Alexander L. Sakhnovich, Lev A. Sakhnovich, Inna Ya. Roitberg
Inverse Problems and Nonlinear Evolution Equations

De Gruyter Studies in Mathematics

Volume 47

Alexander L. Sakhnovich, Lev A. Sakhnovich,
Inna Ya. Roitberg

Inverse Problems and Nonlinear Evolution Equations

Solutions, Darboux Matrices and Weyl–Titchmarsh Functions

DE GRUYTER

Mathematics Subject Classification 2010
Primary: 34A55, 34B20, 34L40, 35G61, 35Q51, 35Q53, 37K15, 37K35, 47A48
Secondary: 34A05, 34L05, 35A01, 35A02, 35F46, 35Q41, 39A12, 93B25

Authors
Alexander Sakhnovich
University of Vienna
Faculty of Mathematics
Nordbergstraße 15
1090 Wien
Austria
oleksandr.sakhnovych@univie.ac.at

Lev Sakhnovich
99 Cove Ave.
Milford CT 06461
United States
lsakhnovich@gmail.com

Inna Roitberg
Universität Leipzig
Faculty of Mathematics & Computer Science
Institute of Mathematics
Augustusplatz 10
04109 Leipzig
Germany
roitberg@math.uni-leipzig.de

ISBN 978-3-11-025860-8
e-ISBN 978-3-11-025861-5
Set-ISBN 978-3-11-220500-6
ISSN 0179-0986

Library of Congress Cataloging-in-Publication Data
A CIP catalog record for this book has been applied for at the Library of Congress.

Bibliographic information published by the Deutsche Nationalbibliothek
The Deutsche Nationalbibliothek lists this publication in the Deutsche Nationalbibliografie; detailed bibliographic data are available in the Internet at http://dnb.dnb.de.

Typesetting: le-tex publishing services GmbH, Leipzig
Printing and binding: Hubert & Co. GmbH & Co. KG, Göttingen
♾ Printed on acid-free paper
Printed in Germany

www.degruyter.com

To the memory of Dora Sakhnovich,
mother of one of the authors and grandmother of another,
with love and gratitude

Preface

This book is based on the method of operator identities and the related theory of S-nodes, both developed by L. A. Sakhnovich. The notion of the transfer matrix function generated by the S-node plays an essential role.

We represent fundamental solutions of various important systems of differential equations using the transfer matrix function, that is, either directly in the form of the transfer matrix function or via the representation in this form of the corresponding Darboux matrix, when Bäcklund–Darboux transformations and explicit solutions are considered. The transfer matrix function representation of the fundamental solution yields, in turn, a solution of an inverse problem, namely, the problem to recover the system from its Weyl function. Weyl theories of self-adjoint and skew-self-adjoint Dirac systems (including the case of rectangular matrix potentials), related canonical systems, discrete Dirac systems, a system auxiliary to the N-wave equation and a system rationally depending on the spectral parameter are obtained in this way.

The results mentioned above on Weyl theory are applied to the study of the initial-boundary value problems for integrable (nonlinear) wave equations via the inverse spectral transformation method. The evolution of the Weyl function is derived for many important nonlinear equations, and some uniqueness and global existence results are proved in detail using these evolution formulas.

Generalized Bäcklund–Darboux transformation (GBDT) is one of the main topics of the book. It is presented in the most general form (i.e. for the case of the linear system of differential equations depending rationally on the spectral parameter). Applications to the Weyl theory and various nonlinear equations are given. Recent results on explicit solutions of the time-dependent Schrödinger equation of dimension $k + 1$ are formulated in order to demonstrate the possibility to apply GBDT to linear systems depending on several variables.

Pseudospectral and Weyl functions of the general-type canonical system are studied in detail in Appendix A.

The last Chapter 9 of the book contains formulations and solutions of the inverse and half-inverse *sliding* problems for radial Schrödinger and Dirac equations, including the case of Coulomb-type potentials. Those results appeared first in 2013.

The reading of the book requires some basic knowledge of linear algebra, calculus and operator theory from standard university courses. All the necessary definitions and results on the method of operator identities and some other additional material is presented in Chapter 1. Moreover, several classical theorems, which are important for the book (e.g. the first Liouville theorem, Phragmen–Lindelöf theorem, Montel's theorem on analytic functions) are formulated in Appendix E.

Chapter 9 was written by L. A. Sakhnovich. Appendix C was written jointly by A. L. Sakhnovich and L. A. Sakhnovich. Chapters 2 and 3 were written by A. L. Sakhnovich and I. Ya. Roitberg. The rest of the book was written by A. L. Sakhnovich.

The book contains results obtained during the last 20 years (or slightly more), and the idea of writing such a book was first considered many years ago. A. L. Sakhnovich is grateful to J. C. Bot for the enthusiastic discussions of the project. The authors are very grateful to F. Gesztesy for his initiation and support of the present (final) version of the book.

A great part of the material, which is presented in the book, appeared during the last 5–6 years as a result of research supported by the Austrian Science Fund (FWF) under grants no. Y330 and no. P24301, and by the German Research Foundation (DFG) under grant no. KI 760/3-1.

Vienna – Milford, CT – Leipzig,
February 2013

Alexander L. Sakhnovich,
Lev A. Sakhnovich, Inna Ya. Roitberg

Notation

A^*	operator adjoint to A
B	de Branges space, see (A.11)
B_1	subspace of B, $B_1 = \widehat{U}L_1$
$\mathfrak{B}$	space, see (A.168)
$\mathfrak{B}_1$	space, see (A.168)
$B(\mathbf{G}, \mathbf{H})$	set of bounded operators acting from **G** into **H**
$B(\Omega) = B^0(\Omega)$	class of bounded on Ω functions (or matrix functions)
$B^N(\Omega)$	class of N times differentiable functions or matrix functions f on the domain Ω such that $\sup \|f^{(N)}\| < \infty$
$\mathbb{C}$	complex plane
$\mathbb{C}_+$	open upper half-plane
$\mathbb{C}_-$	open lower half-plane
$\overline{\mathbb{C}}_\pm$	closed upper (lower) half-planes
$\mathbb{C}_M$	open half-plane $\Im z > M$
$\mathbb{C}_M^-$	open half-plane $\Im z < -M$
$C(\Omega)$	class of continuous on Ω functions (or matrix functions)
$C^N(\Omega)$	class of N times differentiable functions or matrix functions f on the domain Ω such that $f^{(N)}$ is continuous
col	column, $\mathrm{col}\begin{bmatrix} g_1 & g_2 \end{bmatrix} = \begin{bmatrix} g_1 \\ g_2 \end{bmatrix}$
diag	diagonal matrix
$\mathrm{diag}\,\{\alpha_1, \alpha_2\}$	diagonal matrix with the entries (or block entries) α_1 and α_2 on the main diagonal
EG	acronym for *explicitly generated*, a class of potentials $\{C_k\}$ determined by the *admissible triples*, see Notation 5.38
$\widetilde{\mathrm{EG}}$	a class of potentials $\{C_k\}$, see Notation 5.48
I	identity operator
I_p	$p \times p$ identity matrix
$\mathrm{Im}\, A$	image of the operator A
$\Im$	imaginary part of either complex number or matrix
$\mathrm{Ker}\, A$	kernel of the operator A
L_0	subspace, see (A.10)
L_1	subspace, see (A.9)
$L^2(H)$	space, see (A.4)

$L^2_p(d\tau)$	Hilbert space with the scalar product $\langle f, g\rangle_\tau = \int_{-\infty}^{\infty} g(z)^* d\tau(z) f(z)$
$\mathfrak{W}(D, \varphi)$	mapping of a GW-function into potential ζ, see Definition 4.7
$\mathfrak{W}(\varphi)$	mapping of a GW-function into potential v, see Definition 3.29
$\widetilde{\mathfrak{W}}(G, x)$	GBDT-type mapping of G into $\widetilde{G}$, see Notation 7.9
$\mathbb{N}_0$	set of nonnegative integer numbers
$\mathcal{N}(r, z)$ or $\mathcal{N}(r)$	set of values (at z) of Möbius transformations, see Notation 5.10
$\mathcal{N}_u(x, z)$	set of values (at z) of Möbius transformations, see Notation 2.13 for the self-adjoint and Notation 3.1 for the skew-self-adjoint cases
$\mathcal{N}(\mathfrak{A})$	set of functions (of Möbius transformations), see (0.8)
$\mathcal{N}(\mathfrak{A}_A)$	set of functions (of Möbius transformations), see Notation 1.25
$\mathcal{N}(n)$	set of functions (of Möbius transformations), see Notation 5.27
$\mathcal{N}(S, \Phi_1)$	set of Herglotz functions, see Notation 1.24
$\mathcal{N}(\mathcal{U})$	set of functions (of Möbius transformations), see Notation 1.46
$\mathbf{P}_M(j)$	class of matrix functions with property-j, see Section 3.1 and Notation 3.1
$\mathbb{R}$	real axis
R_α	operator, see (A.23)
$\mathfrak{R}$	real part of either complex number or matrix
$\mathcal{S}_{m_2\times m_1}(\Omega)$	class of Schur matrix functions, see Notation 3.2
span	linear span
$\overline{\mathrm{span}}$	closed linear span
Tr	operator (matrix, in particular) trace
$\hat{U}$	operator, see (A.5)
$\overline{z}$	complex conjugate to z
δ_{ij}	Kronecker delta
$\sigma(A)$	spectrum of an operator A
$[D, \mathcal{M}]$	commutator $D\mathcal{M} - \mathcal{M}D$
$(\cdot, \cdot)_{\mathbf{H}}$	scalar product in $\mathbf{H}$
$\wedge$	vector product
$\Vert \cdot \Vert$	l^2 vector norm or the induced matrix norm

Contents

0 Introduction

In recent years, the interplay between inverse spectral methods and gauge transformation techniques to solve nonlinear evolution equations has greatly benefited both areas. The notions of the Weyl and scattering functions, Möbius (linear-fractional) and Bäcklund–Darboux transformations, Darboux matrices and nonlinear integrable equations are all interrelated. The purpose of this book is to treat this interaction by actively using various aspects of the *method of operator identities* (and S-nodes theory, in particular). Thus, Weyl functions that have been initially introduced in the self-adjoint case also prove very useful for solving non-self-adjoint inverse problems and Goursat problems for nonlinear integrable equations. The Darboux matrix can be presented in the form of the transfer matrix function from the system theory, and the Bäcklund–Darboux transformation can be fruitfully applied in the multidimensional case of $k > 1$ space variables. (We write matrix function and vector function meaning matrix-valued function and vector-valued function, respectively.) Some simple examples (as well as several important results) can already be seen in Chapter 1, where basic definitions and statements to make the book self-contained are also presented.

The famous Schrödinger (Sturm–Liouville) equation is usually considered in the form

$$-\frac{d^2}{dx^2}y(x,z) + v(x)y(x,z) = zy(x,z). \tag{0.1}$$

A great number of fundamental notions and results of analysis has been first introduced and obtained for this equation. The list includes the Weyl disc and point, Weyl solution, spectral and Weyl (or Weyl–Titchmarsh) functions, Bäcklund–Darboux transformation, transformation operator and solutions of the inverse problems, and so on. One also has to mention its connections with the Lax pairs, the method of the inverse scattering transform, and nonlinear integrable equations. We can refer to the already classical books [39, 195, 205, 209, 213, 217, 323], though numerous new important papers, surveys and books appear regularly. Among the more recent developments are the notions of bispectrality and $\mathcal{PT}$-symmetry (see [93] and [41], respectively). It is of special interest that practically all of these notions are related in one or another way.

For each continuous $v(x)$ $(0 \le x < \infty)$ and $z \ne \overline{z}$, there is a so called Weyl solution y_w of the Schrödinger equation such that $\int_0^\infty |y_w(x,z)|^2 dx < \infty$ (see, for instance, p. 60 in [196]). Here, we denote by $\overline{z}$ the complex number conjugate to z. Let y_1 and y_2 satisfy the Schrödinger equation and initial conditions

$$y_1(0) = \sin c, \; y_1'(0) = -\cos c; \quad y_2(0) = \cos c, \; y_2'(0) = \sin c \quad \left(y' := \frac{d}{dx}y\right).$$

Then, y_w admits representation $y_w(x,z) = \varphi(z)y_1(x,z) + y_2(x,z)$. Function φ is called the Weyl or Weyl–Titchmarsh function and is extremely important in spectral theory. The Weyl–Titchmarsh approach can be developed for a much wider class of

Schrödinger equations, for matrix Schrödinger equations, and for various other important systems. For the case of the Hamiltonian systems, some basic results have been obtained by Hinton and Shaw [141, 142]. *Canonical system*

$$dy(x,z)/dx = \mathrm{i}zJH(x)y(x,z), \quad J = \begin{bmatrix} 0 & I_p \\ I_p & 0 \end{bmatrix}, \quad H(x) \geq 0 \tag{0.2}$$

with locally summable *Hamiltonian* H, is a particular case of the Hamiltonian systems and a classical object of analysis. Here, J and $H(x)$ are $m \times m$ matrices, I_p is the $p \times p$ identity matrix, $2p = m$, and inequality $H(x) \geq 0$ means that

$$H(x) = H(x)^*$$

($H(x)$ is self-adjoint) and the spectrum of $H(x)$ is nonnegative. (In general, inequality $S_1 \geq S_2$ means that $S_1 = S_1^*$, $S_2 = S_2^*$ and $S_1 - S_2 \geq 0$.) The summability of a matrix function means that its entries belong to L^1 (the entries are summable). The spectral theory of the general-type canonical systems is studied in Appendix A.

The study of the canonical system includes, in turn, such particular cases as the Schrödinger matrix equation (0.1), where *potential* v is a $p \times p$ matrix function, and the well-known self-adjoint Dirac-type (also called Dirac, Zakharov–Shabat or AKNS) system:

$$\frac{d}{dx}y(x,z) = \mathrm{i}(zj + jV(x))y(x,z), \tag{0.3}$$

$$j = \begin{bmatrix} I_p & 0 \\ 0 & -I_p \end{bmatrix}, \quad V = \begin{bmatrix} 0 & v \\ v^* & 0 \end{bmatrix}. \tag{0.4}$$

Usually, we consider systems (0.1)–(0.3) on the finite intervals $[0, l]$ or semiaxis $[0, \infty)$. One can find further references and results on these systems, for instance, in [24, 84, 85, 101, 118, 133, 137, 228, 234, 289, 290], and on the Weyl (Weyl–Titchmarsh or M-) functions for these systems in [118, 195, 205, 289, 290]. Transformations of the Dirac and Schrödinger systems into canonical are given in Subsections 1.1.1 and 1.1.2 of Chapter 1. We note that Dirac (Dirac-type) systems differ from the radial Dirac systems which appeared earlier (and were also called Dirac). Radial Dirac systems are discussed in detail in Chapter 9 (see also some results from Section 8.2).

We recall that fundamental solutions of the first order differential systems are square nondegenerate matrix functions, which satisfy these systems (and generate in an apparent way all other solutions). The fundamental $m \times m$ solution of the canonical system (0.2) is normalized by the condition

$$W(0,z) = I_m. \tag{0.5}$$

Parameter matrix functions $\mathcal{P}_1(z)$, $\mathcal{P}_2(z)$ of order p, such that

$$\mathcal{P}_1(z)^*\mathcal{P}_1(z) + \mathcal{P}_2(z)^*\mathcal{P}_2(z) > 0, \quad \begin{bmatrix} \mathcal{P}_1(z)^* & \mathcal{P}_2(z)^* \end{bmatrix} J \begin{bmatrix} \mathcal{P}_1(z) \\ \mathcal{P}_2(z) \end{bmatrix} \geq 0 \tag{0.6}$$

play an essential role in our book. Next, follow three basic definitions [281, 289, 290].

Definition 0.1. A pair of $p \times p$ matrix functions $\mathcal{P}_1(z)$, $\mathcal{P}_2(z)$, meromorphic in the upper half-plane $\mathbb{C}_+$, is called *nonsingular with property-J* if the first inequality from (0.6) holds in one point (at least) of $\mathbb{C}_+$ and the second inequality from (0.6) holds in all the points of analyticity of $\mathcal{P}_1(z)$, $\mathcal{P}_2(z)$ in $\mathbb{C}_+$.

Remark 0.2. It is apparent that if the first inequality from (0.6) holds in one point of $\mathbb{C}_+$, it holds everywhere, except, possibly, some set of isolated points.

Put

$$\mathfrak{A}(l, z) = \mathfrak{A}(z) = W(l, \overline{z})^* = \begin{bmatrix} a(z) & b(z) \\ c(z) & d(z) \end{bmatrix}. \tag{0.7}$$

Definition 0.3 ([290, p. 7]). Matrix functions $\varphi(z)$, which are obtained via the transformation

$$\varphi(z) = \mathrm{i}\left(a(z)\mathcal{P}_1(z) + b(z)\mathcal{P}_2(z)\right)\left(c(z)\mathcal{P}_1(z) + d(z)\mathcal{P}_2(z)\right)^{-1} \tag{0.8}$$

(where the pairs $\mathcal{P}_1(z)$, $\mathcal{P}_2(z)$ of "parameter" matrix functions are nonsingular with property-J, $\det(c\mathcal{P}_1 + d\mathcal{P}_2) \not\equiv 0$), are called Weyl functions of the canonical system on the interval $[0, l]$.

We denote the class of the Weyl functions of the canonical system on $[0, l]$ by the acronym $\mathcal{N}(\mathfrak{A})$ or simply by $\mathcal{N}(l)$. The discs (Weyl discs) $\mathcal{N}(l)$ are embedded in one another, that is, $\mathcal{N}(l_1) \subseteq \mathcal{N}(l_2)$ for $l_1 > l_2$. It is shown in Appendix A that under a rather weak "positivity" condition on H, we have $\bigcap_{l<\infty} \mathcal{N}(l) \neq \emptyset$, and matrix function $\varphi \in \bigcap_{l<\infty} \mathcal{N}(l)$ satisfies inequality

$$\int_0^\infty \begin{bmatrix} I_p & \mathrm{i}\varphi(z)^* \end{bmatrix} W(x, z)^* H(x) W(x, z) \begin{bmatrix} I_p \\ -\mathrm{i}\varphi(z) \end{bmatrix} dx < \infty \quad (z \in \mathbb{C}_+). \tag{0.9}$$

In other words, the entries of $W(x, z)\begin{bmatrix} I_p \\ -\mathrm{i}\varphi(z) \end{bmatrix}$ belong $L^2(0, \infty)$, which is similar to the definition of the Weyl function of the Schrödinger equation on $[0, \infty)$. Hence, the definition below.

Definition 0.4. Holomorphic functions φ such that inequality (0.9) holds are called Weyl functions of system (0.2) on $[0, \infty)$.

Most of the direct problems in this book consist of the description of the set of Weyl functions, construction of Weyl functions and the study of their existence and uniqueness, whereas inverse problems, which we consider here, usually deal with the recovery of systems from their Weyl functions. Direct and inverse problems are considered either for systems determined on the finite interval $[0, l]$ *or for systems on the semiaxis* $[0, \infty)$.

Though all the notions mentioned in the Introduction are important for our book, *the Weyl functions, Bäcklund–Darboux transformations and Darboux matrices are the principal ones*. Various explicit formulas in the book are obtained through a new version of the well known in the spectral theory and integrable nonlinear equations (see, for instance, [10, 29, 70–72, 83, 84, 140, 192, 199, 206, 209, 213, 215, 231, 335] and references therein) Bäcklund–Darboux transformation (BDT). The BDT transforms the initial equation or system into another one from the same class and also transforms solutions of the initial equation into solutions of the transformed one. Let us illustrate this by the oldest and most popular example, that is, by the Schrödinger equation (0.1), where the scalar potential v is real valued (i.e. $v = \overline{v}$) and $z = \overline{z}$. Assume that $h(x) = \overline{h}(x)$ satisfies (0.1), when $z = c$, that is, $-h'' + vh = ch$ ($h'' := \frac{d^2}{dx^2}h$). Then, one can rewrite (0.1) in the form

$$(\mathcal{A}^*\mathcal{A} + cI)\, y(x,z) = zy(x,z), \tag{0.10}$$

where I is the identity operator, and $\mathcal{A}$ and $\mathcal{A}^*$ are first order differential expressions:

$$\mathcal{A}f = \left(\frac{d}{dx} - \frac{h'}{h}\right) f, \quad \mathcal{A}^* f = -\left(\frac{d}{dx} + \frac{h'}{h}\right) f.$$

The transformed equation is given by the formula

$$(\mathcal{A}\mathcal{A}^* + cI)\, y(x,z) = zy(x,z). \tag{0.11}$$

It easy to see that (0.11) is again the Schrödinger equation, but potential v is transformed into $\widetilde{v} = v - 2(\frac{h'}{h})'$. Notice further that

$$(\mathcal{A}\mathcal{A}^* + (c-z)I)\,\mathcal{A} = \mathcal{A}\,(\mathcal{A}^*\mathcal{A} + (c-z)I)\,.$$

Hence, it follows that if $y(x,z)$ satisfies (0.10), then $\widetilde{y} := \mathcal{A}y$ satisfies (0.11). All solutions of the transformed equations can be constructed in this way. Under rather weak conditions, the spectra of operators A^*A and AA^* may differ only at zero, and so under certain conditions, the spectra of Schrödinger operators L and $\widetilde{L}$ associated with differential expressions $-\frac{d^2}{dx^2} + v$ and $-\frac{d^2}{dx^2} + \widetilde{v}$ may only differ at c. The Bäcklund–Darboux-type (and related commutation) methods of inserting and removing eigenvalues of Schrödinger operators go back historically to Jacobi, Bäcklund and Darboux [29, 84, 146] with decisive later contributions by Crum, Deift and Gesztesy [82, 87, 88, 117, 123]. (See a detailed account in Appendix G in [118].)

One can apply the BDT again to the already transformed equation (0.11) and so on (iterated BDT). There is also a somewhat more complicated binary BDT ([2, 209]). It proves that if v satisfies a nonlinear integrable equation, then $\widetilde{v}$ often satisfies it too, and so the BDT is used to construct solutions of the nonlinear equations.

Elementary Bäcklund–Darboux transformations for Dirac-type and more general AKNS systems one can find, for instance, in [66, 172]. Given first order initial and transformed systems $u' = G(x,z)u$ and $\widetilde{u}' = \widetilde{G}(x,z)\widetilde{u}$, their solutions are connected via the so called Darboux matrix w such that $w' = \widetilde{G}w - wG$. Clearly, if u satisfies the initial system $u' = Gu$, then wu satisfies the transformed one

$\widetilde{u}' = \widetilde{G}\widetilde{u}$. Darboux matrix or gauge transformation is of great interest in this theory (see [172, 193, 209, 301, 318, 332] and references therein). For an interesting comparative analysis of different constructions of Darboux matrices, see the review [71].

As mentioned above, this book is dedicated to the important classical (canonical, Dirac and Schrödinger) and nonclassical systems. Their Weyl functions and Darboux matrices as well as some spectral and scattering matrix functions are studied, and related direct and inverse problems and applications are considered.

Our book is closely related to the books [287, 289, 290], *where the method of operator identities and S-nodes theory are presented and applied to the inversion of operators and the spectral theory of the canonical systems.* S-nodes theory is basic for both our approach to the general solutions of the inverse problems and our version of the BDT. Darboux matrix and fundamental solutions are represented in terms of the S-nodes via transfer matrix function w_A (see formula (0.17) below). Here, we shall briefly discuss S-nodes theory. Basic facts and some applications are given in Chapter 1 and further theory and applications are then found in the following chapters. Suppose that the five operators

$$A,\ B,\ S \in B(\mathbf{H}); \quad \Pi_k,\ \Gamma_k \in B(\mathbf{G}, \mathbf{H}) \quad (k = 1, 2); \tag{0.12}$$

are given, where $\mathbf{G}$ and $\mathbf{H}$ are Hilbert spaces. By $B(\mathbf{H})$, we denote the set of bounded operators acting in $\mathbf{H}$ while $B(\mathbf{G}, \mathbf{H})$ stands for the set of bounded operators acting from $\mathbf{G}$ into $\mathbf{H}$. For simplicity, we shall always assume in this book that $\dim \mathbf{G} < \infty$. We say that *the operators* (0.12) *form an S-node* if the relations

$$AS - SB = \Pi_1 \Pi_2^*, \tag{0.13}$$

$$S\Gamma_1 = \Pi_1, \quad \Gamma_2^* S = \Pi_2^* \tag{0.14}$$

hold (see[290]). Note that the operator identity (0.13) gave name to the *Method of Operator Identities* developed by L. A. Sakhnovich and this method is actively used in our book. The notion of the S-node is a generalization of the earlier notion of a node [59, 200, 201] (see interesting further developments and references in [31, 32, 203]). The S-node reduces to the latter when the relations

$$S = I, \quad B = A^*, \quad \Pi_2^* = \mathrm{i} J \Pi_1^* \tag{0.15}$$

hold. Here, the operator J belongs to the set $B\,(\mathbf{G}, \mathbf{G})$ and

$$J = J^* = J^{-1}. \tag{0.16}$$

Transfer matrix function corresponding to the S-node (transfer matrix function in the L. Sakhnovich form) is given by the equality

$$w_A(\lambda) = I - \Gamma_2^* (A - \lambda I)^{-1} \Pi_1. \tag{0.17}$$

Indeed, this function is the transfer matrix function for system

$$\frac{d}{dt}x = Ax + \Pi_1 u, \quad y = \Gamma_2^* x + u$$

(see Introductions in [276, 280, 290]). At the same time, w_A is a generalization of the Livšic–Brodskii characteristic matrix function and reduces to it under condition (0.15). *The S-multinode generalization of the S-node (and of the Livšic colligations from* [203]*) was applied to the nonstationary multidimensional Schrödinger equation in* [266] *and is discussed in Section* 8.2.

Self-adjoint Dirac-type systems can be treated as a particular case of the canonical systems. In Chapter 2, we consider (in much greater detail than in [289, 290]) the Weyl functions for this particular case. Seminal publications by F. Gesztesy and B. Simon [121, 122, 307] gave rise to a whole series of interesting papers and results on the high energy asymptotics of the Weyl functions (i.e. asymptotics of the Weyl functions when $|z| \to \infty$) and local Borg–Marchenko-type uniqueness theorems (e.g. [74, 75, 77–79, 106, 114, 187, 259] and references therein). For the system (0.3), we present an interesting development of those results in Chapter 2 (see also [252]). Finally, using S-node theory and the high energy asymptotics of the Weyl functions, an inverse problem is solved directly in terms of the Weyl functions.

The Weyl function approach is also fruitfully applied to important non-self-adjoint cases: skew-self-adjoint Dirac-type system (see Chapter 3), a system auxiliary to the nonlinear optics (N-wave) equation (see Chapter 4) and a system depending rationally on the spectral parameter (see Section 8.1). The direct and inverse problems for those cases are solved in the book. Furthermore, the corresponding high energy asymptotics of the Weyl functions and local Borg–Marchenko-type uniqueness theorems are derived.

Discrete self-adjoint and skew-self-adjoint Dirac systems are considered in Chapter 5. Direct and inverse problems are solved. Using GBDT, we also construct explicit solutions of the isotropic Heisenberg magnet model and describe the evolution of the Weyl function of the corresponding discrete skew-self-ajoint Dirac system.

Chapter 6 is dedicated to the initial-boundary value problems for integrable nonlinear equations. This chapter is essentially based on the solutions of the direct and inverse problems from the previous chapters. The Inverse Spectral Transformation method (ISpT) [44, 45, 150, 238, 249, 281, 282, 289] proves very fruitful for nonlinear problems in a quarter-plane. *In terms of the Weyl functions, we consider in this chapter several initial-boundary value problems for nonlinear equations* (for the focusing modified Korteweg–de Vries equation, sine-Gordon and complex sine-Gordon equations and the second harmonic generation (SHG) system). In particular, solutions of the Goursat problems for the second harmonic generation model and sine-Gordon equation in the light cone coordinates follow. We note that the second harmonic generation model has remained an open problem during several decades [166] and was finally solved in the recent paper [256].

In order to demonstrate the *Inverse Spectral Transformation approach* (more precisely, its Weyl function transformation version), let us discuss the initial-boundary value problem for the SHG system

$$\frac{\partial}{\partial x}\psi_1 = -2\overline{\psi_1}\psi_2, \quad \frac{\partial}{\partial t}\psi_2 = \psi_1^2 \tag{0.18}$$

on the semistrip $\{(x,t) : 0 \le x, 0 \le t < T\}$. The Cauchy problem for SHG has no physical meaning, and the Goursat problem for this outwardly simple and physically important integrable equation has remained unsolved in spite of many interesting results (see [12, 166, 168, 314] and references therein). One of the reasons for this situation is connected with the continuously interacting nature of SHG [166]. (The solutions of (0.18) do not necessarily tend to zero at infinity.)

The SHG equation can be presented [162] as the compatibility condition (zero curvature equation)

$$\frac{\partial}{\partial t}G(x,t,z) - \frac{\partial}{\partial x}F(x,t,z) + [G(x,t,z), F(x,t,z)] = 0, \quad [G,F] := GF - FG \tag{0.19}$$

for two auxiliary linear systems $\frac{\partial}{\partial x}w = Gw$ and $\frac{\partial}{\partial t}w = Fw$. The first of these systems:

$$\frac{\partial}{\partial x}w(x,t,z) = G(x,t,z)w(x,t,z), \quad G(x,t,z) = \mathrm{i}\,(zj + jV(x,t)) \tag{0.20}$$

is Dirac-type, and depends on the additional parameter t. Another system is canonical

$$\frac{\partial}{\partial t}w(x,t,z) = F(x,t,z)w(x,t,z), \quad F(x,t,z) = \frac{\mathrm{i}}{z}jH(x,t), \tag{0.21}$$

depending on the additional parameter x. Here, we have $p = 1$,

$$v(x,t) = -2\mathrm{i}\psi_2(x,t), \quad H(x,t) = \vartheta(x,t)\vartheta(x,t)^*, \quad \vartheta(x,t) = \begin{bmatrix} \psi_1(x,t) \\ \overline{\psi_1}(x,t) \end{bmatrix}.$$

The Goursat problem is solved in three steps. First, we find Weyl function $\varphi(0,z)$ of system (0.20) at $t = 0$. Then, similar to [289], we find *evolution* $\varphi(t,z)$ - Weyl functions of system (0.20) for $0 < t < T$. Finally, we solve the inverse problem and recover $V(x,t)$ and hence $\psi_2(x,t)$. The matrix function ψ_1 is now easily expressed via ψ_2.

A general version of GBDT for linear and integrable nonlinear equations is considered in Chapter 7. (Some particular cases are dealt with before in Chapters 1, 2 and 4.) *We present Darboux matrix w in the form of the transfer matrix function, that is, we set $w(x,z) = w_A(x,z)$, where w_A is given by* (0.17). The so called properties-J of w_A are important for spectral and scattering results. In the GBDT case, we construct and use S-nodes differently from the “general inverse problem” situation. Namely, matrix functions $\Pi_k(x)$ are introduced now in terms of the generalized eigenfunctions (with the generalized matrix eigenvalues) of the initial system (see Subsection 1.1.3 and Chapter 7 for details). Recall that eigenfunctions are essential for the classical Bäcklund–Darboux transformation. Since Π_k are generalized eigenfunctions, this approach is called Generalized Bäcklund–Darboux Transformation (or, shorter, GBDT). Matrix function $S(x)$ is expressed directly via $\Pi_k(x)$. Representation (0.17) of the Darboux matrix is treated in this book for $G = \mathrm{i}(zj + jV)$, $G = \mathrm{i}zJH$ as well as in much more general situations. The cases of G polynomially and rationally depending on z are dealt with in Chapter 7. Notice that operator identities have been successfully used for the construction of the explicit solutions of nonlinear equations in [156, 206]. (One can find further developments of the Marchenko scheme in [56] and papers of the "Jena group" [9, 62, 302–304].) Our approach grants an additionally explicit expression for the Darboux matrix, allows one to avoid the stages of the construction of the high order matrix solutions of the nonlinear equation and their reduction to the required order, and minimizes the order of the matrix that has to be inverted (matrix S in our case).

Several further results closely related to various chapters of the book are formulated and discussed (and references to the proofs are given) in Chapter 8.

The last Chapter 9 is dedicated to the special inverse and half-inverse problems: *sliding problems*. The lack of a well-posed half-inverse problem for a radial Dirac system was discussed long before (see [279]). Such a problem (namely, a sliding half-inverse problem for the radial Dirac system) was formulated and solved quite recently [300]. The approach is used to recover several physically important equations from the quantum defect. In particular, sliding problems are solved for a radial Schrödinger equation, a radial Dirac system and a multidimensional Schrödinger equation. Systems with Coulomb-type potentials are considered as well.

Appendices A–E are placed at the end of the book and contain many auxiliary explanations and results. In particular, Appendix A is dedicated to the detailed study of the general-type canonical system.

The results and methods of the book strongly rely on the earlier research of the mathematical schools of Odessa and also Leipzig. The authors are grateful to their coauthors and colleagues from Vienna and Leipzig for the cooperation and discussions. They would like to thank Yu. Yu. Popov for his photos of Odessa in this book.

We already introduced some of the notations used in the book. Recall that by $\mathbb{R}$ and $\mathbb{C}$ we denote the real axis and the complex plane, respectively. By $\mathbb{C}_+$ and $\mathbb{C}_-$, we denote the open upper and lower half-planes, and $\overline{\mathbb{C}}_\pm$ denote the closed half-planes. $\mathbb{N}_0$ stands for the set of nonnegative integer numbers. I_p stands for the $p \times p$ identity matrix and I is the identity operator.

We say operator, meaning *linear operator* (usually, a bounded linear operator). For nonlinear operators, we use the term *mapping*. The spectrum of an operator A is denoted by $\sigma(A)$, Im means the image, Ker means the kernel of the operator, and Tr means the operator (matrix, in particular) trace.

The notation col stands for column, for instance, $\mathrm{col}\,[g_1 \quad g_2] = \left[\begin{smallmatrix} g_1 \\ g_2 \end{smallmatrix}\right]$; diag stands for a diagonal matrix (e.g diag $\{\alpha_1, \alpha_2\}$ is the diagonal matrix with the entries, or block entries, α_1 and α_2 on the main diagonal); δ_{ij} is the Kronecker delta and $[D, \mathcal{M}]$ stands for the commutator $D\mathcal{M} - \mathcal{M}D$. Notation A^* stands for the operator adjoint to A and $\overline{z}$ stands for the number, which is the complex conjugate to z. (When it goes about some function, which takes values complex conjugate to $f(x)$, we write either $\overline{f(x)}$ or $\overline{f}(x)$.) $\Re$ and $\Im$ stand for the real and imaginary parts (of either complex number or matrix), respectively.

We write $S \geq 0$ for the linear operator $S = S^*$ acting in some Hilbert space $\mathbf{H}$ and such that $(Sf, f)_{\mathbf{H}} \geq 0$, where $(\cdot, \cdot)_{\mathbf{H}}$ stands for the scalar product in $\mathbf{H}$. We say that the operator S is positive-definite (or positive) and write $S > 0$ if $(Sf, f)_{\mathbf{H}} > 0$ for each $f \neq 0$, we say that S is strictly positive-definite (or strictly positive) if there is some $\varepsilon > 0$ such that $S \geq \varepsilon I$, that is, $S - \varepsilon I \geq 0$, and we say that S is boundedly invertible if S has a bounded inverse operator. An $m_2 \times m_1$ matrix α is said to be nonexpansive (or contractive) if $\alpha^*\alpha \leq I_{m_1}$ (or, equivalently, if $\alpha\alpha^* \leq I_{m_2}$).

We assume that the spaces $\mathbb{C}^{m_i}$ are equipped with the l^2-norm and use $\|\cdot\|$ to denote this vector norm or the induced matrix norm. We use $B^N(\Omega)$ $(C^N(\Omega))$ to denote the set of N times differentiable functions or matrix functions f on the domain (a closed interval, mostly) Ω such that $\sup \|f^{(N)}\| < \infty$ ($f^{(N)}$ is continuous). Correspondingly, $B(\Omega) = B^0(\Omega)$ stands for the class of bounded functions and $C(\Omega)$ stands for the class of continuous functions (or matrix functions). We say that functions $f \in B^1(\Omega)$ $(f \in C^1(\Omega))$ are boundedly differentiable (continuously differentiable). When we consider functions that are not continuous or piecewise continuous (e.g. bounded functions), we usually regard functions that coincide almost everywhere as one and the same function, and so we use "sup" instead of "ess sup" in the text. We always assume that bounded functions in our considerations are also measurable (i.e. we write "bounded" meaning "bounded and measurable"). Finally, span is the linear span and $\overline{\mathrm{span}}$ is the closed linear span. These and other important notations are explained (or the corresponding references are given) in the glossary "Notation".

Notations J and j are often used in the book, and J either satisfies only (0.16) or is given by (0.2), which is a particular case of matrices satisfying (0.16), or the subcase $p = 1$ of matrices given by (0.2) is dealt with. Depending on the section, matrix j also takes either a more general or a particular form:

$$j = \begin{bmatrix} I_{m_1} & 0 \\ 0 & -I_{m_2} \end{bmatrix}, \quad j = \begin{bmatrix} I_p & 0 \\ 0 & -I_p \end{bmatrix} \quad \text{or} \quad j = \begin{bmatrix} 1 & 0 \\ 0 & -1 \end{bmatrix}.$$

The symbol ":="("=:") is written for the case that the equality defines the corresponding notation given on the left- (right-) hand side. Sometimes, we write that an operator A is applied to some matrix function $F(x)$. The corresponding formula AF or $(AF)(x)$ means that A is applied to F column-wise.

Odessa

Figure 0.1: Odessa, Lighthouse (reproduced by kind permission of Yuri Popov).

Figure 0.2: Odessa, Primorsky Boulevard (reproduced by kind permission of Yuri Popov).

Leipzig

Figure 0.3: Leipzig Tower in the place of the former observatory.

Figure 0.4: Leipzig University, Red College.

1 Preliminaries

“Preliminaries” is an essential component of this book. Within this chapter, we explain, in a more detailed way, the connections between Dirac-type, Sturm–Liouville, and canonical systems (and their Weyl functions), which were mentioned in the Introduction. Furthermore, some basic properties of the S-nodes are explained and interconnections between the S-nodes and canonical systems are treated. Some other fundamental preceding results (e.g. operator similarity and factorization results) are presented to make the book self-sufficient. Several important approaches which will be used throughout the book are demonstrated and interesting results on inverse problems and the asymptotics of the Weyl functions are obtained.

1.1 Simple transformations and examples

1.1.1 Dirac-type systems as a subclass of canonical systems

Consider the self-adjoint Dirac (Dirac-type) system (0.3), where j and V are given by (0.4). Suppose $u(x,z)$ is its *normalized fundamental solution* (further, simply *fundamental solution*). More precisely, u is an $m \times m$ matrix solution of system (0.3) normalized by the condition

$$u(0,z) = I_m \quad (m = 2p). \tag{1.1}$$

In particular, we have

$$\begin{aligned} &\frac{d}{dx}u(x,0) = \mathrm{i}jV(x)u(x,0), \\ &\frac{d}{dx}u(x,0)^{-1} = -u(x,0)^{-1}\left(\frac{d}{dx}u(x,0)\right)u(x,0)^{-1} = -\mathrm{i}u(x,0)^{-1}jV(x). \end{aligned} \tag{1.2}$$

From (1.2), we see that the matrix function

$$W(x,z) = \Theta u(x,0)^{-1}u(x,z)\Theta^*, \tag{1.3}$$

where

$$\Theta := \frac{1}{\sqrt{2}}\begin{bmatrix} I_p & -I_p \\ I_p & I_p \end{bmatrix}, \quad \Theta^* = \Theta^{-1} \tag{1.4}$$

satisfies the equation

$$\begin{aligned} \frac{d}{dx}W(x,z) &= \Theta u(x,0)^{-1}\left(-\mathrm{i}jV(x) + \mathrm{i}zj + \mathrm{i}jV(x)\right)u(x,z)\Theta^* \\ &= \mathrm{i}z\Theta u(x,0)^{-1}ju(x,0)\Theta^*W(x,z). \end{aligned} \tag{1.5}$$

Notice that according to (1.1) and the first equality in (1.2), we have

$$u(x,0)^*ju(x,0) \equiv u(x,0)ju(x,0)^* \equiv j. \tag{1.6}$$

It follows from (1.4) that

$$\Theta j \Theta^* = J, \quad \Theta^* J \Theta = j. \tag{1.7}$$

Using (1.6) and (1.7), we rewrite (1.5) as a canonical system

$$\frac{d}{dx} W(x,z) = \mathrm{i}zJH(x)W(x,z), \tag{1.8}$$

where

$$H(x) = J\Theta u(x,0)^{-1} j u(x,0)\Theta^* = \Theta u(x,0)^* u(x,0)\Theta^* > 0. \tag{1.9}$$

Thus, relations (1.9) and (1.3) transform the Dirac system and its fundamental solution into a canonical system (with positive-definite Hamiltonian H) and its fundamental solution. Furthermore, by putting

$$W_d(x,z) = \exp\left\{-\mathrm{i}z\frac{x}{2}\right\} W\left(\frac{x}{2}, z\right), \quad H_d(x) = \frac{1}{2}\left(H\left(\frac{x}{2}\right) - J\right), \tag{1.10}$$

we obtain the canonical system

$$\frac{d}{dx} W_d(x,z) = \mathrm{i}zJH_d(x)W_d(x,z). \tag{1.11}$$

In view of (1.6), (1.7), (1.9) and (1.10), the Hamiltonian H_d is nonnegative of rank p:

$$H_d(x) = \gamma(x)^*\gamma(x), \tag{1.12}$$

$$\gamma(x) := [0 \quad I_p] u(x/2, 0)\,\Theta^*. \tag{1.13}$$

From the first relation in (1.2) and formula (1.13), it follows that

$$\gamma'(x) = -\frac{\mathrm{i}}{2}\left[v\left(\frac{x}{2}\right)^* \quad 0 \right] u\left(\frac{x}{2}, 0\right)\Theta^*, \quad \gamma' := \frac{d}{dx}\gamma. \tag{1.14}$$

Now, formulas (1.6), (1.7), (1.13) and (1.14) yield

$$\gamma'(x)J\gamma(x)^* \equiv 0, \quad \gamma(0) = \frac{1}{\sqrt{2}}[-I_p \quad I_p]. \tag{1.15}$$

In particular, we have $\gamma(x)J\gamma(x)^* \equiv -I_p$.

Proposition 1.1. *Formula* (1.13) *defines a one-to-one correspondence between Dirac systems with summable potentials $v(x)$ on $[0, l/2]$ and canonical systems where Hamiltonians have the form* (1.12) *and $p \times m$ matrix functions γ in* (1.12) *have summable derivatives γ' on $[0, l]$ and satisfy* (1.15).

Proof. We have already shown that γ given by (1.13) satisfies (1.15). Put

$$\beta(x) := [I_p \quad 0] u(x/2, 0)\,\Theta^*. \tag{1.16}$$

According to (1.6), (1.7), (1.14) and (1.16), we can recover v from γ and β:

$$v(x) = 2\mathrm{i}\beta'(2x)J\gamma(2x)^*. \tag{1.17}$$

Now, we shall show that γ and β are uniquely recovered from H_d. Indeed, in view of (1.12), the matrix function γ admits representation

$$\gamma(x) = \theta(x)\tilde{\gamma}(x), \tag{1.18}$$

where

$$\tilde{\gamma}(x) = [\tilde{\gamma}_1(x) \quad \tilde{\gamma}_2(x)] := [0 \quad I_p]H_d(x), \quad \theta(x) := (\gamma_2(x)^*)^{-1}, \tag{1.19}$$

γ_k and $\tilde{\gamma}_k$ are $p \times p$ blocks of γ and $\tilde{\gamma}$, respectively. Since $\gamma J \gamma^* \equiv -I_p$, from (1.18), we have

$$\begin{aligned} \gamma'(x)J\gamma(x)^* &= \theta'(x)\theta(x)^{-1}\gamma(x)J\gamma(x)^* + \theta(x)\tilde{\gamma}'(x)J\gamma(x)^* \\ &= -\theta'(x)\theta(x)^{-1} + \theta(x)\tilde{\gamma}'(x)J\tilde{\gamma}(x)^*\theta(x)^*. \end{aligned} \tag{1.20}$$

Using (1.19), we obtain $\theta^*\theta = \tilde{\gamma}_2(x)^{-1}$. Hence, taking into account (1.15), we rewrite (1.20) as a linear differential equation on θ:

$$\theta'(x) = \theta(x)\tilde{\gamma}'(x)J\tilde{\gamma}(x)^*\tilde{\gamma}_2(x)^{-1}, \quad \theta(0) = \sqrt{2}I_p. \tag{1.21}$$

Thus, γ and θ are uniquely defined by (1.18), (1.21) and the first relation in (1.19).

The next step is to recover β from γ. According to (1.6), (1.7) and the definitions of γ and β, the equality $\gamma J \beta^* \equiv 0$ is true, that is, β admits representation

$$\beta(x) = \vartheta(x)\tilde{\beta}(x), \quad \tilde{\beta}(x) := [\gamma_1(x)\gamma_1(x)^*(\gamma_2(x)^*)^{-1} \quad -\gamma_1(x)]. \tag{1.22}$$

In the same way in which we derived (1.15), we obtain

$$\beta'(x)J\beta(x)^* = 0, \quad \beta(0) = \frac{1}{\sqrt{2}}[I_p \quad I_p]. \tag{1.23}$$

The second equality in (1.15) and formulas (1.22), (1.23) imply that

$$\vartheta(0) = I_p, \quad \beta(x)J\beta(x)^* \equiv I_p. \tag{1.24}$$

From the first relations in (1.22)–(1.24), we obtain

$$\vartheta'(x) + \vartheta(x)(\tilde{\beta}'(x)J\tilde{\beta}(x)^*)(\tilde{\beta}(x)J\tilde{\beta}(x)^*)^{-1} = 0, \quad \vartheta(0) = I_p. \tag{1.25}$$

The differential equation and initial condition (1.25) uniquely define ϑ. After that, we apply (1.22) to recover β. Therefore, v is uniquely recovered from H_d, that is, different Dirac systems are mapped via (1.13) into different canonical systems (and so the mapping is injective).

It remains to prove that each canonical system of the form (1.11), (1.12), where γ satisfies (1.15), is equivalent to the Dirac system. Indeed, let H_d be given and introduce v by the formula (1.17), where γ is recovered via (1.18), (1.19), (1.21) and β is recovered via (1.22) and (1.25). Then, γ and β satisfy formulas (1.15) and (1.23), respectively. From (1.15), (1.22) and (1.23), it is apparent that

$$\gamma(x)J\gamma(x)^* = -I_p, \quad \gamma(x)J\beta(x)^* = 0, \quad \beta(x)J\beta(x)^* = I_p. \tag{1.26}$$

In particular, the second relation in (1.26) yields $\gamma' J\beta^* = -(\beta' J\gamma^*)^*$. Therefore, from (1.15), (1.17), and (1.23), we derive

$$2\mathrm{i}\begin{bmatrix}\beta'(x)\\ \gamma'(x)\end{bmatrix} J\begin{bmatrix}\beta(x)^* & \gamma(x)^*\end{bmatrix} = V\left(\frac{x}{2}\right). \tag{1.27}$$

Taking into account (1.26), we see that

$$\begin{bmatrix}\beta(x)\\ \gamma(x)\end{bmatrix} J\begin{bmatrix}\beta(x)^* & \gamma(x)^*\end{bmatrix} \equiv j. \tag{1.28}$$

In view of (1.28), we rewrite (1.27) as

$$\begin{bmatrix}\beta'(x)\\ \gamma'(x)\end{bmatrix} = \frac{\mathrm{i}}{2}jV\left(\frac{x}{2}\right)\begin{bmatrix}\beta(x)\\ \gamma(x)\end{bmatrix}. \tag{1.29}$$

Moreover, the second relations in (1.15) and (1.23) yield

$$\begin{bmatrix}\beta(0)\\ \gamma(0)\end{bmatrix} = \Theta^*. \tag{1.30}$$

Formulas (1.29) and (1.30) imply that

$$\begin{bmatrix}\beta(x)\\ \gamma(x)\end{bmatrix} = u\left(\frac{x}{2}, 0\right)\Theta^*. \tag{1.31}$$

In particular, relation (1.13) holds. In other words, formula (1.13) maps the Dirac system (with v constructed in (1.17)) into our initial system (1.11). It shows that the mapping is also surjective. □

Remark 1.2. Our proof also shows that formula (1.13) defines a one-to-one mapping of the Dirac systems with bounded on $[0, l/2]$ potentials into canonical systems of the form (1.11), (1.12), where $\gamma \in B^1([0, l])$ and (1.15) holds.

Remark 1.3. Requirement (1.15) in Proposition 1.1 and in Remark 1.2 can be substituted by the requirement

$$\gamma(x)J\gamma(x)^* = -I_p, \quad \gamma(0) = \frac{1}{\sqrt{2}}[-c \quad c], \quad c^*c = I_p. \tag{1.32}$$

In other words, if y satisfies (1.32) and is summable or bounded, we can construct $\hat{y}$ as also summable or bounded, respectively, and such that

$$H(x) = y(x)^* y(x) = \hat{y}(x)^* \hat{y}(x), \quad \hat{y}'(x) J \hat{y}(x)^* = 0, \quad \hat{y}(0) = \frac{1}{\sqrt{2}} [-I_p \quad I_p]. \tag{1.33}$$

For that purpose, we define $p \times p$ matrix function θ by the relations

$$\theta'(x) = \theta(x) y'(x) J y(x)^*, \quad \theta(0) = c^*. \tag{1.34}$$

From $yJy^* = -I_p$, it follows that $y'Jy^* + (y'Jy^*)^* = 0$, and so θ is unitary. Now, in view of (1.32) and (1.34), equalities (1.33) for $\hat{y} = \theta y$ are straightforward.

See also [252] for the transformations from this subsection. Earlier, some discussion of the connections between Krein and Dirac-type systems was given in [176] and the transformation of the Krein system into the canonical was discussed in [281].

Consider, finally, the skew-self-adjoint Dirac (Dirac-type) system

$$\frac{d}{dx} y(x,z) = (\mathrm{i} z j + jV(x)) y(x,z), \tag{1.35}$$

which is no less important than the self-adjoint system (0.3). (Notice that the term $\mathrm{i}jV$ on the right-hand side of (0.3) is substituted by jV in (1.35).) Skew-self-adjoint Dirac systems are equivalent [238] to a certain subclass of *modified* canonical systems $W' = \mathrm{i}zHW$, more precisely, of the systems:

$$\frac{d}{dx} W_d(x,z) = \mathrm{i} z H_d(x) W_d(x,z), \quad H_d(x) = y(x)^* y(x). \tag{1.36}$$

Indeed, put

$$W_d(x, 2z) = \exp(\mathrm{i}zx) u(x,0)^{-1} u(x,-z), \tag{1.37}$$

where u is the fundamental solution of (1.35) normalized by $u(0,z) = I_m$. Then, we have

$$\begin{aligned} \frac{d}{dx} W_d(x,2z) &= \mathrm{i}z W_d(x,2z) + \exp(\mathrm{i}zx) u(x,0)^{-1} (-\mathrm{i}zj) u(x,-z) \\ &= 2\mathrm{i}z u(x,0)^{-1} \begin{bmatrix} 0 & 0 \\ 0 & I_p \end{bmatrix} u(x,0) W_d(x,2z). \end{aligned} \tag{1.38}$$

By virtue of (1.35), we have $u(x,0)^* u(x,0) = I_m$, and thus equation (1.38) can be rewritten as (1.36), where

$$y(x) := [0 \quad I_p] u(x,0), \quad y(x) y(x)^* = I_p. \tag{1.39}$$

1.1.2 Schrödinger systems as a subclass of canonical systems

Following [281] (see also [290, Ch. 11]), consider connections between canonical and Schrödinger (Sturm–Liouville) systems. Denote by y_1 and y_2 the $p \times p$ matrix solutions of the Schrödinger system (0.1) satisfying initial conditions

$$y_1(0,z) = I_p, \quad y_1'(0,z) = 0, \quad y_2(0,z) = 0, \quad y_2'(0,z) = I_p. \tag{1.40}$$

It is easy to see that the matrix function

$$w(x,z) = \begin{bmatrix} y_1(x,z) & y_2(x,z) \\ y_1'(x,z) & y_2'(x,z) \end{bmatrix} \tag{1.41}$$

satisfies the system

$$\frac{d}{dx} w(x,z) = G(x,z)w(x,z), \quad G(x,z) = \begin{bmatrix} 0 & I_p \\ v(x) - zI_p & 0 \end{bmatrix}, \tag{1.42}$$

and $w(0,z) = I_m$. Now, introduce

$$W(x,z) = \Theta_1^* w(x,0)^{-1} w(x,z)\Theta_1, \quad \Theta_1 := J\Theta \operatorname{diag}\{\mathrm{i}I_p, I_p\}, \tag{1.43}$$

where Θ is defined in (1.4). In view of (1.4) and (1.43), we obtain

$$\Theta_1^* = \Theta_1^{-1}, \quad \Theta_1^* jJ\Theta_1 = \mathrm{i}J, \quad W(0,z) = I_m. \tag{1.44}$$

From (1.42), (1.43) and the first relation in (1.44), we derive

$$\frac{d}{dx} W(x,z) = \Theta_1^* w(x,0)^{-1} \begin{bmatrix} 0 & 0 \\ -zI_p & 0 \end{bmatrix} w(x,0)\Theta_1 W(x,z). \tag{1.45}$$

Assume that $v = v^*$, that is, the differential expression in (0.1) is self-adjoint. According to (1.42), we have

$$w(x,0)^* Jjw(x,0) = Jj, \quad \text{i.e.} \quad w(x,0)^{-1} = jJw(x,0)^* Jj. \tag{1.46}$$

From (1.44)–(1.46), it follows that

$$\frac{d}{dx} W(x,z) = \mathrm{i}zJH(x)W(x,z), \quad H(x) = \Theta_1^* w(x,0)^* \begin{bmatrix} I_p & 0 \\ 0 & 0 \end{bmatrix} w(x,0)\Theta_1. \tag{1.47}$$

In other words, we have

$$H(x) = \beta(x)^*\beta(x), \quad \beta(x) := \begin{bmatrix} I_p & 0 \end{bmatrix} w(x,0)\Theta_1, \quad \beta(x)J\beta(x)^* = 0. \tag{1.48}$$

Formulas (1.41), (1.43) and (1.47) transform solutions of the Schrödinger system into the fundamental solution and Hamiltonian of the corresponding canonical system.

1.1.3 Gauge transformations of the Dirac systems

In this subsection, we consider systems of the form $y' + (zq_1 + q_0)y = 0$, where $y' := dy/dx$. More precisely, we consider self-adjoint Dirac, skew-self-adjoint Dirac and the auxiliary to the N-wave equation systems on some interval $\mathcal{D}$ that contains 0. Using these simple examples, we demonstrate our GBDT approach to the construction of the Darboux matrices.

Dirac system (0.3) can be rewritten in the form

$$y'(x,z) + (zq_1 + q_0(x))\,y(x,z) = 0, \quad q_1 \equiv -\mathrm{i}j, \quad q_0(x) = -\mathrm{i}jV(x). \tag{1.49}$$

Fix an integer $n > 0$ and an $n \times n$ parameter matrix A, and define an $n \times m$ matrix function $\Pi(x)$ by the linear differential equation

$$\Pi'(x) = A\Pi(x)q_1 + \Pi(x)q_0(x) \tag{1.50}$$

and initial value $\Pi(0)$. Here, two other parameter matrices, $\Pi(0)$ and an $n \times n$ matrix $S(0) = S(0)^*$, are chosen so that the operator (matrix) identity

$$AS(0) - S(0)A^* = \mathrm{i}\Pi(0)j\Pi(0)^* \tag{1.51}$$

holds (i.e. A, $S(0)$ and $\Pi(0)$ form a symmetric S-node, see Definition 1.12). Finally, $S(x)$ is given by the equality

$$S' = \Pi\Pi^*. \tag{1.52}$$

From (1.50) and (1.52), we derive $(AS - SA^*)' = A\Pi\Pi^* - \Pi\Pi^*A^*$ and $(\mathrm{i}\Pi j\Pi^*)' = (\mathrm{i}\Pi' j\Pi^*) - (\mathrm{i}\Pi' j\Pi^*)^* = A\Pi\Pi^* - \Pi\Pi^*A^*$, i.e.,

$$(AS - SA^*)' = (\mathrm{i}\Pi j\Pi^*)'. \tag{1.53}$$

Formulas (1.51) and (1.53) imply the identity

$$AS(x) - S(x)A^* = \mathrm{i}\Pi(x)j\Pi(x)^* \tag{1.54}$$

for all $x \in \mathcal{D}$. Now, from (0.14) and (0.17) (putting $\Pi_1 = \Pi$ and $\Pi_2^* = \mathrm{i}j\Pi^*$), we obtain

$$w_A(x,z) = I_m - \mathrm{i}j\Pi(x)^*S(x)^{-1}(A - zI_n)^{-1}\Pi(x) \tag{1.55}$$

in the points of invertibility of $S(x)$. The matrix function w_A, which is constructed in this way, is a Darboux matrix (a gauge transformation) of the Dirac system. In order to show this, we differentiate w_A using (1.50), (1.52) and (1.54). First, we calculate the derivative of Π^*S^{-1}:

$$\left(\Pi^*S^{-1}\right)' = \mathrm{i}j\Pi^*A^*S^{-1} + \mathrm{i}Vj\Pi^*S^{-1} - \Pi^*S^{-1}\Pi\Pi^*S^{-1}. \tag{1.56}$$

Since (1.54) yields $A^*S^{-1} = S^{-1}A - \mathrm{i}S^{-1}\Pi\Pi^*S^{-1}$, we rewrite (1.56) as

$$\left(\Pi^*S^{-1}\right)' = \mathrm{i}j\Pi^*S^{-1}A + \left(\mathrm{i}Vj + j\Pi^*S^{-1}\Pi j - \Pi^*S^{-1}\Pi\right)\Pi^*S^{-1}. \tag{1.57}$$

Taking into account (1.50), (1.55) and (1.57), we derive

$$\begin{aligned}\frac{d}{dx}w_A = &-\mathrm{i}j\Big(\big(\mathrm{i}Vj + j\Pi^*S^{-1}\Pi j - \Pi^*S^{-1}\Pi\big)\Pi^*S^{-1}(A - zI_n)^{-1}\Pi \\ &+ \mathrm{i}j\Pi^*S^{-1}(A - zI_n + zI_n)(A - zI_n)^{-1}\Pi\Big) \\ &- \mathrm{i}j\Pi^*S^{-1}(A - zI_n)^{-1}\left(-\mathrm{i}(A - zI_n + zI_n)\Pi j - \mathrm{i}\Pi jV\right). \qquad (1.58)\end{aligned}$$

From (1.55) and (1.58), it follows that

$$\begin{aligned}\frac{d}{dx}w_A &= \big(\mathrm{i}jV + \Pi^*S^{-1}\Pi - j\Pi^*S^{-1}\Pi j\big)(w_A - I_m) + \Pi^*S^{-1}\Pi + \mathrm{i}zj(w_A - I_m) \\ &\quad - j\Pi^*S^{-1}\Pi j - (w_A - I_m)(\mathrm{i}zj + \mathrm{i}jV) \\ &= \big(\mathrm{i}zj + \mathrm{i}jV + \Pi^*S^{-1}\Pi - j\Pi^*S^{-1}\Pi j\big)w_A - w_A(\mathrm{i}zj + \mathrm{i}jV).\end{aligned}$$

Finally, rewrite the relation above as

$$\frac{d}{dx}w_A(x,z) = \mathrm{i}\left(zj + j\widetilde{V}(x)\right)w_A(x,z) - \mathrm{i}w_A(x,z)\left(zj + jV(x)\right), \qquad (1.59)$$

where

$$\widetilde{V} = \begin{bmatrix} 0 & \widetilde{v} \\ \widetilde{v}^* & 0 \end{bmatrix} = V + \mathrm{i}(\Pi^*S^{-1}\Pi j - j\Pi^*S^{-1}\Pi), \qquad (1.60)$$

$$\widetilde{v} = v - 2\mathrm{i}[I_p \quad 0]\Pi^*S^{-1}\Pi\begin{bmatrix} 0 \\ I_p \end{bmatrix}. \qquad (1.61)$$

Thus, the following proposition is proved.

Proposition 1.4. *Let Dirac system* (0.3) *and the triple of parameter matrices* A, $S(0) = S(0)^*$ *and* $\Pi(0)$ *be given, and let the equality* (1.51) *hold. Then, the matrix function* w_A *defined by* (1.55), *where* $\Pi(x)$ *and* $S(x)$ *are obtained using formulas* (1.50) *and* (1.52), *is a Darboux matrix (gauge transformation) of the Dirac system and satisfies equation* (1.59). *The fundamental solution* $\widetilde{u}$ $(\widetilde{u}(0,z) = I_m)$ *of the transformed Dirac system* $\widetilde{y}' = \mathrm{i}(zj + j\widetilde{V})\widetilde{y}$ *is given by the formula*

$$\widetilde{u}(x,z) = w_A(x,z)u(x,z)w_A(0,z)^{-1},$$

where u *is the fundamental solution of the initial system* (0.3).

Remark 1.5. Usually, when we talk about constructed fundamental solutions of the transformed systems, it is assumed that $\det S(x) \neq 0$ on the corresponding interval (or semiaxis). The case where $\det S(x)$ may turn to zero (i.e. the case of potentials with singularities) is considered in Chapter 8.

Remark 1.6. In the simplest but quite important case $V \equiv 0$ (i.e. $q_0 \equiv 0$), the matrix function Π is immediately recovered explicitly and is given by (2.53). After that, the matrix functions S, the potential $\widetilde{v}$ and the fundamental solution $\widetilde{u}$ are also constructed explicitly. This case (even for a somewhat more general class of matrices j) is dealt with at the beginning of Subsection 2.2.2.

The skew-self-adjoint Dirac system (1.35) can be presented as the first relation in (1.49), where $q_1 \equiv -\mathrm{i}j$, $q_0(x) = -jV(x)$, that is, q_1 coincides with the q_1 in (1.49), but q_0 is different. For the case of the skew-self-adjoint Dirac system, we define Π via (1.50), where $q_0(x) = -jV(x)$, and we define S by $S' = \Pi j \Pi^*$. The *matrix identity* requirement on parameter matrices takes the form

$$AS(0) - S(0)A^* = \mathrm{i}\Pi(0)\Pi(0)^*. \tag{1.62}$$

Then, the identity

$$AS(x) - S(x)A^* = \mathrm{i}\Pi(x)\Pi(x)^* \quad (x \in \mathcal{D}) \tag{1.63}$$

follows from (1.50), where $q_0(x) = -jV(x)$, from $S' = \Pi j \Pi^*$ and from (1.62). Now, the proof of our next proposition is quite similar to the proof of Proposition 1.4 and we omit it.

Proposition 1.7. *Let the skew-self-adjoint Dirac system* (1.35) *be given. Then, the matrix function*

$$w_A(x,z) = I_m - \mathrm{i}\Pi(x)^* S(x)^{-1}(A - zI_n)^{-1}\Pi(x), \tag{1.64}$$

where Π and S are defined above, is a Darboux matrix of (1.35) *and satisfies equation*

$$\frac{d}{dx} w_A(x,z) = \left(\mathrm{i}zj + j\widetilde{V}(x)\right) w_A(x,z) - w_A(x,z)\left(\mathrm{i}zj + jV(x)\right). \tag{1.65}$$

Here, we have

$$\widetilde{V} = \begin{bmatrix} 0 & \widetilde{v} \\ \widetilde{v}^* & 0 \end{bmatrix} = V + \Pi^* S^{-1}\Pi - j\Pi^* S^{-1}\Pi j. \tag{1.66}$$

Proposition 1.7 admits a generalization for the case of the linear system auxiliary to the N-wave (also called nonlinear optics) equation

$$\frac{d}{dx} y(x,z) = (\mathrm{i}zD - [D, \varrho(x)])\, y(x,z), \quad \varrho^* = B\varrho B, \quad [D,\varrho] := D\varrho - \varrho D, \tag{1.67}$$

$$D = \mathrm{diag}\,\{d_1, d_2, \ldots, d_m\} = D^*, \quad B = \mathrm{diag}\,\{b_1, b_2, \ldots, b_m\} \quad (b_k = \pm 1). \tag{1.68}$$

For the case of system (1.67), we have

$$q_1 \equiv -\mathrm{i}D, \quad q_0(x) = [D, \varrho(x)], \tag{1.69}$$

and we define Π via (1.50) and (1.69), whereas S is defined by its derivative $S' = \Pi D B \Pi^*$. The matrix identity takes the form $AS(x) - S(x)A^* = \mathrm{i}\Pi(x)B\Pi(x)^*$.

Proposition 1.8. *Let system* (1.67) *be given. Then, the matrix function*

$$w_A(x,z) = I_m - \mathrm{i}B\Pi(x)^* S(x)^{-1}(A - zI_n)^{-1}\Pi(x) \tag{1.70}$$

is a Darboux matrix of (1.67) *and satisfies equation*

$$\frac{d}{dx} w_A(x,z) = (\mathrm{i}zD - [D, \widetilde{\varrho}(x)])\, w_A(x,z) - w_A(x,z)\, (\mathrm{i}zD - [D, \varrho(x)])\,. \quad (1.71)$$

Here, we have

$$\widetilde{\varrho} = \varrho - B\Pi^* S^{-1}\Pi, \quad \widetilde{\varrho}^* = B\widetilde{\varrho}B. \quad (1.72)$$

Propositions 1.4–1.8 are particular subcases of the general GBDT version (of the Bäcklund–Darboux transformation) for systems with rational dependence on the spectral parameter z, which will be presented in Chapter 7.

Remark 1.9. When A is a scalar (i.e. $n = 1$), we see from (1.50) that A is an eigenvalue and Π is an eigenfunction of the dual to (1.49) system $\widehat{y}' = (zq_1 + q_0)\widehat{y}$. In the general situation, we call A a generalized matrix eigenvalue and Π - a generalized eigenfunction.

1.2 S-nodes and high energy asymptotics of the Weyl functions

In this section, we shall formulate some basic results from [276, 280, 289, 290] on the S-nodes (see also references therein) to make the book self-sufficient to a great extent. The results will be actively used in this and following chapters. In particular, in the subsection 1.2.3, using S-node theory, we obtain the high energy asymptotics of the Weyl functions for canonical systems corresponding to the S-nodes ([236, 252]). Various other results on the high energy asymptotics of the Weyl functions and the procedure to solve inverse problems follow in the next chapters.

1.2.1 Elementary properties of S-nodes

Recall that the linear operators A, B, S, Π_k, Γ_k $(k = 1, 2)$, which form an S-node, satisfy (0.12) (i.e. they are bounded) and equalities

$$AS - SB = \Pi_1\Pi_2^*, \quad S\Gamma_1 = \Pi_1, \quad \Gamma_2^* S = \Pi_2^*. \quad (1.73)$$

In this section, we assume that (0.12) and (1.73) hold. Introduce the operator-valued transfer matrix functions from [153] in the L. Sakhnovich (e.g. [276]) form:

$$w_A(\lambda) = I - \Gamma_2^*(A - \lambda I)^{-1}\Pi_1, \quad (1.74)$$

$$w_B(\lambda) = I + \Pi_2^*(B - \lambda I)^{-1}\Gamma_1. \quad (1.75)$$

Representation (1.74) takes roots in the Livšic–Brodskii characteristic matrix function (recall Chapter 0). A characteristic matrix function has been (and still is) extremely fruitfully used in the problems of the unitary and linear operator similarity, factorization and scattering problems, and applications [8, 23, 32, 59, 201, 203, 317]. Transfer matrix function (1.74) is also successfully used in these domains as well as in the inversion, interpolation, and direct and inverse spectral problems and nonlinear integrable equations (see results and references in [235–238, 240–249, 251–254, 256–265, 267, 268, 270, 271, 276, 277, 280–287, 289, 290]).

Theorem 1.10 ([290]). *Assume that z and ζ do not belong to the spectrum of A and B, respectively. Then, we have*

$$w_A(z)w_B(\zeta) = I - (z-\zeta)\Gamma_2^*(A-zI)^{-1}S(B-\zeta I)^{-1}\Gamma_1. \tag{1.76}$$

Proof. Using (1.74) and (1.75), we obtain

$$\begin{aligned} w_A(z)w_B(\zeta) &= I - \Gamma_2^*(A-zI)^{-1}\Pi_1 + \Pi_2^*(B-\zeta I)^{-1}\Gamma_1 \\ &\quad - \Gamma_2^*(A-zI)^{-1}\Pi_1\Pi_2^*(B-\zeta I)^{-1}\Gamma_1. \end{aligned} \tag{1.77}$$

It follows from the operator identity $AS - SB = \Pi_1\Pi_2^*$ in (1.73) that

$$\Pi_1\Pi_2^* = (A-zI)S - S(B-\zeta I) + (z-\zeta)S. \tag{1.78}$$

Finally, we substitute (1.78) into (1.77) and use the last two equalities in (1.73) in order to derive (1.76). □

The next theorem is similar to Theorem 1.10.

Theorem 1.11. *Let operator S have a bounded inverse and assume that z and ζ do not belong to the spectrum of A and B, respectively. Then, we have*

$$w_B(\zeta)w_A(z) = I - (z-\zeta)\Pi_2^*(B-\zeta I)^{-1}S^{-1}(A-zI)^{-1}\Pi_1. \tag{1.79}$$

Proof. Using definitions (1.74) and (1.75) and the equalities

$$\Gamma_1 = S^{-1}\Pi_1, \quad \Gamma_2^* = \Pi_2^*S^{-1}, \tag{1.80}$$

which are immediate from (1.73), we derive the formula

$$\begin{aligned} w_B(\zeta)w_A(z) &= I + \Pi_2^*(B-\zeta I)^{-1}S^{-1}\Pi_1 - \Pi_2^*S^{-1}(A-zI)^{-1}\Pi_1 \\ &\quad - \Pi_2^*(B-\zeta I)^{-1}S^{-1}\Pi_1\Pi_2^*S^{-1}(A-zI)^{-1}\Pi_1, \end{aligned} \tag{1.81}$$

which is similar to (1.77). Next, taking into account (1.78) and (1.81), we obtain (1.79). □

When $z = \zeta$, we have

$$w_A(z)w_B(z) = I. \tag{1.82}$$

Definition 1.12. An S-node is called symmetric if

$$S = S^*, \ B = A^*, \ \Pi_2^* = \mathrm{i}J\Pi_1^*, \ \Gamma_2^* = \mathrm{i}J\Gamma_1^* \quad (J = J^* = J^{-1}). \tag{1.83}$$

When S has a bounded inverse, we also say that the considered triple A, S, $\Pi := \Pi_1$ (or $\{A, S, \Pi\}$ if the triple is written more precisely) forms a symmetric S-node.

Let us rewrite (1.82) for the case that the S-node is symmetric.

Corollary 1.13 ([276]). *In the case of the symmetric S-nodes, the reflection principle holds for $w_A(\lambda)$, that is,*

$$w_A(\lambda)Jw_A(\overline{\lambda})^* = w_A(\overline{\lambda})^*Jw_A(\lambda) = J. \tag{1.84}$$

If the S-node is symmetric, we omit the index in Π_1. In view of (1.80) and (1.83), for the symmetric S-node with invertible operator S, relations (1.74) and (1.75) take the form

$$w_A(\lambda) = I - \mathrm{i}J\Pi^*S^{-1}(A - \lambda I)^{-1}\Pi \quad (\Pi := \Pi_1) \tag{1.85}$$

$$w_B(\lambda) = I + \mathrm{i}J\Pi^*(A^* - \lambda I)^{-1}S^{-1}\Pi = Jw_A(\overline{\lambda})^*J. \tag{1.86}$$

Remark 1.14. Recall that *matrix* symmetric S-nodes were used already in Subsection 1.1.3 to construct Darboux matrices of the form (1.85). Darboux matrices are essential in the construction of fundamental solutions for the case that explicit formulas are available. Furthermore, we shall also use (in a different way) *operator* symmetric S-nodes for the construction of fundamental solutions in general situations (see e.g. Theorem 1.20).

Taking into account (1.86), we rewrite (1.76) and (1.79) (see the corollary below).

Corollary 1.15. *In the case of the symmetric S-nodes, where S has a bounded inverse, we have*

$$w_A(z)Jw_A(\overline{\zeta})^* = J - \mathrm{i}(z-\zeta)J\Pi^*S^{-1}(A - zI)^{-1}S(A-\zeta I)^{-1}S^{-1}\Pi J, \tag{1.87}$$

$$w_A(\overline{\zeta})^*Jw_A(z) = J - \mathrm{i}(z-\zeta)\Pi^*(A^* - \zeta I)^{-1}S^{-1}(A - zI)^{-1}\Pi. \tag{1.88}$$

1.2.2 Continual factorization

Continual factorization results are based on the factorization theorem [276] (see also, e.g. [290, Theorem 2.1, p. 21]). As before, we assume that the operators A, B, S, Π_k ($k = 1, 2$) form an S-node (i.e. they satisfy (0.12) and the equalities (1.73)). Assume also that the Hilbert space $\mathbf{H}$ from (0.12) is an orthogonal sum of Hilbert subspaces $\mathbf{H}_1$ and $\mathbf{H}_2$ ($\mathbf{H} = \mathbf{H}_1 \oplus \mathbf{H}_2$) and denote by P_k the orthogonal projector acting from $\mathbf{H}$ onto $\mathbf{H}_\mathbf{k}$. It is apparent that P_k^* is the natural embedding of $\mathbf{H}_\mathbf{k}$ in $\mathbf{H}$. We put

$$R_{jk} = P_jRP_k^* \in B(\mathbf{H}_\mathbf{k}, \mathbf{H}_\mathbf{j}) \quad \text{for } R \in B(\mathbf{H}) \quad (j, k = 1, 2).$$

Theorem 1.16 ([276]). *Suppose that the operator-valued function $w(\lambda)$ admits a realization (1.74) such that (1.73) holds and the bounded operator S is also boundedly invertible, $T := S^{-1}$. If S_{11} is boundedly invertible on $\mathbf{H}_1$ and if*

$$A_{12} = P_1AP_2^* = 0, \quad P_2BP_1^* = B_{21} = 0, \tag{1.89}$$

then

$$w(\lambda) = w_2(\lambda) w_1(\lambda), \tag{1.90}$$

where

$$w_1(\lambda) = I - \Pi_2^* P_1^* S_{11}^{-1} (A_{11} - \lambda I)^{-1} P_1 \Pi_1, \tag{1.91}$$

$$w_2(\lambda) = I - \Gamma_2^* P_2^* (A_{22} - \lambda I)^{-1} T_{22}^{-1} P_2 \Gamma_1. \tag{1.92}$$

As applications of Theorem 1.16, the continuous factorization Theorems 1.1 and 1.2 from [290, pp. 40–42] are derived, see also [280]. Here, we will only need their reduction below for the symmetric S-node case ([290, p. 54]). Recall also that we assume $\dim \mathbf{G} < \infty$ in this book.

Theorem 1.17. *Suppose that the operator-valued function $w(\lambda)$ satisfies the following conditions:*

(i) *$w(\lambda)$ admits a realization of the form* (1.85), *where the operator identity*

$$AS - SA^* = \mathrm{i}\Pi J \Pi^* \tag{1.93}$$

holds, and the operator S is a bounded and strictly positive operator.

(ii) *The spectrum $\sigma(A)$ of the operator A is located at zero (i.e. $\sigma(A) = 0$). Moreover, there exists a family $\{P_\xi\}$ of orthogonal projectors (acting from $\mathbf{H}$ onto the subspaces $\mathbf{H}_\xi \in \mathbf{H}$):*

$$P_\xi \quad (a \le \xi \le b), \quad P_a = 0, \quad P_b = I,$$

so that the family H_ξ is increasing,

$$(I - P_\xi) A^* P_\xi^* = 0, \quad a \le \xi \le b, \tag{1.94}$$

and for the projectors $(P_{\xi+\Delta\xi} - P_\xi) \in B(\mathbf{H}, \mathbf{H}_{\xi+\Delta\xi} \ominus \mathbf{H}_\xi)$ and $(P_\xi - P_{\xi-\Delta\xi}) \in B(\mathbf{H}, \mathbf{H}_\xi \ominus \mathbf{H}_{\xi-\Delta\xi})$, where $\Delta\xi > 0$, we have

$$\|(P_{\xi+\Delta\xi} - P_\xi) A (P_{\xi+\Delta\xi} - P_\xi)^*\| \le M\Delta\xi, \tag{1.95}$$

$$\|(P_{\xi\pm\Delta\xi} - P_\xi) f\| \to 0 \quad \text{for } \Delta\xi \to 0 \quad (\forall f \in \mathbf{H}). \tag{1.96}$$

Then, $w(\lambda)$ admits the following representation:

$$w(\lambda) = \overset{\frown}{\int_a^b} \exp\left\{\frac{1}{\lambda} d\mathcal{B}(\xi)\right\}, \tag{1.97}$$

where $\overset{\frown}{\int}$ denotes the multiplicative integral, and $\mathcal{B}(\xi)$ is a continuous matrix function on $[a, b]$, which is given by

$$\mathcal{B}(\xi) = \mathrm{i} J \Pi^* P_\xi^* S_\xi^{-1} P_\xi \Pi, \quad S_\xi := P_\xi S P_\xi^* \in B(P_\xi \mathbf{H}), \tag{1.98}$$

and whose variation is bounded in norm.

Example 1.18. In the space $L^2_r(0, l)$ of vector functions

$$f(t) = \mathrm{col}\,[f_1(t), f_2(t), \dots, f_r(t)] \quad (\text{col means column}),$$

consider the family of orthogonal projectors

$$(P_\xi f)(x) = f(x) \quad (0 < x < \xi), \quad P_\xi \in B(L^2_r(0, l), L^2_r(0, \xi)). \tag{1.99}$$

Then, for

$$Af = \mathrm{i}D\int_0^x f(t)dt, \tag{1.100}$$

$$D = \mathrm{diag}\,\{d_1, d_2, \dots, d_m\}, \tag{1.101}$$

condition (ii) of Theorem 1.17 is valid.

When the matrix function $\Pi^* P^*_\xi S^{-1}_\xi P_\xi \Pi$ is absolutely continuous, we obtain

$$w(x,\lambda) := \int_a^{\curvearrowleft x} \exp\left\{\frac{1}{\lambda} d\mathcal{B}(\xi)\right\} = \int_a^{\curvearrowleft x} \exp\left\{\frac{\mathrm{i}}{\lambda} JH(\xi)d\xi\right\}, \tag{1.102}$$

where

$$H(\xi) := \frac{d}{d\xi}\left(\Pi^* P^*_\xi S^{-1}_\xi P_\xi \Pi\right). \tag{1.103}$$

In other words, $w(\lambda) = w(b, \lambda)$ is the so called matrizant of the system

$$\frac{d}{dx} w(x,\lambda) = (\mathrm{i}/\lambda)JH(x)w(x,\lambda). \tag{1.104}$$

In particular, we will be interested in the important subcases $J = I_{2p}$ and

$$J = \begin{bmatrix} 0 & I_p \\ I_p & 0 \end{bmatrix}. \tag{1.105}$$

Remark 1.19. According to [290, p. 42], where Theorem 1.2 is proved, the matrix function $-\mathrm{i}J\mathcal{B}(\xi)$ is nondecreasing and so $H(\xi)$ in (1.103) satisfies the inequality $H(\xi) \geq 0$ $(a \leq \xi \leq b)$.

From Theorem 1.17, the next theorem easily follows.

Theorem 1.20. *Suppose that the triple $\{A, S, \Pi\}$ forms a symmetric S-node and satisfies three groups of additional conditions:*

(i) *The operator S is a bounded and strictly positive operator.*

(ii) *The conditions (ii) of Theorem 1.17 (where $a = 0$ and $b = l$) are fulfilled.*

(iii) *The matrix function $\Pi^* P^*_\xi S^{-1}_\xi P_\xi \Pi$ is absolutely continuous.*

Then, the fundamental solution W of the system

$$\frac{d}{dx}y(x,z) = \mathrm{i}zJH(x)y(x,z), \quad H(x) \geq 0, \quad 0 \leq x \leq l, \tag{1.106}$$

where H has the form (1.103), *admits a representation*

$$W(\xi,z) = I_m + \mathrm{i}zJ\Pi^* P_\xi^* S_\xi^{-1}(I - zA_\xi)^{-1} P_\xi \Pi. \tag{1.107}$$

We recall that $J = J^* = J^{-1}$, S_ξ denotes in (1.98) the reduction $P_\xi S P_\xi^*$ of S and the corresponding reductions of other operators are denoted in a similar way (e.g. $A_\xi := P_\xi A P_\xi^*$ in the formula above).

Remark 1.21. In order to prove Theorem 1.20, we should recall the definition (1.85) of w_A and rewrite (1.107) in the equivalent form

$$W(\xi,z) = w_A(\xi, 1/z), \tag{1.108}$$

where $w_A(\xi,\cdot)$ is the transfer matrix function generated by the triple $\{A_\xi,\ S_\xi,\ P_\xi\Pi\}$. Now, Theorem 1.20 is apparent from Theorem 1.17.

Remark 1.22. We note that the right-hand side of (1.107) makes no sense when $\xi = 0$. However, condition (1.96) means that W, given by (1.107), has the property

$$\lim_{x\to 0} W(x,z) = I_m. \tag{1.109}$$

Hence, we can assume that $W(0,z) = I_m$. Therefore, indeed, W given by (1.107) is the fundamental solution of (1.106), which is normalized by I_m at $x = 0$.

1.2.3 Canonical systems and representation of the corresponding S-nodes

In this and the next subsections, we assume that the matrix J is given by (1.105). Without loss of generality, we put $a = 0$ and $b = l$, that is, we consider systems in this subsection (and, usually, further in the book) on the interval $[0,\ l]$ instead of the interval $[a,\ b]$.

The spectral theory of system (1.106), where J is given by (1.105), that is, of the canonical system

$$\frac{d}{dx}y(x,z) = \mathrm{i}z\begin{bmatrix} 0 & I_p \\ I_p & 0 \end{bmatrix} H(x)y(x,z), \quad H(x) \geq 0, \quad x \geq 0, \tag{1.110}$$

has been treated in [236, 280, 290] using the S-node theory approach. For that, we consider symmetric S-nodes where $\mathbf{G} = \mathbb{C}^m$ and $m = 2p$. We partition the operator Π into two blocks $\Pi = [\Phi_1 \quad \Phi_2]$, acting from $\mathbb{C}^p$ into $\mathbf{H}$. The operator identity (1.93), which the triple A, S, Π should satisfy, takes the form

$$AS - SA^* = \mathrm{i}\Pi J\Pi^* = \mathrm{i}(\Phi_1\Phi_2^* + \Phi_2\Phi_1^*) \quad (\Pi = [\Phi_1 \quad \Phi_2]). \tag{1.111}$$

General-type canonical systems (1.110) (without the requirement for the S-nodes to exist) were studied in [138, 240] (see also Appendix A and references therein). However, far less results may be obtained for the general case. The fundamental solution W of system (1.110) is normalized by the condition

$$W(0,z) = I_m. \tag{1.112}$$

Definition 1.23. We say that the symmetric S-node (or the triple $\{A, S, \Pi\}$, which forms this S-node) corresponds to a canonical system and that the system corresponds to a symmetric S-node (or triple) if S is strictly positive, J has the form (1.105), $\sigma(A) = 0$ and for some family of orthoprojectors P_ξ (satisfying condition (ii) of Theorem 1.20), the equalities (1.103) and (1.107) hold.

The solutions of inverse problems treated in [280, 289] are based on the recovery of the corresponding S-nodes from the fixed operators A and Φ_2 and the spectral function τ. This problem is closely related to the interpolation problem of representation of the operators S and Φ_1 in the form

$$S = \int_{-\infty}^{\infty} (I - tA)^{-1}\Phi_2(d\tau(t))\Phi_2^*(I - tA^*)^{-1}, \tag{1.113}$$

$$\Phi_1 = \mathrm{i}\left(\Phi_2\nu - \int_{-\infty}^{\infty}\left(A(I - tA)^{-1} + \frac{t}{t^2+1}I\right)\Phi_2 d\tau(t)\right). \tag{1.114}$$

Notation 1.24. *We denote by $\mathcal{N}(S,\Phi_1)$ the set of functions φ of the form*

$$\varphi(z) = \nu + \int_{-\infty}^{\infty}\left(\frac{1}{t-z} - \frac{t}{1+t^2}\right) d\tau(t) \quad (\nu = \nu^*), \tag{1.115}$$

where the nondecreasing matrix function τ satisfies the inequality

$$\int_{-\infty}^{\infty}\frac{d\tau(t)}{1+t^2} < \infty, \tag{1.116}$$

the integrals in the right-hand sides of (1.113) *and* (1.114) *weakly converge, and the equalities* (1.113) *and* (1.114) *hold.*

The set $\mathcal{N}(S,\Phi_1)$ is described in Theorem 2.4 [290, Ch. 4], using the set of linear-fractional (Möbius) transformations $\mathcal{N}(\mathfrak{A}_A)$ (see Notation 1.25 below).

Notation 1.25. *We denote by $\mathcal{N}(\mathfrak{A}_A)$ the set of functions φ of the form*

$$\varphi(z) = \mathrm{i}\,(a(z)\mathcal{P}_1(z) + b(z)\mathcal{P}_2(z))\,(c(z)\mathcal{P}_1(z) + d(z)\mathcal{P}_2(z))^{-1}, \tag{1.117}$$

$$\mathfrak{A}_A(z) = \begin{bmatrix} a(z) & b(z) \\ c(z) & d(z)\end{bmatrix} := (I_m + \mathrm{i}\overline{z}J\Pi^*S^{-1}(I - \overline{z}A)^{-1}\Pi)^*, \tag{1.118}$$

where the blocks a, b, c, d are $p \times p$ matrix functions and the pairs $\mathcal{P}_1, \mathcal{P}_2$ are so called nonsingular pairs with property-J.

Recall that according to Definition 0.1, the nonsingular pair $\mathcal{P}_1, \mathcal{P}_2$ with property-J is a pair of the $p \times p$ matrix functions $\mathcal{P}_1, \mathcal{P}_2$, which are meromorphic in $\mathbb{C}_+$ and satisfy (excluding, possibly, isolated points) relations

$$\begin{bmatrix} \mathcal{P}_1(z)^* & \mathcal{P}_2(z)^* \end{bmatrix} \begin{bmatrix} \mathcal{P}_1(z) \\ \mathcal{P}_2(z) \end{bmatrix} > 0, \quad \begin{bmatrix} \mathcal{P}_1(z)^* & \mathcal{P}_2(z)^* \end{bmatrix} J \begin{bmatrix} \mathcal{P}_1(z) \\ \mathcal{P}_2(z) \end{bmatrix} \geq 0. \tag{1.119}$$

Remark 1.26. Under conditions of Theorem 1.20, it follows from (1.107) and (1.118) that

$$\mathfrak{A}_A(z) = W(l, \overline{z})^* = \mathfrak{A}(z). \tag{1.120}$$

Hence, the set $\mathcal{N}(\mathfrak{A}_A)$, which is given by (1.117) and (1.118), coincides with the set $\mathcal{N}(\mathfrak{A})$ of the Weyl functions of canonical system (1.110) on $[0, l]$ (Definition 0.3).

Theorem 1.27 ([290, Ch. 4]). *Let a symmetric S-node (where J is given by* (1.105)*) satisfy conditions:*

(i) *The operator S is a bounded and strictly positive operator.*

(ii) $\sigma(A) = \{0\}$ *and zero is not the eigenvalue of* A.

(iii) *The equalities* $\mathrm{Im}\,(A) \cap \mathrm{Im}\,(\Phi_2) = 0$ *and* $\mathrm{Ker}\,(\Phi_2) = 0$ *hold.*

Then, we have $\mathcal{N}(S, \Phi_1) = \mathcal{N}(\mathfrak{A}_A)$.

Let a symmetric S-node, some matrix $\nu = \nu^*$ and a nondecreasing matrix function τ satisfy relations (1.113), (1.114) and (1.116). Now, define matrix function φ by the Herglotz (or Nevanlinna) formula (1.115) in the domain $\mathbb{C}_+ \cup \mathbb{C}_-$. (More often, φ is considered, in this book, in $\mathbb{C}_+$.) We have

$$\frac{\varphi(z) - \varphi(z)^*}{z - \overline{z}} = \int_{-\infty}^{\infty} \frac{d\tau(t)}{(t - z)(t - \overline{z})} \quad (z \neq \overline{z}). \tag{1.121}$$

In view of (1.114) and (1.115), we obtain

$$\begin{aligned} \omega(z) :&= (I - zA)^{-1}(\Phi_2 \varphi(z) + \mathrm{i}\Phi_1) \\ &= (I - zA)^{-1} \int_{-\infty}^{\infty} \left(A(I - tA)^{-1} + \frac{1}{t - z} I \right) \Phi_2 d\tau(t). \end{aligned} \tag{1.122}$$

Taking into account the equality

$$(I - tA)^{-1} - (I - zA)^{-1} = (t - z)(I - zA)^{-1} A (I - tA)^{-1}, \tag{1.123}$$

we rewrite (1.122) in the form

$$\omega(z) = \int_{-\infty}^{\infty} \frac{(I - tA)^{-1}}{t - z} \Phi_2 d\tau(t). \tag{1.124}$$

Using (1.113), (1.121) and (1.124), we get a Potapov-type (compare with [212, 224] and FMI in Appendix A) inequality

$$\begin{bmatrix} S & \omega(z) \\ \omega(z)^* & \frac{\varphi(z)-\varphi(z)^*}{z-\overline{z}} \end{bmatrix} \geq 0. \tag{1.125}$$

The next inequality follows directly from (1.125):

$$||\omega(z)|| \leq ||S||^{\frac{1}{2}} \left\| \frac{\varphi(z) - \varphi(z)^*}{z - \overline{z}} \right\|^{\frac{1}{2}} \quad (z \neq \overline{z}). \tag{1.126}$$

Suppose Ker $\Phi_2 = 0$ and thus $\det(\Phi_2^*(I - zA)^{-1}\Phi_2) \not\equiv 0$. Then, (1.122) yields

$$\varphi(z) = \left(\Phi_2^*(I - zA)^{-1}\Phi_2\right)^{-1} \left(\Phi_2^*\omega(z) - \mathrm{i}\Phi_2^*(I - zA)^{-1}\Phi_1\right). \tag{1.127}$$

According to (1.116), (1.121) and (1.126), for any $\delta > 0$, we have

$$||\omega(z)|| \to 0 \quad (|z| \to \infty, \quad |\Im(z)/\Re(z)| > \delta).$$

Therefore, from (1.127), we derive

$$\begin{aligned} \varphi(z) = &- \mathrm{i}\left(\Phi_2^*(I - zA)^{-1}\Phi_2\right)^{-1} \Phi_2^*(I - zA)^{-1}\Phi_1 \\ &+ o(1)\left(\Phi_2^*(I - zA)^{-1}\Phi_2\right)^{-1} \end{aligned} \tag{1.128}$$

for $|z| \to \infty$, $|\Im(z)/\Re(z)| > \delta$. Since $\varphi(\overline{z}) = \varphi(z)^*$, we rewrite (1.128) in the equivalent form

$$\begin{aligned} \varphi(z)^* = &- \mathrm{i}\left(\Phi_2^*(I - \overline{z}A)^{-1}\Phi_2\right)^{-1} \Phi_2^*(I - \overline{z}A)^{-1}\Phi_1 \\ &+ o(1)\left(\Phi_2^*(I - \overline{z}A)^{-1}\Phi_2\right)^{-1}. \end{aligned} \tag{1.129}$$

Hence, using Theorem 1.27, we obtain our next theorem.

Theorem 1.28. *Let an S-node satisfy the conditions of Theorem* 1.27. *Then, for any $\varphi \in \mathcal{N}(\mathfrak{A}_A)$ and for any $\delta > 0$, the equalities* (1.128) *and* (1.129) *hold, when $|z| \to \infty$ in the domain $\Im(z)/|\Re(z)| > \delta$.*

Theorem 1.28 is basic in our approach to the high energy asymptotics of the Weyl functions and Borg–Marchenko-type results. (Compare it with [236, Lemma 1].)

1.2.4 Asymptotics of the Weyl functions, a special case

Consider now the subclass of the canonical systems (1.110), where the Hamiltonians H are degenerate (of rank $p = m/2$), having the form

$$H(x) = \gamma(x)^*\gamma(x), \tag{1.130}$$

and y are differentiable $p \times m$ matrix functions, which satisfy relations

$$y(x)Jy(x)^* = -I_p; \quad \sup_{0<x<l} \|y'(x)\| < \infty; \quad y(0) = \frac{1}{\sqrt{2}}[-c \quad c], \quad c^*c = I_p. \tag{1.131}$$

This subclass of the canonical systems is equivalent to the class of Dirac-type systems with bounded potentials (Remarks 1.2 and 1.3). The S-node, which corresponds to system (1.110) (where the Hamiltonian satisfies (1.130) and (1.131)), is constructed [236, 281, 290] using the operator

$$K = \mathrm{i}y(x)J\int_0^x y(t)^* \cdot dt \in B(L_p^2(0, l)). \tag{1.132}$$

A more general class of operators

$$K = \int_0^x K(x,t) \cdot dt, \quad K(x,t) := F(x)G(t) \tag{1.133}$$

is also of interest. Here, $F(x)$ is a $p \times r$ matrix function, $G(t)$ is an $r \times p$ matrix function, K is acting in $L_p^2(0, l)$ and the identity

$$\mathrm{i}F(x)G(x) \equiv I_p \tag{1.134}$$

is valid. The following proposition is a particular case of Theorem 1 [273].

Proposition 1.29. *Let F and G be boundedly differentiable and let* (1.134) *hold. Then, the operator K, which is given by* (1.133), *is similar to the operator of integration. More precisely, we have*

$$K = \mathcal{E}A\mathcal{E}^{-1}, \quad A := -\mathrm{i}\int_0^x \cdot\, dt \in B(L_p^2(0, l)), \tag{1.135}$$

where $\mathcal{E}$ is a lower triangular operator of the form

$$(\mathcal{E}f)(x) = \kappa(x)f(x) + \int_0^x \mathcal{E}(x,t)f(t)dt, \quad \frac{d}{dx}\kappa = \mathrm{i}F'G\kappa, \quad \det\kappa(0) \neq 0. \tag{1.136}$$

Moreover, the operators $\mathcal{E}^{\pm 1}$ map functions from $B^1([0,l])$ into $B^1([0,l])$.

Remark 1.30. We call operator $\mathcal{E}$ (from the similarity transformation (1.135)) a transformation operator.

An analog of Proposition 1.29 for the case that $F, G \in C^1([0,l])$ is stated as Proposition 4.1 in [20]. We recall that $B^N([0, l])$ ($C^N([0, l])$) denotes the set of N

times differentiable functions or matrix functions on $[0, l]$ such that $\sup f^{(N)} < \infty$ ($f^{(N)}$ is continuous). Substitute

$$F = \gamma, \quad G = \mathrm{i}J\gamma^* \tag{1.137}$$

into (1.133) and (1.134) to see that K of the form (1.132) satisfies conditions of Proposition 1.29. Therefore, taking into account (1.132) and (1.135), we obtain

$$K - K^* = \mathcal{E}A\mathcal{E}^{-1} - (\mathcal{E}^{-1})^*A^*\mathcal{E}^* = \mathrm{i}\gamma(x)J\int_0^l \gamma(t)^* \cdot dt. \tag{1.138}$$

The second equality in (1.138) implies that the operators given by

$$A = -\mathrm{i}\int_0^x \cdot dt, \quad S = \mathcal{E}^{-1}(\mathcal{E}^{-1})^*, \quad \Pi g = -\mathcal{E}^{-1}\gamma g \tag{1.139}$$

form a symmetric S-node, that is, the operator identity (1.93) holds.

The choice of $\mathcal{E}$ in (1.135) is not unique and so $\mathcal{E}$ admits normalization [274, pp. 103, 104] (see also papers [106, 236]), which is discussed in the proposition below.

Proposition 1.31. *Let K be given by* (1.132) *and let $\mathcal{E}$ be a transformation operator from Proposition* 1.29. *Suppose $\mathcal{E}_0 \in B(L_p^2(0, l))$ is determined by the equalities*

$$(\mathcal{E}_0 f)(x) = \kappa(0)^{-1}\gamma_2(0)f(x) + \int_0^x \mathcal{E}_0(x-t)f(t)dt, \quad \gamma_2 := \gamma\begin{bmatrix} 0 \\ I_p \end{bmatrix}, \tag{1.140}$$

$$\mathcal{E}_0(x) := \left(\mathcal{E}^{-1}\gamma_2\right)'(x). \tag{1.141}$$

Then, the operator $E := \mathcal{E}\mathcal{E}_0$ is another transformation operator from Proposition 1.29 *(i.e. $K = EAE^{-1}$ and $E^{\pm 1}$ map functions from $B^1([0, l])$ into $B^1([0, l])$). It is a lower triangular operator and admits representation*

$$Ef = \gamma_2(0)f + \int_0^x E(x, t)f(t)dt. \tag{1.142}$$

Moreover, E satisfies the additional condition

$$E^{-1}\gamma_2 \equiv I_p. \tag{1.143}$$

Proof. The next identity can be easily shown directly (and follows also from the fact that $\mathcal{E}_0$ is a lower triangular convolution operator):

$$A\mathcal{E}_0 = \mathcal{E}_0 A, \quad A := -\mathrm{i}\int_0^x \cdot dt. \tag{1.144}$$

Furthermore, because of (1.131), we have $\det y_2(0) \neq 0$, and so $\mathcal{E}_0$ is invertible. Hence, equalities (1.144) and $K = \mathcal{E}A\mathcal{E}^{-1}$ imply that

$$K = EAE^{-1} \quad \text{for } E = \mathcal{E}\mathcal{E}_0. \tag{1.145}$$

Formulas (1.140) and (1.141) lead us to the equality

$$\begin{aligned}\left(\mathcal{E}_0 I_p\right)(x) &= \kappa(0)^{-1}y_2(0) + \int_0^x \mathcal{E}_0(x-t)dt = \kappa(0)^{-1}y_2(0) + \int_0^x \mathcal{E}_0(t)dt \\ &= \kappa(0)^{-1}y_2(0) + \left(\mathcal{E}^{-1}y_2\right)(x) - \left(\mathcal{E}^{-1}y_2\right)(0).\end{aligned} \tag{1.146}$$

Using (1.136), we obtain $(\mathcal{E}^{-1}y_2)(0) = \kappa(0)^{-1}y_2(0)$ and rewrite (1.146) as

$$\left(\mathcal{E}_0 I_p\right)(x) = \left(\mathcal{E}^{-1}y_2\right)(x). \tag{1.147}$$

Hence, equality (1.143) follows. It remains to be shown that $\mathcal{E}_0^{\pm 1}$ maps $B^1([0,l])$ into $B^1([0,l])$. First, note that the integral operators $\mathcal{E}_0$, $\mathcal{E}_0^{-1}$ have bounded kernels and map bounded functions into bounded functions. In particular, for $\mathcal{E}_0^{-1}$, it follows from a series representation of the operator of the form $(I + \int_0^x k(x,t)\ \cdot\ dt)^{-1}$ (where the kernel k is bounded). Now, taking into account that any $f \in B^1([0,l])$ admits representation $f = f(0) + \mathrm{i}Af'$, we see that formulas (1.144) and (1.147) yield the fact that $\mathcal{E}_0$ maps $B^1([0,l)]$ into $B^1([0,l])$.

The identity $A\mathcal{E}_0^{-1} = \mathcal{E}_0^{-1}A$ is immediate from (1.144). Thus, to prove that $\mathcal{E}_0^{-1}$ maps $B^1([0,l])$ into $B^1([0,l])$, we need only to show that $\mathcal{E}_0^{-1}I_p \in B^1([0,l])$. For that purpose, we represent $f = \mathcal{E}^{-1}y_2$ as $f(0) + \mathrm{i}Af'$, that is,

$$\begin{aligned}\left(\mathcal{E}^{-1}y_2\right)(x) &= \left(\mathcal{E}^{-1}y_2\right)(0) + \mathrm{i}\left(A\left(\mathcal{E}^{-1}y_2\right)'\right)(x) \\ &= \kappa(0)^{-1}y_2(0) + \mathrm{i}\left(A\left(\mathcal{E}^{-1}y_2\right)'\right)(x).\end{aligned} \tag{1.148}$$

Applying $\mathcal{E}_0^{-1}$ to the left- and right-hand sides of (1.148) and taking into account (1.147), we derive

$$\mathcal{E}_0^{-1}I_p = \left(I_p - \mathrm{i}\mathcal{E}_0^{-1}A\left(\mathcal{E}^{-1}y_2\right)'\right)y_2(0)^{-1}\kappa(0) \in B^1([0,l]). \qquad \square$$

Substitute E instead of $\mathcal{E}$ into (1.139) and use (1.143) to obtain our next proposition.

Proposition 1.32. *Let y satisfy* (1.131). *Then, the triple given by*

$$A = -\mathrm{i}\int_0^x \cdot\, dt; \quad S = E^{-1}(E^{-1})^*; \quad \Pi = \begin{bmatrix}\Phi_1 & \Phi_2\end{bmatrix}; \tag{1.149}$$

$$\Phi_1 g = -s(x)g, \quad \Phi_2 g = g; \quad s(x) := -E^{-1}y_1(x), \quad y_1(x) = y(x)\begin{bmatrix} I_p \\ 0\end{bmatrix} \tag{1.150}$$

forms a symmetric S-node.

The second relation in (1.150) (i.e. $\Phi_2 g = g$) means that Φ_2 acts as $I_p g$ after the natural embedding of $g \in \mathbb{C}^p$ into $L_p^2(0, l)$.

The set of projectors which we fix for this S-node is given by the natural projections from $L_p^2(0, l)$ onto $L_p^2(0, \xi)$, see (1.99) (see also Example 1.18, putting there $r = p$ and $D = -I_p$).

Lemma 1.33. *Let the triple* $\{A, S, \Pi\}$ *be given by* (1.149) *and* (1.150), *and introduce the projectors* P_ξ *by* (1.99). *Then,* $\{A, S, \Pi\}$ *forms a symmetric* S*-node and the conditions (i)–(iii) of Theorems* 1.20 *and* 1.27 *are fulfilled.*

Proof. According to Proposition 1.32, the triple A, S, Π forms a symmetric S-node. Since $S = E^{-1}(E^{-1})^*$, the condition (i) of Theorems 1.20 and 1.27 is fulfilled. The conditions (ii) of Theorem 1.20 and (ii), (iii) of Theorem 1.27 easily follow from the definitions of A and Φ_2 in (1.149) and (1.150), respectively. It remains to be shown that the matrix function $\Pi^* P_\xi^* S_\xi^{-1} P_\xi \Pi$ is absolutely continuous. Indeed, formula (1.149) implies

$$S_\xi = P_\xi E^{-1}(E^*)^{-1} P_\xi^*, \tag{1.151}$$

and, since $E^{\pm 1}$ are lower triangular operators, we also have

$$(E^*)^{-1} P_\xi^* = P_\xi^* P_\xi (E^*)^{-1} P_\xi^*, \quad P_\xi (E^*)^{-1} P_\xi^* = (P_\xi E^* P_\xi^*)^{-1}, \tag{1.152}$$

$$P_\xi E = P_\xi E P_\xi^* P_\xi. \tag{1.153}$$

We easily derive from (1.151) and (1.152) that

$$S_\xi^{-1} = \left(P_\xi E^{-1} P_\xi^* P_\xi (E^*)^{-1} P_\xi^*\right)^{-1} = P_\xi E^* P_\xi^* P_\xi E P_\xi^*. \tag{1.154}$$

Finally, it follows from (1.150), (1.153) and (1.154) that

$$\begin{aligned} \Pi^* P_\xi^* S_\xi^{-1} P_\xi \Pi g &= (P_\xi E \Pi)^* (P_\xi E \Pi) g \\ &= \int_0^\xi \left(E\begin{bmatrix} -s & I_p \end{bmatrix} g\right)(t)^* \left(E\begin{bmatrix} -s & I_p \end{bmatrix} g\right)(t)\,dt \end{aligned} \tag{1.155}$$

for the vectors $g \in \mathbb{C}^m$, and the right-hand side of (1.155) is absolutely continuous. □

Thus, we see that the triple given by (1.149) and (1.150) corresponds to the canonical system (1.110), (1.130), (1.131) (see Definition 1.23 of the *corresponding* triple). More precisely, we have the following theorem.

Theorem 1.34. *Let* γ *satisfy* (1.131). *Then, the triple given by* (1.149) *and* (1.150), *where* E *is constructed in Proposition* 1.31, *corresponds to the canonical system* (1.110), *where* H *has the form* (1.130).

Proof. According to Proposition 1.32, our triple forms a symmetric S-node. In view of Lemma 1.33, this S-node satisfies conditions of Theorem 1.20, and so it corresponds to the canonical system where H has the form (1.103). Using (1.155), we rewrite (1.103):

$$H(x) = \hat{y}(x)^* \hat{y}(x), \quad \hat{y}(x) := \left(E\left[-s \quad I_p\right]\right)(x). \tag{1.156}$$

Relations (1.143) and (1.150) imply that $\hat{y}$ from formula (1.156) equals y, that is, (1.130) holds. □

Recall that the set of Weyl functions of the canonical system on $[0, l]$ is denoted by $\mathcal{N}(\mathfrak{A})$. Next, we solve the inverse problem to recover the canonical system, where Hamiltonian H satisfies (1.130) and (1.131), from its Weyl function.

Theorem 1.35. *Let φ be a Weyl function of some canonical system* (1.110) *on* $[0, l]$, *where the Hamiltonian H is expressed via $p \times m$ matrix function y in* (1.130) *and y satisfies* (1.131). *Then, system* (1.110) *(or, equivalently, the Hamiltonian H) is uniquely recovered from φ using the following procedure.*

First, we recover via formulas (1.149) *and* (1.150) *(for A and Φ_2, respectively) and formulas* (1.113) *and* (1.114) *(for S and Φ_1, respectively) an S-node which corresponds to our system, whereas τ and ν in* (1.113) *and* (1.114) *are given in Herglotz representation* (1.115) *of φ. Next, we recover H from the formula* (1.103).

Proof. An S-node, which corresponds to the system of the form (1.110), (1.130), (1.131), was already constructed above (Theorem 1.34). Recall that this S-node and projectors P_ξ given (1.99) satisfy (according to Lemma 1.33) conditions of Theorem 1.20, and so Remark 1.26 implies that the Weyl functions of the system belong to $\mathcal{N}(\mathfrak{A}_A)$. Moreover, according to Lemma 1.33, conditions of Theorem 1.27 are also fulfilled. Thus, we have $\mathcal{N}(S, \Phi_1) = \mathcal{N}(\mathfrak{A}_A)$, which implies that our Weyl function φ belongs to $\mathcal{N}(S, \Phi_1)$. In view of Notation 1.24, it means that we, indeed, recover S and Φ_1 from the representations (1.113) and (1.114). Finally, we see from Definition 1.23 (of the "corresponding" S-node) that the Hamiltonian of our system can be recovered using formula (1.103). □

In order to derive asymptotics of the Weyl functions, we note that the following equalities hold for operators A and Φ_2 given by (1.149) and (1.150), respectively:

$$(I - zA)^{-1} = I - \mathrm{i}z \int_0^x \mathrm{e}^{\mathrm{i}z(t-x)} \cdot dt, \tag{1.157}$$

$$\Phi_2^*(I - zA)^{-1} = \mathrm{e}^{-\mathrm{i}zl} \int_0^l \mathrm{e}^{\mathrm{i}zt} \cdot dt, \quad \Phi_2^*(I - zA)^{-1}\Phi_2 = \frac{\mathrm{i}}{z}(\mathrm{e}^{-\mathrm{i}zl} - 1)I_p, \tag{1.158}$$

which is easily checked directly.

Theorem 1.36. *Let φ be a Weyl function of the canonical system* (1.110) *on* $[0, l]$, *where the Hamiltonian H is expressed via the $p\times m$ matrix function y in* (1.130) *and y satisfies* (1.131).

Then, in the angle $\Im z/|\Re z| > \delta$ (for any $\delta > 0$), we have

$$\varphi(z) = z\int_0^l \mathrm{e}^{\mathrm{i}zt}s(t)dt + o(z\mathrm{e}^{\mathrm{i}zl}), \quad |z| \to \infty, \tag{1.159}$$

where s is constructed in Proposition 1.32 *(see* (1.150)*).*

Proof. As mentioned in the proof of Theorem 1.35, the S-node constructed in Proposition 1.32 satisfies conditions of Theorem 1.27 and $\varphi \in \mathcal{N}(\mathfrak{A})$. Therefore, Theorem 1.28 implies representation (1.128) of φ. We substitute (1.158) into (1.128) to derive (1.159). □

1.2.5 Factorization of the operators S

Let us consider S-node (1.149), (1.150), which corresponds to the canonical system of the form (1.110), (1.130), (1.131), in greater detail. In view of (1.150), the operator identity (1.111) takes for this case the form

$$AS - SA^* = -\mathrm{i}\int_0^l \left(s(x) + s(t)^*\right) \cdot dt. \tag{1.160}$$

The operator S satisfying (1.160) is given in Corollary D.4 and the next proposition is immediate.

Proposition 1.37. *The S-node* (1.149), (1.150) *can be presented in the form*

$$A = A_l := -\mathrm{i}\int_0^x \cdot\, dt, \quad S = S_l := \frac{d}{dx}\int_0^l s(x-t) \cdot dt, \quad s(-x) = -s(x)^*, \tag{1.161}$$

where $A, S \in B(L^2_p(0, l))$. The operator $\Pi \in B(\mathbb{C}^m, L^2_p(0, l))$ is defined via its blocks by

$$\Pi = \Pi_l := [\Phi_{l,1} \quad \Phi_{l,2}], \quad \Phi_1 g = \Phi_{l,1} g = -s(x)g, \quad \Phi_2 g = \Phi_{l,2} g = g. \tag{1.162}$$

In accordance with (1.161) and (1.162), we add index "l"in our notations when the length of the interval, where the system is treated, may vary. It is apparent that the expressions for the reductions $A_\xi = P_\xi A P_\xi^*$, $S_\xi = P_\xi S P_\xi^*$ and $\Pi_\xi := P_\xi \Pi$ are obtained by the substitution of ξ instead of l in (1.161) and (1.162).

We note that according to Theorem 1.2 [287, p. 10], the operator identity (1.111) holds in the case of the arbitrary triple given by (1.161) and (1.162), that is, A, S and Π, which are given by (1.161) and (1.162), form a symmetric S-node. (See also Appendix D on operator identities.) Therefore, it is possible (and sometimes useful) to start research on canonical systems directly from the matrix function $s(x)$ and the S-node of the form (1.161), (1.162), which is determined by s (or from the operator S with the difference kernel), see [20, 176, 280, 290] and references therein. We recall that the important operator E, which was studied in the previous subsection, is connected with S by the relation $S^{-1} = E^*E$. Therefore, operator factorization is essential in the studied domain and below we provide some basic statements.

Factorization results from [137, pp. 184–186] yield the following proposition.

Proposition 1.38.

(a) *Let an operator S act in $L^2_p(0, l)$ and have the form*

$$S = I + \int_0^l k(x,t) \cdot dt, \tag{1.163}$$

where $k(x,t)$ is a Hilbert–Schmidt (i.e. square-integrable) kernel. Assume that S is boundedly invertible together with all the reductions S_ξ $(0 < \xi < l)$.
Then, the operator S^{-1} admits a triangular factorization

$$S^{-1} = E_+E_-, \quad E_+ = I + \int_x^l E_+(x,t) \cdot dt, \quad E_- = I + \int_0^x E_-(x,t) \cdot dt, \tag{1.164}$$

where $E_\pm(x,t)$ are also square-integrable.

(b) *Furthermore, if the matrix function $k(x,t)$ is not only Hilbert–Schmidt but continuous with respect to x and t, we can choose continuous matrix functions $E_\pm(x,t)$. These matrix functions are uniquely defined by the relations:*

$$E_+(x,\xi) = \Gamma(x,\xi,\xi), \quad E_-(\xi,t) = \Gamma(\xi,t,\xi), \tag{1.165}$$

where $\Gamma(x,t,\xi)$ is the continuous (with respect to x,t,ξ) kernel of S_ξ^{-1}, that is,

$$S_\xi^{-1} = I + \int_0^\xi \Gamma(x,t,\xi) \cdot dt. \tag{1.166}$$

A generalization of Proposition 1.38 for the case that $k(x,t)$ is piecewise continuous (with a possible jump discontinuity at $x = t$) is given in [135, p. 273].

When $S = S^*$, it is apparent from the uniqueness of the factorization (of the form (1.164)) that the factorizations E_+E_- and $E_-^*E_+^*$ coincide and we have $E_+ = E_-^*$.

Corollary 1.39. *If $S > 0$ has the form* (1.163) *and k is continuous, then S is boundedly invertible and S^{-1} admits factorization*

$$S^{-1} = E^*E, \qquad E = I + \int_0^x E(x,t) \cdot dt, \tag{1.167}$$

where $E(x,t) = E_-(x,t)$ is a continuous matrix function given by the second equality in (1.165).

For the case where $S = S^*$ and $k(x,t)$ is a Hilbert–Schmidt kernel, we have a similar corollary.

Corollary 1.40. *If $S > 0$ has the form* (1.163) *and $k(x,t)$ is a Hilbert–Schmidt kernel, then S is boundedly invertible and S^{-1} admits factorization*

$$S^{-1} = E^*E, \qquad E = I + \int_0^x E(x,t) \cdot dt, \tag{1.168}$$

where $E(x,t) = E_-(x,t)$ is square integrable.

Proof. According to (1.164), the operator E_-S is an upper triangular operator. Hence, the operator $E_-SE_-^*$ is an upper triangular operator. On the other hand, $E_-SE_-^*$ is self-adjoint, and so the integral part of $E_-SE_-^*$ equals zero. That is, $E_-SE_-^* = I$, which yields (1.168). □

Using the factorization of S, we can modify Lemma 1.33 (without any essential changes in the proof) in the following way.

Lemma 1.41. *Let the triple $\{A, S, \Pi\}$ be given by* (1.161) *and* (1.162), *and introduce the projectors P_ξ by* (1.99). *Assume that S is bounded and strictly positive. Then, A, S, Π form a symmetric S-node and the conditions (i)–(iii) of Theorems* 1.20 *and* 1.27 *are fulfilled.*

1.2.6 Weyl functions of Dirac and Schrödinger systems

Finally, we introduce Weyl functions of the Dirac (and Schrödinger) system on $[0, l]$ and $[0, \infty)$ and show that they coincide with the Weyl functions of the corresponding canonical system. First, note that Definition 0.1 admits an easy generalization.

Definition 1.42. Let an $m \times m$ matrix J satisfy equalities $J = J^* = J^{-1}$ and have $m_1 > 0$ positive eigenvalues (counted in accordance with their multiplicity). Then, an $m \times m_1$ matrix function $\mathcal{P}(z)$, which is meromorphic in

$$\mathbb{C}_M = \{z \in \mathbb{C} : \Im(z) > M \geq 0\}, \tag{1.169}$$

is called nonsingular with property-J if the inequalities

$$\mathcal{P}(z)^*\mathcal{P}(z) > 0, \quad \mathcal{P}(z)^*J\mathcal{P}(z) \geq 0 \tag{1.170}$$

hold in $\mathbb{C}_M$ (excluding, possibly, isolated points).

Since $\mathcal{P}$ is meromorphic, the fact that the first inequality in (1.170) holds in one point implies that it holds in all $\mathbb{C}_M$ (excluding, possibly, isolated points). It is also apparent that $\mathcal{P}$ satisfies the second inequality in (1.170) in all its points of analyticity.

The next proposition will be useful to show that Möbius transformations of $\mathcal{P}$ are well-defined.

Proposition 1.43. *Let the $m \times m$ matrix J satisfy equalities $J = J^* = J^{-1}$ and have $m_1 > 0$ positive eigenvalues. Let $m \times m_1$ matrices ϑ and $\widetilde{\vartheta}$ satisfy inequalities*

$$\vartheta^*\vartheta > 0, \quad \vartheta^*J\vartheta > 0, \quad \widetilde{\vartheta}^*\widetilde{\vartheta} > 0, \quad \widetilde{\vartheta}^*J\widetilde{\vartheta} \geq 0. \tag{1.171}$$

Then, we have

$$\det \vartheta^*J\widetilde{\vartheta} \neq 0. \tag{1.172}$$

Proof. Recall that a J-positive (J-nonnegative) subspace is such a subspace of $\mathbb{C}^m$ that for any of its vectors $g \neq 0$, the inequality $g^*Jg > 0$ ($g^*Jg \geq 0$) holds. (See, for instance, [51, 223] on some foundations of so called J-theory.) Clearly, the maximal J-positive (J-nonnegative) subspaces are m_1-dimensional, and all the m_1-dimensional J-positive (J-nonnegative) subspaces are maximal [51]. From (1.171), we see that $\operatorname{Im} \vartheta$ is a maximal J-positive subspace and $\operatorname{Im} \widetilde{\vartheta}$ is a maximal J-nonnegative subspace.

Now, we prove (1.172) by contradiction. Suppose that the inequality (1.172) does not hold, that is, there is a vector g such that

$$\vartheta^*J\widetilde{\vartheta}g = 0, \quad g \in \mathbb{C}^{m_1}, \quad g \neq 0. \tag{1.173}$$

It is apparent from (1.173) and J-nonnegativity of $\operatorname{Im} \vartheta$ and $\operatorname{Im} \widetilde{\vartheta}$ that the subspace span $(\operatorname{Im} \vartheta \cup \operatorname{Im} \widetilde{\vartheta}g)$ is J-nonnegative. Hence, from the fact that $\operatorname{Im} \vartheta$ is already a maximal J-nonnegative subspace, we derive

$$\widetilde{\vartheta}g \in \operatorname{Im} \vartheta. \tag{1.174}$$

Since $\widetilde{\vartheta}$ is nondegenerate and $g \neq 0$, we obtain $\widetilde{\vartheta}g \neq 0$. Recall also that $\operatorname{Im} \vartheta$ is not only J-nonnegative, but it is a J-positive subspace. Therefore, (1.174) contradicts (1.173). □

We also need another proposition which is proved in a similar way.

Proposition 1.44. *Let the $m \times m$ matrix J satisfy equalities $J = J^* = J^{-1}$ and have $m_1 > 0$ positive eigenvalues. Let $m \times m_1$ matrix ϑ and $m \times \widetilde{m}$ ($\widetilde{m} \leq m_1$) matrix $\widetilde{\vartheta}$ satisfy inequalities*

$$\vartheta^*\vartheta > 0, \quad \vartheta^*J\vartheta > 0, \quad \widetilde{\vartheta}^*\widetilde{\vartheta} > 0, \quad \vartheta^*J\widetilde{\vartheta} = 0. \tag{1.175}$$

Then, we have

$$\widetilde{\vartheta}^* J\widetilde{\vartheta} < 0. \tag{1.176}$$

Proof. Suppose that the inequality (1.176) does not hold, that is, there is a vector g such that

$$g^*\widetilde{\vartheta}^* J\widetilde{\vartheta}g \geq 0, \quad g \in \mathbb{C}^{\widetilde{m}}, \quad g \neq 0. \tag{1.177}$$

It is apparent from formula (1.177) and the second and the last relations in (1.175) that the subspace span $(\mathrm{Im}\,\vartheta \cup \mathrm{Im}\,\widetilde{\vartheta}g)$ is J-nonnegative. Hence, from the fact that $\mathrm{Im}\,\vartheta$ is already a maximal J-nonnegative subspace, we derive

$$\widetilde{\vartheta}g \in \mathrm{Im}\,\vartheta. \tag{1.178}$$

Since $\widetilde{\vartheta}$ is nondegenerate and $g \neq 0$, we obtain $\widetilde{\vartheta}g \neq 0$. Thus, taking into account the second relation in (1.175) and formula (1.178), we derive $g^*\widetilde{\vartheta}^* J\widetilde{\vartheta}g > 0$. On the other hand, the last relation in (1.175) and formula (1.178) imply $g^*\widetilde{\vartheta}^* J\widetilde{\vartheta}g = 0$. Thus, we arrive at a contradiction. □

Remark 1.45. The proof above is easily modified to show that under conditions of Proposition 1.44, with $\vartheta^* J\vartheta \geq 0$ substituted for $\vartheta^* J\vartheta > 0$, we have $\widetilde{\vartheta}^* J\widetilde{\vartheta} \leq 0$.

Further in this subsection, we still assume that $\mathbb{C}_M = \mathbb{C}_+$, $m_1 = p$ and $m = 2p$. Recall that u is the normalized by (1.1) fundamental solution of (0.3) and put

$$\mathcal{U}(z) = \mathcal{U}(l,z) := \Theta u(l,\overline{z})^*, \quad \mathcal{U}(z) =: \{\mathcal{U}_{ij}(z)\}_{i,j=1}^2, \tag{1.179}$$

where Θ is given by (1.4) and $\mathcal{U}_{ij}$ are the $p \times p$ blocks of $\mathcal{U}$. Below, we present Möbius transformation, which is similar to (0.8), in a slightly more compact form.

Notation 1.46. *Denote by $\mathcal{N}(\mathcal{U})$ the set of Möbius transformations*

$$\varphi(z) = \varphi(z,l,\mathcal{P}) = \mathrm{i}\begin{bmatrix} I_p & 0 \end{bmatrix}\mathcal{U}(z)\mathcal{P}(z)\left(\begin{bmatrix} 0 & I_p \end{bmatrix}\mathcal{U}(z)\mathcal{P}(z)\right)^{-1}, \tag{1.180}$$

where the matrix functions $\mathcal{P}$ are nonsingular matrix functions with property-j and j is given by (0.4) *(which implies, in particular, that $m_1 = p$ and $\mathcal{P}$ are $m \times p$ matrix functions).*

In the next chapter, we also study the case that $j = \mathrm{diag}\,\{I_{m_1}, -I_{m_2}\}$, where $m_1 + m_2 = m$ and m_1 may differ from m_2. Useful general results on linear-fractional transformations are contained in [181] (see also [94] for some related results and references).

Proposition 1.47. *If* (1.170) *holds, then we have*

$$\det\left(\begin{bmatrix} 0 & I_p \end{bmatrix}\mathcal{U}(z)\mathcal{P}(z)\right) \neq 0, \tag{1.181}$$

and so the Möbius transformation (1.180) *is always well-defined.*

Proof. Let us fix some $z \in \mathbb{C}_+$. From (0.3), it is immediately apparent that

$$\frac{d}{dx}\left(u(x,\overline{z})^* j u(x,\overline{z})\right) = \mathrm{i}(\overline{z} - z) u(x,\overline{z})^* u(x,\overline{z}) > 0. \tag{1.182}$$

According to (1.1) and (1.182), we have

$$\aleph(x,\overline{z}) > j \quad (x > 0), \quad \aleph(x,z) := u(x,z)^* j u(x,z) \tag{1.183}$$

since $\aleph(x,\overline{z})$ is strictly monotonic and $\aleph(0,\overline{z}) = j$. It is well known (Corollary E.4) that the inequality in (1.183) is equivalent to

$$u(x,\overline{z}) j u(x,\overline{z})^* > j. \tag{1.184}$$

Taking into account (1.184) and inequalities (1.170) for $\mathcal{P}$, we obtain

$$\widetilde{\mathcal{P}}(z)^* j \widetilde{\mathcal{P}}(z) > 0 \quad \text{for } \widetilde{\mathcal{P}}(z) := u(l,\overline{z})^* \mathcal{P}(z). \tag{1.185}$$

In view of (1.4) and (1.185), the matrices $\vartheta = \widetilde{\mathcal{P}}(z)$ and $\widetilde{\vartheta}^* = [0 \;\; I_p]\Theta j$ satisfy conditions of Proposition 1.43. Therefore, we have $\det(\vartheta^* j \widetilde{\vartheta}) \neq 0$ and so we also have $\det(\widetilde{\vartheta}^* j \vartheta) \neq 0$. Finally, we note that according to (1.179) and (1.185), the left-hand side of (1.181) coincides with $\det(\widetilde{\vartheta}^* j \vartheta)$. □

Definition 1.48. Möbius (also called linear-fractional) transformations $\varphi \in \mathcal{N}(\mathcal{U})$ are called Weyl functions of the Dirac system (0.3) on the interval $[0, l]$.

Proposition 1.49. *The set of Weyl functions of the Dirac system* (0.3) *on the interval* $[0, l]$ *coincides with the sets of Weyl functions of the canonical system* (1.8), (1.9) *on* $[0, l]$ *and of the canonical system* (1.11) *on* $[0, 2l]$.

Proof. According to (1.6) and (1.7), we have

$$\Theta u(l,0)^{-1} j \left(u(l,0)^*\right)^{-1} \Theta^* = J, \quad u(l,0)\Theta^* J \Theta u(l,0)^* = j.$$

Thus, the transformations

$$\mathcal{P} = \left(u(l,0)^*\right)^{-1} \Theta^* \widehat{\mathcal{P}}, \quad \widehat{\mathcal{P}} = \Theta u(l,0)^* \mathcal{P} \tag{1.186}$$

map, respectively, nonsingular matrix functions with property-J onto nonsingular matrix functions with property-j and nonsingular matrix functions with the property-j onto nonsingular matrix functions with property-J (where J is given in (0.2)). Hence, equalities (0.7), (1.3) and (1.179) imply that the sets of Möbius transformations (0.8) and (1.180) coincide. Therefore, according to Definitions 0.3 and 1.48, the first statement of the proposition is proved. Moreover, Definition 0.3 and the first equality in (1.10) imply that the sets of Weyl functions of canonical systems (1.8) on $[0, l]$ and (1.11) on $[0, 2l]$ coincide too. □

Remark 1.50. It is shown in Appendix A that Weyl functions $\varphi(z) \in \mathcal{N}(\mathfrak{A})$ of the canonical systems belong to the Nevanlinna (also called Herglotz) class, that is, $\mathrm{i}(\varphi(z)^* - \varphi(z) \geq 0$ in $\mathbb{C}_+$, and so φ admits a unique Herglotz representation

$$\varphi(z) = \mu z + \nu + \int_{-\infty}^{\infty} \left(\frac{1}{t-z} - \frac{t}{1+t^2} \right) d\tau(t), \tag{1.187}$$

where $\mu \geq 0$, $\nu = \nu^*$ and τ is a nondecreasing $p \times p$ distribution matrix function, which is normalized by the conditions

$$\tau(-\infty) = 0, \quad \tau(t-0) = \tau(t). \tag{1.188}$$

The famous Stieltjes–Perron inversion formula [11, Ch. 3] (see [124] for further references and developments) constructs τ via φ. If t_1 and t_2 are the points of continuity of τ, we have

$$\tau(t_1) - \tau(t_2) = \frac{\mathrm{i}}{2\pi} \lim_{\eta \to +0} \int_{t_2}^{t_1} (\varphi(\xi + \mathrm{i}\eta)^* - \varphi(\xi + \mathrm{i}\eta)) d\xi \quad (\xi \in \mathbb{R}, \ \eta > 0). \tag{1.189}$$

Definition 1.51. A holomorphic function φ such that

$$\int_0^{\infty} \begin{bmatrix} I_p & \mathrm{i}\varphi(z)^* \end{bmatrix} \Theta u(x,z)^* u(x,z) \Theta^* \begin{bmatrix} I_p \\ -\mathrm{i}\varphi(z) \end{bmatrix} dx < \infty, \quad z \in \mathbb{C}_+ \tag{1.190}$$

is called a Weyl function of the Dirac system (0.3) on $[0, \infty)$.

Proposition 1.52. *The Weyl functions of the Dirac system* (0.3) *and canonical system* (1.8), (1.9) *on the semiaxis* $[0, \infty)$ *coincide.*

Proof. Because of (1.3) and (1.9), the equivalence of (0.9) and (1.190) is immediately apparent. □

Following [238, 242] (see also [76, 108, 259]), we define Weyl functions of the skew-self-adjoint Dirac systems (1.35) by the inequality

$$\int_0^{\infty} \begin{bmatrix} I_p & \varphi(z)^* \end{bmatrix} u(x,z)^* u(x,z) \begin{bmatrix} I_p \\ \varphi(z) \end{bmatrix} dx < \infty, \quad \mathrm{Im}\,(z) > M, \tag{1.191}$$

which is similar to (1.190). Here, u is the normalized (by $u(0,z) = I_m$) fundamental solution of the skew-self-adjoint Dirac system (1.35) where the potential v is bounded by M:

$$||v(x)|| \leq M, \quad 0 < x < \infty.$$

The existence and uniqueness of the Weyl function defined above are studied in Chapter 3.

Weyl function of the Schrödinger (Sturm–Liouville) system (0.1) is defined ([281] and [290, Ch. 11]) via the $p \times m$ solution Y of (0.1) satisfying initial conditions

$$Y(0,z) = (\sqrt{2})^{-1}[\mathrm{i}I_p \quad I_p], \quad \left(\frac{d}{dx}Y\right)(0,z) = (\sqrt{2})^{-1}[\mathrm{i}I_p \quad -I_p]. \tag{1.192}$$

Definition 1.53. A holomorphic function φ such that

$$\int_0^\infty \begin{bmatrix} I_p & \mathrm{i}\varphi(z)^* \end{bmatrix} Y(x,z)^* Y(x,z) \begin{bmatrix} I_p \\ -\mathrm{i}\varphi(z) \end{bmatrix} dx < \infty, \quad z \in \mathbb{C}_+ \tag{1.193}$$

is called a Weyl function of the Schrödinger system (0.1) on $[0, \infty)$.

Proposition 1.54. *Weyl functions of the Schrödinger systems* (0.1) *on* $[0, \infty)$ *coincide with Weyl functions of the corresponding canonical systems* (1.47) *on* $[0, \infty)$.

Proof. From (1.40), (1.41) and (1.192), we derive

$$Y(x,z) = \begin{bmatrix} I_p & 0 \end{bmatrix} w(x,z)\Theta_1, \tag{1.194}$$

where Θ_1 is given in (1.43). Therefore, using (1.43) and (1.47), we can rewrite (0.9) in the form (1.193). □

2 Self-adjoint Dirac system: rectangular matrix potentials

Dirac (also called a ZS-AKNS) system is very well known in mathematics and applications, especially in mathematical physics (see, e.g. books [26, 68, 179, 195, 290], recent publications [19, 20, 25, 74, 76, 113, 114, 293] and numerous references therein). The Dirac system (0.3), (0.4) is used, in particular, in the study of transmission lines and acoustic problems [333]. The most interesting applications are, however, caused by the fact that the Dirac system is an auxiliary linear system for many important integrable nonlinear wave equations and as such, it was studied, for instance, in [5–7, 101, 120, 162, 166, 256, 337]. (The name ZS-AKNS system is connected with these applications.) Nonlinear Schrödinger equations, modified Korteweg–de Vries equations and the second harmonic generation model, which describe various wave processes (including, e.g. photoconductivity and nonlinear wave processes in water, waveguides, nonlinear optics and on silicon surfaces), are only some of the well-known examples.

Chapter 2 is dedicated to self-adjoint Dirac systems with the potentials v, which are rectangular matrix functions. At first, the case of square matrix functions v is studied. We introduce the important notion of the spectral functions and solve inverse problems to recover a system from its spectral or Weyl functions. The spectral and Weyl theories of Dirac systems with scalar or square matrix potentials v were actively studied before, mostly under the restriction that v is continuous (see [20, 24, 25, 74, 76, 95, 176, 179, 195, 290, 293, 295] and various references therein). Following [252], we develop these results under the conditions that v is locally summable or locally bounded (in norm) for direct and inverse problems, respectively. In particular, we further develop the so called high energy asymptotics results (on asymptotics of Weyl functions for $|z| \to \infty$) from the seminal paper [74].

Next, we treat the main topic of this chapter, that is, the Weyl theory of Dirac systems with $m_1 \times m_2$ potentials v where m_1 is not necessarily equal to m_2 and so v is not necessarily a square matrix function. The "nonclassical" Weyl theory for the case $m_1 \neq m_2$ is essential for the study of coupled, multicomponent, and matrix nonlinear equations. This topic is new but the method of operator identities and the approach (which is based on it) from [252] can be successfully adapted for our case as shown in the recent papers [106, 107, 110]. We note also that a discrete self-adjoint Dirac system was dealt with in [105] and [109], respectively, for the cases of square and rectangular matrix potentials (depending on discrete variable $k \in \mathbb{N}_0$).

We always assume here that the potential v is locally summable, that is, summable on any finite interval.

2.1 Square matrix potentials: spectral and Weyl theories

2.1.1 Spectral and Weyl functions: direct problem

In this section, we study Dirac system (0.3), where j is given by (0.4). The operator $\mathcal{H}_D$ in $L^2_m(0, \infty)$, which corresponds to system (0.3) (so that (0.3) may be rewritten as $\mathcal{H}_D f = zf$), is determined by its differential expression and domain $\mathcal{D}(\mathcal{H}_D)$:

$$\mathcal{H}_D f = -\left(\mathrm{i}j\frac{d}{dx} + V(x)\right) f, \tag{2.1}$$

$$\mathcal{D}(\mathcal{H}_D) := \left\{f : -\left(\mathrm{i}j\frac{d}{dx} + V(x)\right) f \in L^2_m(0, \infty), \quad [I_p \quad \omega]f(0) = 0\right\}. \tag{2.2}$$

We assume that the matrix ω in the initial condition

$$\begin{bmatrix} I_p & \omega \end{bmatrix} f(0) = 0 \tag{2.3}$$

is unitary, which is necessary and sufficient for $\mathcal{H}_D$ being self-adjoint [211]. For the case of the continuous potential and in a somewhat more general setting than system (0.3), the problems of self-adjointness have also been studied in [190] (see [191] for some further developments).

Put

$$U_D = \frac{1}{\sqrt{2}}\int_0^\infty \begin{bmatrix} I_p & -\omega \end{bmatrix} u(x,\overline{z})^* \cdot dx \quad (\omega\omega^* = I_p). \tag{2.4}$$

For functions $f \in \mathcal{D}(\mathcal{H}_D)$ with compact support, we easily derive

$$(U_D\mathcal{H}_D f)(z) = z(U_D f)(z), \tag{2.5}$$

that is, U_D diagonalizes $\mathcal{H}_D$.

Definition 2.1. A nondecreasing $p \times p$ matrix function τ on the real axis $\mathbb{R}$ is called a spectral function of the system (0.3) on $[0, \infty)$ with the initial condition (2.3) if U_D, which is defined by (2.4) for f with compact support, extends to an isometry, also denoted by U_D, from $L^2_m(0, \infty)$ into $L^2_p(d\tau)$. Here, $L^2_p(d\tau)$ is a Hilbert space of functions with the scalar product $\langle f, g\rangle_\tau = \int_{-\infty}^\infty g(z)^* d\tau(z) f(z)$.

Correspondingly, a nondecreasing matrix function τ is called a spectral function of the system (0.3) on the interval $[0, l]$ $(0 < l < \infty)$ with the initial condition (2.3) if U_D, defined by (2.4), restricts to an isometry from $L^2_m(0, l)$ into $L^2_p(d\tau)$.

Put now

$$\Theta_2 = \begin{bmatrix} I_p & 0 \\ 0 & -\omega \end{bmatrix}, \quad \hat{u}(x,z) = \Theta_2 u(x,z)\Theta_2^*, \quad \hat{V}(x) = \Theta_2 V(x)\Theta_2^*, \tag{2.6}$$

where Θ_2 is unitary. According to (0.3), (1.1)) and (2.6), $\hat{u}$ is the normalized fundamental solution of a new Dirac system

$$\frac{d}{dx}\hat{u}(x,z) = \mathrm{i}(zj + j\hat{V}(x))\hat{u}(x,z) \quad (x \geq 0), \quad \hat{u}(0,z) = I_m. \tag{2.7}$$

Substitute the initial condition (2.3) by

$$\begin{bmatrix} I_p & -I_p \end{bmatrix} \hat{f}(0) = 0. \tag{2.8}$$

Then, formulas (2.6)–(2.8) yield

$$\mathcal{D}(\hat{\mathcal{H}}_D) = \Theta_2 \mathcal{D}(\mathcal{H}_D), \quad \hat{U}_D = U_D \Theta_2^*, \tag{2.9}$$

where $\hat{\mathcal{H}}_D$ corresponds to system (2.7) with the initial condition (2.8). In view of (2.9) and Definition 2.1, we obtain

Proposition 2.2. *The spectral functions of systems* (0.3) *and* (2.7) *with initial conditions* (2.3) $(\omega\omega^* = I_p)$ *and* (2.8), *respectively, coincide.*

Remark 2.3. Thus, without loss of generality we suppose further that $\omega = -I_p$ and we usually omit references to the initial condition, speaking about the spectral functions of system (0.3), as this condition remains the same. The corresponding initial condition for the canonical system (1.8) is

$$\begin{bmatrix} I_p & 0 \end{bmatrix} f(0) = 0. \tag{2.10}$$

The operators

$$U_{C,r} := \begin{bmatrix} 0 & I_p \end{bmatrix} \int_0^r W(x, \overline{z})^* H(x) \cdot dx \quad (r \leq \infty) \tag{2.11}$$

correspond to (1.8) (or to (0.2), which is the same) on the finite intervals $[0, l]$ $(l = r < \infty)$ and the semiaxis $[0, \infty)$(for the case that $r = \infty$), respectively.

Definition 2.4. A nondecreasing $p \times p$ matrix function τ on $\mathbb{R}$ is called a spectral function of the canonical system (0.2) given on $[0, l]$ $(l < \infty)$ if $U_{C,l}$, defined by (2.11), is an isometry from $L^2_m(H, l)$ into $L^2_p(d\tau)$, where $L^2_m(H, l)$ is a Hilbert space of functions with the scalar product

$$(f, g)_H = \int_0^l g(x)^* H(x) f(x) dx.$$

Matrix function τ is called a spectral function of the system (0.2) on $[0, \infty)$ if this τ is a spectral function of (0.2), considered on $[0, l]$, for all $0 < l < \infty$. In other words, τ is called a spectral function if $U_C := U_{C,\infty}$, defined by (2.11)) on functions from $L^2_m(H, \infty)$ with compact support, extends to isometry, also denoted by U_C, on the entire space $L^2_m(H) := L^2_m(H, \infty)$.

From (1.3), (1.4) and (1.9), we obtain

$$[0 \quad I_p] W(x, \overline{z})^* H(x) = \frac{1}{\sqrt{2}} [I_p \quad I_p] u(x, \overline{z})^* u(x, 0) \Theta^*. \tag{2.12}$$

According to (2.4), (2.11) and (2.12), we have

$$U_{C,l} = U_D u(x,0)\Theta^* \tag{2.13}$$

Finally, notice that in view of (1.9), the multiplication by $u(x,0)\Theta^*$ unitarily maps $L^2_m(H,\, l)$ onto $L^2_m(0,\, l)$ and thus equality (2.13) yields the proposition below.

Theorem 2.5. *The spectral functions of the Dirac system* (0.3) *on* $[0,\, l]$ ($[0,\, \infty)$) *coincide with the spectral functions of the canonical system* (1.8) *on* $[0,\, l]$ ($[0,\, \infty)$), *assuming that H is given by* (1.9) *and the initial conditions for these systems are*

$$\begin{bmatrix} I_p & -I_p \end{bmatrix} f(0) = 0 \tag{2.14}$$

and (2.10), *respectively.*

The spectral theory of the general-type canonical systems on a finite interval is studied in Appendix A. Our next statement on canonical systems with H of the form (1.9) follows from Theorem A.23 and the apparent fact that $\det H \neq 0$ in (1.9).

Theorem 2.6. *Assume that H is given by* (1.9). *Then, the set of spectral functions of the canonical system* (1.8) *on* $[0,\, l]$ (*where the initial condition is given by* (2.10)) *coincides with the set of distribution functions for the Weyl functions* $\varphi \in \mathcal{N}(\mathfrak{A})$.

Corollary 2.7. *The set of spectral functions of the Dirac system* (0.3) *on* $[0,\, l]$ (*where the initial condition is given by* (2.14)) *coincides with the set of distribution functions for the Weyl functions* $\varphi \in \mathcal{N}(\mathcal{U})$.

Proof. For the case that H is given by (1.9), the equality $\mathcal{N}(\mathcal{U}) = \mathcal{N}(\mathfrak{A})$ follows from Proposition 1.49. Now, Theorems 2.5 and 2.6 yield the statement of our corollary. □

Theorem 2.8. *Let the Hamiltonian H of the canonical system* (1.8) *be given by* (1.9), *where u is defined by* (1.1) *and* (1.2) *with v locally summable. Then, there exists a unique spectral function of* (1.8) *on* $[0,\, \infty]$.

Proof. It is immediate from $\det H(x) \neq 0$ that the "positivity type" condition (A.205) is fulfilled. Therefore, we can use Theorem A.26, which yields the existence of the matrix function $\varphi \in \cap_{l<\infty}\mathcal{N}(l)$, where $\mathcal{N}(l) := \mathcal{N}(\mathfrak{A}(l))$. According to Theorem 2.6, the distribution function in the Herglotz representation of this φ is a spectral function of our canonical system on all the finite intervals $[0,\, l]$, and so on $[0,\, \infty)$.

Using Theorem 2.6 and the Montel's theorem approach from the proof of Theorem A.26 (see also Appendix E for the Montel's theorem), it is easy to show that for each spectral function τ of the canonical system (1.8), (1.9) on $[0,\, \infty)$ there is a function from the intersection $\cap_{l<\infty}\mathcal{N}(l)$ such that τ is its distribution function. Thus, the uniqueness of τ follows from the uniqueness of $\varphi \in \cap_{l<\infty}\mathcal{N}(l)$. In turn, the uniqueness of $\varphi \in \cap_{l<\infty}\mathcal{N}(l)$ is implied by Remark A.29 and Theorem A.31. We note that the inequality $H(x) \geq -J$ is apparent from (1.6), (1.7), (1.9), and so the conditions of Theorem A.31 are satisfied. □

Clearly, the proof of Theorem 2.8 yields the corollary below.

Corollary 2.9. *Let the Hamiltonian H of the canonical system* (1.8) *be given by* (1.9), *where u is defined by* (1.1) *and* (1.2) *with v locally summable. Then, there exists a unique Weyl function of* (1.8) *on* $[0, \infty)$.

Proposition 1.52 (Theorem 2.5) states that the Weyl (spectral) functions of the Dirac system (0.3) and canonical system (1.8), (1.9) on $[0, \infty)$ coincide. Hence, Theorem 2.8 and Corollary 2.9 yield our next result.

Corollary 2.10. *There exist unique spectral and Weyl functions of the Dirac system* (0.3) *on* $[0, \infty)$ *(with v locally summable).*

2.1.2 Spectral and Weyl functions: inverse problem

Here, we recover the potential v of the Dirac system from its spectral and Weyl functions. Clearly, it suffices to be able to recover v on each finite interval $[0, l/2]$, and so we consider system (0.3) on an interval.

Theorem 2.11. *A Dirac system on $[0, l/2]$ with a bounded (in norm) potential is uniquely recovered from its spectral function. Namely, the potential v is recovered from γ by the formula* (1.17), *where J is given in* (0.2) *and β is given by* (1.22) *and* (1.25). *The matrix function γ is recovered from H_d using formulas* (1.18), (1.19) *and* (1.21). *Finally, H_d is recovered from the spectral function τ via formulas* (1.103) *(where $H_d = H$),* (1.113), (1.114) *and equality*

$$\nu = \mathrm{i}I_p + \left(\int_{-\infty}^{\infty} \left(A(I - tA)^{-1} + \frac{t}{1+t^2} I \right) \Phi_2 d\tau(t) I_p \right)(+0). \tag{2.15}$$

Proof. According to Proposition 1.49 and Corollary 2.7, our spectral function is a distribution function in the Herglotz representation of some Weyl function of the canonical system given by (1.11), (1.12) and (1.13) on $[0, l]$. Hence, in view of (1.15), we can apply Theorem 1.35 to recover H_d once we recover ν. The matrix functions γ, β and, finally, v are recovered from H_d using the proof of Proposition 1.1. Moreover, the recovery procedures discussed above coincide with the corresponding formulas in the theorem that we are proving now.

It remains to be shown that ν satisfies (2.15). First, we recall that Theorem 1.35 implies representation (1.114) for the operator Φ_1 from the S-node constructed in Theorem 1.34. On the other hand, formula (1.150) yields the equality

$$\Phi_1 I_p = \left(E^{-1} \gamma_1 \right)(x), \tag{2.16}$$

where (according to Proposition 1.31) the entries of $E^{-1}y_1$ are absolutely continuous. Hence, from (1.15) and (1.142), we see that

$$(\Phi_1 I_p)(+0) = -I_p. \tag{2.17}$$

Taking into account (1.114) and (2.17), we obtain (2.15). □

Our next corollary is apparent from the proof of Theorem 2.11.

Corollary 2.12. *A Dirac system on* $[0, l/2]$ *with a bounded (in norm) potential is uniquely recovered from its Weyl function* φ. *Namely, the potential* v *is recovered from* y *by the formula* (1.17), *where* J *is given in* (0.2) *and* β *is given by* (1.22) *and* (1.25). *The matrix function* y *is recovered from* H_d *using formulas* (1.18), (1.19) *and* (1.21). *The Hamiltonian* H_d *is recovered from the matrix* ν *and matrix function* τ *via formulas* (1.103) *(where* $H_d = H$), (1.113) *and* (1.114). *In turn,* ν *and* τ *are the matrix and distribution matrix function, respectively, from the Herglotz representation* (1.187) *of* φ.

We note that the S-node and, finally, the potential v can also be recovered directly from φ (without using the Herglotz representation of φ). More precisely, one can use formula (1.159) instead of the Herglotz representation, see [252]. Such an approach is applied further in this chapter for the case that the potential v is rectangular.

2.2 Weyl theory for Dirac system with a rectangular matrix potential

2.2.1 Direct problem

The form of the self-adjoint Dirac system

$$\frac{d}{dx}y(x,z) = \mathrm{i}(zj + jV(x))y(x,z) \quad (x \geq 0), \quad V = \begin{bmatrix} 0 & v \\ v^* & 0 \end{bmatrix}, \tag{2.18}$$

which we consider in this section, coincides with (0.3). However, j has a somewhat more general structure, that is,

$$j = \begin{bmatrix} I_{m_1} & 0 \\ 0 & -I_{m_2} \end{bmatrix} \tag{2.19}$$

and, correspondingly, $v(x)$ is an $m_1 \times m_2$ matrix function.

System (2.18) is an auxiliary linear system for many important nonlinear integrable wave equations, including nonlinear Schrödinger and modified Korteweg–de Vries equations. (As such, it was studied, e.g. in [5–7, 101, 120, 256, 337].) Hence, the case that $m_1 \neq m_2$ is important for the coupled, multicomponent and $m_1 \times m_2$ matrix nonlinear equations, also much less results were derived, till recently, for this case.

In this subsection, we consider Dirac system (2.18) on the semiaxis $x \in [0, \infty)$, assume that v is measurable and locally summable (i.e. summable on all the finite intervals) and study the direct problem. Following [110], a definition of the nonexpansive generalized Weyl function is given, its existence and uniqueness are proved, and some basic properties are derived.

We denote $m_1 + m_2$ by m and also preserve other notations from the previous sections. Thus, the fundamental solution of (2.18) is denoted by $u(x, z)$, and this solution is normalized by the condition

$$u(0, z) = I_m. \tag{2.20}$$

Putting $J = j$ in Definition 1.42, we use a class of nonsingular $m \times m_1$ matrix functions $\mathcal{P}(z)$ with property-j. We recall that the fundamental solutions are invertible and, moreover, in our case, we have $u(x, \overline{z})^* j u(x, z) = j$ (as follows from formula (2.22), which will appear later).

Notation 2.13. *The set $\mathcal{N}_u(x, z)$ of Möbius transformations is the set of values at x, z of matrix functions*

$$\varphi(x, z, \mathcal{P}) = \begin{bmatrix} 0 & I_{m_2} \end{bmatrix} u(x, z)^{-1} \mathcal{P}(z) \left(\begin{bmatrix} I_{m_1} & 0 \end{bmatrix} u(x, z)^{-1} \mathcal{P}(z) \right)^{-1}, \tag{2.21}$$

where $\mathcal{P}(z)$ are nonsingular matrix functions with property-j.

Remark 2.14. We recall that the notations $\mathcal{N}(\mathfrak{A})$ and $\mathcal{N}(\mathcal{U})$ given in the Introduction and Notation 1.46, respectively, stand for the sets of functions depending on z. However, it is also convenient to consider sets of values of such functions pointwise like the sets $\mathcal{N}_u(x, z)$ in Notation 2.13 above. In particular, the uniqueness of the intersection of sets is proved pointwise.

Differently from Subsection 1.2.6 (where Möbius transformations (1.180) belonged to Nevanlinna class), it is more convenient for the present case (that the potentials are rectangular) to deal with nonexpansive Möbius transformations. However, many considerations in this subsection are similar to considerations in Subsection 1.2.6.

Proposition 2.15. *Let Dirac system (2.18) be given on $[0, l]$ and assume that v is summable. Then, the set $\mathcal{N}_u(z) := \mathcal{N}_u(l, z)$ is well-defined.*

Proof. The proof that $\mathcal{N}_u(z)$ is well-defined is similar to the proof of Proposition 1.47. It is immediate from (2.18) that for $0 \le x \le l$, we have

$$\frac{d}{dx}\left(u(x, z)^* j u(x, z)\right) = \mathrm{i}(z - \overline{z}) u(x, z)^* u(x, z) < 0, \quad z \in \mathbb{C}_+. \tag{2.22}$$

In view of (2.20) and (2.22), we obtain

$$\aleph(x, z) := u(x, z)^* j u(x, z) < j \quad (x > 0, \quad z \in \mathbb{C}_+). \tag{2.23}$$

Inequality (2.23) yields

$$(u(x,z)^*)^{-1} j u(x,z)^{-1} > j \quad (x > 0, \quad z \in \mathbb{C}_+). \tag{2.24}$$

Thus, using Proposition 1.43, we derive

$$\det\left(\begin{bmatrix} I_{m_1} & 0 \end{bmatrix} u(l,z)^{-1}\mathcal{P}(z)\right) \neq 0 \tag{2.25}$$

for $l > 0$, and so $\mathcal{N}_u$ is well-defined via (2.21) for $l > 0$.

It is easy to see that (2.25) also holds for $l = 0$. Indeed, if $\det([I_{m_1} \;\; 0]\mathcal{P}(z)) = 0$, there is a vector $g \in \mathbb{C}^{m_1}$, $g \neq 0$ such that $[I_{m_1} \;\; 0]\mathcal{P}(z)g = 0$. Hence, in view of the second inequality in (1.170), we derive $\mathcal{P}(z)g = 0$, which contradicts the first inequality in (1.170). □

Since transformations (2.21) are well-defined, the following definition is also well-posed.

Definition 2.16. Möbius transformations $\varphi(z)$ $(z \in \mathbb{C}_+)$ of the form (2.21), where $x = l$ and $\mathcal{P}(z)$ are nonsingular matrix functions with property-j, are called Weyl functions of the Dirac system (2.18) on the interval $[0, l]$.

Below, we write sometimes $\varphi(z)$ meaning the set $\{\varphi(z)\}$ consisting of the matrix $\varphi(z)$.

Proposition 2.17. *Let Dirac system* (2.18) *be given on* $[0, \infty)$ *and assume that* v *is locally summable. Then, there is a unique matrix function* $\varphi(z)$ *in* $\mathbb{C}_+$ *such that*

$$\varphi(z) = \bigcap_{x<\infty} \mathcal{N}_u(x,z). \tag{2.26}$$

This function is analytic and nonexpansive.

Proof. We rewrite (2.21) in the equivalent form

$$\begin{bmatrix} I_{m_1} \\ \varphi(x,z,\mathcal{P}) \end{bmatrix} = u(x,z)^{-1}\mathcal{P}(z)\left(\begin{bmatrix} I_{m_1} & 0 \end{bmatrix} u(x,z)^{-1}\mathcal{P}(z)\right)^{-1}. \tag{2.27}$$

In view of (1.170), (2.27) and of the definition of $\aleph$ in (2.23), formula

$$\hat{\varphi}(z) \in \mathcal{N}_u(x,z) \tag{2.28}$$

is equivalent to

$$\begin{bmatrix} I_{m_1} & \hat{\varphi}(z)^* \end{bmatrix} \aleph(x,z) \begin{bmatrix} I_{m_1} \\ \hat{\varphi}(z) \end{bmatrix} \geq 0. \tag{2.29}$$

From (2.22), we see that $\aleph(x_1,z) < \aleph(x_2,z)$ for $x_1 > x_2$. Therefore, using the equivalence of (2.28) and (2.29), in a standard way, we obtain

$$\mathcal{N}_u(x_1,z) \subset \mathcal{N}_u(x_2,z) \quad \text{for } x_1 > x_2. \tag{2.30}$$

Moreover, (2.29) at $x = 0$ means that

$$\mathcal{N}_u(0, z) = \{\widehat{\varphi}(z) : \widehat{\varphi}(z)^* \widehat{\varphi}(z) \leq I_{m_1}\}. \tag{2.31}$$

Hence, in the same way as in the proof of Theorem 2.8, we can use Montel's theorem. This time we use Montel's theorem in order to show the existence of an analytic and nonexpansive matrix function $\varphi(z)$ such that

$$\varphi(z) \in \bigcap_{x<\infty} \mathcal{N}_u(x, z). \tag{2.32}$$

Indeed, because of (2.30) and (2.31), we see that the set of functions $\varphi(x, z, \mathcal{P})$ of the form (2.21) is uniformly bounded in $\mathbb{C}_+$. So, Montel's theorem is applicable and there is an analytic matrix function which we denote by $\varphi_\infty(z)$ and which is a uniform limit of some sequence

$$\varphi_\infty(z) = \lim_{i\to\infty} \varphi(x_i, z, \mathcal{P}_i) \quad (i \in \mathbb{N}, \quad x_i \uparrow, \quad \lim_{i\to\infty} x_i = \infty) \tag{2.33}$$

on all the bounded and closed subsets of $\mathbb{C}_+$. Since $x_i \uparrow$ and equalities (2.27) and (2.30) hold, it follows that the matrix functions

$$\mathcal{P}_{ik}(z) := u(x_i, z) \begin{bmatrix} I_{m_1} \\ \varphi(x_k, z, \mathcal{P}_k) \end{bmatrix} \quad (k \geq i)$$

satisfy relations (1.170) ($J = j$). Therefore, using (2.33), we derive that (1.170) holds for

$$\mathcal{P}_{i,\infty}(z) := u(x_i, z) \begin{bmatrix} I_{m_1} \\ \varphi_\infty(z) \end{bmatrix},$$

which implies that we can substitute $\mathcal{P} = \mathcal{P}_{i,\infty}$ and $x = x_i$ into (2.27) to obtain

$$\varphi_\infty(z) \in \mathcal{N}_u(x_i, z). \tag{2.34}$$

Since (2.34) holds for all $i \in \mathbb{N}$, we see that (2.32) is valid for $\varphi(z) = \varphi_\infty(z)$.

Next, following [218, 224] (see also [92] and references therein), we show that $\mathcal{N}_u$ is a matrix ball (disc). Namely, we partition $\aleph$ of the form (2.23) into four blocks $\aleph_{ij}$ $(i, j = 1, 2)$, where $\aleph_{22}$ is an $m_2 \times m_2$ block, and derive from (2.23) that

$$\aleph_{22} \leq -I_{m_2}.$$

Moreover, it follows from (2.22) and (2.23) that $\frac{d}{dx}\aleph \leq \mathrm{i}(\overline{z} - z)\aleph \leq \mathrm{i}(\overline{z} - z)j$. The inequalities above imply that

$$-\aleph_{22}(x, z) \geq (1 + \mathrm{i}(\overline{z} - z)x) I_{m_2}. \tag{2.35}$$

Note also that (2.23) yields the inequality $\aleph(x, z)^{-1} \geq j$ for $z \in \mathbb{C}_+$ (Corollary E.3), and so $(\aleph^{-1})_{11} \geq I_{m_1}$. Using a simple Lemma 2.1 from [280, Ch.1] (and formulas from its proof), we express $(\aleph^{-1})_{11}$ via the blocks of $\aleph$ and the last inequality takes the form

$$\left(\aleph^{-1}\right)_{11} = \left(\aleph_{11} - \aleph_{12}\aleph_{22}^{-1}\aleph_{21}\right)^{-1} \geq I_{m_1}. \tag{2.36}$$

Since $-\aleph_{22} > 0$, the square root $\Upsilon = (-\aleph_{22})^{1/2}$ is well-defined. Thus, we rewrite (2.29) in the form

$$\aleph_{11} - \aleph_{12}\aleph_{22}^{-1}\aleph_{21} - \left(\widehat{\varphi}^*\Upsilon - \aleph_{12}\Upsilon^{-1}\right)\left(\Upsilon\widehat{\varphi} - \Upsilon^{-1}\aleph_{21}\right) \geq 0,$$

where $\aleph_{12} = \aleph_{21}^*$. Equivalently, we have

$$\widehat{\varphi} = \rho_l \omega \rho_r - \aleph_{22}^{-1}\aleph_{21}, \quad \omega^*\omega \leq I_{m_2}, \tag{2.37}$$

$$\rho_l := \Upsilon^{-1} = (-\aleph_{22})^{-1/2}, \quad \rho_r := \left(\aleph_{11} - \aleph_{12}\aleph_{22}^{-1}\aleph_{21}\right)^{1/2}. \tag{2.38}$$

Here, ρ_l and ρ_r are the left and right semi-radii and ω is an $m_2 \times m_1$ matrix function. Since (2.28) is equivalent to (2.37), the sets $\mathcal{N}_u(x,z)$ (where the values of x and z are fixed) are matrix balls, indeed. According to (2.35), (2.36) and (2.38), the next formula holds:

$$\rho_l(x,z) \to 0 \quad (x \to \infty), \quad \rho_r(x,z) \leq I_{m_1}. \tag{2.39}$$

Finally, relations (2.32), (2.37) and (2.39) imply (2.26).

□

Relations (2.30) and (2.31) immediately yield the next corollary.

Corollary 2.18. *Matrices $\varphi(z) \in \mathcal{N}_u(x,z)$ are nonexpansive, that is, we have*

$$\varphi(z)^*\varphi(z) \leq I_{m_1}. \tag{2.40}$$

Remark 2.19. If $\varphi(z) \in \mathcal{N}_u(l,z)$ for all $z \in \mathbb{C}_+$ and, in addition, φ is analytic in $\mathbb{C}_+$, then it admits representation (2.21), where $x = l$ and $\mathcal{P}$ satisfies (1.170). This statement is immediately apparent since we can choose

$$\mathcal{P}(z) = u(l,z)\begin{bmatrix} I_{m_1} \\ \varphi(z) \end{bmatrix}.$$

The property-j of this $\mathcal{P}$ follows from the equivalence of (2.28) and (2.29).

Now, from Definition 2.16, we see that our φ is a Weyl function of system (2.18).

Our next definition of Weyl functions is similar to Definitions 0.9 and 1.51, although (for the sake of convenience) we omit the factor "i" before φ and the matrix Θ in the formula (1.190) from the Definition 1.51. (As we shall see, in this way, we switch from the Weyl functions belonging to Nevanlinna class to the contractive Weyl functions.)

Definition 2.20. A holomorphic function φ such that

$$\int_0^\infty \begin{bmatrix} I_{m_1} & \varphi(z)^* \end{bmatrix} u(x,z)^* u(x,z) \begin{bmatrix} I_{m_1} \\ \varphi(z) \end{bmatrix} dx < \infty \tag{2.41}$$

is called a Weyl function of the Dirac system (2.18) (where j is given by (2.19)) on $[0, \infty)$.

Corollary 2.21. *Let Dirac system* (2.18) *be given on* $[0, \infty)$ *and assume that* v *is locally summable. Then, there is a unique Weyl function of this system. This Weyl function is given by* (2.26).

Proof. According to Proposition 2.17, the function φ given by (2.26) is well-defined. Using the equalities in (2.22) and (2.23), and inequality (2.29), we derive that our φ satisfies relations

$$\begin{aligned} &\int_0^r \begin{bmatrix} I_{m_1} & \varphi(z)^* \end{bmatrix} u(x,z)^* u(x,z) \begin{bmatrix} I_{m_1} \\ \varphi(z) \end{bmatrix} dx \\ &= \frac{\mathrm{i}}{z - \overline{z}} \begin{bmatrix} I_{m_1} & \varphi(z)^* \end{bmatrix} (\aleph(0,z) - \aleph(r,z)) \begin{bmatrix} I_{m_1} \\ \varphi(z) \end{bmatrix} \\ &\le \frac{\mathrm{i}}{z - \overline{z}} \begin{bmatrix} I_{m_1} & \varphi(z)^* \end{bmatrix} \aleph(0,z) \begin{bmatrix} I_{m_1} \\ \varphi(z) \end{bmatrix}. \end{aligned} \tag{2.42}$$

Inequality (2.41) is immediate from (2.42).

It remains to be shown that the Weyl function is unique. Since $u^* u \ge -\aleph$, the inequality (2.35) yields

$$\int_0^r \begin{bmatrix} 0 & I_{m_2} \end{bmatrix} u(x,z)^* u(x,z) \begin{bmatrix} 0 \\ I_{m_2} \end{bmatrix} dx \ge r I_{m_2}. \tag{2.43}$$

In view of (2.43), the function satisfying (2.41) is unique. Indeed, let some matrices $\widehat{\varphi}$ and $\widetilde{\varphi}$ ($\widehat{\varphi} \ne \widetilde{\varphi}$) satisfy (2.41) at some $z \in \mathbb{C}_+$. Then, we obtain

$$\int_0^\infty g^* u(x,z)^* u(x,z) g dx < \infty \quad \text{for } g \in L_\varphi := \operatorname{span}\left(\operatorname{Im} \begin{bmatrix} I_{m_1} \\ \widehat{\varphi} \end{bmatrix} \cup \operatorname{Im} \begin{bmatrix} I_{m_1} \\ \widetilde{\varphi} \end{bmatrix} \right).$$

On the other hand, (2.43) yields

$$\int_0^\infty g^* u(x,z)^* u(x,z) g dx = \infty \quad \text{for } g \in L(\infty) := \operatorname{Im} \begin{bmatrix} 0 \\ I_{m_2} \end{bmatrix}, \ g \ne 0.$$

Clearly, we have $\dim(L_\varphi) > m_1$ and $\dim(L(\infty)) = m_2$. Since $L_\varphi \cap L(\infty) = 0$ and $L_\varphi + L(\infty) \subseteq \mathbb{C}^m$, we arrive at a contradiction. □

Remark 2.22. From Proposition 2.17 and Corollary 2.21, we see that formula (2.26) can be used as an equivalent definition of the Weyl function of system on the semi-axis. However, the definition of the form (2.41) is a more classical one and deals with solutions of (2.18) which belong to $L^2(0, \infty)$. Definition (2.41) is close to definitions of the Weyl–Titchmarsh or M-functions for discrete and continuous systems in [76, 195, 205, 238, 242, 290, 320, 322] (see also references therein).

Remark 2.23. It is also shown in the proof of Corollary 2.21 that for any $z \in \mathbb{C}_+$, there is a unique matrix $\varphi(z)$ such that (2.41) holds at this z. That is, we have uniqueness even without the requirement that the Weyl function is holomorphic.

Remark 2.24. Let the conditions of Corollary 2.21 hold. Then, formulas (2.37) and (2.39) yield the equality

$$\varphi(z) = \lim_{b \to \infty} \varphi_b(z) \quad (z \in \mathbb{C}_+) \tag{2.44}$$

for the Weyl function $\varphi(z)$ and any set of functions $\varphi_b(z) \in \mathcal{N}_u(b, z)$.

From Corollary 2.21, Proposition 2.17 and Remark 2.19, we immediately obtain the next corollary.

Corollary 2.25. *The Weyl function of each Dirac system* (2.18) *on* $[0, \infty)$ *(where the potential* v *is locally summable) is also a Weyl function of the same Dirac system on all the finite intervals* $[0, l]$.

The last proposition in this subsection is dedicated to a property of the Weyl function, an analog of which may be used as a definition of generalized Weyl functions in a more complicated non-self-adjoint cases (see, e.g. [113, 242, 249]).

Proposition 2.26. *Let the Dirac system* (2.18) *be given on* $[0, \infty)$, *let its potential* v *be locally summable, and assume that* φ *is its Weyl function. Then, the inequality*

$$\sup_{x \le l,\, z \in \mathbb{C}_+} \|\hat{u}(x,z)\| < \infty, \quad \hat{u}(x,z) := \mathrm{e}^{-\mathrm{i}zx} u(x,z) \begin{bmatrix} I_{m_1} \\ \varphi(z) \end{bmatrix} \tag{2.45}$$

holds on any finite interval $[0, l]$.

Proof. We fix some l and choose x such that $0 < x \le l < \infty$. Because of (2.26), the Weyl function φ admits representations (2.21) (i.e. $\varphi(z) = \varphi(x, z, \mathcal{P})$). Hence, using (1.170) (where $J = j$) and (2.27), we obtain

$$\hat{u}(x,z)^* j \hat{u}(x,z) \ge 0. \tag{2.46}$$

On the other hand, equation (2.18) and the definition of $\hat{u}$ in (2.45) imply that

$$\begin{aligned} &\frac{d}{dx}\left(\mathrm{e}^{-2xM}\hat{u}(x,z)^*(I_m+j)\hat{u}(x,z)\right) \\ &= \mathrm{e}^{-2xM}\hat{u}(x,z)^*\left(\mathrm{i}\left((I_m+j)jV - Vj(I_m+j)\right) - 2M(I_m+j)\right)\hat{u}(x,z) \\ &= 2\mathrm{e}^{-2xM}\hat{u}(x,z)^*\begin{bmatrix} -2MI_{m_1} & \mathrm{i}v(x) \\ -\mathrm{i}v(x)^* & 0 \end{bmatrix}\hat{u}(x,z), \quad M := \sup_{x<l}\|V(x)\|. \end{aligned} \tag{2.47}$$

Using (2.46) and (2.47), we derive

$$\begin{aligned} &\frac{d}{dx}\left(\mathrm{e}^{-2xM}\hat{u}(x,z)^*(I_m+j)\hat{u}(x,z)\right) \\ &\leq 2\mathrm{e}^{-2xM}\hat{u}(x,z)^*\left(\begin{bmatrix} & \mathrm{i}v(x) \\ -\mathrm{i}v(x)^* & 0 \end{bmatrix} - MI_m\right)\hat{u}(x,z) \leq 0. \end{aligned} \tag{2.48}$$

Finally, inequalities (2.46) and (2.48) lead us to

$$\hat{u}(x,z)^*\hat{u}(x,z) \leq \hat{u}(x,z)^*(I_m+j)\hat{u}(x,z) \leq 2\mathrm{e}^{2xM}I_m, \tag{2.49}$$

and (2.45) follows. □

2.2.2 Direct and inverse problems: explicit solutions

In this subsection, we consider Dirac systems with the so called *generalized pseudo-exponential* potentials (Definition 2.28). Direct and inverse problems for such systems on the semiaxis $[0, \infty)$ are solved explicitly. For that purpose, we again follow [110]. The same scheme was applied earlier in [111, 129, 133, 245]. See also some references therein. In particular, explicit solutions of the direct and inverse problems (for spectral functions and for a smaller class of Dirac systems with square potentials) were obtained in [17]. We use our GBDT version of the Bäcklund–Darboux transformation, some classical results from system theory [153] and several results on Riccati equations [185]. Basic notions from system theory and one of the related theorems on the Riccati equation are given in Appendix B.

GBDT transformations of the fundamental solutions and potentials of Dirac systems, where $m_1 = m_2 = p$, were constructed in Subsection 1.1.3. Precisely in the same way, the analog (see below) of Proposition 1.4 is proved for the case, where j has the form (2.19).

Proposition 2.27. *Let Dirac system* (2.18), (2.19) *and a triple of parameter matrices* A, $\Pi(0)$ *and* $S(0) = S(0)^*$ *(satisfying* (1.51)*) be given. Then, the matrix function* w_A *defined by* (1.55)*, where* $\Pi(x)$ *and* $S(x)$ *are obtained by using formulas* (1.50) *and* (1.52), *respectively, is a Darboux matrix (gauge transformation) of the Dirac system and satisfies equation* (1.59). *The fundamental solution* $\widetilde{u}$ *of the transformed system*

$$\widetilde{u}_x = \mathrm{i}(zj + j\widetilde{V})\widetilde{u}, \quad \widetilde{V}(x) = \begin{bmatrix} 0 & \widetilde{v}(x) \\ \widetilde{v}(x)^* & 0 \end{bmatrix} \tag{2.50}$$

is given by the formula

$$\widetilde{u}(x,z) = w_A(x,z)u(x,z)w_A(0,z)^{-1}. \tag{2.51}$$

Here, u *is the fundamental solution of the initial system* (2.18) *and* $\widetilde{V}$ *is given by* (1.60).

Recall that A and $S(0)$ are $n \times n$ matrices, $\Pi(0)$ is an $n \times m$ matrix, $n \in \mathbb{N}$ and

$$w_A(x,z) = I_m - \mathrm{i}j\Pi(x)^* S(x)^{-1}(A - zI_n)^{-1}\Pi(x). \tag{2.52}$$

In this subsection, we deal with the case where the initial system is trivial, that is,

$$V(x) \equiv 0.$$

For that case, the matrix functions $\widetilde{u}$ and $\widetilde{V}$ can be constructed explicitly. Namely, we easily derive from (1.50) that

$$\Pi(x) = \begin{bmatrix} \Lambda_1(x) & \Lambda_2(x) \end{bmatrix}, \quad \Lambda_1(x) = \mathrm{e}^{-\mathrm{i}xA}\varkappa_1, \quad \Lambda_2(x) = \mathrm{e}^{\mathrm{i}xA}\varkappa_2, \tag{2.53}$$

where $\varkappa_1$ and $\varkappa_2$ are $n \times m_1$ and $n \times m_2$, respectively, blocks of $\Pi(0)$. Rewriting (1.51) in terms of $\varkappa_1$ and $\varkappa_2$, we have

$$AS(0) - S(0)A^* = \mathrm{i}(\varkappa_1\varkappa_1^* - \varkappa_2\varkappa_2^*). \tag{2.54}$$

Next, we derive from (1.52) the equality

$$S(x) = S(0) + \int_0^x \Pi(t)\Pi(t)^* dt \tag{2.55}$$

and assume that

$$S(0) > 0, \tag{2.56}$$

so that $S(x) > 0$ for all $x \geq 0$. Clearly, $u(x,z) = \mathrm{e}^{\mathrm{i}zxj}$ for the case that $V = 0$, and therefore formula (2.51) takes the form

$$\widetilde{u}(x,z) = w_A(x,z)\mathrm{e}^{\mathrm{i}zxj}w_A(0,z)^{-1}. \tag{2.57}$$

Finally, since $V = 0$ and $\Pi = \begin{bmatrix} \Lambda_1 & \Lambda_2 \end{bmatrix}$, formula (1.60) is equivalent to the equality $\widetilde{v} = -2\mathrm{i}\Lambda_1^* S^{-1}\Lambda_2$. Hence, in view of (2.53), we obtain

$$\widetilde{v}(x) = -2\mathrm{i}\varkappa_1^* \mathrm{e}^{\mathrm{i}xA^*} S(x)^{-1}\mathrm{e}^{\mathrm{i}xA}\varkappa_2. \tag{2.58}$$

Taking into account the invertibility of $S(x)$ for $x \geq 0$, we see that the potential $\widetilde{v}$ in (2.58) is well-defined.

Definition 2.28. The $m_1 \times m_2$ potential $\widetilde{v}$ of the form (2.58), where relations (2.54)–(2.56) hold, is called a generalized pseudoexponential potential. It is said that $\widetilde{v}$ is generated by the parameter matrices A, $S(0)$, $\varkappa_1$ and $\varkappa_2$.

We note that the case $m_1 = m_2$ (i.e. the case of the pseudoexponential potentials) was treated in greater detail in [133] (see [133] and references therein for the term *pseudoexponential*, itself, too).

The next theorem provides an explicit solution of the direct problem.

Theorem 2.29. *Let Dirac system* (2.50) *be given on* $[0, \infty)$. *Assume that* $\widetilde{v}$ *is a generalized pseudoexponential potential which is generated by the matrices* A, $S(0)$, $\varkappa_1$ *and* $\varkappa_2$. *Then, the Weyl function* φ *of system* (2.50) *has the form*

$$\varphi(z) = \mathrm{i}\varkappa_2^* S(0)^{-1}(\alpha - zI_n)^{-1}\varkappa_1, \quad \alpha := A - \mathrm{i}\varkappa_1\varkappa_1^* S(0)^{-1}. \tag{2.59}$$

Proof. First, we partition w_A into blocks and show that

$$w_{21}(z)w_{11}(z)^{-1} = \mathrm{i}\varkappa_2^* S(0)^{-1}(\alpha - zI_n)^{-1}\varkappa_1, \quad w_A(0,z) =: \{w_{ij}(z)\}_{i,j=1}^2. \tag{2.60}$$

Here, w_{11} and w_{21} are $m_1 \times m_1$ and $m_2 \times m_1$ blocks, respectively. In order to prove (2.60), we use standard calculations from system theory, see Appendix B. First, we take into account (2.52) and write realizations

$$w_{11}(z) = I_{m_1} - \mathrm{i}\varkappa_1^* S(0)^{-1}(A - zI_n)^{-1}\varkappa_1, \tag{2.61}$$

$$w_{21}(z) = \mathrm{i}\varkappa_2^* S(0)^{-1}(A - zI_n)^{-1}\varkappa_1. \tag{2.62}$$

Next, from (2.61) and (B.5), we obtain

$$w_{11}(z)^{-1} = I_{m_1} + \mathrm{i}\varkappa_1^* S(0)^{-1}(\alpha - zI_n)^{-1}\varkappa_1, \tag{2.63}$$

where α is given in (2.59) and $z \notin (\sigma(A) \cup \sigma(\alpha))$. In view of the equality $\mathrm{i}\varkappa_1\varkappa_1^* S(0)^{-1} = A - \alpha$, relations (2.62) and (2.63) yield (2.60).

According to (2.57) and (2.60), the function φ given by (2.59) satisfies the equality

$$\widetilde{u}(x,z)\begin{bmatrix} I_{m_1} \\ \varphi(z) \end{bmatrix} = \mathrm{e}^{\mathrm{i}zx} w_A(x,z)\begin{bmatrix} I_{m_1} \\ 0 \end{bmatrix} w_{11}(z)^{-1}. \tag{2.64}$$

Let us consider the factor $\mathrm{e}^{\mathrm{i}xz} w_A(x,z)$ on the right-hand side of (2.64). We note that the entries of the matrix function $\Pi^* S^{-1}$, which appears in the definition (2.52) of w_A, belong to $L^2(0, \infty)$, that is,

$$\Pi^* S^{-1} \in L^2_{m\times n}(0, \infty). \tag{2.65}$$

Indeed, we derive from (2.55) that

$$S(x)^{-1}\Pi(x)\Pi(x)^* S(x)^{-1} = -\frac{d}{dx} S(x)^{-1}. \tag{2.66}$$

Since $S(x) > 0$, using (2.66), we obtain

$$\int_0^\infty S(t)^{-1}\Pi(t)\Pi(t)^* S(t)^{-1} dt \le S(0)^{-1}, \tag{2.67}$$

and so (2.65) holds. Furthermore, (2.53) implies that

$$\sup_{\Im z > \|A\| + \varepsilon} \|e^{izx}\Pi(x)\| < M_\varepsilon \quad (\varepsilon > 0). \tag{2.68}$$

It follows from (2.52), (2.63), (2.65) and (2.68) that the entries of the right-hand side of (2.64) are well-defined and uniformly bounded in the $L^2(0, \infty)$ norm with respect to x for all z such that $\Im z \ge \max(\|A\|, \|\alpha\|) + \varepsilon$ $(\varepsilon > 0)$. Hence, taking into account (2.64), we see that (2.41) holds for z from the above-mentioned domain. Thus, according to the uniqueness statement in Remark 2.23, $\varphi(z)$ of the form (2.59) coincides with the Weyl function in that domain. In view of analyticity, the matrix function φ coincides with the Weyl function in $\mathbb{C}_+$ (i.e. φ is the Weyl function, indeed). □

When $\widetilde{v}$ is a generalized pseudoexponential potential, our Weyl function coincides with the reflection coefficient from [111], more precisely, with the reflection coefficient from [111] for the subcase $S(0) > 0$ since the singular case $S(0) \not> 0$ is studied there too ([111, Theorems 3.3 and 4.1]).

From [185, Theorems 21.1.3, 21.2.1], an auxiliary lemma easily follows.

Lemma 2.30. *Let $\varphi(z)$ be a strictly proper rational $m_2 \times m_1$ matrix function, which is nonexpansive on $\mathbb{R}$ and has no poles in $\mathbb{C}_+$. Assume that*

$$\varphi(z) = -C(\mathcal{A} - zI_n)^{-1}B \tag{2.69}$$

is a minimal realization of φ. Then, the Riccati equation

$$XC^*CX + i(X\mathcal{A}^* - \mathcal{A}X) + BB^* = 0 \tag{2.70}$$

has a positive solution $X > 0$. Furthermore, all the Hermitian solutions of (2.70) *are positive.*

Using the lemma above, we explicitly solve the corresponding inverse problem.

Theorem 2.31. *Let $\varphi(z)$ satisfy conditions of Lemma* 2.30. *Assume that* (2.69) *is its minimal realization and that $X > 0$ is a solution of* (2.70).

Then, $\varphi(z)$ is the Weyl function of the Dirac system on the semiaxis $[0, \infty)$, the potential of which is given by (2.58), *where S is defined in* (2.55) *(via* (2.53)*) and*

$$A = \mathcal{A} + iBB^*X^{-1}, \quad S(0) = X, \quad \varkappa_1 = B, \quad \varkappa_2 = -iXC^*. \tag{2.71}$$

Proof. From (2.71), we see that

$$AS(0) - S(0)A^* = \mathcal{A}X - X\mathcal{A}^* + 2iBB^*, \quad i(\varkappa_1\varkappa_1^* - \varkappa_2\varkappa_2^*) = iBB^* - iXC^*CX.$$

Taking into account (2.70) and the equalities above, we easily derive (2.54). Since (2.54) holds, we apply Theorem 2.29. Theorem 2.29 states that the Weyl function of the Dirac system, where $\widetilde{v}$ is given by (2.58), has the form (2.59). Next, we substitute (2.71) into (2.59) to derive that the right-hand sides of (2.69) and the first equality in (2.59) coincide. In other words, the Weyl function of the constructed system admits representation (2.69). □

It is shown in Section 2.3 that the solution of the inverse problem is unique in the class of Dirac systems with the locally bounded potentials.

Because of the operator (matrix) identity (2.54) and the second equality in (2.59), the matrices α and $S(0)$ satisfy another matrix identity

$$\alpha S(0) - S(0)\alpha^* = -\mathrm{i}(\varkappa_1\varkappa_1^* + \varkappa_2\varkappa_2^*). \tag{2.72}$$

Since $S(0)$ is invertible, identity (2.72) yields

$$S(0)^{-1}\alpha - \alpha^* S(0)^{-1} = -\mathrm{i}S(0)^{-1}(\varkappa_1\varkappa_1^* + \varkappa_2\varkappa_2^*)S(0)^{-1}. \tag{2.73}$$

If $g \neq 0$ is an eigenvector of α (i.e. $\alpha g = \lambda g$), identity (2.73) implies that

$$(\lambda - \overline{\lambda})g^* S(0)^{-1} g = -\mathrm{i}g^* S(0)^{-1}(\varkappa_1\varkappa_1^* + \varkappa_2\varkappa_2^*)S(0)^{-1}g. \tag{2.74}$$

Taking into account the inequality $S(0) > 0$, we derive from (2.74) that

$$\sigma(\alpha) \subset \mathbb{C}_- \cup \mathbb{R}. \tag{2.75}$$

Real eigenvalues of α play a special role in the spectral theory of an operator $\widetilde{\mathcal{H}}$ which corresponds to the Dirac system with a generalized pseudoexponential potential (see, e.g. [132] for the case of square potentials). This operator is determined by the differential expression like in (2.1), but on the subspace $\mathcal{D}(\widetilde{\mathcal{H}})$ (of the set of absolutely continuous functions), which is smaller than the subspace $\mathcal{D}(\mathcal{H}_D)$ given by (2.2). Namely, we have

$$\widetilde{\mathcal{H}}f = -\left(\mathrm{i}j\frac{d}{dx} + \widetilde{V}(x)\right)f, \tag{2.76}$$

$$\mathcal{D}(\widetilde{\mathcal{H}}) := \left\{f : -\left(\mathrm{i}j\frac{d}{dx} + \widetilde{V}(x)\right)f \in L^2_m(0,\infty), \quad f(0) = 0\right\}. \tag{2.77}$$

Proposition 2.32. *Let the conditions of Theorem 2.29 hold, let α be given by the second relation in (2.59) and let λ be a real eigenvalue of α, that is,*

$$\alpha g = \lambda g, \quad g \neq 0, \quad \lambda \in \mathbb{R}. \tag{2.78}$$

Then, the matrix function

$$f(x) := j\Pi(x)^* S(x)^{-1} g \tag{2.79}$$

is a bounded state of $\widetilde{\mathcal{H}}$ and $\widetilde{\mathcal{H}}f = \lambda f$.

Proof. We note that formulas (2.74) and (2.78) yield

$$\varkappa_1^* S(0)^{-1} g = 0, \quad \varkappa_2^* S(0)^{-1} g = 0, \quad A g = \lambda g. \tag{2.80}$$

Indeed, the first two equalities in (2.80) easily follow from (2.74) for the case that $\lambda = \overline{\lambda}$. The equality $Ag = \lambda g$ is immediate from $\alpha g = \lambda g$, definition of α in (2.59) and equality $\varkappa_1^* S(0)^{-1} g = 0$.

It is apparent from the proof of (1.57) that this formula holds for j of the form (2.19) as well. (In fact, (1.57) admits much more essential generalizations, see Chapter 7.) For the case that $V = 0$, we rewrite (1.57) in the form

$$\left(j\Pi^* S^{-1}\right)' = \mathrm{i}\Pi^* S^{-1} A + \left(\Pi^* S^{-1}\Pi - j\Pi^* S^{-1}\Pi j\right) j\Pi^* S^{-1}. \tag{2.81}$$

Recall that according to Proposition 2.27, the matrix function $\widetilde{V}$ is given by (1.60), and so we have

$$\Pi^* S^{-1}\Pi - j\Pi^* S^{-1}\Pi j = \mathrm{i} j \widetilde{V}. \tag{2.82}$$

We substitute (2.82) into (2.81) and then apply both sides of (2.81) to f to derive

$$\left(j\Pi^* S^{-1} g\right)' = \mathrm{i} j^2 \Pi^* S^{-1} A g + \mathrm{i} j \widetilde{V} j \Pi^* S^{-1} g. \tag{2.83}$$

Using (2.79) and (2.80), we rewrite (2.83) as

$$f' = \mathrm{i}(\lambda j + j\widetilde{V}) f. \tag{2.84}$$

In view of (2.84), it remains to be shown that $f \in \mathcal{D}(\widetilde{\mathcal{H}})$, more precisely, that $f \in L^2_m(0, \infty)$ and $f(0) = 0$. From (2.67), we see that $f \in L^2_m(0, \infty)$. Finally, the equalities

$$f(0) = \begin{bmatrix} \varkappa_1 & -\varkappa_2 \end{bmatrix}^* S(0)^{-1} g = 0 \tag{2.85}$$

are immediate from (2.53), (2.79) and (2.80). Hence, the initial condition is fulfilled. □

2.3 Recovery of the Dirac system: general case

The topic of this section is closely related to the topic of Subsection 2.1.2, where inverse problems for locally bounded square potentials are treated. However, in this section the potentials are not necessarily square and it would be difficult to talk about spectral functions. Hence, the approach is modified correspondingly (similar to [238, 259] and following [106], where precisely this case was dealt with).

The first Subsection 2.3.1 is dedicated to representation of the fundamental solution. In Subsection 2.3.2 we study the high energy asymptotics of Weyl functions. The solution of the inverse problem to recover the potential (and system) from the Weyl function is given in Subsection 2.3.3. Borg–Marchenko-type uniqueness results are contained in that subsection as well. Finally, Subsection 2.3.4 is dedicated to conditions for an analytic matrix function to be the Weyl function of some Dirac system.

2.3.1 Representation of the fundamental solution

The results of this subsection can be formulated for a Dirac system on a fixed final interval $[0, l]$. We assume that v is bounded on this interval, that is,

$$\sup_{x<l} \|v(x)\| < \infty. \tag{2.86}$$

Modifying formulas (1.13) and (1.16), we put

$$\beta(x) = \begin{bmatrix} I_{m_1} & 0 \end{bmatrix} u(x,0), \quad \gamma(x) = \begin{bmatrix} 0 & I_{m_2} \end{bmatrix} u(x,0). \tag{2.87}$$

It follows from (2.18) and (2.87) that

$$\sup_{x<l} \|\gamma'(x)\| < \infty. \tag{2.88}$$

Clearly, normalization (2.20) and the equality in (2.22) yield

$$u(x,z)^* j u(x,z) = j = u(x,z) j u(x,z)^*, \quad z \in \mathbb{R}. \tag{2.89}$$

Therefore, (2.87) implies

$$\beta j \beta^* \equiv I_{m_1}, \quad \gamma j \gamma^* \equiv -I_{m_2}, \quad \beta j \gamma^* \equiv 0. \tag{2.90}$$

Next, we again use Proposition 1.29 from Chapter 1, where we put (this time) $p = m_2$ and $r = m$, that is, F and G are $m_2 \times m$ and $m \times m_2$ matrix functions, respectively. More precisely, we set

$$F = \gamma, \quad G = \mathrm{i} j \gamma^*, \quad \gamma j \gamma^* \equiv -I_{m_2}, \quad \gamma \in B^1([0, l]), \tag{2.91}$$

$$K = \mathrm{i}\gamma(x) j \int_0^x \gamma(t)^* \cdot dt, \quad K \in B\left(L^2_{m_2}(0, l)\right). \tag{2.92}$$

It is apparent from (2.91) that F and G, which are given there, satisfy conditions of Proposition 1.29. We partition $\gamma(x)$ into two blocks $\gamma = \begin{bmatrix} \gamma_1 & \gamma_2 \end{bmatrix}$, where γ_1 and γ_2, respectively, are $m_2 \times m_1$ and $m_2 \times m_2$ matrix functions. Our next proposition is the direct analog of Proposition 1.31.

Proposition 2.33. *Let K be given by* (2.92) *(where γ satisfies two last equalities in* (2.91)*) and let $\mathcal{E}$ be a transformation operator from Proposition* 1.29*. Introduce $\mathcal{E}_0 \in B(L^2_{m_2}(0, l))$ by the equalities*

$$(\mathcal{E}_0 f)(x) = \kappa(0)^{-1}\gamma_2(0) f(x) + \int_0^x \mathcal{E}_0(x-t) f(t) dt, \quad \mathcal{E}_0(x) := \left(\mathcal{E}^{-1}\gamma_2\right)'(x). \tag{2.93}$$

Then, the operator $E := \mathcal{E}\mathcal{E}_0$ is another transformation operator from Proposition 1.29*, which satisfies the additional condition*

$$E^{-1}\gamma_2 \equiv I_{m_2}. \tag{2.94}$$

The proof of the Proposition 2.33 is quite similar to the proof of Proposition 1.31, which is formulated for the case that $m_1 = m_2$.

According to Proposition 2.33, we have

$$K = EAE^{-1}, \quad A := -\mathrm{i}\int_0^x \cdot\, dt, \quad A, E, E^{-1} \in B\left(L^2_{m_2}(0, l)\right), \tag{2.95}$$

and (2.94) holds.

Remark 2.34. The kernels of the operators $\mathcal{E}^{\pm 1}$, which are given in [274], as well as the kernels of the operators $\mathcal{E}_0^{\pm 1}$ from Proposition 1.31 are bounded by construction. Therefore, without loss of generality, we always assume further that the kernels of $E^{\pm 1}$ are bounded.

The next lemma easily follows from Proposition 2.33 and will be used to construct the fundamental solution.

Lemma 2.35. *Let y be an $m_2 \times m$ matrix function which satisfies the last two relations in* (2.91), *and set*

$$S := E^{-1}\left(E^*\right)^{-1}, \quad \Pi := \begin{bmatrix} \Phi_1 & \Phi_2 \end{bmatrix}, \quad \Phi_k \in B\left(\mathbb{C}^{m_k}, L^2_{m_2}(0, l)\right); \tag{2.96}$$

$$(\Phi_1 g)(x) = \Phi_1(x) g, \quad \Phi_1(x) := \left(E^{-1} y_1\right)(x); \quad \Phi_2 g = I_{m_2} g \equiv g; \tag{2.97}$$

where E is constructed (for the given y) in Proposition 2.33. *Then, A, S and Π form an S-node, that is, the operator identity*

$$AS - SA^* = \mathrm{i}\Pi j \Pi^* \tag{2.98}$$

holds.

Proof. Because of (2.92) and (2.95), we obtain

$$EAE^{-1} - \left(E^{-1}\right)^* A^* E^* = K - K^* = \mathrm{i} y(x) j \int_0^l y(t)^* \cdot\, dt. \tag{2.99}$$

Formulas (2.94), (2.96) and (2.97) lead us to the equality

$$\Pi f = \left(E^{-1} y\right)(x) f. \tag{2.100}$$

Now, the operator identity (2.98) follows from (2.99), (2.100) and the first equality in (2.96). □

Remark 2.36. The condition on the sum of images

$$\overline{\sum_{i=0}^{\infty} \mathrm{Im}\,\left((A^*)^i S^{-1} \Pi\right)} = L^2_{m_2}(0, l), \tag{2.101}$$

which appears in some important theorems in [290], is also satisfied. Recall that the sum $\mathbf{H}_1 + \mathbf{H}_2$ of linear subspaces $\mathbf{H}_1$, $\mathbf{H}_2$ stands for the span $(\mathbf{H}_1 \cup \mathbf{H}_2)$.

In order to show that (2.101) holds, rewrite (2.98) as $S^{-1}A = A^*S^{-1} + \mathrm{i}S^{-1}\Pi j \Pi^* S^{-1}$. Hence, for $N_1, N_2 \geq 0$, we obtain

$$\operatorname{Im}\left((A^*)^{N_1+1} S^{-1} A^{N_2}\Pi\right) + \sum_{i=0}^{N_1+N_2} \operatorname{Im}\left((A^*)^i S^{-1}\Pi\right) \supseteq \operatorname{Im}\left((A^*)^{N_1} S^{-1} A^{N_2+1}\Pi\right). \tag{2.102}$$

Using (2.102), we derive (by induction) the formula

$$\sum_{i=0}^{N} \operatorname{Im}\left((A^*)^i S^{-1}\Pi\right) \supseteq \sum_{i=0}^{N} \operatorname{Im}\left(S^{-1}A^i\Pi\right) = S^{-1}\sum_{i=0}^{N} \operatorname{Im}\left(A^i\Pi\right). \tag{2.103}$$

In view of (2.103), the equality (2.101) follows from

$$\overline{\sum_{i=0}^{\infty} \operatorname{Im}\,(A^i\Pi)} = L^2_{m_2}(0, l), \tag{2.104}$$

which, in turn, follows from (2.97).

Putting in (1.99) $r = m_2$, we introduce a family of orthoprojectors P_ξ:

$$\left(P_\xi f\right)(x) = f(x) \quad (0 < x < \xi), \quad P_\xi \in B\left(L^2_{m_2}(0, l), L^2_{m_2}(0, \xi)\right) \tag{2.105}$$

and set

$$A_\xi := P_\xi A P_\xi^*, \quad S_\xi := P_\xi S P_\xi^*. \tag{2.106}$$

We recall that for A of the form (2.95), identity (2.98) implies identities

$$A_\xi S_\xi - S_\xi A_\xi^* = \mathrm{i}P_\xi \Pi j \Pi^* P_\xi^*, \tag{2.107}$$

that is, the operators A_ξ, S_ξ and $P_\xi\Pi$ form an S-node. The transfer matrix function corresponding to an S-node satisfying (2.107) has the form

$$w_A(\xi, \lambda) = I - \mathrm{i}j\Pi^* S_\xi^{-1}(A_\xi - \lambda I)^{-1} P_\xi \Pi, \tag{2.108}$$

where j is given by (2.19).

Theorem 2.37. *Let y be an $m_2 \times m$ matrix function which satisfies the last two relations in* (2.91). *Then, the fundamental solution W of the system*

$$\frac{d}{dx} W(x, z) = \mathrm{i}zjy(x)^* y(x) W(x, z) \tag{2.109}$$

admits representation

$$W(\xi, z) = w_A(\xi, 1/z). \tag{2.110}$$

Here, W is normalized by the condition $W(0, z) = I_m$, w_A has the form (2.108) *and the triple $\{A, S, \Pi\}$ is given by* (2.95), (2.96) *and* (2.97).

Proof. The statement of the theorem follows from Theorem 1.20. Namely, quite similar to the proof of Theorem 1.34, we can show that the triple $\{A, S, \Pi\}$ given by (2.95), (2.96) and (2.97) corresponds to system (2.109). □

Remark 2.38. We note that the proofs of Theorem 2.37 above and Theorem 1.34 are similar in spite of the fact that the system treated in Theorem 1.34 is canonical (see also Definition 1.23 of the triple corresponding to a canonical system) and system (2.109) is not even equivalent to canonical when $m_1 \neq m_2$.

Now, again consider the case of Dirac system with bounded potential. In other words, the case of γ given by (2.87). Because of (2.87) and (2.90), we immediately obtain

$$u(x,0)j\gamma(x)^*\gamma(x)u(x,0)^{-1} = -\begin{bmatrix} 0 & 0 \\ 0 & I_{m_2} \end{bmatrix}. \tag{2.111}$$

Hence, direct differentiation shows that the next corollary is valid.

Corollary 2.39. *Let $u(x,z)$ be the fundamental solution of a Dirac system with the bounded potential v and let γ be given by* (2.87)*. Then, $u(x,z)$ admits representation*

$$u(x,z) = \mathrm{e}^{\mathrm{i}zx}u(x,0)W(x,2z), \tag{2.112}$$

where W has the form (2.110) *and the S-node generating the transfer matrix function w_A is recovered from γ in Lemma 2.35 (see* (2.96) *and* (2.97)*).*

Remark 2.40. For the case that γ is given by (2.87), it follows from (2.18), (2.87) and (2.90) that

$$\gamma'(x)j\gamma(x)^* = -\mathrm{i}\begin{bmatrix} v(x)^* & 0 \end{bmatrix} u(x,0)j\gamma(x)^* = -\mathrm{i}v(x)^*\beta(x)j\gamma(x)^* \equiv 0. \tag{2.113}$$

Thus, from (2.91) and (2.113), we see that κ in (1.136) is a constant matrix. Moreover, we recall equality (2.100), where E (according to Proposition 2.33) belongs to the family of transformation operators (1.136). Therefore, since $\gamma_2(0) = I_{m_2}$, equality (2.100) implies that $\kappa(x) \equiv I_{m_2}$ and E has the form

$$(Ef)(x) = f(x) + \int_0^x E(x,t)f(t)dt, \quad E \in B\left(L^2_{m_2}(0,l)\right). \tag{2.114}$$

Proposition 2.41. *Let $\Pi = [\Phi_1 \quad \Phi_2]$ be constructed in Lemma 2.35 where γ is given by* (2.87)*. Then, there is a unique solution $S \in B\left(L^2_{m_2}(0,l)\right)$ of the operator identity* (2.98)*, this S is strictly positive and is defined by the equalities*

$$(Sf)(x) = f(x) - \int_0^l s(x,t)f(t)dt, \; s(x,t) := \int_0^{\min(x,t)} \Phi_1'(x-\zeta)\Phi_1'(t-\zeta)^*d\zeta. \tag{2.115}$$

Proof. It is immediate from (2.96) and (2.97) that S is strictly positive and Φ_1 is boundedly differentiable. Since $y_1(0) = 0$, formulas (2.97) and (2.114) yield

$$\Phi_1(+0) = 0. \tag{2.116}$$

Now, the statement of the proposition follows from Appendix D, see Theorem D.1. □

Continuous operator kernels (with a possible jump at $x = t$) from (2.115) were considered in [135, Section 2.4], where they were called "close to displacement kernels" (see also references therein).

Next, we use (2.113) to recover y from $y_2^{-1}y_1$. It is immediate from $yjy^* \equiv -I_{m_2}$ that $\det y_2(x) \neq 0$ and so $y_2^{-1}y_1$ is well-defined. Hence, we rewrite (2.113) as

$$y_2' = y_1'(y_2^{-1}y_1)^*. \tag{2.117}$$

We also have

$$(y_2^{-1}y_1)' = -y_2^{-1}y_2'y_2^{-1}y_1 + y_2^{-1}y_1', \quad \text{i.e.} \quad y_1' = y_2'(y_2^{-1}y_1) + y_2(y_2^{-1}y_1)'.$$

In view of (2.87), (2.117) and the formula above, we see that

$$y_2'\left(I_{m_2} - (y_2^{-1}y_1)(y_2^{-1}y_1)^*\right) = y_2(y_2^{-1}y_1)'(y_2^{-1}y_1)^*, \quad y_2(0) = I_{m_2}.$$

Therefore, taking into account that $I_{m_2} - (y_2^{-1}y_1)(y_2^{-1}y_1)^* = y_2^{-1}(y_2^{-1})^*$ is invertible, we obtain

$$y_2' = y_2(y_2^{-1}y_1)'(y_2^{-1}y_1)^*\left(I_{m_2} - (y_2^{-1}y_1)(y_2^{-1}y_1)^*\right)^{-1}, \quad y_2(0) = I_{m_2}, \tag{2.118}$$

which recovers y_2. Then, y_1 is immediately recovered from $y_2^{-1}y_1$ and y_2. Recall that in a somewhat similar way, y is recovered from $H_d = y^*y$ in Subsection 1.1.1.

Remark 2.42. If the conditions of Remark 2.40 hold, then y_2, y and Hamiltonian $H = y^*y$ are consecutively recovered from $y_2^{-1}y_1$ via the differential equation (2.118). Furthermore, from the third relation in (2.91), we see that $(y_2^{-1}y_1)(y_2^{-1}y_1)^* < I_{m_2}$. Thus, the matrix function $y_2^{-1}y_1$ is a continuous analog of Schur coefficients for system (2.109). (Compare with Remark 3.1 from the paper [114] on the canonical systems.)

2.3.2 Weyl function: high energy asymptotics

As we already mentioned in the Introduction, the recent surge of interest in the high energy asymptotics of the Weyl functions (i.e. asymptotics of the Weyl functions for the case that $|z| \to \infty$) was caused by the important works [121, 122, 307]. Here, we further develop the S-node scheme from the proof of Theorem 1.36 to derive the high energy asymptotics of the Weyl functions of Dirac system (2.18) (see also closely related papers [106, 236, 238, 252]).

Theorem 2.43. *Let the Dirac system* (2.18) *be given on* $[0, l]$, *let its potential* v *be bounded and assume that* φ *is its Weyl function. Then (uniformly with respect to* $\Re(z)$*), we have*

$$\varphi(z) = 2\mathrm{i}z \int_0^l \mathrm{e}^{2\mathrm{i}zx}\Phi_1(x)dx + O\left(2z\mathrm{e}^{2\mathrm{i}zl}\Big/\sqrt{\Im(z)}\right), \quad \Im(z) \to \infty, \tag{2.119}$$

where the corresponding triple $\{A, S, \Pi\}$ *and* Φ_1, *in particular, are constructed in Lemma* 2.35 *via* y *of the form* (2.87).

Proof. To prove the theorem, we consider the matrix function

$$\beth(z) = \begin{bmatrix} I_{m_1} & \varphi(z)^* \end{bmatrix} (j - w_A(l, 1/(2z))^* j w_A(l, 1/(2z))) \begin{bmatrix} I_{m_1} \\ \varphi(z) \end{bmatrix}. \tag{2.120}$$

Because of (2.110), (2.112) and (2.120), we have

$$\beth(z) = I_{m_1} - \varphi(z)^*\varphi(z) - \mathrm{e}^{\mathrm{i}l(\overline{z}-z)} \begin{bmatrix} I_{m_1} & \varphi(z)^* \end{bmatrix} u(l,z)^* j u(l,z) \begin{bmatrix} I_{m_1} \\ \varphi(z) \end{bmatrix}.$$

Taking into account (2.27), we rewrite the formula above as

$$\begin{aligned} \beth(z) &= I_{m_1} - \varphi(z)^*\varphi(z) - \mathrm{e}^{\mathrm{i}l(\overline{z}-z)} \\ &\times \left(\left(\begin{bmatrix} I_{m_1} & 0 \end{bmatrix} u(l,z)^{-1}\mathcal{P}(z)\right)^{-1}\right)^* \mathcal{P}(z)^* j \mathcal{P}(z) \left(\begin{bmatrix} I_{m_1} & 0 \end{bmatrix} u(l,z)^{-1}\mathcal{P}(z)\right)^{-1}. \end{aligned} \tag{2.121}$$

From (1.170) (where $J = j$ in our case) and (2.121), we see that

$$\beth(z) \leq I_{m_1}. \tag{2.122}$$

It easily follows from (1.88) that

$$w_A(l, 1/z)^* j w_A(l, 1/z) = j + \mathrm{i}(z - \overline{z})\Pi^*(I - \overline{z}A^*)^{-1}S_l^{-1}(I - zA)^{-1}\Pi. \tag{2.123}$$

In view of (2.40), formulas (2.120), (2.122) and (2.123) imply

$$2\mathrm{i}(\overline{z} - z) \begin{bmatrix} I_{m_1} & \varphi(z)^* \end{bmatrix} \Pi^*(I - 2\overline{z}A^*)^{-1}S_l^{-1}(I - 2zA)^{-1}\Pi \begin{bmatrix} I_{m_1} \\ \varphi(z) \end{bmatrix} \leq I_{m_1}. \tag{2.124}$$

Since S is strictly positive and bounded, inequality (2.124) yields

$$\left\| (I - 2zA)^{-1}\Pi \begin{bmatrix} I_{m_1} \\ \varphi(z) \end{bmatrix} \right\| \leq C/\sqrt{\Im z} \quad \text{for some} \quad C > 0. \tag{2.125}$$

After applying $-\mathrm{i}\Phi_2^*$ to the operator on the left-hand side of (2.125), we derive

$$-\mathrm{i}\Phi_2^*(I - 2zA)^{-1}\Phi_2\varphi(z) = \mathrm{i}\Phi_2^*(I - 2zA)^{-1}\Phi_1 + O\left(\frac{1}{\sqrt{\Im(z)}}\right). \tag{2.126}$$

Recall that according to (1.158), we have

$$\Phi_2^*(I-2zA)^{-1}f=\int_0^l e^{2iz(x-l)}f(x)dx, \quad \Phi_2^*(I-2zA)^{-1}\Phi_2=\frac{i}{2z}\left(e^{-2izl}-1\right)I_{m_2}.$$

Hence, from (2.126), we see that

$$\frac{1}{2z}\left(e^{-2izl}-1\right)\varphi(z)=ie^{-2izl}\int_0^l e^{2izx}\Phi_1(x)dx+O\left(\frac{1}{\sqrt{\Im(z)}}\right). \tag{2.127}$$

Since φ is nonexpansive, we obtain from (2.127) that (2.119) holds. □

Now, consider the Dirac system given on the semiaxis.

Corollary 2.44. *Let φ be the Weyl function of Dirac system* (2.18) *on* $[0,\infty)$, *where the potential v is locally bounded. Then, we have*

$$\varphi(z)=2iz\int_0^\infty e^{2izx}\Phi_1(x)dx, \quad \Im(z)>0. \tag{2.128}$$

Proof. Since φ is analytic and nonexpansive in $\mathbb{C}_+$, for any $\varepsilon>0$, it admits (Theorem E.11) a representation

$$\varphi(z)=2iz\int_0^\infty e^{2izx}\Phi(x)dx, \quad \Im(z)>\varepsilon>0, \tag{2.129}$$

where $e^{-2\varepsilon x}\Phi(x)\in L^2_{m_2\times m_1}(0,\infty)$. Because of (2.119) and (2.129), we obtain

$$\begin{aligned}\omega_d(z):&=\int_0^l e^{2iz(x-l)}\left(\Phi_1(x)-\Phi(x)\right)dx\\ &=\int_l^\infty e^{2iz(x-l)}\Phi(x)dx+O\left(1\Big/\sqrt{\Im(z)}\right).\end{aligned} \tag{2.130}$$

From (2.130), we see that $\omega_d(z)$ is bounded in some half-plane $\Im(z)\geq\eta_0>0$. Clearly, $\omega_d(z)$ is also bounded in the half-plane $\Im(z)<\eta_0$. Since ω_d is analytic and bounded in $\mathbb{C}$ and tends to zero on some rays, according to the first Liouville theorem (Theorem E.5), we have

$$\omega_d(z)=\int_0^l e^{2iz(x-l)}\left(\Phi_1(x)-\Phi(x)\right)dx\equiv 0. \tag{2.131}$$

It follows from (2.131) that $\Phi_1(x)\equiv\Phi(x)$ on all finite intervals $[0,l]$. Hence, (2.129) implies (2.128). □

Remark 2.45. Since $\Phi_1 \equiv \Phi$, we see that $\Phi_1(x)$ does not depend on l for $l > x$. Compare this with the proof of Proposition 4.1 in [114], where the fact that $E(x,t)$ (and so Φ_1) does not depend on l follows from the uniqueness of the factorizations of operators S_l^{-1}. See also Section 3 in [20] on the uniqueness of the accelerant. Furthermore, since $\Phi_1 \equiv \Phi$, the proof of Corollary 2.44 also implies that $e^{-\varepsilon x}\Phi_1(x) \in L^2_{m_2\times m_1}(0,\infty)$ for any $\varepsilon > 0$.

Remark 2.46. From (2.128), we see that Φ_1' is an analog (for the case of Dirac system) of A-amplitude, which was studied in [122, 307]. On the other hand, Φ_1' is closely related to the so called accelerant, which appears for the case that $m_1 = m_2$ in papers by M. Krein (see, e.g. [20, 176, 179]).

2.3.3 Inverse problem and Borg–Marchenko-type uniqueness theorem

Taking into account the Plancherel theorem and Remark 2.45, we apply the inverse Fourier transform to formula (2.128) and derive

$$\Phi_1\left(\frac{x}{2}\right) = \frac{1}{\pi}e^{\eta x}\mathrm{l.i.m.}_{a\to\infty}\int_{-a}^{a} e^{-i\xi x}\frac{\varphi(\xi+i\eta)}{2i(\xi+i\eta)}d\xi, \quad \eta > 0. \tag{2.132}$$

Here, l.i.m. stands for the entrywise limit in the norm of $L^2(0,r)$, $0 < r \le \infty$. (Note that if we additionally put $\Phi_1(x) = 0$ for $x < 0$, equality (2.132) holds for l.i.m. as the entrywise limit in $L^2(-r,r)$.) Thus, operators S and Π are recovered from φ.

According to Theorem 1.20 (see also Theorem 2.37 and its proof), the Hamiltonian $H = \gamma^*\gamma$ is recovered from S and Π via formula (1.103). Next, in order to recover γ, we first recover the Schur coefficient (see Remark 2.42 for the motivation of the term “Schur coefficient”):

$$\left(\begin{bmatrix}0 & I_{m_2}\end{bmatrix}H\begin{bmatrix}0\\ I_{m_2}\end{bmatrix}\right)^{-1}\begin{bmatrix}0 & I_{m_2}\end{bmatrix}H\begin{bmatrix}I_{m_1}\\ 0\end{bmatrix} = (\gamma_2^*\gamma_2)^{-1}\gamma_2^*\gamma_1 = \gamma_2^{-1}\gamma_1. \tag{2.133}$$

Next, we recover γ_2 from $\gamma_2^{-1}\gamma_1$ using (2.118) and, finally, we easily recover γ from γ_2 and $\gamma_2^{-1}\gamma_1$.

To recover β from γ, we could directly follow the scheme from the proof of Proposition 1.1. However, we slightly modify it in order to better use the Schur coefficient. Partition β into two blocks $\beta = \begin{bmatrix}\beta_1 & \beta_2\end{bmatrix}$, where β_i $(i = 1,2)$ is an $m_1 \times m_i$ matrix function. We put

$$\widetilde{\beta} = \begin{bmatrix}I_{m_1} & \gamma_1^*(\gamma_2^*)^{-1}\end{bmatrix}. \tag{2.134}$$

Because of (2.90) and (2.134), we have $\beta j\gamma^* = \widetilde{\beta} j\gamma^* = 0$, and so

$$\beta(x) = \beta_1(x)\widetilde{\beta}(x). \tag{2.135}$$

It follows from (2.18) and (2.87) that

$$\beta'(x) = \mathrm{i}v(x)\gamma(x), \tag{2.136}$$

which (in view of (2.90)) implies

$$\beta' j\beta^* = 0, \quad \beta' j\gamma^* = -\mathrm{i}v. \tag{2.137}$$

Formula (2.135) and the first relations in (2.90) and (2.137) lead us to

$$\widetilde{\beta} j\widetilde{\beta}^* = \beta_1^{-1}(\beta_1^*)^{-1}, \quad \beta' j\beta^* = \beta_1'\beta_1^{-1} + \beta_1(\widetilde{\beta}' j\widetilde{\beta}^*)\beta_1^* = 0. \tag{2.138}$$

According to (2.20) and (2.138), β_1 satisfies the first order differential equation

$$\beta_1' = -\beta_1(\widetilde{\beta}' j\widetilde{\beta}^*)(\widetilde{\beta} j\widetilde{\beta}^*)^{-1}, \quad \beta_1(0) = I_{m_1}. \tag{2.139}$$

Thus, we recover β from (2.134), (2.135) and (2.139). Our considerations are summarized in the next theorem.

Theorem 2.47. *Let φ be the Weyl function of Dirac system* (2.18) *on $[0, l]$, where the potential v is bounded. Then, v can be uniquely recovered from φ via the formula*

$$v(x) = \mathrm{i}\beta'(x)j\gamma(x)^*. \tag{2.140}$$

Here, β is recovered from γ using (2.134), (2.135) *and* (2.139); *γ is recovered from the Hamiltonian H using* (2.133) *and equation* (2.118); *H is given by* (1.103); *operators $\Pi = [\Phi_1 \ \ \Phi_2]$ and S in* (1.103) *are expressed via $\Phi_1(x)$ in formulas* (2.97) *and* (2.115), *respectively. Finally, $\Phi_1(x)$ is recovered from φ using* (2.132).

In view of Corollary 2.25, Theorem 2.47 also yields the procedure to solve the inverse problem on $[0, \infty)$.

Corollary 2.48. *Let φ be the Weyl function of Dirac system* (2.18) *on $[0, \infty)$, where the potential v is locally bounded. Then, the potential v is uniquely recovered from φ, using $\Phi_1(x)$ given on $[0, \infty)$ by* (2.132), *via the procedure described in Theorem* 2.47.

Remark 2.49. There is another way to recover β and γ. Namely, we can recover β directly from Π and S as described in the proposition below, and then recover γ from β in the same way that β is recovered from γ at the beginning of this subsection.

Proposition 2.50. *Let Dirac system* (2.18) *be given on $[0, \infty]$ (on $[0, l]$). Assume that v is locally bounded. Then, the matrix function β, which is defined in* (2.87), *satisfies the equality*

$$\beta(x) = \begin{bmatrix} I_{m_1} & 0 \end{bmatrix} + \int_0^x \left(S_x^{-1}\Phi_1'\right)(t)^* \begin{bmatrix} \Phi_1(t) & I_{m_2} \end{bmatrix} dt, \tag{2.141}$$

where $0 < x < \infty$ $(0 < x \leq l)$. Here, $\Phi_1(x)$ and S_x are given by (2.132) *and* (2.115) *(after putting $l = x$), respectively.*

Proof. First, we fix an arbitrary l (denote the end of the interval by l if the finite interval is dealt with) and rewrite (2.100) (for $x \leq l$) in the form

$$\gamma(x) = \left(E\begin{bmatrix}\Phi_1 & I_{m_2}\end{bmatrix}\right)(x). \tag{2.142}$$

It follows that

$$\mathrm{i}\left(EAE^{-1}E\Phi_1'\right)(x) = \gamma_1(x) - \gamma_2(x)\Phi_1(+0). \tag{2.143}$$

Since (2.116) holds, we rewrite (2.143) as

$$\mathrm{i}\left(EAE^{-1}E\Phi_1'\right)(x) = \gamma_1(x). \tag{2.144}$$

Next, we substitute the equality $K = EAE^{-1}$ from (2.95) into (2.144) and (using (2.92)) we obtain

$$\gamma_1(x) = -\gamma(x)j\int_0^x \gamma(t)^*\left(E\Phi_1'\right)(t)dt. \tag{2.145}$$

Formulas (2.142) and (2.145) imply

$$\gamma_1(x) = -\gamma(x)j\int_0^x \left(E\begin{bmatrix}\Phi_1 & I_{m_2}\end{bmatrix}\right)(t)^*\left(E\Phi_1'\right)(t)dt. \tag{2.146}$$

We rewrite (2.146) in the form

$$\gamma(x)j\widehat{\beta}(x)^* \equiv 0, \tag{2.147}$$

where

$$\widehat{\beta}(x) := \begin{bmatrix}I_{m_1} & 0\end{bmatrix} + \int_0^x \left(E\Phi_1'\right)(t)^*\left(E\begin{bmatrix}\Phi_1 & I_{m_2}\end{bmatrix}\right)(t)dt. \tag{2.148}$$

Recall that formula (1.154) follows from the first relation in (2.96). Taking into account (1.153) and (1.154), we rewrite the definition of $\widehat{\beta}$ above as

$$\widehat{\beta} = \begin{bmatrix}I_{m_1} & 0\end{bmatrix} + \int_0^x \left(S_x^{-1}\Phi_1'\right)(t)^*\begin{bmatrix}\Phi_1(t) & I_{m_2}\end{bmatrix}dt. \tag{2.149}$$

On the other hand, in view of (2.142) and (2.148), we have

$$\widehat{\beta}'(x) = \left(E\Phi_1'\right)(x)^*\gamma(x). \tag{2.150}$$

Therefore, (2.90) leads us to

$$\beta(x)j\widehat{\beta}'(x)^* \equiv 0. \tag{2.151}$$

Furthermore, compare (2.90) with (2.147) to see that

$$\widehat{\beta}(x) = \widehat{\vartheta}(x)\beta(x), \tag{2.152}$$

where $\widehat{\vartheta}(x)$ is an $m_1 \times m_1$ matrix function, which is boundedly differentiable on $[0, l]$. Using equality (2.152) and the first relations in (2.90) and (2.137) to rewrite (2.151), we see that $\widehat{\vartheta}' \equiv 0$ (i.e. $\widehat{\vartheta}$ is a constant). It follows from the equalities (2.87), (2.148) and (2.152) (considered at $x = 0$) that $\widehat{\vartheta}(0) = I_{m_1}$, and so $\widehat{\vartheta} \equiv I_{m_1}$, that is,

$$\widehat{\beta} \equiv \beta. \tag{2.153}$$

Thus, (2.141) is immediate from (2.149). □

Remark 2.51. Because of (2.136), (2.150) and (2.153), we see that the potential v can be recovered via the formula

$$v(x) = \left(\mathrm{i}E\Phi_1'\right)(x)^*. \tag{2.154}$$

Moreover, after we choose (following (1.154)) a continuous kernel for operators S_x^{-1}, it follows from (1.154), (2.114) and (2.154) that

$$v(x) = \left(\mathrm{i}S_x^{-1}\Phi_1'\right)(x)^*. \tag{2.155}$$

The last statement in this section is a Borg–Marchenko-type uniqueness theorem. Such results are closely related to the high energy asymptotics of Weyl functions and (as we already mentioned) the study of both theories gained momentum after the publishing of the papers [121, 122, 307]. We derive our theorem from Theorems 2.43 and 2.47.

Theorem 2.52. *Let φ and $\widehat{\varphi}$ be Weyl functions of two Dirac systems on $[0, l]$ (on $[0, \infty)$) with bounded (locally bounded) potentials, which are denoted by v and $\widehat{v}$, respectively. Suppose that on some ray $\Re z = c\Im z$, where $c \in \mathbb{R}$ and $\Im z > 0$, the equality*

$$\|\varphi(z) - \widehat{\varphi}(z)\| = O(\mathrm{e}^{2\mathrm{i}zr}) \quad (\Im z \to \infty) \tag{2.156}$$

holds for all $0 < r < \widetilde{l}$ ($\widetilde{l} < l < \infty$). Then, we have

$$v(x) = \widehat{v}(x), \quad 0 < x < \widetilde{l}. \tag{2.157}$$

Proof. Since Weyl functions are nonexpansive, we obtain

$$\|\mathrm{e}^{-2\mathrm{i}zr}\left(\varphi(z) - \widehat{\varphi}(z)\right)\| \leq c_1\mathrm{e}^{2|z|r}, \quad \Im z \geq c_2 > 0 \tag{2.158}$$

for some c_1 and c_2. It is also apparent that the matrix function $\mathrm{e}^{-2\mathrm{i}zr}(\varphi(z) - \widehat{\varphi}(z))$ is bounded on the line $\Im z = c_2$. Furthermore, formula (2.156) implies that $\mathrm{e}^{-2\mathrm{i}zr}(\varphi(z) - \widehat{\varphi}(z))$ is bounded on the ray $\Re z = c\Im z$. Therefore, applying the

Phragmen–Lindelöf theorem (see Corollary E.7) in the angles generated by the line $\Im z = c_2$ and the ray $\Re z = c\Im z$ ($\Im z \geq c_2$), we see that

$$\|e^{-2izr}\left(\varphi(z) - \widehat{\varphi}(z)\right)\| \leq c_3, \quad \Im z \geq c_2 > 0. \tag{2.159}$$

Functions associated with $\widehat{\varphi}$ are written with a hat (e.g. $\widehat{v}$, $\widehat{\Phi}_1$). Because of formula (2.119), its analog for $\widehat{\varphi}$, $\widehat{\Phi}_1$ and the inequality (2.159), we have

$$\left\| \int_0^r e^{2iz(x-r)}\left(\Phi_1(x) - \widehat{\Phi}_1(x)\right)dx \right\| \leq c_4, \quad \Im z \geq c_2 > 0. \tag{2.160}$$

Clearly, the left-hand side of (2.160) is bounded in the semiplane $\Im z < c_2$ and tends to zero on some rays. Thus, we derive

$$\int_0^r e^{2iz(x-r)}\left(\Phi_1(x) - \widehat{\Phi}_1(x)\right)dx \equiv 0, \quad \text{i.e.} \quad \Phi_1(x) \equiv \widehat{\Phi}_1(x) \quad (0 < x < r). \tag{2.161}$$

Since (2.161) holds for all $r < \widetilde{l}$, we obtain $\Phi_1(x) \equiv \widehat{\Phi}_1(x)$ for $0 < x < \widetilde{l}$. In view of Theorem 2.47, the last identity implies (2.157). □

2.3.4 Weyl function and positivity of S

In this (last) subsection of Chapter 2, we discuss some sufficient conditions for a nonexpansive matrix function φ, which is analytic in $\mathbb{C}_+$, to be a Weyl function of Dirac system on the semiaxis. For the case that v is a square matrix function (or scalar function) and Weyl functions φ are the so called Nevanlinna functions (i.e. $\Im\varphi \geq 0$), a sufficient condition for φ to be a Weyl function can be given in terms of spectral function [195, 205], which is connected with φ via Herglotz representation (1.187). Furthermore, a positive operator S is also recovered from the spectral function, see formula (1.113) and Theorem 2.11. (See also [290, Chapters 4,10] and [252].) The invertibility of the convolution operators (analogs of our operator S), which is required in [19, 20, 176], provides their positivity too, and the spectral problem is treated in this way. Here, we formulate conditions on the $m_2 \times m_1$ nonexpansive matrix functions, again in terms of S.

First, consider a useful procedure to recover y from β. We note that β is recovered from y in the Subsection 2.3.3. The scheme to solve the inverse problem using another way around, that is, using the recovery of y from β, is discussed in Remark 2.49.

Proposition 2.53. *Let a given $m_1 \times m$ matrix function $\beta(x)$ $(0 \leq x \leq l)$ be boundedly differentiable and satisfy relations*

$$\beta(0) = \begin{bmatrix} I_{m_1} & 0 \end{bmatrix}, \quad \beta' j \beta^* \equiv 0. \tag{2.162}$$

Then, there is a unique $m_2 \times m$ matrix function γ which is boundedly differentiable and satisfies relations

$$\gamma(0) = \begin{bmatrix} 0 & I_{m_2} \end{bmatrix}, \quad \gamma' j \gamma^* \equiv 0, \quad \gamma j \beta^* \equiv 0. \tag{2.163}$$

This γ is given by the formula

$$\gamma = \gamma_2 \tilde{\gamma}, \quad \tilde{\gamma} := \begin{bmatrix} \tilde{\gamma}_1 & I_{m_2} \end{bmatrix}, \quad \tilde{\gamma}_1 := \beta_2^* (\beta_1^*)^{-1}, \tag{2.164}$$

where γ_2 is recovered via a differential system and initial condition, namely,

$$\gamma_2' = \gamma_2 \tilde{\gamma}_1' \tilde{\gamma}_1^* \left(I_{m_2} - \tilde{\gamma}_1 \tilde{\gamma}_1^*\right)^{-1}, \quad \gamma_2(0) = I_{m_2}. \tag{2.165}$$

Moreover, the procedure above is well-defined since

$$\det \beta_1(x) \neq 0, \quad \det\left(I_{m_2} - \tilde{\gamma}_1(x)\tilde{\gamma}_1(x)^*\right) \neq 0. \tag{2.166}$$

Proof. Because of (2.162), we have $\beta j \beta^* \equiv I_{m_1}$ (and so $\det \beta_1 \neq 0$). On the other hand, formula (2.164) implies $\tilde{\gamma}\tilde{\gamma}^* > 0$ and $\tilde{\gamma} j \beta^* = 0$. Therefore, equality $\tilde{\gamma} = \begin{bmatrix} \tilde{\gamma}_1 & I_{m_2} \end{bmatrix}$ and Proposition 1.44 yield

$$\tilde{\gamma}_1 \tilde{\gamma}_1^* - I_{m_2} = \tilde{\gamma} j \tilde{\gamma}^* < 0. \tag{2.167}$$

In particular, we see that $\det(I_{m_2} - \tilde{\gamma}_1 \tilde{\gamma}_1^*) \neq 0$. Both inequalities in (2.166) are proved.

Next, we show that γ given by (2.164) and (2.165) satisfies (2.163). Indeed, the equality $\gamma(0) = \begin{bmatrix} 0 & I_{m_2} \end{bmatrix}$ is apparent from $\beta(0) = \begin{bmatrix} I_{m_1} & 0 \end{bmatrix}$, (2.164) and (2.165). The identity $\gamma j \beta^* \equiv 0$ follows from $\tilde{\gamma} j \beta^* = 0$. Finally, we rewrite $\gamma' j \gamma^* = 0$ in the equivalent form

$$\left(\gamma_2' \tilde{\gamma} + \begin{bmatrix} \gamma_2 \tilde{\gamma}_1' & 0 \end{bmatrix}\right) j \tilde{\gamma}^* = 0, \tag{2.168}$$

which, in turn, is equivalent to the first equality in (2.165). Thus, $\gamma' j \gamma^* = 0$ is also proved, and so all the relations in (2.163) hold.

On the other hand, for each γ satisfying (2.163), we have inequality $\det \gamma_2 \neq 0$ (since $\gamma j \gamma^* = -I_{m_2}$) and representation (2.164). Moreover, as mentioned above, relations (2.165) follow from (2.163) and (2.164). Hence, $\tilde{\gamma}_1$ is uniquely recovered from (2.164) and (after that) $\tilde{\gamma}_2$ is uniquely recovered from (2.165), which grants the uniqueness of γ. □

Now, we formulate the main statement in this section.

Theorem 2.54. *Let an $m_2 \times m_1$ matrix function $\varphi(z)$ be analytic and nonexpansive in $\mathbb{C}_+$. Furthermore, let matrix function $\Phi_1(x)$ given on $\mathbb{R}$ by (2.132) be boundedly differentiable on each finite interval $[0, l]$ and satisfy equality*

$$\Phi_1(x) = 0 \quad \text{for } x \leq 0.$$

Assume that operators S_l, which are expressed via Φ_1 in (2.115), are boundedly invertible for all $0 < l < \infty$.

Then, φ is the Weyl function of some Dirac system on $[0, \infty)$. The operators S_l^{-1} admit unique factorizations

$$S_l^{-1} = E_{\Phi,l}^* E_{\Phi,l}, \quad E_{\Phi,l} = I + \int_0^x E_\Phi(x,t) \cdot dt \in B\left(L_2^{m_2}(0,l)\right), \tag{2.169}$$

where $E_\Phi(x,t)$ is continuous (with respect to x and t) and does not depend on l, and the potential of the Dirac system is constructed via formula

$$v(x) = \left(\mathrm{i}E_{\Phi,l}\Phi_1'\right)(x)^*, \quad 0 < x < l. \tag{2.170}$$

Clearly, formula (2.170) in Theorem 2.54 is similar to (2.154). In order to prove Theorem 2.54, we use several results from Appendix D. First, note that the existence and uniqueness of the factorization (2.169) and properties of $E_\Phi(x,t)$ follow directly from Theorem D.7. To proceed with our proof, we need the lemma below.

Lemma 2.55. *Let a matrix function $\Phi_1(x)$ be boundedly differentiable on each finite interval $[0, l]$ and satisfy equality $\Phi_1(0) = 0$. Assume that the operators S_l, which are expressed via Φ_1 in (2.115), are boundedly invertible for all $0 < l < \infty$. Then, the matrix functions*

$$\beta_\Phi(x) := \begin{bmatrix} I_{m_1} & 0 \end{bmatrix} + \int_0^x \left(S_x^{-1}\Phi_1'\right)(t)^* \begin{bmatrix} \Phi_1(t) & I_{m_2} \end{bmatrix} dt, \tag{2.171}$$

$$\gamma_\Phi(x) := \begin{bmatrix} \Phi_1(x) & I_{m_2} \end{bmatrix} + \int_0^x E_\Phi(x,t) \begin{bmatrix} \Phi_1(t) & I_{m_2} \end{bmatrix} dt \tag{2.172}$$

are boundedly differentiable and satisfy conditions

$$\beta_\Phi(0) := \beta_\Phi(+0) = \begin{bmatrix} I_{m_1} & 0 \end{bmatrix}, \quad \beta_\Phi' j \beta_\Phi^* \equiv 0; \tag{2.173}$$

$$\gamma_\Phi(0) = \begin{bmatrix} 0 & I_{m_2} \end{bmatrix}, \quad \gamma_\Phi' j \gamma_\Phi^* \equiv 0; \quad \gamma_\Phi j \beta_\Phi^* \equiv 0. \tag{2.174}$$

Proof. Step 1. The first equalities in (2.173) and (2.174) are immediate from (2.171) and (2.172), respectively. Furthermore, (2.172) is equivalent to the equalities

$$\gamma_\Phi(x) = \left(E_{\Phi,l}\begin{bmatrix} \Phi_1 & I_{m_2} \end{bmatrix}\right)(x), \quad 0 \le x \le l \quad (\text{for all } l < \infty). \tag{2.175}$$

Next, fix any $0 < l < \infty$ and recall that according to Theorem D.1, the operator identity

$$AS_l - S_l A^* = \mathrm{i}\Pi j \Pi^*, \tag{2.176}$$

where A is given in (2.95) and Π is given by the second and third relations in (2.96) (and the first and third relations in (2.97)), holds. Hence, taking into account (2.169) and (2.175), and turning around the proof of (2.176) in Lemma 2.35, we get

$$E_\Phi A E_\Phi^{-1} - \left(E_\Phi^{-1}\right)^* A^* E_\Phi^* = \mathrm{i}\gamma_\Phi(x) j \int_0^l \gamma_\Phi(t)^* \cdot dt, \quad \text{i.e.}$$

$$E_\Phi A E_\Phi^{-1} = \mathrm{i}\gamma_\Phi(x) j \int_0^x \gamma_\Phi(t)^* \cdot dt \quad (\text{for } E_\Phi = E_{\Phi,l}). \tag{2.177}$$

Introducing the resolvent kernel Γ_Φ of $E_\Phi^{-1} = I + \int_0^x \Gamma_\Phi(x,t) \cdot dt$, we rewrite (2.177) in the form of an equality for kernels:

$$\begin{aligned} I_{m_2} &+ \int_t^x (E_\Phi(x,r) + \Gamma_\Phi(r,t))\, dr + \int_t^x \int_\xi^x E_\Phi(x,r) dr \Gamma_\Phi(\xi,t) d\xi \\ &= -\gamma_\Phi(x) j \gamma_\Phi(t)^*. \end{aligned} \tag{2.178}$$

In particular, formula (2.178) for the case that $x = t$ implies

$$\gamma_\Phi(x) j \gamma_\Phi(x)^* \equiv -I_{m_2}. \tag{2.179}$$

In a way quite similar to the first part of the proof of Proposition 2.50, we use equalities (2.175), (2.177) and $\Phi_1(0) = 0$ to derive

$$\gamma_\Phi(x) j \beta_\Phi(x)^* \equiv 0 \tag{2.180}$$

(compare (2.180) with (2.147)). Because of (2.169), (2.171) and (2.175), we have

$$\beta_\Phi'(x) = \left(E_\Phi \Phi_1'\right)(x)^* \gamma_\Phi(x). \tag{2.181}$$

In view of (2.180) and (2.181), β_Φ is boundedly differentiable and the last relation in (2.173) holds. Thus, the properties of β_Φ are proved.

Step 2. It remains to be shown that γ_Φ is boundedly differentiable and the identity $\gamma_\Phi' j \gamma_\Phi^* \equiv 0$ holds. For that purpose, note that β_Φ satisfies conditions of Proposition 2.53, and so there is a boundedly differentiable matrix function $\hat{\gamma}$ such that

$$\hat{\gamma}(0) = \begin{bmatrix} 0 & I_{m_2} \end{bmatrix}, \quad \hat{\gamma}' j \hat{\gamma}^* \equiv 0; \quad \hat{\gamma} j \beta_\Phi^* \equiv 0. \tag{2.182}$$

Formulas (2.173) and (2.182) yield

$$\beta_\Phi j \beta_\Phi^* \equiv I_{m_1}, \quad \hat{\gamma} j \hat{\gamma}^* \equiv -I_{m_2}, \tag{2.183}$$

respectively. That is, the rows of β_Φ (the rows of $\hat{\gamma}$) are linearly independent. Hence, the last relations in (2.182) and (2.183) and formulas (2.179) and (2.180) imply that there is a unitary matrix function ω such that

$$\gamma_\Phi(x) = \omega(x)\hat{\gamma}(x), \quad \omega(x)^* = \omega(x)^{-1}. \tag{2.184}$$

Moreover, formulas (2.173), (2.182) and (2.183) lead us to the relations

$$\hat{u}' j\hat{u}^* j = \mathrm{i}j \begin{bmatrix} 0 & \hat{v} \\ \hat{v}^* & 0 \end{bmatrix}, \quad \hat{u}(x) := \begin{bmatrix} \beta_\Phi(x) \\ \hat{y}(x) \end{bmatrix}, \quad \hat{v} := \mathrm{i}\beta_\Phi' j\hat{y}^*; \tag{2.185}$$

$$\hat{u} j\hat{u}^* j \equiv I_m, \quad \hat{u}(0) = I_m. \tag{2.186}$$

According to (2.185) and (2.186), the matrix function $\hat{u}(x)$ is the normalized by the $\hat{u}(0) = I_m$ fundamental solution of the Dirac system, where $\hat{v}$ is the bounded potential and the spectral parameter z equals zero. In other words, $\beta = \beta_\Phi$ and $\gamma = \hat{y}$ correspond to $\hat{v}$ via equalities (2.87), and we can apply the results of Subsection 2.3.1. Therefore, according to Proposition 2.33 and Remark 2.40, there is an operator E of the form (2.114) such that

$$EA = \mathrm{i}\hat{y}(x)j \int_0^x \hat{y}(t)^* \cdot dt\, E, \quad \hat{y}_2 = EI_{m_2}. \tag{2.187}$$

It follows from (2.184) and (2.187) that

$$\omega EA = \mathrm{i}\gamma_\Phi(x)j \int_0^x \gamma_\Phi(t)^* \cdot dt\, \omega E, \quad \gamma_{2,\Phi} = \omega EI_{m_2}, \tag{2.188}$$

where ω denotes the operator of multiplication by the matrix function $\omega(x)$. On the other hand, formulas (2.175) and (2.177) lead us to

$$E_\Phi A = \mathrm{i}\gamma_\Phi(x)j \int_0^x \gamma_\Phi(t)^* \cdot dt\, E_\Phi, \quad \gamma_{2,\Phi} = E_\Phi I_{m_2}. \tag{2.189}$$

It is easy to see that

$$\overline{\mathrm{span}}\left(\bigcup_{i=0}^{\infty} \mathrm{Im}\left(A^i I_{m_2}\right)\right) = L^2_{m_2}(0,l).$$

Hence, equalities (2.188) and (2.189) imply $E_\Phi = \omega E$. Finally, compare (2.114) and (2.169) to see that

$$\omega(x) \equiv I_{m_2}, \quad E_\Phi = E. \tag{2.190}$$

Furthermore, because of (2.184) and (2.190), we have $\gamma_\Phi = \hat{y}$, and so γ_Φ is boundedly differentiable and satisfies (2.174). □

Proof of Theorem 2.54. Using Theorem D.7, we already proved the theorem's statements about operators S_l and $E_{\Phi,l}$. From the assumptions of Theorem 2.54, we see that the conditions of Lemma 2.55 are also fulfilled. It is shown in the proof of Lemma 2.55 that starting from the Dirac system characterized by $\beta = \beta_\Phi$, $\gamma = \gamma_\Phi$ and $v = \hat{v}$ (given by (2.140)) and applying the procedure from Subsection 2.3.1 (see Lemma 2.35), we recover our Φ_1. That is, Φ_1, which is constructed in Lemma 2.35, coincides with Φ_1 given in the theorem. (Here, we took into account the equalities (2.175), (2.185), (2.186), $E_\Phi = E$ and $y_\Phi = \hat{y}$ from the proof of Lemma 2.55.) Moreover, in view of (2.154) and $E_\Phi = E$, we derive that our $v = \hat{v}$ has the form (2.170). We also note that $v = \hat{v}$ is bounded on $[0, l]$.

In view of Corollaries 2.21 and 2.44, there is a unique Weyl function φ_W of the Dirac system with a locally bounded on $[0, \infty)$ potential, and this Weyl function is given by (2.128). It remains to be shown that φ_W equals the function φ which generates Φ_1 via (2.132). Recalling that (2.132) also holds for φ_W, we see that

$$\text{l.i.m.}_{a\to\infty} \int_{-a}^{a} \frac{e^{-ix\xi}}{(\xi + i\eta)} \left(\varphi(\xi + i\eta) - \varphi_W(\xi + i\eta)\right) d\xi \equiv 0, \quad \eta > 0, \tag{2.191}$$

where l.i.m. stands for the entrywise limit in the norm of $L^2(-r, r)$ $(0 < r \leq \infty)$. Therefore, we get $\varphi_W = \varphi$, that is, φ is the Weyl function of the Dirac system, where the potential is given by (2.170). □

3 Skew-self-adjoint Dirac system: rectangular matrix potentials

Let us consider the skew-self-adjoint Dirac (also called ZS or AKNS) system

$$\frac{d}{dx}y(x,z) = (\mathrm{i}zj + jV(x))y(x,z) \quad (x \geq 0, \quad z \in \mathbb{C}), \tag{3.1}$$

$$j = \begin{bmatrix} I_{m_1} & 0 \\ 0 & -I_{m_2} \end{bmatrix}, \quad V = \begin{bmatrix} 0 & v \\ v^* & 0 \end{bmatrix}, \tag{3.2}$$

where $v(x)$ is an $m_1 \times m_2$ locally summable matrix function, which is called the *potential* (often V is called the potential as well). System (3.1) can be rewritten in the form $(-\mathrm{i}j\frac{d}{dx} + \mathrm{i}V(x))y = zy$, where the term $\mathrm{i}V$ is skew-self-adjoint. The system above differs from the more classical self-adjoint Dirac system $(-\mathrm{i}j\frac{d}{dx} - V(x))y = zy$ by the factor i (instead of -1) before V.

Skew-self-adjoint Dirac systems are just like self-adjoint Dirac systems well known in analysis and applications, and actively studied as auxiliary linear systems for important integrable nonlinear equations, though the Weyl theory of such systems is nonclassical. For the case that $m_1 \neq m_2$, systems of the form (3.1), (3.2) are, in particular, auxiliary linear systems for the coupled, multicomponent, and $m_1 \times m_2$ matrix versions of the focusing nonlinear Schrödinger equations. We note that the above-mentioned case $m_1 \neq m_2$ was much less studied. We present recent results from [108], where earlier works [238, 259] (on the case $m_1 = m_2$), are generalized for the case of rectangular potentials. We solve direct and inverse problems, that is, we construct Weyl functions and recover the $m_1 \times m_2$ potential v from the Weyl function, respectively.

As in Chapter 2, the fundamental solution of system (3.1) is denoted by $u(x,z)$, and this solution is normalized by the condition

$$u(0,z) = I_m. \tag{3.3}$$

Usually, we assume that

$$\|v(x)\| \leq M \quad \text{for } 0 < x < r \leq \infty. \tag{3.4}$$

It is immediately apparent from (3.1) and (3.4) that

$$\frac{d}{dx}\left(u(x,z)^* j u(x,z)\right) = -2u(x,z)^*(\Im(z)I_m - V(x))u(x,z) < 0, \quad z \in \mathbb{C}_M, \tag{3.5}$$

where $\mathbb{C}_M$ stands for the open half-plane $\{z : \Im(z) > M > 0\}$. Taking into account (3.3) and (3.5), we derive

$$\aleph(x,z) := u(x,z)^* j u(x,z) < j \quad (x > 0, \quad z \in \mathbb{C}_M). \tag{3.6}$$

It follows from (3.6) that many of the techniques that are valid for the self-adjoint Dirac system (Chapter 2) can be adapted to study the skew-self-adjoint system (3.1). Sections 3.1 and 3.2 present analogs of various results from Sections 2.2 and 2.3 in Chapter 2.

The case of system (3.1) on $[0, \infty)$, where the potential v is only locally bounded, is studied in the last Section 3.3. Generalized Weyl functions (GW-functions) are introduced there and a corresponding inverse problem is solved.

3.1 Direct problem

As mentioned above, using inequality (3.6), we can derive analogs to many results on a self-adjoint Dirac system from Chapter 2. Therefore, in this section, we mostly restrict ourselves to the more important case of the semiaxis ($x \in [0, \infty)$) and omit the case of the finite interval. For a fixed $M > 0$ and $J = j$, we denote the class of $m \times m_1$ matrix functions, which are nonsingular and have the property-j as introduced in Definition 1.42, by $\mathbf{P}_M(j)$. We recall that the fundamental solutions are invertible. Our next notation coincides with Notation 2.13 and only u is different there.

Notation 3.1. *The set $\mathcal{N}_u(x, z)$ of Möbius transformations is the set of values (at the fixed points $x \in [0, \infty)$, $z \in \mathbb{C}_M$) of matrix functions*

$$\varphi(x, z, \mathcal{P}) = \begin{bmatrix} 0 & I_{m_2} \end{bmatrix} u(x, z)^{-1}\mathcal{P}(z) \left(\begin{bmatrix} I_{m_1} & 0 \end{bmatrix} u(x, z)^{-1}\mathcal{P}(z) \right)^{-1}, \quad z \in \mathbb{C}_M, \tag{3.7}$$

where $\mathcal{P}(z)$ are nonsingular matrix functions with property-j, that is, $\mathcal{P}(z) \in \mathbf{P}_M(j)$.

We also need another notation.

Notation 3.2. *The class of $m_2 \times m_1$ nonexpansive matrix functions (Schur matrix functions) on some domain Ω is denoted by $S_{m_2 \times m_1}(\Omega)$.*

For simplicity, we write $\varphi(z)$ below, meaning the set $\{\varphi(z)\}$ consisting of $\varphi(z)$.

Proposition 3.3. *Let Dirac system* (3.1) *be given on $[0, \infty)$ and assume that $\|v\|$ is bounded by M, that is,*

$$\|v(x)\| \leq M \quad \text{for } x \in [0, \infty). \tag{3.8}$$

Then, the sets $\mathcal{N}_u(x, z)$ are well-defined in $\mathbb{C}_M$. There is a unique matrix function $\varphi(z)$ such that

$$\varphi(z) = \bigcap_{x < \infty} \mathcal{N}_u(x, z). \tag{3.9}$$

This function is analytic and nonexpansive in $\mathbb{C}_M$ (i.e. $\varphi \in S_{m_2 \times m_1}(\mathbb{C}_M)$).

Proof. The proof is similar to the proof of Proposition 2.17, but z is considered now in the half-plane $\mathbb{C}_M$ instead of the half-plane $\mathbb{C}_+$, where z is considered in Chapter 2. We present, however, the main details of this proof since some formulas contained in it will be also used afterwards. (Later, we will skip various details of the proofs similar to those in Chapter 2.) First, note that inequality (3.6) yields

$$(u(x,z)^*)^{-1} ju(x,z)^{-1} > j \quad (x > 0, \quad z \in \mathbb{C}_M). \tag{3.10}$$

In view of (1.170) and (3.10), we can use Proposition 1.43 to derive

$$\det\left(\begin{bmatrix} I_{m_1} & 0 \end{bmatrix} u(x,z)^{-1}\mathcal{P}(z)\right) \neq 0 \quad (z \in \mathbb{C}_M) \tag{3.11}$$

for $x > 0$. It is immediate from (1.170) and (3.3) that (3.11) holds for $x = 0$. Thus, $\mathcal{N}_u$ is well-defined via (3.7).

Next, we rewrite (3.7) in the equivalent form

$$\begin{bmatrix} I_{m_1} \\ \varphi(x,z,\mathcal{P}) \end{bmatrix} = u(x,z)^{-1}\mathcal{P}(z)\left(\begin{bmatrix} I_{m_1} & 0 \end{bmatrix} u(x,z)^{-1}\mathcal{P}(z)\right)^{-1}, \tag{3.12}$$

which, in turn, is equivalent to

$$u(x,z)\begin{bmatrix} I_{m_1} \\ \varphi(x,z,\mathcal{P}) \end{bmatrix} = \mathcal{P}(z)\left(\begin{bmatrix} I_{m_1} & 0 \end{bmatrix} u(x,z)^{-1}\mathcal{P}(z)\right)^{-1}. \tag{3.13}$$

Taking into account the definition of $\aleph$ in (3.6) and relations (1.170) and (3.13), we see that the formula

$$\varphi(z) \in \mathcal{N}_u(x,z) \tag{3.14}$$

is equivalent to

$$\begin{bmatrix} I_{m_1} & \varphi(z)^* \end{bmatrix} \aleph(x,z) \begin{bmatrix} I_{m_1} \\ \varphi(z) \end{bmatrix} \geq 0. \tag{3.15}$$

Using formula (3.5) and the equivalence of (3.14) and (3.15), we easily obtain

$$\mathcal{N}_u(x_1,z) \subset \mathcal{N}_u(x_2,z) \quad \text{for } x_1 > x_2. \tag{3.16}$$

Moreover, (3.15) at $x = 0$ means that

$$\mathcal{N}_u(0,z) = \{\varphi(z) : \varphi(z)^*\varphi(z) \leq I_{m_1}\}. \tag{3.17}$$

By virtue of Montel's theorem, formulas (3.16) and (3.17) imply the existence of an analytic and nonexpansive matrix function $\varphi(z)$ such that

$$\varphi(z) \in \bigcap_{x<\infty} \mathcal{N}_u(x,z), \tag{3.18}$$

the proof of which completely coincides with the proof of (2.32) in Chapter 2.

It follows from (3.5) and (3.6) that

$$\frac{d}{dx}\aleph \leq -2(\Im(z) - M)u^*u \leq 2(\Im(z) - M)\aleph \leq 2(\Im(z) - M)j. \tag{3.19}$$

We partition $\aleph$ into four blocks so that $\aleph_{ii}$ is an $m_i \times m_i$ matrix function ($i = 1, 2$). Taking into account the inequalities above and the equality $\aleph_{22}(0, z) = -I_{m_2}$, we obtain

$$-\aleph_{22}(x, z) \geq (1 + 2(\Im(z) - M)x)\, I_{m_2} \quad (z \in \mathbb{C}_M). \tag{3.20}$$

Note also that (3.10) implies $\aleph(x, z)^{-1} \geq j$ for $z \in \mathbb{C}_M$ (see Corollary E.3). Thus, we have

$$\left(\aleph^{-1}\right)_{11} = \left(\aleph_{11} - \aleph_{12}\aleph_{22}^{-1}\aleph_{21}\right)^{-1} \geq I_{m_1}. \tag{3.21}$$

Further arguments completely coincide with the arguments from Proposition 2.17. Fixing x and z, we see that $\mathcal{N}_u$ is a matrix disc (ball)

$$\varphi = \rho_l \omega \rho_r - \aleph_{22}^{-1}\aleph_{21}, \quad \omega^*\omega \leq I_{m_2}, \tag{3.22}$$

where ω is an $m_2 \times m_1$ nonexpansive matrix, ρ_l and ρ_r are the left and right semiradii of the Weyl disc and

$$\rho_l = (-\aleph_{22})^{-1/2}, \quad \rho_r = \left(\aleph_{11} - \aleph_{12}\aleph_{22}^{-1}\aleph_{21}\right)^{1/2}. \tag{3.23}$$

Moreover, according to (3.20), (3.21) and (3.23), the next formula holds:

$$\rho_l(x, z) \to 0 \quad (x \to \infty), \quad \rho_r(x, z) \leq I_{m_1}. \tag{3.24}$$

Finally, relations (3.18), (3.22) and (3.24) imply (3.9). □

Remark 3.4. It is convenient sometimes to partition $\mathcal{P} \in \mathbf{P}_M(j)$ into blocks $\mathcal{P}_1$ and $\mathcal{P}_2$ where $\mathcal{P}_1$ is an $m_1 \times m_1$ matrix function (so that pairs like in Definition 0.1 appear) and to normalize $\mathcal{P}$, that is,

$$\mathcal{P} = \begin{bmatrix} \mathcal{P}_1 \\ \mathcal{P}_2 \end{bmatrix}, \quad \widetilde{\mathcal{P}} = \mathcal{P}\mathcal{P}_1^{-1} = \begin{bmatrix} I_{m_1} \\ \mathcal{P}_2\mathcal{P}_1^{-1} \end{bmatrix}. \tag{3.25}$$

Here, the invertibility of $\mathcal{P}_1$ and the property $\mathcal{P}_2\mathcal{P}_1^{-1} \in S_{m_2 \times m_1}$ are immediately apparent from (1.170). Furthermore, formulas (3.7) and (3.25) yield $\varphi(x, z, \mathcal{P}) = \varphi(x, z, \widetilde{\mathcal{P}})$, and so the set of functions $\varphi(x, z, \mathcal{P})$ can also be described in terms of the parameter Schur functions. In the latter way, various linear-fractional transformations were treated, for instance, in [22, 92]. The results from [22, 92] could be used to derive formulas (3.22) and (3.23) as well.

Remark 3.5. From the proof of Proposition 3.3, we see that the inequality (3.4) for $r = l < \infty$ implies that the embedded matrix balls (discs) $\mathcal{N}_u(x, z)$ are well-defined for $z \in \mathbb{C}_M$ and x on the interval $[0, l]$, which is useful for the study of a Dirac system on intervals.

Remark 3.6. If $\varphi(z) \in \mathcal{N}_u(l, z)$ for all $z \in \mathbb{C}_M$ and, in addition, φ is analytic in $\mathbb{C}_M$, then it admits representation (3.7), where $x = l$ and $\mathcal{P}$ satisfies (1.170). This statement is immediately apparent since we can choose

$$\mathcal{P}(z) = u(l, z) \begin{bmatrix} I_{m_1} \\ \varphi(z) \end{bmatrix}.$$

The property-j of this $\mathcal{P}$ follows from the equivalence of (3.14))and (3.15).

We define Weyl function for (3.1) like in the self-adjoint case (Definition 2.20).

Definition 3.7. An $m_2 \times m_1$ matrix function φ, which is holomorphic in $\mathbb{C}_M$ (for some $M > 0$) and satisfies the inequality

$$\int_0^\infty \begin{bmatrix} I_{m_1} & \varphi(z)^* \end{bmatrix} u(x, z)^* u(x, z) \begin{bmatrix} I_{m_1} \\ \varphi(z) \end{bmatrix} dx < \infty, \quad z \in \mathbb{C}_M \tag{3.26}$$

is called a Weyl function of the skew-self-adjoint Dirac system (3.1).

From Proposition 3.3 and Definition 3.7, the next analog of Corollary 2.21 follows, which is proved quite like Corollary 2.21 also (and so this proof is omitted).

Corollary 3.8. *Let Dirac system* (3.1) *be given on* $[0, \infty)$ *and assume that* (3.8) *holds. Then, there is a unique Weyl function of this system. This Weyl function is determined in* $\mathbb{C}_M$ *via* (3.9).

Remark 3.9. Let the conditions of Corollary 3.8 hold. Then, formulas (3.22) and (3.24) yield the equality

$$\varphi(z) = \lim_{b \to \infty} \varphi_b(z) \quad (z \in \mathbb{C}_M) \tag{3.27}$$

for the Weyl function $\varphi(z)$ and any set of functions $\varphi_b(z) \in \mathcal{N}_u(b, z)$.

3.2 The inverse problem on a finite interval and semiaxis

Definition 3.10. Weyl functions of the skew-self-adjoint Dirac system (3.1), which is given on $[0, l]$ and satisfies the inequality

$$\|v(x)\| \leq M \quad \text{for } 0 < x < l, \tag{3.28}$$

are the functions of the form (3.7), where $x = l$ and $\mathcal{P} \in \mathbf{P}_M(j)$.

Recall that, in view of Remark 3.5, the set $\mathcal{N}_u(l,z)$ is well-defined only if (3.28) holds. Further in the text, we assume that (3.28) holds and put

$$\beta(x) = \begin{bmatrix} I_{m_1} & 0 \end{bmatrix} u(x,0), \quad \gamma(x) = \begin{bmatrix} 0 & I_{m_2} \end{bmatrix} u(x,0). \tag{3.29}$$

It follows from (3.1), (3.28) and (3.29) that

$$\sup_{x<l} \|\gamma'(x)\| < \infty, \quad \gamma' := \frac{d}{dx}\gamma. \tag{3.30}$$

Moreover, from (3.1)–(3.3), we obtain

$$u(x,\overline{z})^* u(x,z) = u(x,z)u(x,\overline{z})^* = I_m, \quad z \in \mathbb{C}. \tag{3.31}$$

Therefore, (3.29) implies

$$\beta\beta^* \equiv I_{m_1}, \quad \gamma\gamma^* \equiv I_{m_2}, \quad \beta\gamma^* \equiv 0. \tag{3.32}$$

As in Chapter 2, we partition $\gamma(x)$ into two blocks $\gamma = \begin{bmatrix} \gamma_1 & \gamma_2 \end{bmatrix}$, where γ_1, γ_2 are, respectively, $m_2 \times m_1$ and $m_2 \times m_2$ matrix functions. Next, we introduce the Volterra operator

$$K = \int_0^x F(x)G(t) \cdot dt, \quad K \in B\left(L^2_{m_2}(0,l)\right), \tag{3.33}$$

$$F = \gamma, \quad G = -\mathrm{i}\gamma^*, \quad \gamma\gamma^* \equiv I_{m_2}, \quad \gamma \in B^1([0,l]), \quad \det \gamma_2(0) \neq 0. \tag{3.34}$$

Relations (3.34) yield

$$\mathrm{i}F(x)G(x) \equiv I_{m_2}. \tag{3.35}$$

Therefore, Proposition 1.29 is applicable. Using Proposition 1.29 and the proof of Proposition 1.31, we immediately obtain the next direct analog of Proposition 2.33.

Proposition 3.11. *Let K be given by (3.33), where F and G satisfy (3.34). Then, there is a triangular operator $E \in B(L^2_{m_2}(0,l))$ of the form*

$$(Ef)(x) = \rho(x)f(x) + \int_0^x E(x,t)f(t)dt, \quad \frac{d}{dx}\rho = \gamma'\gamma^*\rho, \quad \rho(0) = \gamma_2(0), \tag{3.36}$$

such that the operators $E^{\pm 1}$ map functions with bounded derivatives into functions with bounded derivatives, the kernel $E(x,t)$, as well as the kernel E^{-1}, is bounded and the following relations hold:

$$K = EAE^{-1}, \quad A := -\mathrm{i}\int_0^x \cdot dt, \tag{3.37}$$

$$E^{-1}\gamma_2 \equiv I_{m_2}. \tag{3.38}$$

The boundedness of the kernels of $E^{\pm 1}$ in Proposition 3.11 follows from the arguments in Remark 2.34. The next lemma is easily derived from Proposition 3.11 and will be used to construct a transfer matrix function representation of the fundamental solution u.

Lemma 3.12. *Let γ be an $m_2 \times m$ matrix function which satisfies the last three relations in* (3.34), *and set*

$$S := E^{-1}\left(E^*\right)^{-1}, \quad \Pi := \begin{bmatrix} \Phi_1 & \Phi_2 \end{bmatrix}, \quad \Phi_k \in B\left(\mathbb{C}^{m_k}, L^2_{m_2}(0, l)\right); \tag{3.39}$$

$$\left(\Phi_1 g\right)(x) = \Phi_1(x) g, \quad \Phi_1(x) := \left(E^{-1}\gamma_1\right)(x); \quad \Phi_2 g = I_{m_2} g \equiv g; \tag{3.40}$$

where E is described (for the given γ) in Proposition 3.11 *and g is used to denote finite-dimensional vectors. Then, A, S and Π form an S-node, that is, the following operator identity is valid:*

$$AS - SA^* = -\mathrm{i}\Pi\Pi^*. \tag{3.41}$$

Proof. Because of (3.33), (3.34) and (3.37), we obtain

$$EAE^{-1} - \left(E^{-1}\right)^* A^* E^* = K - K^* = -\mathrm{i}\gamma(x) \int_0^l \gamma(t)^* \cdot dt. \tag{3.42}$$

The second equality in (3.39) and formulas (3.38) and (3.40) lead us to the equalities

$$\Pi g = \left(E^{-1}\gamma\right)(x) g, \quad \Pi^* f = \int_0^l \left(E^{-1}\gamma\right)(t)^* f(t) dt = \int_0^l \gamma(t)^* \left((E^{-1})^* f\right)(t) dt. \tag{3.43}$$

Using the first equality in (3.39) and relations (3.43), we rewrite (3.42) as the operator identity (3.41). □

Remark 3.13. For the symmetric S-node constructed above, we have $J = -I_m$ in (1.83) and so formula (1.85) for the transfer matrix function takes the form

$$w_A(\lambda) = I_m + \mathrm{i}\Pi^* S^{-1}(A - \lambda I)^{-1}\Pi. \tag{3.44}$$

As in Chapter 2, the family of orthoprojectors $\{P_\xi\}$ given by (2.105) and the triple $\{A, S, \Pi\}$ generate a family of S-nodes such that

$$A_\xi := P_\xi A P_\xi^*, \quad S_\xi := P_\xi S P_\xi^* \quad (0 < \xi \le l); \quad A_\xi S_\xi - S_\xi A_\xi^* = -\mathrm{i} P_\xi \Pi\Pi^* P_\xi^*. \tag{3.45}$$

The corresponding transfer matrix function (depending now on ξ and λ) has the form

$$w_A(\xi, \lambda) = I_m + \mathrm{i}\Pi^* S_\xi^{-1}(A_\xi - \lambda I)^{-1} P_\xi \Pi, \quad 0 < \xi \le l. \tag{3.46}$$

Next, we construct the fundamental solution W of the *pseudocanonical* system

$$\frac{d}{dx}W(x,z) = -\mathrm{i}z\gamma(x)^*\gamma(x)W(x,z), \quad W(0,z) = I_m. \tag{3.47}$$

Theorem 3.14. *Let γ be an $m_2 \times m$ matrix function which satisfies the last three relations in* (3.34). *Then, the fundamental solution W of the system* (3.47) *admits the representation*

$$W(\xi,z) = w_A(\xi, 1/z), \tag{3.48}$$

where w_A is defined in Remark 3.13.

Theorem 3.14 is an analog of Theorems 1.34 and 2.37. Its statement follows from Theorem 1.20, and the proof of this is quite similar to the proof of Theorem 1.34.

Now, again consider the case of a Dirac system. It is apparent that

$$u'(x,0)u(x,0)^{-1} = jV(x). \tag{3.49}$$

Because of (3.29) and (3.32), we obtain

$$u(x,0)\gamma(x)^*\gamma(x)u(x,0)^{-1} = \begin{bmatrix} 0 & 0 \\ 0 & I_{m_2} \end{bmatrix}. \tag{3.50}$$

In view of (3.47)–(3.50), direct differentiation shows that the following corollary of Theorem 3.14 holds.

Corollary 3.15. *Let $u(x,z)$ be the fundamental solution of a skew-self-adjoint Dirac system where the potential v is bounded, and let γ be given by* (3.29). *Then, $u(x,z)$ admits the representation*

$$u(x,z) = \mathrm{e}^{\mathrm{i}zx}u(x,0)w_A\left(x, 1/(2z)\right). \tag{3.51}$$

Remark 3.16. If γ is given by (3.29), it follows from (3.1) and (3.32) that

$$\gamma'(x)\gamma(x)^* = -\begin{bmatrix} v(x)^* & 0 \end{bmatrix} u(x,0)\gamma(x)^* = -v(x)^*\beta(x)\gamma(x)^* \equiv 0. \tag{3.52}$$

Thus, from (3.34) and (3.52), we see that $F'G \equiv 0$, and so ρ in (3.36) is a constant matrix, that is, $\rho(x) \equiv \gamma_2(0) = I_{m_2}$. Therefore, formula (3.36) can be rewritten in the form

$$(Ef)(x) = f(x) + \int_0^x E(x,t)f(t)dt, \quad E \in B\left(L^2_{m_2}(0,l)\right). \tag{3.53}$$

Next, we use the method of operator identities to derive a high energy asymptotics result for the Dirac system (3.1) (see also [108, 259]).

Theorem 3.17. *Assume that φ is a Weyl function of system* (3.1) *on* $[0, l]$ *and* (3.28) *holds. Then (uniformly with respect to $\Re(z)$), we have*

$$\varphi(z) = 2\mathrm{i}z \int_0^l \mathrm{e}^{2\mathrm{i}zx}\Phi_1(x)dx + O\left(2z\mathrm{e}^{2\mathrm{i}zl}\Big/\sqrt{\Im(z)}\right), \quad \Im(z) \to \infty. \tag{3.54}$$

Proof. Step 1. As in the self-adjoint case, we first show that

$$\left\|(I - 2zA)^{-1}\Pi\begin{bmatrix} I_{m_1} \\ \varphi(z)\end{bmatrix}\right\| \le C/\sqrt{\Im z} \quad \text{for some} \quad C > 0. \tag{3.55}$$

For this purpose, we consider the matrix function

$$\beth(z) := \begin{bmatrix} I_{m_1} & \varphi(z)^*\end{bmatrix}\left(w_A(l, 1/(2z))^* w_A(l, 1/(2z)) - I_m\right)\begin{bmatrix} I_{m_1} \\ \varphi(z)\end{bmatrix}. \tag{3.56}$$

However, the function itself and the proof that this function is bounded are different from those in the self-adjoint case (compare with the proof of Theorem 2.43). More specifically, because of (3.31) and (3.51), we have

$$\beth(z) = \mathrm{e}^{\mathrm{i}(\overline{z}-z)l}\begin{bmatrix} I_{m_1} & \varphi(z)^*\end{bmatrix} u(l,z)^* u(l,z)\begin{bmatrix} I_{m_1} \\ \varphi(z)\end{bmatrix} - I_{m_1} - \varphi(z)^*\varphi(z).$$

We substitute l for x and $\widetilde{\mathcal{P}}$ (of the form (3.25)) for $\mathcal{P}$ into (3.12), after which we substitute the result into the formula above. Thus, we obtain

$$\begin{aligned}\beth(z) &= \mathrm{e}^{\mathrm{i}(\overline{z}-z)l}\left(\left(\begin{bmatrix} I_{m_1} & 0\end{bmatrix} u(l,z)^{-1}\widetilde{\mathcal{P}}(z)\right)^{-1}\right)^* \widetilde{\mathcal{P}}(z)\widetilde{\mathcal{P}}(z)^* \\ &\quad \times \left(\begin{bmatrix} I_{m_1} & 0\end{bmatrix} u(l,z)^{-1}\widetilde{\mathcal{P}}(z)\right)^{-1} - I_{m_1} - \varphi(z)^*\varphi(z). \end{aligned}\tag{3.57}$$

Step 2. To derive (3.55), we also need to examine the asymptotics of $u(l, z)$ ($z \to \infty$). This is achieved by using the procedure for constructing the transformation operator for Dirac system. We show that u admits an integral representation

$$u(x,z) = \mathrm{e}^{\mathrm{i}zxj} + \int_{-x}^{x} \mathrm{e}^{\mathrm{i}zt}N(x,t)dt, \quad \sup \|N(x,t)\| < \infty \quad (0 < |t| < x < l). \tag{3.58}$$

Indeed, it is easily checked that

$$u(x,z) = \sum_{i=0}^{\infty} \nu_i(x,z), \tag{3.59}$$

$$\nu_0(x,z) := \mathrm{e}^{\mathrm{i}zxj}, \quad \nu_i(x,z) := \int_0^x \mathrm{e}^{\mathrm{i}z(x-t)j} jV(t)\nu_{i-1}(t,z)dt \quad \text{for } i > 0 \tag{3.60}$$

since the right-hand side of (3.59) satisfies both equation (3.1) and the normalization condition at $x = 0$. Moreover, using induction, we derive in a standard way that

$$\nu_i(x,z) = \int_{-x}^{x} \mathrm{e}^{\mathrm{i}zt} N_i(x,t)dt, \quad \sup \|N_i(x,t)\| \le (2M)^i x^{i-1}/(i-1)!, \tag{3.61}$$

where $i > 0$ and M is the same as in (3.28). We see that formulas (3.59) and (3.61) yield (3.58). Because of (3.31) and (3.58), we obtain

$$u(l,z)^{-1} = u(l,\overline{z})^* = \mathrm{e}^{-\mathrm{i}zl}\left(\begin{bmatrix} I_{m_1} & 0 \\ 0 & \mathrm{e}^{2\mathrm{i}zl} I_{m_2}\end{bmatrix} + o(1)\right) \quad (z \in \mathbb{C}_+,\ z \to \infty). \tag{3.62}$$

Step 3. In view of (3.7), (3.57), (3.62) and Remark 3.4, for any $\varepsilon > 0$, there are numbers $\widetilde{M} > M > 0$ and $\widetilde{C} > 0$ such that for all z satisfying $\Im z > \widetilde{M}$, we have

$$\|\varphi(z)\| \le \varepsilon, \quad \|\beth(z)\| \le \widetilde{C}. \tag{3.63}$$

From (1.88), we see that

$$\begin{aligned} w_A(l, 1/(2z))^* w_A(l, 1/(2z)) = I_m &+ 2\mathrm{i}(\overline{z} - z)\Pi^*(I - 2\overline{z}A^*)^{-1} \\ &\times S_l^{-1}(I - 2zA)^{-1}\Pi. \end{aligned} \tag{3.64}$$

We substitute (3.64) into (3.56) to rewrite the second inequality in (3.63) in the form

$$2\mathrm{i}(\overline{z} - z)\begin{bmatrix} I_{m_1} & \varphi(z)^* \end{bmatrix}\Pi^*(I - 2\overline{z}A^*)^{-1}S_l^{-1}(I - 2zA)^{-1}\Pi\begin{bmatrix} I_{m_1} \\ \varphi(z)\end{bmatrix} \le \widetilde{C} I_{m_1}. \tag{3.65}$$

Since S is strictly positive, inequality (3.65) yields (3.55).

The remaining part of the proof is similar to the proof of Theorem 2.43. Namely, we apply $-\mathrm{i}\Phi_2^*$ to the operator on the left-hand side of (3.55) and use (1.158) to obtain (uniformly with respect to $\Re(z)$) the equality

$$\frac{1}{2z}\left(\mathrm{e}^{-2\mathrm{i}zl} - 1\right)\varphi(z) = \mathrm{i}\mathrm{e}^{-2\mathrm{i}zl}\int_0^l \mathrm{e}^{2\mathrm{i}zx}\Phi_1(x)dx + O\left(\frac{1}{\sqrt{\Im(z)}}\right). \tag{3.66}$$

Because of (3.66) (and the first inequality in (3.63)), we see that (3.54) holds. □

The integral representation below follows from the high energy asymptotics of φ and is essential in interpolation and inverse problems.

Corollary 3.18. *Let φ be a Weyl function of a Dirac system on $[0, l]$ and let (3.28) hold. Then, we have*

$$\Phi_1\left(\frac{x}{2}\right) = \frac{1}{\pi}\mathrm{e}^{\eta x}\mathrm{l.i.m.}_{a\to\infty}\int_{-a}^{a} \mathrm{e}^{-\mathrm{i}\xi x}\frac{\varphi(\xi + \mathrm{i}\eta)}{2\mathrm{i}(\xi + \mathrm{i}\eta)}d\xi, \quad \eta > M, \tag{3.67}$$

where l.i.m. stands for the entrywise limit in the norm of $L^2(0, 2l)$.

Proof. According to Definition 3.10 and relations (3.16) and (3.17), the matrix function φ is nonexpansive in $\mathbb{C}_M$. Since φ is analytic and nonexpansive in $\mathbb{C}_M$, it admits (see, e.g. Theorem E.11) a representation

$$\varphi(z) = 2\mathrm{i}z \int_0^\infty \mathrm{e}^{2\mathrm{i}zx}\Phi(x)dx, \quad z = \xi + \mathrm{i}\eta, \quad \eta > M > 0, \tag{3.68}$$

where $\mathrm{e}^{-2\eta x}\Phi(x) \in L^2_{m_2 \times m_1}(0, \infty)$. (Here, $L^2_{m_2\times m_1}$ stands for the space $L^2_{m_2 m_1}$, where vector functions are presented in the $m_2 \times m_1$ matrix function form.) Because of (3.54) and (3.68), we derive

$$\int_0^l \mathrm{e}^{2\mathrm{i}z(x-l)}(\Phi_1(x) - \Phi(x))dx = \int_l^\infty \mathrm{e}^{2\mathrm{i}z(x-l)}\Phi(x)dx + O\left(1\Big/\sqrt{\Im(z)}\right)$$

for $\Im(z) \to +\infty$. Taking into account that $\mathrm{e}^{-2(M+\varepsilon)x}\Phi(x) \in L^2_{m_2\times m_1}(0, \infty)$ for $\varepsilon > 0$, we rewrite the formula above as

$$\int_0^l \mathrm{e}^{2\mathrm{i}z(x-l)}(\Phi_1(x) - \Phi(x))dx = O\left(1\Big/\sqrt{\Im(z)}\right), \quad \Im(z) \to +\infty, \tag{3.69}$$

and the equality (3.69) is uniform with respect to $\Re(z)$. Clearly, the left-hand side of (3.69) is bounded in the domains, where $\Im(z)$ is bounded from above. Hence, in view of (3.69), its left-hand side is bounded in all of $\mathbb{C}$ and tends to zero on some rays. Now, the following identities are immediate:

$$\int_0^l \mathrm{e}^{2\mathrm{i}z(x-l)}(\Phi_1(x) - \Phi(x))dx \equiv 0, \quad \text{i.e.} \quad \Phi_1(x) \equiv \Phi(x). \tag{3.70}$$

Using the Plancherel theorem, we apply the inverse Fourier transform to formula (3.68) and derive a representation of the form (3.67) for Φ. Since according to (3.70), we have $\Phi_1 \equiv \Phi$, the formula (3.67) is valid. □

Directly from (3.1) and (3.29), we easily obtain a useful formula

$$\beta'(x) = v(x)\gamma(x), \tag{3.71}$$

which, because of (3.32), implies

$$\beta'\beta^* = 0, \quad \beta'\gamma^* = v. \tag{3.72}$$

Thus, we apply (3.72) and solve the inverse problem once we recover β and γ from φ. The recovery of Φ_1 from φ is studied in Corollary 3.18. The next step to solve the inverse problem is to recover β from Φ_1 in the proposition below.

Proposition 3.19. *Let the potential v of Dirac system* (3.1) *on* $[0, l]$ *satisfy* (3.28), *let β and γ be given by* (3.29), *and let Π and S be the operators determined by γ in Lemma* 3.12. *Then, we have*

$$\beta(x) = \begin{bmatrix} I_{m_1} & 0 \end{bmatrix} - \int_0^x \left(S_x^{-1}\Phi_1'\right)(t)^* \begin{bmatrix} \Phi_1(t) & I_{m_2} \end{bmatrix} dt. \tag{3.73}$$

Proof. First, we use (3.38) and (3.40) to express γ in the form

$$\gamma(x) = \left(E\begin{bmatrix} \Phi_1 & I_{m_2} \end{bmatrix}\right)(x). \tag{3.74}$$

It follows that

$$\mathrm{i}\left(EAE^{-1}E\Phi_1'\right)(x) = \gamma_1(x) - \gamma_2(x)\Phi_1(+0). \tag{3.75}$$

Since $\gamma_1(0) = 0$ and E has the form (3.53), formula (3.74) also yields the equality

$$\Phi_1(+0) = 0. \tag{3.76}$$

Therefore, we rewrite (3.75) as

$$\mathrm{i}\left(EAE^{-1}E\Phi_1'\right)(x) = \gamma_1(x). \tag{3.77}$$

Next, we substitute $K = EAE^{-1}$ from (3.37) into (3.77) and (using (3.33) and (3.34)) we obtain

$$\gamma_1(x) = \gamma(x)\int_0^x \gamma(t)^* \left(E\Phi_1'\right)(t)dt. \tag{3.78}$$

Formulas (3.74) and (3.78) imply

$$\gamma_1(x) = \gamma(x)\int_0^x \left(E\begin{bmatrix} \Phi_1 & I_{m_2} \end{bmatrix}\right)(t)^* \left(E\Phi_1'\right)(t)dt. \tag{3.79}$$

Because of (1.153), (1.154) and (3.79), we see that

$$\gamma(x)\widehat{\beta}(x)^* \equiv 0, \tag{3.80}$$

where

$$\begin{aligned}\widehat{\beta}(x) :&= \begin{bmatrix} I_{m_1} & 0 \end{bmatrix} - \int_0^x \left(E\Phi_1'\right)(t)^* \left(E\begin{bmatrix} \Phi_1 & I_{m_2} \end{bmatrix}\right)(t)dt \\ &= \begin{bmatrix} I_{m_1} & 0 \end{bmatrix} - \int_0^x \left(S_x^{-1}\Phi_1'\right)(t)^* \begin{bmatrix} \Phi_1(t) & I_{m_2} \end{bmatrix} dt.\end{aligned} \tag{3.81}$$

Consider $\widehat{\beta}$ in greater detail. In view of (3.74) and (3.81), we have

$$\widehat{\beta}'(x) = -\left(E\Phi_1'\right)(x)^*\gamma(x). \tag{3.82}$$

Therefore, (3.32) leads us to

$$\beta(x)\widehat{\beta}'(x)^* \equiv 0. \tag{3.83}$$

We furthermore compare (3.32) with (3.80) to see that

$$\widehat{\beta}(x) = \widehat{\vartheta}(x)\beta(x), \tag{3.84}$$

where $\widehat{\vartheta}(x)$ is an $m_1 \times m_1$ matrix function. According to (3.29) and (3.82), respectively, β and $\widehat{\beta}$ are boundedly differentiable on $[0, l]$, and so $\widehat{\vartheta}$ is also boundedly differentiable. Now, equalities (3.83), (3.84) and the first relations in (3.32), (3.72) yield that $\widehat{\vartheta}' \equiv 0$ (i.e. $\widehat{\vartheta}$ is a constant). It follows from (3.29), (3.81) and (3.84) that $\widehat{\vartheta}(0) = I_{m_1}$, and therefore $\widehat{\vartheta} \equiv I_{m_1}$, that is, $\widehat{\beta} \equiv \beta$. Thus, (3.73) follows directly from (3.81). □

Proposition 3.19 is an analog of Proposition 2.50 from Chapter 2 (only the signs before the integral term in the expressions for β differ). As in Chapter 2, we need a procedure here to recover γ from β. Directly from (3.1) and the last equalities in (3.29) and (3.32), we obtain

$$\gamma(0) = \begin{bmatrix} 0 & I_{m_2} \end{bmatrix}, \quad \gamma' = -v^*\beta, \quad \gamma'\gamma^* = 0. \tag{3.85}$$

From (3.29) and (3.71), we also see that $\beta(0) = [I_{m_1} \ \ 0]$ and β is boundedly differentiable. Therefore, in view of (3.72), the proof of the proposition below will provide us with the required procedure.

Proposition 3.20. *Let a given $m_1 \times m$ matrix function $\beta(x)$ $(0 \le x \le l)$ be boundedly differentiable and satisfy relations*

$$\beta(0) = \begin{bmatrix} I_{m_1} & 0 \end{bmatrix}, \quad \beta'\beta^* \equiv 0. \tag{3.86}$$

Then, there is a unique $m_2 \times m$ matrix function γ, which is boundedly differentiable and satisfies relations

$$\gamma(0) = \begin{bmatrix} 0 & I_{m_2} \end{bmatrix}, \quad \gamma'\gamma^* \equiv 0, \quad \gamma\beta^* \equiv 0. \tag{3.87}$$

Proof. Since $\beta\beta^* \equiv I_{m_1}$, there are m_1 columns of $\beta(x)$ (for each x), which are linearly independent. We partition $[0, l]$ into a finite number of closed subintervals such that for each interval, some set of m_1 numbers of such columns of β can be fixed. After that, we easily construct (successively on all subintervals) an $m_2 \times m$ matrix function $\widetilde{\gamma}$ which is boundedly differentiable on $[0, l]$ and satisfies relations

$$\beta\widetilde{\gamma}^* \equiv 0, \quad \widetilde{\gamma}\widetilde{\gamma}^* > 0, \quad \widetilde{\gamma}(0) = \begin{bmatrix} 0 & I_{m_2} \end{bmatrix}. \tag{3.88}$$

That is, we construct such a matrix function $\widetilde{y}$ using β.

Next, we need some heuristics. For that, we write down certain properties of y and $\widetilde{y}$ following from (3.87) and (3.88):

$$\beta y^* = \beta \widetilde{y}^* = 0, \quad y y^* = I_{m_2}, \quad \widetilde{y}\widetilde{y}^* > 0, \quad \widetilde{y}(0) = y(0), \tag{3.89}$$

and compare them. In view of (3.89), the matrix function y admits representation

$$y = \widetilde{\vartheta}\widetilde{y}, \quad \widetilde{\vartheta}\widetilde{\vartheta}^* > 0, \quad \widetilde{\vartheta}(0) = I_{m_2}, \tag{3.90}$$

where $\widetilde{\vartheta}$ is boundedly differentiable. Because of (3.90), the second equality in (3.87) can be rewritten as

$$\widetilde{\vartheta}' = -\widetilde{\vartheta}\widetilde{y}'\widetilde{y}^*(\widetilde{y}\widetilde{y}^*)^{-1}, \quad \widetilde{\vartheta}(0) = I_{m_2}, \tag{3.91}$$

which uniquely defines $\widetilde{\vartheta}$.

It is apparent from (3.88) that y given by the first relation in (3.90) (and formula (3.91)) satisfies (3.87). Moreover, if y and $\widetilde{y}$ satisfy both (3.87), then we have $y = \widetilde{\vartheta}\widetilde{y}$, where (3.91) yields $\widetilde{\vartheta} \equiv I_{m_2}$ (i.e. y is unique). □

Hence, taking into account Corollary 3.18, Propositions 3.19 and 3.20 and formula (3.72), we see that we have a procedure to solve the inverse problem.

Theorem 3.21. *Let the potential v of a Dirac system* (3.1) *on* $[0, l]$ *satisfy* (3.28). *Then, v can be uniquely recovered from a Weyl function φ of this system, which is done in the following way. First, we recover Φ_1 using formula* (3.67). *Next, we use operator identity* (3.41), *where Π has the form $\Pi\left[\begin{smallmatrix} g_1 \\ g_2 \end{smallmatrix}\right] = \Phi_1(x)g_1 + g_2$, to recover S and we use formula* (3.73), *where $S_x = P_x S P_x^*$, to recover β. We recover (from β) the matrix functions $\widetilde{y}$ and $\widetilde{\vartheta}$ via formulas* (3.88) *and* (3.91), *respectively. The matrix function y is given by $y = \widetilde{\vartheta}\widetilde{y}$. Finally, we obtain v from the equality*

$$v(x) = \beta'(x)y(x)^*. \tag{3.92}$$

We note that (3.92) coincides with the second equality in (3.72).

Remark 3.22. In order to recover S from (3.41), we use Theorem D.5. Thus, taking into account (3.76), we obtain

$$S = I + \int_0^l s(x,t) \cdot dt, \quad s(x,t) = \int_0^{\min(x,t)} \Phi_1'(x-\xi)\Phi_1'(t-\xi)^* d\xi. \tag{3.93}$$

The next local Borg–Marchenko-type uniqueness theorem, which follows from Theorems 3.17 and 3.21, is an analog of Theorem 2.52. Weyl functions φ, and $\widehat{\varphi}$ of two Dirac systems are considered. The notations of matrix functions associated with $\widehat{\varphi}$ are written with a "hat" (e.g. $\widehat{v}$, $\widehat{\Phi}_1$).

Theorem 3.23. *Let φ and $\widehat{\varphi}$ be Weyl functions of two Dirac systems on $[0, l]$, the potentials of which are denoted, respectively, by v and $\widehat{v}$, and assume that*

$$\max(\|v(x)\|, \|\widehat{v}(x)\|) \le M, \quad 0 < x < l. \tag{3.94}$$

Suppose that on some ray $\Re z = c\Im z$ $(c \in \mathbb{R}, \Im z > 0)$, the equality

$$\|\varphi(z) - \widehat{\varphi}(z)\| = O(\mathrm{e}^{2izr}) \quad (\Im z \to \infty) \tag{3.95}$$

holds for all $0 < r < \widetilde{l} \le l$. Then, we have

$$v(x) = \widehat{v}(x), \quad 0 < x < \widetilde{l}. \tag{3.96}$$

Proof. Since according to (3.17) Weyl functions are nonexpansive, we see that the matrix function $\mathrm{e}^{-2izr}(\varphi(z) - \widehat{\varphi}(z))$ is bounded on each line $\Im z = M + \varepsilon$ $(\varepsilon > 0)$ and the inequality

$$\|\mathrm{e}^{-2izr}\left(\varphi(z) - \widehat{\varphi}(z)\right)\| \le 2\mathrm{e}^{2|z|r} \quad (\Im z > M > 0) \tag{3.97}$$

holds. Furthermore, formula (3.95) implies that $\mathrm{e}^{-2izr}(\varphi(z) - \widehat{\varphi}(z))$ is bounded on the ray $\Re z = c\Im z$. Therefore, applying the Phragmen–Lindelöf theorem to $\mathrm{e}^{-2izr} \cdot (\varphi(z) - \widehat{\varphi}(z))$ in the angles generated by the line $\Im z = M + \varepsilon$ and the ray $\Re z = c\Im z$ $(\Im z \ge M + \varepsilon)$, we derive

$$\|\mathrm{e}^{-2izr}\left(\varphi(z) - \widehat{\varphi}(z)\right)\| \le M_1 \quad \text{for } \Im z \ge M + \varepsilon. \tag{3.98}$$

Because of formula (3.54), of the formula's analog for $\widehat{\varphi}$ and $\widehat{\Phi}_1$ and of the inequality (3.98), we have

$$\int_0^r \mathrm{e}^{2iz(x-r)}\left(\Phi_1(x) - \widehat{\Phi}_1(x)\right) dx = O(1/\Im(z)) \quad (\Im(z) \to +\infty), \tag{3.99}$$

uniformly with respect to $\Re(z)$. Hence, the left-hand side of (3.99) is bounded in $\mathbb{C}$ and tends to zero on some rays. Thus, we obtain

$$\int_0^r \mathrm{e}^{2iz(x-r)}\left(\Phi_1(x) - \widehat{\Phi}_1(x)\right) dx \equiv 0, \quad \text{i.e.,} \quad \Phi_1(x) \equiv \widehat{\Phi}_1(x) \quad (0 < x < r). \tag{3.100}$$

Since (3.100) holds for all $r < \widetilde{l}$, it yields $\Phi_1(x) \equiv \widehat{\Phi}_1(x)$ for $0 < x < \widetilde{l}$. In view of Theorem 3.21, the last identity implies (3.96). □

Finally, we again consider Dirac system (3.1) on the semiaxis $[0, \infty)$. According to Corollary 3.8, the Weyl function of system (3.1), such that (3.8) holds, is unique and satisfies (3.9). Moreover, it follows from the proof of Corollary 3.18 that

$$\varphi(z) = 2iz\int_0^\infty \mathrm{e}^{2ixz}\Phi(x)dx, \quad \mathrm{e}^{-2(M+\varepsilon)x}\Phi(x) \in L^2_{m_2\times m_1}, \tag{3.101}$$

and for any $0 < l < \infty$, we have

$$\Phi(x) \equiv \Phi_1(x) \quad (0 < x < l). \tag{3.102}$$

Since $\Phi_1 \equiv \Phi$, we see that $\Phi_1(x)$ does not depend on l for $l > x$. Our last result in this section is immediate from (3.101), (3.102) and Theorem 3.21.

Theorem 3.24. *Let a skew-self-adjoint Dirac system* (3.1) *be given on* $[0, \infty)$ *and satisfy* (3.8). *Then, its Weyl function is unique and admits representation*

$$\varphi(z) = 2\mathrm{i}z \int_0^\infty \mathrm{e}^{2\mathrm{i}xz}\Phi_1(x)dx, \quad \mathrm{e}^{-2\eta x}\Phi_1(x) \in L^2_{m_2\times m_1} \quad (\eta > M). \tag{3.103}$$

The procedure to recover Φ_1 *and* v *from* φ *is given in Theorem* 3.21.

3.3 System with a locally bounded potential

We start with several results for the system (3.1) satisfying (3.8). An analog of Corollary 2.25 is immediate from Remark 3.6, Corollary 3.8 and Definition 3.10.

Corollary 3.25. *Let a skew-self-adjoint Dirac system* (3.1) *be given on* $[0, \infty)$ *and satisfy* (3.8). *Then, its Weyl function is also a Weyl function of the same Dirac system on all the finite intervals* $[0, l]$.

Using Corollary 3.25, we obtain an analog of Proposition 2.26 precisely following the lines of its proof (hence, we omit the proof of this analog).

Proposition 3.26. *Let a system* (3.1) *be given on* $[0, \infty)$, *let its potential* v *satisfy* (3.8), *and assume that* φ *is its Weyl function. Then, the inequality*

$$\sup_{x\leq l,\, z\in\mathbb{C}_M} \left\| \mathrm{e}^{-\mathrm{i}zx}u(x,z)\begin{bmatrix} I_{m_1} \\ \varphi(z) \end{bmatrix} \right\| < \infty \tag{3.104}$$

holds on any finite interval $[0, l]$.

We already mentioned in Chapter 2 that the property (3.104) can be used as a definition of a generalized Weyl function. Here, we use it to study the case of the locally bounded potentials v, that is, potentials v such that $\sup \|v(x)\| < \infty$ on all the finite intervals $[0, l]$.

Definition 3.27. A generalized Weyl function (GW-function) of the system (3.1), where v is locally bounded on $[0, \infty)$, is an $m_2 \times m_1$ matrix function φ such that, for some $M > 0$, it is analytic in $\mathbb{C}_M$ and the inequalities (3.104) also hold for each $l < \infty$.

Proposition 3.28. *For any system* (3.1), *where* v *is locally bounded on* $[0, \infty)$, *there is no more than one GW-function.*

Proof. Let $\varphi_1 \neq \varphi_2$ be two GW-functions of the system (3.1). Then, according to (3.58) and (3.104), the inequality

$$\sup_{z \in \mathbb{C}_{\widetilde{M}}} \left\| \mathrm{e}^{-2\mathrm{i}zl}(\varphi_1(z) - \varphi_2(z)) \right\| < \infty \tag{3.105}$$

is valid for any $l > 0$ and some $\widetilde{M}(l) > M > 0$. The equality $\varphi_1(z) \equiv \varphi_2(z)$ easily follows. (To see this, we could consider, e.g. an integral representation of $(\varphi_1(z) - \varphi_2(z))/z$, which follows from Theorem E.11, and apply some arguments from the proof of Corollary 3.18.) □

Definition 3.29. The inverse spectral problem (ISpP) for system (3.1), where v is locally bounded on $[0, \infty)$, is the problem to recover (from a GW-function φ) a potential $v(x)$, which satisfies (3.104).

The notation $\mathfrak{W}$ stands for the operator mapping $\varphi(z)$ into v (i.e. $\mathfrak{W}(\varphi) = v$).

The following theorem is necessary in the process of solving a Goursat problem for the sine-Gordon equation in Subsection 6.2.2.

Theorem 3.30. *Let an $m_2 \times m_1$ matrix function $\varphi(z)$ be holomorphic in $\mathbb{C}_M$ (for some $M > 0$) and satisfy condition*

$$\sup \|z^2(\varphi(z) - \phi_0/z)\| < \infty \quad (z \in \mathbb{C}_M), \tag{3.106}$$

where ϕ_0 is an $m_2 \times m_1$ matrix. Then, φ is a GW-function of a skew-self-adjoint Dirac system. This Dirac system is uniquely recovered from φ by successively using relations (3.67) (in order to recover $\Phi_1(x)$), (3.93) (in order to recover operators S_l), (3.73) (in order to recover β), (3.88) and (3.91) (in order to recover $\gamma = \widetilde{\vartheta}\widetilde{\gamma}$) and, finally, using formula (3.92) in order to obtain $v = \mathfrak{W}(\varphi)$.

Proof. Step 1. Clearly, the matrix function $\Phi_1(x)$ does not depend on l and, moreover, in view of (3.106), formula (3.67) can be rewritten as a pointwise limit

$$\Phi_1\left(\frac{x}{2}\right) = \frac{1}{\pi}\mathrm{e}^{\eta x}\int_{-\infty}^{\infty} \mathrm{e}^{-\mathrm{i}\xi x}\frac{\varphi(\xi + \mathrm{i}\eta)}{2\mathrm{i}(\xi + \mathrm{i}\eta)}d\xi, \quad \eta > M, \tag{3.107}$$

which does not depend on the choice of $\eta > M$. Rewriting (3.107) in the form

$$\Phi_1\left(\frac{x}{2}\right) = \mathrm{i}x\phi_0 + \frac{1}{2\pi\mathrm{i}}\mathrm{e}^{\eta x}\int_{-\infty}^{\infty} \mathrm{e}^{-\mathrm{i}\xi x}\left(\frac{\varphi(\xi + \mathrm{i}\eta)}{(\xi + \mathrm{i}\eta)} - \frac{\phi_0}{(\xi + \mathrm{i}\eta)^2}\right)d\xi, \tag{3.108}$$

we immediately see that Φ_1 is locally boundedly differentiable. We also have

$$\Phi_1(0) = 0, \quad \Phi_1 \in C^1([0, \infty)); \qquad \sup_{x \in [0, \infty)} \|\mathrm{e}^{-2(M+\varepsilon)x}\Phi_1(x)\| < \infty \tag{3.109}$$

for any $\varepsilon > 0$. Now, we will construct via Φ_1 the fundamental solution of the Dirac system, the potential of which is described in our theorem.

Step 2. Let us fix an arbitrary $l > 0$. According to Remark 3.22 and Theorem D.5, the operator S_l, given by (3.93), satisfies the operator identity (3.41) and the inequality $S_l \geq I$. In particular, the triple $\{A, S_l, \Pi = [\Phi_1 \quad \Phi_2]\}$, where $\Phi_1 g = \Phi_1(x)g$ and $\Phi_2 g = g$, forms a symmetric S-node. (We note that the corresponding operators $S_\xi = P_\xi S_l P_\xi^*$ have the form (3.93) with ξ in place of l.) It is apparent that S_l satisfies the conditions of Corollary 1.39 (i.e. $S_l^{-1} = E^*E$) and that (1.154) is valid too. Therefore, we see that the matrix function $\Pi^* P_\xi^* S_\xi^{-1} P_\xi \Pi$ is continuously differentiable and the equality

$$\frac{d}{d\xi}(\Pi^* P_\xi^* S_\xi^{-1} P_\xi \Pi) = \widetilde{\gamma}(\xi)^* \widetilde{\gamma}(\xi), \quad \widetilde{\gamma}(x) := E\left[\Phi_1(x) \quad I_{m_2}\right] \tag{3.110}$$

holds. Like in the proof of Lemma 2.55, we derive from the operator identity (3.41) and triangular factorization $S_l^{-1} = E^*E$ that

$$K := EAE^{-1} = -\mathrm{i}\widetilde{\gamma}(x) \int_0^x \widetilde{\gamma}(t)^* \cdot dt. \tag{3.111}$$

In view of (3.111) and second equality in (3.110), we can precisely follow the lines of the proof of Proposition 3.19. In this way, taking into account that we determine β via (3.73), we substitute β in place of $\widehat{\beta}$ and $\widetilde{\gamma}$ in place of γ into (3.80)–(3.82) and derive

$$\widetilde{\gamma}(x)\beta(x)^* = 0, \quad \beta'(x) = -\left(E\Phi_1'\right)(x)^* \widetilde{\gamma}(x). \tag{3.112}$$

Since $\beta(0) = \left[I_{m_1} \quad 0\right]$, equalities (3.112) yield

$$\beta'(x)\beta(x)^* \equiv 0, \quad \beta(x)\beta(x)^* \equiv I_{m_1}, \tag{3.113}$$

and so β satisfies (3.86). On the other hand, rewriting the second equality in (3.111) in terms of the equality of the kernels of the corresponding triangular integral operators, we obtain the relation

$$I_{m_2} + \int_t^x (E(x,\xi) + \Upsilon_E(\xi,t))d\xi + \int_t^x E(x,\xi) \int_t^\xi \Upsilon_E(r,t)dr d\xi = \widetilde{\gamma}(x)\widetilde{\gamma}(t)^*, \tag{3.114}$$

where $x \geq t$ and Υ_E stands for the kernel of E^{-1}. In particular, (3.114) implies

$$\widetilde{\gamma}(x)\widetilde{\gamma}(x)^* \equiv I_{m_2}. \tag{3.115}$$

From the definition of $\widetilde{\gamma}$ in (3.110) and the equality $\Phi_1(0) = 0$, we have $\widetilde{\gamma}(0) = [0 \ I_{m_2}]$. Since equalities (3.112) and (3.115) also hold, $\widetilde{\gamma}$ satisfies (3.88). Therefore, according to the proof of Proposition 3.20, the matrix function $\gamma = \widetilde{\vartheta}\widetilde{\gamma}$ satisfies (3.87), and $\widetilde{\vartheta}$ given by (3.91) is unitary. Taking into account that $\gamma\beta^* \equiv 0$ and v is determined in our theorem via (3.92), we easily see that

$$\beta'\gamma^* = v, \quad \gamma'\beta^* = -v^*. \tag{3.116}$$

Therefore, setting

$$u(x,0) = \begin{bmatrix} \beta(x) \\ \gamma(x) \end{bmatrix} \tag{3.117}$$

and using $\beta\beta^* \equiv I_{m_1}$, $\gamma\gamma^* \equiv I_{m_2}$ and $\gamma\beta^* \equiv 0$, we have $u(x,0)^*u(x,0) = I_m$. Hence, formulas $\beta'\beta^* \equiv 0$, $\gamma'\gamma^* \equiv 0$, (3.116) and (3.117) yield

$$u'(x,0) = u'(x,0)u(x,0)^*u(x,0) = jV(x)u(x,0). \tag{3.118}$$

In view of (3.110), the S-node $\{A, S_l, \Pi\}$ satisfies conditions of Theorem 1.20 for the case $J = -I_m$. Thus, introducing $u(x,z)$ via (3.51), where w_A is given by (3.46), and using (3.110) and (3.118), we derive

$$u'(x,z) = jV(x)u(x,z) + \left(\mathrm{i}zI_m - 2\mathrm{i}zu(x,0)\tilde{\gamma}(x)^*\tilde{\gamma}(x)u(x,0)^*\right)u(x,z). \tag{3.119}$$

Recall that $\gamma = \tilde{\vartheta}\tilde{\gamma}$, where $\tilde{\vartheta}$ is unitary, and so $\tilde{\gamma}^*\tilde{\gamma} = \gamma^*\gamma$. Hence, according to (3.117), the equality

$$I_m - 2u(x,0)\tilde{\gamma}^*\tilde{\gamma}u(x,0)^* = j \tag{3.120}$$

is valid. In view of (3.51), (3.119) and (3.120), the constructed u is the fundamental solution of system (3.1), that is,

$$u'(x,z) = (\mathrm{i}zj + jV(x))u(x,z), \quad u(0,z) = I_m. \tag{3.121}$$

Step 3. Since we already proved that u is the fundamental solution of system (3.1), that the potential v of this system is constructed in accordance with the procedure given in the formulation of our theorem and does not depend on l, and that l is an arbitrary positive number, it only remains to be shown that (3.104) holds for this l. Moreover, we can prove an analog of (3.104) in $\mathbb{C}_{\widetilde{M}}$:

$$\sup_{x\le l,\, z\in\mathbb{C}_{\widetilde{M}}} \left\| \mathrm{e}^{-\mathrm{i}zx}u(x,z)\begin{bmatrix} I_{m_1} \\ \varphi(z) \end{bmatrix} \right\| < \infty \tag{3.122}$$

for some arbitrary $\widetilde{M} > M$, and according to (3.58), the inequality (3.104) will follow. In view of (3.51), it suffices to show that

$$\sup_{z\in\mathbb{C}_{\widetilde{M}}} \left\| z(I - 2zA)^{-1}\Pi \begin{bmatrix} I_{m_1} \\ \varphi(z) \end{bmatrix} \right\|_2 < \infty, \tag{3.123}$$

where the norm $\|f\|_2$ of $f \in L^2_{m_2\times m_1}(0,l)$ can be introduced via the trend $\mathrm{Tr}(f^*f)$ using the relation

$$\|f\|_2 := \left(\int_0^l \mathrm{Tr}\left(f(x)^*f(x)\right)dx \right)^{\frac{1}{2}}. \tag{3.124}$$

Taking into account (1.157), we rewrite (3.123) in an equivalent form

$$\sup_{z\in\mathbb{C}_{\widetilde{M}}} \left\| z\left(\Phi_1(x) + \mathrm{e}^{-2\mathrm{i}zx}\left(\varphi(z) - 2\mathrm{i}z\int_0^x \mathrm{e}^{2\mathrm{i}zt}\Phi_1(t)dt\right)\right)\right\|_2 < \infty. \tag{3.125}$$

In order to express φ via Φ_1 (in formula (3.125)), we note first that similar to the move from l.i.m. in the norm of $L^2(0, 2l)$ (considered in (3.67)) to the pointwise limit in (3.107), we can deal with l.i.m. in the norm of $L^2(-2l, 0)$. In that case (i.e. the case $x < 0$), formula (3.107) implies $\Phi_1(x) = 0$. Therefore, returning to (3.67) and applying the inverse Fourier transform, we obtain

$$\frac{\varphi(\xi + \mathrm{i}\eta)}{\mathrm{i}(\xi + \mathrm{i}\eta)} = \mathrm{l.i.m.}_{a\to\infty} \int_0^a \mathrm{e}^{\mathrm{i}\xi x}\mathrm{e}^{-\eta x}\Phi_1\left(\frac{x}{2}\right) dx, \quad \eta > M. \tag{3.126}$$

Taking into account that the last relation in (3.109) holds for any $\varepsilon > 0$, we turn again to a pointwise limit and rewrite (3.126):

$$\varphi(z) = 2\mathrm{i}z \int_0^\infty \mathrm{e}^{2\mathrm{i}zt}\Phi_1(t)dt, \quad z \in \mathbb{C}_M, \tag{3.127}$$

which coincides with the representation of the Weyl function in (3.103). It follows from (3.127) that

$$\varphi(z) - 2\mathrm{i}z \int_0^x \mathrm{e}^{2\mathrm{i}zt}\Phi_1(t) = 2\mathrm{i}z \int_x^\infty \mathrm{e}^{2\mathrm{i}zt}\Phi_1(t)dt. \tag{3.128}$$

Moreover, using (3.106) and (3.108), we easily check directly that

$$\lim_{x\to\infty} \mathrm{e}^{-2\eta x}\Phi_1(x) = \lim_{x\to\infty} \mathrm{e}^{-2\eta x}\Phi_1'(x) = 0, \tag{3.129}$$

and that Φ_1 is two times differentiable, that is,

$$2\pi\mathrm{i}\left(\mathrm{e}^{-\eta x}(\Phi_1(x/2) + \mathrm{i}x\phi_0)\right)'' \tag{3.130}$$

$$= \mathrm{l.i.m.}_{a\to\infty} - \int_{-a}^a \xi^2\mathrm{e}^{-\mathrm{i}\xi x}\left(\frac{\varphi(\xi + \mathrm{i}\eta)}{(\xi + \mathrm{i}\eta)} - \frac{\phi_0}{(\xi + \mathrm{i}\eta)^2}\right) d\xi, \tag{3.131}$$

$$\mathrm{e}^{-2\eta x}\Phi_1'(x),\ \mathrm{e}^{-2\eta x}\Phi_1''(x) \in L^2_{m_2\times m_1}(0, \infty) \tag{3.132}$$

for all $\eta > M$. (Here, (3.129) is also immediate from the last relation in (3.109) and a similar relation for Φ_1'.)

Integrating by parts and taking into account (3.129) and (3.132), we obtain

$$2\mathrm{i}z\mathrm{e}^{-2\mathrm{i}zx} \int_x^\infty \mathrm{e}^{2\mathrm{i}zt}\Phi_1(t)dt = \frac{1}{2\mathrm{i}z}\left(\mathrm{e}^{-2\mathrm{i}zx}\int_x^\infty \mathrm{e}^{2\mathrm{i}zt}\Phi_1''(t)dt + \Phi_1'(x)\right) - \Phi_1(x). \tag{3.133}$$

It is apparent from (3.128), (3.132) and (3.133) that (3.125) holds. Thus, φ is a GW-function of the system (3.1) with the prescribed v.

Step 4. Concluding this proof we show by contradiction that the solution of the inverse problem is unique. Suppose that there are two solutions of the inverse problem and attach, to the notations corresponding to these solutions the indices, "1" and "2," respectively (e.g. v_1, u_1, v_2, u_2,...). Fix some l such that the subset of $[0, l]$, where $v_1 \neq v_2$, is not a null set. Then, we have $u_1(l,z) \not\equiv u_2(l,z)$ since otherwise the Weyl functions of both systems on $[0,l]$ coincide and according to Theorem 3.21, the potentials are uniquely recovered from the Weyl functions, that is, $v_1 \equiv v_2$.

Let us consider the matrix function $u_1(l,z)u_2(l,z)^{-1}$. We rewrite it in the form

$$u_1(l,z)u_2(l,z)^{-1} = \hat{u}_1(l,z)\hat{u}_2(l,z)^{-1}, \quad \hat{u}_i(l,z) := u_i(l,z)\begin{bmatrix} I_{m_1} & 0 \\ \varphi(z) & I_{m_2} \end{bmatrix} \mathrm{e}^{-izlj}. \tag{3.134}$$

Relations (3.58) and (3.104) show that the matrix functions $\hat{u}_i(l,z)$ $(i = 1,2)$ are bounded in $\mathbb{C}_M$. Moreover, formulas (3.58) and (3.106) yield $\hat{u}_i(l,\xi + \mathrm{i}\eta) - I_m \in L^2_{m\times m}(-\infty, \infty)$ for any fixed $\eta > M$, that is, $\hat{u}_i(l,\xi + \mathrm{i}\eta) - I_m$ satisfy conditions of Theorem E.11 and admit representations (E.9), where we put $M + \varepsilon$ $(\varepsilon > 0)$ in place of M. In particular, it follows from (E.9) that

$$\lim_{\eta\to\infty} \hat{u}_i(l,\xi + \mathrm{i}\eta) = I_m, \tag{3.135}$$

and that $\hat{u}_i(l,z)$ are invertible and their inverse are uniformly bounded in some half-plane $\mathbb{C}_{\widetilde{M}}$ $(\widetilde{M} > M)$. Since the matrix functions $\hat{u}_i(l,z)$ are bounded and boundedly invertible, the matrix functions $u_1(l,z)u_2(l,z)^{-1}$ and $u_2(l,z)u_1(l,z)^{-1}$ are also bounded in $\mathbb{C}_{\widetilde{M}}$. From (3.31), we see that $u_1(l,\overline{z})u_2(l,\overline{z})^{-1} = (u_2(l,z)u_1(l,z)^{-1})^*$ and so the boundedness of $u_2(l,z)u_1(l,z)^{-1}$ in $\mathbb{C}_{\widetilde{M}}$ is equivalent to the boundedness of $u_1(l,z)u_2(l,z)^{-1}$ in the half-plane $\mathbb{C}^-_{\widetilde{M}} := \{z : \Im(z) < -\widetilde{M}\}$. Finally, according to (3.58), the matrix functions $u_1(l,z)$ and $u_2(l,\overline{z})^* = u_2(l,z)^{-1}$ are bounded in the strip

$$\Omega(-\widetilde{M},\widetilde{M}) = \{z : -\widetilde{M} \le \Im(z) \le \widetilde{M}\}.$$

Thus, $u_1(l,z)u_2(l,z)^{-1}$ is bounded in $\mathbb{C}_{\widetilde{M}}$, $\mathbb{C}^-_{\widetilde{M}}$ and $\Omega(-\widetilde{M},\widetilde{M})$, that is,

$$\sup_{z\in\mathbb{C}} \|u_1(l,z)u_2(l,z)^{-1}\| < \infty, \quad \text{i.e.} \quad u_1(l,z)u_2(l,z)^{-1} \equiv \text{const}. \tag{3.136}$$

Relations (3.134)–(3.136) imply that $u_1(l,z) \equiv u_2(l,z)$, which contradicts our assumption. □

Remark 3.31. In Theorem 3.30, differently from Theorems 2.47 and 3.24, where inverse problems for Dirac systems are also dealt with, we don't assume a priori that φ is a GW-function (or Weyl function).

Remark 3.32. Explicit solutions of the direct and inverse problems for the skew-self-adjoint Dirac system are easily constructed similar to the corresponding considerations for the self-adjoint case, see Subsection 2.2.2. See also [130] on the skew-self-adjoint Dirac systems with square potentials.

4 Linear system auxiliary to the nonlinear optics equation

We shall consider the system

$$\frac{dy(x,z)}{dx} = (\mathrm{i}zD - \zeta(x))\, y(x,z), \quad x \geq 0; \tag{4.1}$$

$$D = \operatorname{diag}\{d_1, d_2, \ldots, d_m\}, \quad d_1 > d_2 > \ldots > d_m > 0; \quad \zeta(x) = -\zeta(x)^*, \tag{4.2}$$

where the *potential* $\zeta(x)$ is an $m \times m$ matrix function. System (4.1) is the auxiliary linear system for the well-known nonlinear optics (N-wave) equation

$$\left[D, \frac{\partial \varrho}{\partial t}\right] - \left[\widehat{D}, \frac{\partial \varrho}{\partial x}\right] = [D, \varrho]\,[\widehat{D}, \varrho] - [\widehat{D}, \varrho]\,[D, \varrho],$$

where $[D, \varrho] := D\varrho - \varrho D$, $\widehat{D}$ is another diagonal matrix,

$$\widehat{D} = \widehat{D}^* = \operatorname{diag}\{\widehat{d}_1, \widehat{d}_2, \ldots, \widehat{d}_m\},$$

and $\varrho(x,t)$ is an $m \times m$ matrix function. In case $D = j$, $\zeta = -jV$, system (4.1) turns into the skew-self-adjoint Dirac system (Chapter 3). Various references on the N-wave problem and its auxiliary system can be found in [7, 334], see also Section 7.1 in this book. An extensive amount of research on the scattering problem for system (4.1) with complex-valued entries d_k of D was done by R. Beals, R. R. Coifman, P. Deift and X. Zhou with coauthors (see [36, 40] and references therein). The unique solvability of the problem on a suitable dense set of scattering data, in particular, was obtained. The case (4.2) of the positive D, which is considered here, is less general and therefore a more explicit description of the class of data, for which the solution of the inverse problem exists, and a procedure, which is quite close to the classical one, to solve this inverse problem are given.

In the following sections, we consider system (4.1), where D and ζ satisfy (4.2). Condition $\zeta = -\zeta^*$ (as well as the other reductions of the form (4.113) below) follows from the physical considerations connected with the N–wave problem. Definitions of the Weyl and generalized Weyl (denoted by the acronym GW-) functions are given in Section 4.1 and solutions of the direct and inverse problems follow. Some interrelations between GW-functions and Weyl functions are also discussed in Section 4.1. Conditions on the potential, which imply some required asymptotics of the GW-functions, are studied in Section 4.2. Section 4.3 is dedicated to the explicit solutions of the direct and inverse problems. This chapter's results were obtained in [239, 241, 249, 262].

4.1 Direct and inverse problems

4.1.1 Bounded potentials

We denote the fundamental solution of system (4.1) (satisfying (4.2)) by w, and normalize this solution by $w(0,z) = I_m$. The Weyl functions of system (4.1), (4.2) are considered in the lower half-plane (in order to demonstrate some small changes that are required for the switch from $\mathbb{C}_+$ to $\mathbb{C}_-$).

Definition 4.1. A Weyl function of system (4.1), (4.2) is an analytic matrix function $\varphi(z) = \left\{\varphi_{ij}(z)\right\}_{i,j=1}^{m}$, satisfying for certain $M > 0$ and $r > 0$, and for all z from the domain $\mathbb{C}_M^- = \{z : \Im(z) < -M\}$ the relation

$$\int_0^\infty \exp\{\mathrm{i}\overline{z}xD\}\varphi(z)^* w(x,z)^* w(x,z)\varphi(z) \exp\{-(\mathrm{i}zD + rI_m)x\}\, dx < \infty, \tag{4.3}$$

and the normalization conditions

$$\varphi_{ij}(z) \equiv 1 \quad \text{for} \quad i = j, \quad \varphi_{ij}(z) \equiv 0 \quad \text{for} \quad i > j. \tag{4.4}$$

Theorem 4.2. *Let the inequality*

$$\sup_{0<x<\infty} ||\zeta(x)|| \le M_0 \tag{4.5}$$

hold. Then, there is a Weyl function of system (4.1), (4.2) *and this Weyl function is unique.*

At first, some notations and lemmas are introduced. We set

$$\aleph(x,z) = w(x,z)^* J_k w(x,z), \tag{4.6}$$

where

$$J_k = P_1 - P_2, \quad P_1 = \begin{bmatrix} I_k & 0 \\ 0 & 0 \end{bmatrix}, \quad P_2 = \begin{bmatrix} 0 & 0 \\ 0 & I_{m-k} \end{bmatrix}, \tag{4.7}$$

and k is usually fixed in our considerations of $\aleph$.

Lemma 4.3. *Let the condition* (4.5) *be valid and some* $0 < k < m$ *be fixed. Put* $M_k = 2M_0/(d_k - d_{k+1})$. *Then, we have*

$$\exp\left\{-(\mathrm{i}(z-\overline{z})d_{k+1} + 2M_0)x\right\}\aleph(x,z) \ge J_k \quad (z \in \mathbb{C}_{M_k}^-). \tag{4.8}$$

Proof. Let w satisfy (4.1). In view of (4.1) and (4.5), it is easy to estimate the derivative of $\aleph$

$$\begin{aligned} \frac{d}{dx}\aleph(x,z) &= w(x,z)^*\left(J_k(\mathrm{i}zD - \zeta(x)) - (\mathrm{i}\overline{z}D + \zeta(x)^*)J_k\right)w(x,z) \\ &\ge w(x,z)^*\left(\mathrm{i}(z-\overline{z})DJ_k - 2M_0I_m\right)w(x,z). \end{aligned} \tag{4.9}$$

Fix now $\varepsilon > 0$ and put $\delta = \varepsilon(d_k - d_{k+1})$. Then, taking into account (4.2), for $z \in \mathbb{C}^-_{M_k+\varepsilon}$, we have

$$\begin{aligned} &\mathrm{i}(z-\overline{z})DJ_k - 2M_0 I_m - (\mathrm{i}(z-\overline{z})d_{k+1} + 2M_0 + \delta)J_k \\ &= \mathrm{i}(z-\overline{z})(D - d_{k+1}I_m)J_k - 4M_0P_1 - \delta(P_1 - P_2) \\ &\geq 2(M_k+\varepsilon)(d_k - d_{k+1})P_1 - 4M_0P_1 - \delta(P_1-P_2) = \varepsilon(d_k - d_{k+1})(P_1+P_2). \end{aligned}$$

Since $P_1 + P_2 = I_m$, the formula above yields

$$\mathrm{i}(z-\overline{z})DJ_k - 2M_0 I_m - (\mathrm{i}(z-\overline{z})d_{k+1} + 2M_0 + \delta)J_k \geq \delta I_m. \tag{4.10}$$

The estimate (4.9) and inequality (4.10) infer

$$\begin{aligned} &\frac{d}{dx}\left(\exp\left\{-(\mathrm{i}(z-\overline{z})d_{k+1} + 2M_0 + \delta)x\right\}\aleph(x,z)\right) \\ &\geq \delta \exp\left\{-(\mathrm{i}(z-\overline{z})d_{k+1} + 2M_0 + \delta)x\right\} w(x,z)^* w(x,z) > 0. \end{aligned} \tag{4.11}$$

It is immediate from (4.11) that

$$\exp\left\{-(\mathrm{i}(z-\overline{z})d_{k+1} + 2M_0 + \delta)x\right\}\aleph(x,z) > J_k \quad (x > 0,\ z \in \mathbb{C}^-_{M_k+\varepsilon}). \tag{4.12}$$

If $\varepsilon \to 0$, we have $\delta \to 0$ and (4.12) implies the inequality (4.8). □

Inequalities (4.8) and (4.12) are analogs of the inequality (3.6) from the previous chapter in the sense that they allow one to apply the J-theory technique in our considerations. Since we already used this technique before, the arguments here are slightly less detailed. Our next definition is a direct analog of Definition 1.42.

Definition 4.4. An $m \times (m-k)$ matrix function $\mathcal{Q}_k$, which is meromorphic in $\mathbb{C}^-_{M_k}$ is called nonsingular with property-J_k if the inequalities

$$\mathcal{Q}_k(z)^*\mathcal{Q}_k(z) > 0, \quad \mathcal{Q}_k(z)^* J_k \mathcal{Q}_k(z) \leq 0 \tag{4.13}$$

hold in $\mathbb{C}^-_{M_k}$ (excluding, possibly, isolated points).

Matrix functions $\mathcal{Q}_k$ are used in the proof of the following lemma.

Lemma 4.5. *Let the condition (4.5) be valid. Then, there exist a value $M > 0$ and a matrix function φ that is analytic in $\mathbb{C}^-_M$, such that equalities (4.4) hold, $\sum_{i=1}^m \left|\varphi_{ij}(z)\right|^2 \leq 2$ $(1 \leq j \leq m)$, and for all $l < \infty$, we have*

$$\sup_{x \leq l,\ \Im(z) < -M} \left\|w(x,z)\varphi(z)\exp\{-\mathrm{i}zxD\}\right\| < \infty, \tag{4.14}$$

where w is the normalized fundamental solution of (4.1).

Proof. According to (4.12) and (4.13), the matrix function

$$\psi_k(x,z) = \begin{bmatrix} I_k & 0 \end{bmatrix} w(x,z)^{-1}\mathcal{Q}_k(z)\left(\begin{bmatrix} 0 & I_{m-k} \end{bmatrix} w(x,z)^{-1}\mathcal{Q}_k(z)\right)^{-1} \tag{4.15}$$

is well-defined for $x \geq 0$ and satisfies the inequality

$$\begin{bmatrix} \psi_k(z)^* & I_{m-k} \end{bmatrix} J_k \begin{bmatrix} \psi_k(z) \\ I_{m-k} \end{bmatrix} \leq 0,$$

that is,

$$\psi_k(z)^*\psi_k(z) \leq I_{m-k}. \tag{4.16}$$

Here, we took into account that

$$\begin{bmatrix} \psi_k(z) \\ I_{m-k} \end{bmatrix} = w(x,z)^{-1}\mathcal{Q}_k(z)\left(\begin{bmatrix} 0 & I_{m-k} \end{bmatrix} w(x,z)^{-1}\mathcal{Q}_k(z)\right)^{-1}. \tag{4.17}$$

From (4.13) and (4.17), we see also that

$$\begin{bmatrix} \psi_k(z)^* & I_{m-k} \end{bmatrix} \aleph(x,z) \begin{bmatrix} \psi_k(z) \\ I_{m-k} \end{bmatrix} \leq 0. \tag{4.18}$$

On the other hand, formula (4.18) infers (4.13) and (4.15), where

$$\mathcal{Q}_k(z) = w(x,z)\begin{bmatrix} \psi_k(z) \\ I_{m-k} \end{bmatrix}.$$

That is, the set of Möbius transformations of the form (4.15), where $\mathcal{Q}_k$ are nonsingular, with property-J_k, coincides with the set of matrix functions satisfying inequality (4.18).

Taking into account (4.6) and Corollary E.2, we rewrite (4.12) as

$$\exp\left\{-(\mathrm{i}(z-\overline{z})d_{k+1} + 2M_0 + \delta)x\right\} J_k > \aleph(x,z)^{-1} \quad (x > 0,\ z \in \mathbb{C}^-_{M_k+\varepsilon}). \tag{4.19}$$

We partition $\aleph$ into four blocks, that is, $\aleph = \{\aleph_{ij}\}_{i,j=1}^2$, where $\aleph_{11}$ is a $k \times k$ matrix and $\aleph_{22}$ is an $(m-k) \times (m-k)$ matrix. Like in (2.36), we can express the blocks of $\aleph^{-1}$ via the blocks of $\aleph$. In particular, we have $(\aleph^{-1})_{22} = (\aleph_{22} - \aleph_{12}^*\aleph_{11}^{-1}\aleph_{12})^{-1}$. Now, from (4.12) and (4.19), we see that

$$\aleph_{11} > 0, \quad \aleph_{22} - \aleph_{12}^*\aleph_{11}^{-1}\aleph_{12} < 0. \tag{4.20}$$

Formula (4.18) is equivalent to the relation

$$\left(\psi_k^*\aleph_{11}^{1/2} + \aleph_{12}^*\aleph_{11}^{-1/2}\right)\left(\aleph_{11}^{1/2}\psi_k + \aleph_{11}^{-1/2}\aleph_{12}\right) \leq \aleph_{12}^*\aleph_{11}^{-1}\aleph_{12} - \aleph_{22}.$$

This relation is equivalent to the representation

$$\psi_k(x,z) = \rho_l(x,z)\omega(x,z)\rho_r(x,z) - \aleph_{11}(x,z)^{-1}\aleph_{12}(x,z); \tag{4.21}$$

$$\omega^*\omega \leq I_{m-k}, \quad \rho_l = \aleph_{11}^{-1/2}, \quad \rho_r = \left(\aleph_{12}^*\aleph_{11}^{-1}\aleph_{12} - \aleph_{22}\right)^{1/2}. \tag{4.22}$$

Here, $\omega(x,z)$ are $k \times (m-k)$ matrix functions. The set of matrix values of functions ψ_k at x, z is denoted by $\mathcal{N}_k(x,z)$. As in previous chapters, we have

$$\mathcal{N}_k(x_1,z) \subset \mathcal{N}_k(x_2,z) \quad \text{for } x_1 > x_2. \tag{4.23}$$

Using again Montel's theorem, after taking into account (4.16) and (4.23), we see that there is an analytic function

$$\widetilde{\psi}_k(z) \in \bigcap_{x<\infty} \mathcal{N}_k(x,z) \quad (z \in \mathbb{C}^-_{M_k}). \tag{4.24}$$

From (4.8), we (similar to the derivation of (4.19)) obtain

$$\aleph(x,z)^{-1} \le \exp\left\{-(\mathrm{i}(z-\overline{z})d_{k+1} + 2M_0)x\right\} J_k,$$

which, by taking into account (4.22) and $(\aleph^{-1})_{22} = (\aleph_{22} - \aleph_{12}^* \aleph_{11}^{-1} \aleph_{12})^{-1}$, implies

$$\rho_r(x,z) \le \exp\left\{(1/2)(\mathrm{i}(z-\overline{z})d_{k+1} + 2M_0)x\right\} I_{m-k}. \tag{4.25}$$

From (4.12) and (4.22), we also derive

$$\rho_l(x,z) \le \exp\left\{-(1/2)(\mathrm{i}(z-\overline{z})d_{k+1} + 2M_0 + \delta)x\right\} I_k. \tag{4.26}$$

Relations (4.21), (4.25) and (4.26) infer the uniqueness of $\widetilde{\psi}_k$. Therefore, we change (4.24) in the following way:

$$\widetilde{\psi}_k(z) = \bigcap_{x<\infty} \mathcal{N}_k(x,z) \quad (z \in \mathbb{C}^-_{M_k}). \tag{4.27}$$

Since (4.21) yields (4.18), $\widetilde{\psi}_k$ satisfies the inequality

$$\begin{bmatrix} \widetilde{\psi}_k(z)^* & I_{m-k} \end{bmatrix} \aleph(x,z) \begin{bmatrix} \widetilde{\psi}_k(z) \\ I_{m-k} \end{bmatrix} \le 0 \quad (0 < x < \infty). \tag{4.28}$$

From (4.1), (4.5) and (4.28), we obtain $f_x(x,z) \le 2M_0 f(x,z)$ for the function

$$f(x,z) = \exp\{-\mathrm{i}(z-\overline{z})d_{k+1}x\} \begin{bmatrix} \widetilde{\psi}_k(z)^* & I_{m-k} \end{bmatrix} w(x,z)^* P_2 w(x,z) \begin{bmatrix} \widetilde{\psi}_k(z) \\ I_{m-k} \end{bmatrix},$$

where P_2 is given in (4.7). Hence, $f(x,z) \le \exp\{2M_0 x I_{m-k}\}$, and, in view of (4.28), we conclude

$$\sup_{\Im(z)<-M_k,\, x \le l} \left\| \exp\{-\mathrm{i}zd_{k+1}x\} w(x,z) \begin{bmatrix} \widetilde{\psi}_k(z) \\ I_{m-k} \end{bmatrix} \right\| < \infty. \tag{4.29}$$

We introduce column vectors $e_j(k)$ and $\{\varphi_{i,k+1}(z)\}_{i=1}^k$ by the equalities

$$\{\varphi_{i,k+1}(z)\}_{i=1}^k = \widetilde{\psi}_k(z) e_1(m-k), \quad e_j(k) := \{\delta_{ij}\}_{i=1}^k. \tag{4.30}$$

Fixing any $\varepsilon > 0$ and setting $M = \max_{1 \le k \le m} (M_k + \varepsilon_0)$, from (4.16) and (4.29), we see that the matrix function φ, the entries of which are defined by (4.4) and (4.30), is the function we are looking for. □

Proof of Theorem 4.2. In view of (4.28) and (4.30), we have

$$e_{k+1}(m)^*\varphi(z)^*\aleph(x,z)\varphi(z)e_{k+1}(m) \le 0. \tag{4.31}$$

Setting

$$r = 2M + \delta, \quad \delta = \varepsilon_0 \min_{1\le k\le m}(d_k - d_{k+1}),$$

from (4.11) and (4.31), we derive

$$\int_0^\infty \exp\{-(\mathrm{i}(z-\overline{z})d_{k+1} + r)x\}\, e_{k+1}(m)^*\varphi(z)^* w(x,z)^* w(x,z)\varphi(z)e_{k+1}(m)dx$$
$$\le -\delta^{-1}e_{k+1}(m)^*\varphi(z)^* J_k\varphi(z)e_{k+1}(m) < \infty \quad (\Im(z) < -M). \tag{4.32}$$

Thus, the existence of a Weyl function is proved. From (4.8), it follows that for any M and r, there is a value $z \in \mathbb{C}_M^-$ such that (for all $0 < l < \infty$) the inequality

$$\int_0^l \exp\{-(\mathrm{i}(z-\overline{z})d_{k+1} + r)x\} \begin{bmatrix} I_k & 0 \end{bmatrix} w(x,z)^* w(x,z) \begin{bmatrix} I_k \\ 0 \end{bmatrix} dx \ge l\, I_k \tag{4.33}$$

is valid. In view of (4.33), we derive the uniqueness of $\varphi(z)$, column by column. □

Remark 4.6. Using easy algebraic transformations and induction on k, from (4.15), (4.27) and (4.28), we obtain

$$\widetilde{\psi}_k(z) = \begin{bmatrix} I_k & 0 \end{bmatrix}\varphi(z)\begin{bmatrix} 0 \\ I_{m-k} \end{bmatrix}\left(\begin{bmatrix} 0 & I_{m-k} \end{bmatrix}\varphi(z)\begin{bmatrix} 0 \\ I_{m-k} \end{bmatrix}\right)^{-1}. \tag{4.34}$$

It is apparent from the proof of Theorem 4.2 that the Weyl function $\varphi(z)$ satisfies (4.14).

4.1.2 Locally bounded potentials

Following similar considerations from Section 3.3, we will also study such a system (4.1), where the potential ζ is only locally bounded.

Definition 4.7. A generalized Weyl function (GW-function) of system (4.1), (4.2), where inequalities

$$\sup_{0<x<l} \|\zeta(x)\| < \infty \tag{4.35}$$

hold for each $l < \infty$, is an $m \times m$ matrix function φ such that for some $M > 0$, it is analytic in $\mathbb{C}_M^-$ and the inequalities (4.14) also hold for each $l < \infty$.

Recall that D is fixed and given by (4.2). The inverse spectral problem (ISpP) for system (4.1) is the problem to recover (from a GW-function φ) a potential $\zeta(x) = -\zeta^*(x)$ ($\zeta_{kk} \equiv 0$), which satisfies (4.35). The notation $\mathfrak{W}$ stands for the operator mapping the pair D and $\varphi(z)$ into ζ (i.e. $\mathfrak{W}(D,\varphi) = \zeta$).

We no longer assume that $\varphi(z)$ is *normalized* via (4.4).

Theorem 4.8. *For any matrix function $\varphi(z)$ which is analytic and bounded in $\mathbb{C}_M^-$ and has the property*

$$\int_{-\infty}^{\infty} (\varphi(z) - I_m)^* \, (\varphi(z) - I_m) \, d\xi < \infty \quad (z = \xi - \mathrm{i}\eta, \quad \eta > M), \tag{4.36}$$

there is at most one solution of the ISpP.

In order to prove the theorem above, we fix some $l > 0$ and construct the representation (4.37) of w below, which is a direct analog of representation (3.58) in Chapter 3 (and so the construction of the representation (4.37) is closely related to the construction of the transformation operators in inverse problems).

Lemma 4.9. *Relations* (4.1) *and* (4.35) *yield the representation*

$$w(x,z) = \exp\{\mathrm{i}zxD\} + \int_{ax}^{bx} \exp\{\mathrm{i}zt\} N(x,t)\, dt, \tag{4.37}$$

where $a = d_m$, $b = d_1$, $\sup \|N(x,t)\| < \infty$ $(x \le l)$.

Proof. The "wave" function w admits a representation in the form

$$w(x,z) = \sum_{k=0}^{\infty} \nu_k(x,z), \tag{4.38}$$

where

$$\begin{aligned} \nu_0(x,z) &= \exp\{\mathrm{i}zxD\}, \\ \nu_k(x,z) &= -\int_0^x \exp\{\mathrm{i}z(x-t)D\}\zeta(t)\,\nu_{k-1}(t,z)\,dt \quad (k > 0). \end{aligned} \tag{4.39}$$

In the case $k = 1$, we have

$$\nu_1(x,z) = \int_{ax}^{bx} \exp\{\mathrm{i}zt\} N_1(x,t)\, dt, \quad N_1 = \left\{N_{1ij}\right\}_{i,j=1}^{m}, \tag{4.40}$$

$$N_{1ij}(x,t) = \frac{1}{d_j - d_i}\, \zeta_{ij}\left(\frac{d_i x - t}{d_i - d_j}\right) \begin{cases} 1 & \text{for } j > i, \quad d_j x < t \le d_i x, \\ -1 & \text{for } j < i, \quad d_i x < t \le d_j x, \\ 0 & \text{otherwise.} \end{cases} \tag{4.41}$$

Using induction, we show that

$$\nu_k(x,z) = \int_{ax}^{bx} \exp\{izt\} N_k(x,t)\,dt \quad (k \ge 1), \quad N_k = \left\{N_{kij}\right\}_{i,j=1}^m, \tag{4.42}$$

$$N_{kij}(x,t) = -\int_{\chi}^{x} \sum_{r=1}^{m} \zeta_{ir}(v)\, N_{k-1,r,j}\,(v, d_i(v-x)+t)\,dv \quad (k \ge 2), \tag{4.43}$$

where

$$\chi = (d_i x - t)(d_i - a)^{-1} \quad (t \le d_i x), \quad \chi = (d_i x - t)(d_i - b)^{-1} \quad (t \ge d_i x). \tag{4.44}$$

Here, $N_k(x,t)$ is defined for $ax < t < bx$, which excludes the cases $d_i = a$ and $d_i = b$ in (4.44). From (4.41) and (4.43), we conclude $\|N_k(x,t)\| \le C^k x^{k-1}/(k-1)!$. Hence, in view of (4.38), (4.39) and (4.42), the statement of the lemma follows. □

Proof of Theorem 4.8. We prove the theorem by contradiction. Suppose there exist two solutions of the ISpP, namely, $\zeta_1, \zeta_2 \in \mathfrak{M}(D,\varphi)$ ($\zeta_1 \ne \zeta_2$ on a set of positive measure), and let w_i ($i = 1,2$) denote normalized fundamental solutions of the nonlinear optics equations with the potentials ζ_1, ζ_2, respectively. Lemma 4.9 and formulas (4.14) and (4.36) yield (see Theorem E.11) representations

$$w_i(x,z)\varphi(z)\exp\{-izxD\} = I_m + \int_0^\infty \exp\{-izt\}\mathfrak{N}_k(x,t)\,dt, \tag{4.45}$$

$$\text{where} \quad z \in \mathbb{C}^-_{M+\varepsilon}, \quad \int_0^\infty \exp\{-2(M+\varepsilon)t\}\mathfrak{N}_i(x,t)^*\mathfrak{N}_i(x,t)dt < \infty$$

for any $\varepsilon > 0$. From (4.45), we see that there is $\widehat{M} > M$ such that

$$\sup_{z\in\mathbb{C}^-_{\widehat{M}}} \left\| w_i(x,z) w_j(x,z)^{-1} \right\| < \infty \quad (i \ne j; \quad i = 1,2; \quad j = 1,2). \tag{4.46}$$

Taking into account (4.1) and (4.2), we derive

$$w(x,\overline{z})^* = w(x,z)^{-1}, \tag{4.47}$$

and therefore $w_1(x,\overline{z})w_2(x,\overline{z})^{-1} = \left(w_2(x,z)w_1(x,z)^{-1}\right)^*$. Using (4.46), we now obtain the boundedness of $w_1 w_2^{-1}$ in both half-planes $\Im(z) \ge \widehat{M}$ and $\Im(z) \le -\widehat{M}$:

$$\sup \left\| w_1(x,z)\, w_2^{-1}(x,z) \right\| < \infty \quad \left(|\Im(z)| \ge \widehat{M}\right). \tag{4.48}$$

In view of (4.1), (4.35) and (4.47), there exist values $c_1 > 0$ and $c_2 > 0$ such that $\sup \|w_1(x,z)\, w_2(x,z)^{-1}\| \le c_1 e^{c_2|z|}$ for all $z \in \mathbb{C}$. Thus, the Phragmen–Lindelöf theorem may be applied (see Corollary E.8) to show the boundedness of $w_1 w_2^{-1}$ in the strip $|\Im(z)| < \widehat{M}$. Since $w_1 w_2^{-1}$ is bounded in the domains $|\Im(z)| < \widehat{M}$ and $|\Im(z)| \geq \widehat{M}$, it is bounded in $\mathbb{C}$. Therefore, according to the first Liouville theorem (Theorem E.5), the equality

$$w_1(x,z)w_2(x,z)^{-1} \equiv \text{const} \tag{4.49}$$

is valid. From Lemma 4.9, it follows that

$$\lim_{\xi \to \infty} w_i(x, \xi - \mathrm{i}\eta) \exp\{-\mathrm{i}(\xi - \mathrm{i}\eta)xD\} = I_m. \tag{4.50}$$

The relations (4.49) and (4.50) mean that $w_1 \equiv w_2$, which contradicts $\zeta_1 \neq \zeta_2$. □

In order to derive the existence of the solutions of ISpP, we need somewhat stronger restrictions, namely,

$$\sup \| z(\varphi(z) - I_m) \| < \infty \quad (\Im(z) < -M); \tag{4.51}$$

$$z\left(\varphi(z) - I_m - \phi_0/z\right) \in L^2_{m\times m}(-\infty,\infty) \quad (z = \xi - \mathrm{i}\eta,\ -\infty < \xi < \infty) \tag{4.52}$$

for some matrix ϕ_0 and all fixed values of $\eta > M$. Without loss of generality, we assume

$$\det \varphi(z) \neq 0. \tag{4.53}$$

Our next theorem is an analog of Theorem 3.30, where an inverse problem for a skew-self-adjoint Dirac system was dealt with.

Theorem 4.10. *Let the analytic matrix function φ satisfy conditions* (4.51)–(4.53). *Then, a solution of the ISpP exists and is unique.*

Proof. Step 1. The uniqueness follows from Theorem 4.8. The procedure of recovering the solution is started by introducing the matrix function

$$\Pi(x) = \frac{1}{2\pi\mathrm{i}} \exp\{\eta x D\} \mathrm{l.i.m.}_{a\to\infty} \int_{-a}^{a} \exp\{\mathrm{i}\xi x D\} \frac{\varphi(\xi - \mathrm{i}\eta)^{-1}}{\xi - \mathrm{i}\eta} d\xi, \tag{4.54}$$

where $\eta > M$ and the norm limit in (4.54) is taken in $L^2_{m\times m}(0,l)$ or $L^2_{m\times m}(0,\infty)$. (Compare (4.54) with formulas (2.132) and (3.67), where Φ_1 is constructed.)

The right-hand side of (4.54) is well-defined since from (4.51)–(4.53), we obtain

$$\sup \left\| z(\varphi(z)^{-1} - I_m)\right\| < \infty \quad (\Im(z) < -M), \tag{4.55}$$

$$z\left(\varphi(z)^{-1} - I_m + \phi_0/z\right) \in L^2_{m\times m}(-\infty,\infty) \quad (z = \xi - \mathrm{i}\eta,\ \eta > M). \tag{4.56}$$

It is apparent that for $\eta > M$, we have

$$\frac{1}{2\pi\mathrm{i}} \exp\{\eta x D\} \mathrm{l.i.m.}_{a\to\infty} \int_{-a}^{a} \exp\{\mathrm{i}\xi x D\} \frac{d\xi}{\xi - \mathrm{i}\eta} \equiv I_m \quad (x \geq 0). \tag{4.57}$$

In view of (4.55) and (4.57), we rewrite (4.54) as the pointwise equality

$$\Pi(x) = I_m + \frac{1}{2\pi i} \int_{-\infty}^{\infty} \exp\{izxD\} \left(\varphi(z)^{-1} - I_m\right) \frac{d\xi}{z}, \quad z = \xi - i\eta \tag{4.58}$$

for $x \geq 0$, where $\Pi(x)$ does not depend on $\eta > M$. According to (4.55), (4.56) and (4.58), $\Pi(x)$ is twice differentiable and has the following properties:

$$\Pi(0) = I_m, \quad \Pi'(0) = -iD\phi_0; \tag{4.59}$$

$$\exp\{-xMD\}\Pi'(x), \ \exp\{-xMD\}\Pi''(x) \in L^2_{m\times m}(0, \infty). \tag{4.60}$$

Step 2. An operator $S = S_l \in B(L^2_m(0, l))$ is introduced via the operator identity

$$AS - SA^* = i\Pi\Pi^*, \tag{4.61}$$

where the operators $A = A_l \in B(L^2_m(0, l))$ and $\Pi = \Pi_l \in B(\mathbb{C}^m, L^2_m(0, l))$, respectively, are determined by the relations

$$A = iD \int_0^x \cdot \, dt, \qquad (\Pi g)(x) = \Pi(x)g, \tag{4.62}$$

and the index "l" in the notations of operators is sometimes omitted. Using Corollary D.12, from (4.61), we derive

$$S_l f = D^{-1} f + \int_0^l s(x,t) f(t)\, dt, \quad s(x,t) = \left\{s_{ij}(x,t)\right\}_{i,j=1}^m, \tag{4.63}$$

$$s_{ij}(x,t) = \int_\chi^x \breve{\Upsilon}_{ij}(\xi, t + d_i d_j^{-1}(\xi - x))\, d\xi \tag{4.64}$$

$$+ \begin{cases} d_i^{-1}\Pi'_{ij}(x - d_j d_i^{-1} t) & \text{for} \quad t \leq d_i d_j^{-1} x, \\ d_j^{-1}\overline{\Pi}'_{ji}(t - d_i d_j^{-1} x) & \text{for} \quad d_i d_j^{-1} x < t; \end{cases}$$

$$\breve{\Upsilon}(x,t) = \left\{\breve{\Upsilon}_{ij}(x,t)\right\}_{i,j=1}^m = \Pi'(x)\Pi'(t)^* D^{-1}, \quad \chi = \max\left(0, x - d_j d_i^{-1} t\right).$$

If we substitute S_l for S_ε into the proof of Theorem D.5 and precisely follow this proof, we see that relations (4.61)–(4.63) imply Ker $S_l = 0$ and $S_l \geq 0$. Therefore, again using (4.63), we obtain

$$S_l \geq \varepsilon(l) I > 0 \tag{4.65}$$

for some $\varepsilon(l) > 0$. As in Chapter 3, the triple $\{A, S, \Pi\}$ and the family of orthoprojectors $\{P_\xi\}$ given by (2.105) generate a family of S-nodes, such that

$$A_\xi = P_\xi A P_\xi^*, \quad S_\xi = P_\xi S P_\xi^* \quad (0 < \xi \leq l); \quad A_\xi S_\xi - S_\xi A_\xi^* = iP_\xi \Pi\Pi^* P_\xi^*, \tag{4.66}$$

and the corresponding transfer matrix function (depending on ξ and λ)

$$w_A(\xi,\lambda) = I_m - \mathrm{i}\Pi^* S_\xi^{-1}(A_\xi - \lambda I)^{-1} P_\xi \Pi, \quad 0 < \xi \le l. \tag{4.67}$$

Different signs in (3.46) and (4.67) are explained by the difference of signs in the right-hand sides of the operator identities (3.41) and (4.61).

From (4.63)–(4.65), we see that $D^{-1/2}SD^{-1/2}$ satisfies conditions of Corollary 1.40, and so S admits the factorization

$$S^{-1} = E^*E, \quad E = D^{1/2} + \int_0^x E(x,t)\,\cdot\,dt, \quad \int_0^l\int_0^x E^*(x,t)E(x,t)\,dt\,dx < \infty. \tag{4.68}$$

In view of (4.68), the equality (1.154) is valid and the matrix function $\Pi^* P_\xi^* S_\xi^{-1} P_\xi \Pi$ is absolutely continuous. Furthermore, using (1.153) and (1.154) as in Theorem 1.34, we obtain

$$H(\xi) := \frac{d}{d\xi}\left(\Pi^* P_\xi^* S_\xi^{-1} P_\xi \Pi\right) = \gamma(\xi)^*\gamma(\xi), \quad \gamma(\xi) := (E\Pi)(\xi). \tag{4.69}$$

From Theorem 1.20, we derive

$$\frac{d}{d\xi}W(\xi,z) = \mathrm{i}zH(\xi)W(\xi,z), \quad W(\xi,z) := w_A(\xi,1/z), \tag{4.70}$$

where H is given by (4.69) and w_A is given by (4.67).

Step 3. Next, we show that $\gamma(x)$ is boundedly differentiable on $[0,\ l]$ and satisfies the equality

$$\gamma(x)\gamma(x)^* \equiv D. \tag{4.71}$$

Indeed, taking into account (4.61), (4.68) and the last equality in (4.69), we obtain

$$EAE^{-1} - (E^*)^{-1}A^*E^* = \mathrm{i}\gamma(x)\int_0^l \gamma(t)^*\,\cdot\,dt,$$

that is, the triangular operator EAE^{-1} has the form

$$EAE^{-1} = \mathrm{i}\gamma(x)\int_0^x \gamma(t)^*\,\cdot\,dt. \tag{4.72}$$

Rewriting (4.72) as $AE^{-1} = \mathrm{i}E^{-1}\gamma(x)\int_0^x \gamma(t)^*\,\cdot\,dt$ and using representation

$$E^{-1} = D^{-1/2} + \int_0^x \Gamma(x,t)\,\cdot\,dt, \tag{4.73}$$

for $x \ge t$, we derive

$$D^{1/2} + D\int_t^x \Gamma(\xi,t)\,d\xi = D^{-1/2}\gamma(x)\,\gamma(t)^* + \int_t^x \Gamma(x,\xi)\gamma(\xi)\,d\xi\,\gamma(t)^*. \tag{4.74}$$

Since $E^{-1} = SE^*$, we can put

$$E(t,x)^* = -S_t^{-1}s(x,t)D^{1/2} \quad (x \leq t), \tag{4.75}$$

$$\Gamma(x,t) = s(x,t)D^{1/2} + \int_0^t s(x,\xi)\,E(t,\xi)^*d\xi \quad (x \geq t). \tag{4.76}$$

Hence, $y(x)$ is continuous in x and the matrix functions

$$\int_t^x \Gamma(x,\xi)y(\xi)d\xi \quad \text{and} \quad \int_t^x \Gamma(\xi,t)d\xi$$

are continuous in x and t. In this way, we see that (4.74) is valid pointwise. In particular, when $x = t$, we obtain (4.71). Since E^{-1} has the form (4.73) and $(E^{-1}y)(x) = \Pi(x)$, formula (4.74) can be rewritten in the form

$$D^{1/2} + D\int_t^x \Gamma(\xi,t)\,d\xi = \Pi(x)y(t)^* - \int_0^t \Gamma(x,\xi)y(\xi)\,d\xi\,y(t)^*. \tag{4.77}$$

Multiplying both sides of (4.77) (from the right) by $D^{-1}y(t)$, in view of (4.71), we have

$$D^{-1/2}y(t) = \Pi(x) - \int_0^t \Gamma(x,\xi)y(\xi)d\xi - D\int_t^x \Gamma(\xi,t)d\xi D^{-1}y(t). \tag{4.78}$$

Differentiating both sides of (4.78) with respect to t and putting $t = x$, we obtain the result

$$y'(x) = D^{1/2}\left(D\Gamma(x,x)D^{-1} - \Gamma(x,x)\right)y(x). \tag{4.79}$$

From (4.75) and (4.76), we see that $\Gamma(x,x)$ is bounded on finite intervals. Therefore, (4.79) yields $y(x) \in B^1([0,\,l])$. We set

$$w(x,z) = D^{-1/2}y(x)w_A(x,1/z). \tag{4.80}$$

According to (4.69)–(4.71), the matrix function w constructed above satisfies (4.1), where

$$\zeta(x) = -\tilde{y}'(x)\,\tilde{y}(x)^*, \quad \tilde{y}(x) := D^{-1/2}y(x). \tag{4.81}$$

Moreover, w is the normalized fundamental solution of (4.1) since

$$w(0,z) = D^{-1/2}y(0) = \Pi(0) = I_m. \tag{4.82}$$

Step 4. Let us show that the potential ζ from (4.81) is the solution of ISpP. Formula (1.157) for the resolvent of A given by (1.149) is easily transformed for the case, where A is given in (4.62), that is,

$$((I - zA)^{-1}\Pi)(x) = \Pi(x) + izD\int_0^x \exp\{iz(x-t)D\}\Pi(t)dt. \tag{4.83}$$

In view of (4.56) and (4.60), taking the transform, which is inverse to (4.54), we have

$$\mathrm{i}zD\int_0^\infty \exp\{-\mathrm{i}ztD\}\Pi(t)dt = \varphi(z)^{-1} \quad (\Im(z) < -M). \tag{4.84}$$

Taking into account (4.59), (4.83) and (4.84), we conclude

$$\begin{aligned}\left((I - zA)^{-1}\Pi\right)(x) \quad & \quad (4.85)\\ = \Pi(x) + \exp\{\mathrm{i}zxD\}\left(\varphi(z)^{-1} - \mathrm{i}zD\int_x^\infty \exp\{-\mathrm{i}ztD\}\Pi(t)\,dt\right)\\ = \exp\{\mathrm{i}zxD\}\varphi(z)^{-1} + (\mathrm{i}/z)D^{-1}\\ \times\left(\Pi'(x) + \int_x^\infty \exp\{\mathrm{i}z(x-t)D\}\Pi''(t)\,dt\right).\end{aligned}$$

It is apparent from (4.85) that

$$\left((I - zA)^{-1}\Pi\right)(x)\varphi(z)\exp\{-\mathrm{i}zlD\} = \exp\{\mathrm{i}z(x-l)D\} + f(z,x,l), \tag{4.86}$$

where (for all $l_0 < \infty$ and $\varepsilon > 0$) the following estimate on f is valid:

$$\sup\|zf(z,x,l)\| < \infty \quad (\Im(z) \le -M - \varepsilon, \quad x \le l \le l_0). \tag{4.87}$$

The transfer matrix function w_A given by (4.67) corresponds to the case $J = I_m$. Hence, formula (1.88) yields

$$w_A(l,1/z)^* w_A(l,1/z) = I_m + \mathrm{i}(z-\overline{z})\Pi^*(I - \overline{z}A^*)^{-1}S_l^{-1}(I - zA)^{-1}\Pi. \tag{4.88}$$

Finally, from (4.71), (4.80) and (4.86)–(4.88), we derive

$$\sup_{\Im(z)<-M-\varepsilon,\ l\le l_0}\|w(l,z)\,\varphi(z)\exp\{-\mathrm{i}zlD\}\| < \infty. \tag{4.89}$$

Substituting in (4.89) M for $M+\varepsilon$, x for l and l for l_0, we obtain (4.14). Thus, φ is a GW-function of the constructed system.

It remains to add that by virtue of (4.71), (4.79) and (4.81), we have

$$\zeta(l) = \left(\Gamma(l,l) - D\Gamma(l,l)D^{-1}\right)D^{1/2}, \tag{4.90}$$

and so $\zeta_{kk} = 0$. □

Remark 4.11. We already mentioned in the proof above that relations (4.75) and (4.76) yield the boundedness of $\Gamma(l,l)$. More detailed calculations which use these relations and the continuous differentiability of Π show that $\Gamma(l,l)$ is continuous, that is, ζ is continuous and $\gamma \in C^1([0,\,l])$. From (4.68), (4.75) and (4.76), we also see that

$$\Gamma(l,l) = D^{-1}\left(S_l^{-1}s(x,l)\right)(l)D^{1/2}, \tag{4.91}$$

$$\gamma(l) = (E\Pi)(l) = D^{1/2}\left(\Pi(l) - \int_0^l \left(S_l^{-1}s(t,l)\right)^*\Pi(t)dt\right). \tag{4.92}$$

Remark 4.12. Under the conditions of Theorem 4.10, equations (4.58), (4.63), (4.64), (4.90) and (4.91) now uniquely define the mapping $\mathfrak{M}(D, \varphi)$ and produce a procedure for solving ISpP.

Example 4.13. Let $\varphi(z) \equiv I_m$. Then, according to (4.58) and (4.64), we have $\Pi(x) \equiv I_m$ and $s(x,t) \equiv 0$. Therefore, from (4.90) and (4.91), we see that the potential ζ is trivial, that is, $\zeta \equiv 0$.

Example 4.14. Let $m = 2$,

$$\varphi(z) = \left(I_2 - \frac{\phi_0}{z}\right)^{-1}, \quad \phi_0 := \begin{bmatrix} 0 & -1 \\ 1 & 0 \end{bmatrix}. \tag{4.93}$$

It follows from (4.58) that $\Pi(x) = I_2 - \mathrm{i}xD\phi_0$. Thus, we have $\Pi'(x) \equiv -\mathrm{i}D\phi_0$ and $\breve{Y}(x,t) \equiv D\phi_0\phi_0^* = D$. Therefore, formula (4.64) implies

$$s_{ii}(x,t) = d_k t \text{ for } t \le x, \quad s_{ii}(x,t) = d_k x \text{ for } t \ge x \quad (i = 1,2), \tag{4.94}$$

$$s_{12}(x,t) = \mathrm{i}, \quad s_{21}(x,t) = -\mathrm{i}, \quad s(x,l) = xD - \mathrm{i}\phi_0. \tag{4.95}$$

In view of (4.63), (4.94) and (4.95), we obtain

$$Sf = D^{-1}f + D\left(\int_0^x tf(t)dt + x\int_x^l f(t)dt\right) - \mathrm{i}\phi_0\int_0^l f(t)dt. \tag{4.96}$$

Putting $h(x) = s(x,l) = xD - \mathrm{i}\phi_0$, we construct $f = S^{-1}h$. Notice, that by virtue of formulas $h(x) = xD - \mathrm{i}\phi_0$ and (4.96), we have

$$\frac{d^2}{dx^2}h = 0 = \frac{d^2}{dx^2}D^{-1}f - Df, \quad h(0) = -\mathrm{i}\phi_0 = D^{-1}f(0) - \mathrm{i}\phi_0\int_0^l f(t)dt, \tag{4.97}$$

$$h'(0) = D = D^{-1}f'(0) + D\int_0^l f(t)dt. \tag{4.98}$$

According to (4.97), the equality $f(x) = \mathrm{e}^{xD}\phi^+ + \mathrm{e}^{-xD}\phi^-$, is valid, where $\phi^\pm$ are constant matrices. Now, it follows from (4.98) that

$$\mathrm{e}^{lD}\phi^+ - \mathrm{e}^{-lD}\phi^- = D. \tag{4.99}$$

From the second equality in (4.97) and formula (4.99), we derive

$$D^{-1}(\phi^+ + \phi^-) + \mathrm{i}\phi_0 D^{-1}(\phi^+ - \phi^-) = 0. \tag{4.100}$$

After some easy calculations, formulas (4.99) and (4.100) yield

$$\phi^+ = c_1(l)D\left(\mathrm{e}^{lD} - \mathrm{i}\phi_0\mathrm{e}^{lD}\right), \quad c_1(l) := \left(\mathrm{e}^{2d_1 l} + \mathrm{e}^{2d_2 l}\right)^{-1}, \tag{4.101}$$

$$\phi^- = D\left(c_1(l)\mathrm{e}^{3lD} - \mathrm{e}^{lD} - \mathrm{i}c_2(l)\mathrm{e}^{lD}\phi_0\right), \quad c_2(l) := c_1(l)\mathrm{e}^{d_1 l + d_2 l}. \tag{4.102}$$

In view of (4.101) and (4.102), the off-diagonal part $f(l)^{od}$ of $f(l)$ has the form

$$f(l)^{od} = \left(\mathrm{e}^{lD}\phi^+ + \mathrm{e}^{-lD}\phi^-\right)^{od} = -2\mathrm{i}c_2(l)D\phi_0.$$

So, recalling that $f(l) = \left(S^{-1}s(x,l)\right)(l)$ and using (4.90) and (4.91), we obtain

$$\zeta(l) = D^{-1}f(l)D - f(l) = -2\mathrm{i}(d_1 - d_2)c_2(l)\begin{bmatrix} 0 & 1 \\ 1 & 0 \end{bmatrix}, \quad \zeta(l) = -\zeta(l)^*, \tag{4.103}$$

where $c_2(l)$ is defined in (4.101) and (4.102).

Further, in Theorem 4.23, which is based on the GBDT transformation, we give a simpler procedure to recover potentials directly from the rational Weyl functions. However, the way to obtain explicit solutions using general results on inverse problems and inversion of semiseparable operators (like in the example above) is sometimes also fruitful (see Subsection D.4 and papers [19, 113, 270]).

4.1.3 Weyl functions

In Subsection 4.1.1, we have shown that, under condition (4.5), the Weyl function is also a GW-function. Now, we show that (4.3) follows from (4.5) and (4.14).

Theorem 4.15. *Let the potential ζ of system (4.1), (4.2) satisfy (4.5). Assume that a GW-function $\varphi(z)$ of this system is normalized (i.e. (4.4) holds). Then, $\varphi(z)$ is not only a GW-function, but also the Weyl function of the system, that is, (4.3) is valid.*

Proof. First, we note that, for each x, there is $M(x) > 0$ such that the inequality

$$e_{k+1}^*\varphi(z)^*\aleph(x,z)\,\varphi(z)e_{k+1} \le 0, \tag{4.104}$$

where $\aleph$ and $e_{k+1} = e_{k+1}(m)$ are defined in (4.6) and (4.30), respectively, is valid in $\mathbb{C}^-_{M(x)}$. Indeed, if (4.104) is not fulfilled for some x_0, there is a sequence $\{z_i\}$ ($\Im(z_i) \to -\infty$) such that the inequalities

$$e_{k+1}^*\varphi(z_i)^*\aleph(x_0,z_i)\,\varphi(z_i)e_{k+1} > 0$$

hold. Hence, from (4.11), we see that

$$e_{k+1}^*\varphi(z_i)^*\aleph(x,z_i)\,\varphi(z_i)e_{k+1} > 0 \quad \text{for} \quad x \ge x_0. \tag{4.105}$$

Inequality $DP_1 \ge d_k P_1$, where $P_1 = (I_m + J_k)/2$, and relations (4.1) and (4.105) imply that for $M_0 = \sup\|\zeta(x)\|$, we have

$$\begin{aligned}
&\left(\exp\{\mathrm{i}(\overline{z_i}-z_i)d_{k+1}x + 4M_0x\}e_{k+1}^*\varphi(z_i)^*w(x,z_i)^*(I_m+J_k)w(x,z_i)\varphi(z_i)e_{k+1}\right)' \\
&\ge \mathrm{i}(z_i - \overline{z_i})(d_k - d_{k+1})\exp\{\mathrm{i}(\overline{z_i} - z_i)d_{k+1}x + 4M_0x\} \\
&\quad\times e_{k+1}^*\varphi(z_i)^*w(x,z_i)^*w(x,z_i)\varphi(z_i)e_{k+1}.
\end{aligned} \tag{4.106}$$

From (4.1) and (4.2), we also obtain

$$\begin{aligned}&\frac{d}{dx}\Big(\exp\{\mathrm{i}(\overline{z}-z)d_{k+1}x+2M_0x\}w(x,z)^*J_{k+1}w(x,z)\Big)\\&=\exp\{\mathrm{i}(\overline{z}-z)d_{k+1}x+2M_0x\}w(x,z)^* \qquad (4.107)\\&\quad\times\left(\mathrm{i}(z-\overline{z})(DJ_{k+1}-d_{k+1}J_{k+1})+2M_0J_{k+1}+\zeta J_{k+1}-J_{k+1}\zeta\right)w(x,z).\end{aligned}$$

We recall that $J_{k+1} = \operatorname{diag}\{I_{k+1}, -I_{m-k-1}\}$. It is apparent that the corresponding upper diagonal block of

$$\mathrm{i}(z-\overline{z})(DJ_{k+1}-d_{k+1}J_{k+1})+2M_0J_{k+1}+\zeta J_{k+1}-J_{k+1}\zeta$$

is positive and nondecreasing, whereas the lower diagonal block tends to infinity when $\Im(z)\to-\infty$. Hence, (4.107) implies that

$$\frac{d}{dx}\Big(\exp\{\mathrm{i}(\overline{z}-z)d_{k+1}x+2M_0x\}w(x,z)^*J_{k+1}w(x,z)\Big)>0 \qquad (4.108)$$

for $z\in\mathbb{C}^-_{\widetilde{M}}$ and sufficiently large $\widetilde{M}$. Since the normalization (4.4) yields

$$e_{k+1}^*\varphi(z_i)^*J_{k+1}\varphi(z_i)e_{k+1}\geq 1,$$

by taking into account (4.106) and (4.108), we derive

$$\begin{aligned}&\Big(\exp\{\mathrm{i}(\overline{z_i}-z_i)d_{k+1}x+4M_0x\}e_{k+1}^*\varphi(z_i)^*w(x,z_i)^*(I_m+J_k)w(x,z_i)\varphi(z_i)e_{k+1}\Big)'\\&\geq\mathrm{i}(z_i-\overline{z_i})(d_k-d_{k+1})\exp\{2M_0x\}.\end{aligned}$$

Therefore, we easily conclude

$$\begin{aligned}&\exp\{\mathrm{i}(\overline{z_i}-z_i)d_{k+1}x+4M_0x\}e_{k+1}^*\varphi(z_i)^*w(x,z_i)^*(I_m+J_k)w(x,z_i)\varphi(z_i)e_{k+1}\\&\geq\mathrm{i}(z_i-\overline{z_i})(d_k-d_{k+1})\left(\exp\{2M_0x\}-\exp\{2M_0x_0\}\right)/(2M_0). \qquad (4.109)\end{aligned}$$

The right-hand side of (4.109) tends to infinity when $\mathrm{i}\to\infty$, and so the left-hand side of (4.109) also tends to infinity, which contradicts (4.14). Hence, (4.104) is valid. Without loss of generality, we assume that $M(x)$ is increasing.

Because of (4.104), the vector function $\varphi(z)e_{k+1}$ can be considered as the first column of some matrix function $\left[\begin{smallmatrix}\psi_k(z,\omega_1)\\ I_{m-k}\end{smallmatrix}\right]$ satisfying (4.18). According to Theorem 4.2, there exists the Weyl function of system (4.1) which we denote by $\widehat{\varphi}(z)$. In view of the construction of $\widehat{\varphi}$ (in Subsection 4.1.1), the vector function $\widehat{\varphi}(z)e_{k+1}$ can be also considered as the first column of some matrix function $\left[\begin{smallmatrix}\psi_k(z,\omega_2)\\ I_{m-k}\end{smallmatrix}\right]$ satisfying (4.18). Therefore, we can estimate $\|(\widehat{\varphi}(z)-\varphi(z))e_{k+1}\|$ using (4.21). For that purpose, we need a better than (4.26) estimate of ρ_l. We substitute k for $k+1$ into (4.108) and obtain

$$\begin{aligned}&\aleph(x,z)>\exp\{\mathrm{i}(z-\overline{z})d_kx-2M_0x\}J_k, \quad \text{and so}\\&\aleph_{11}(x,z)^{-1/2}<\exp\{-(\mathrm{i}/2)(z-\overline{z})d_kx+M_0x\}I_k. \qquad (4.110)\end{aligned}$$

Since $\rho_l = \aleph_{11}(x,z)^{-1/2}$, from (4.21), (4.25) and (4.110), we see that

$$\|(\widehat{\varphi}(z) - \varphi(z))e_{k+1}\| \le 2\exp\{(\Im(z)(d_k - d_{k+1}) + 2M_0)x\}, \quad \Im z < -\widehat{M} \tag{4.111}$$

where $\widehat{M} = \widehat{M}(x) = \max(M(x), \widetilde{M})$. The remaining part of the proof is somewhat similar to the proof of Corollary 2.44. From the boundedness of $(\widehat{\varphi}(z) - \varphi(z))e_{k+1}$, we derive representation

$$(\widehat{\varphi}(z) - \varphi(z))e_{k+1} = z\int_0^\infty \exp\{-\mathrm{i}zt\}f(t)\,dt, \quad \Im z < -\widehat{M}(x_0), \tag{4.112}$$

where some x_0 is fixed and $\exp\{-t\widehat{M}(x_0)\}f(t) \in L^2_m(0,\infty)$. For $\xi = (d_k - d_{k+1})x$ $(x > x_0)$, relations (4.111) and (4.112) imply the inequality

$$\sup\left\|\int_0^\xi \exp\{\mathrm{i}z(\xi - t)\}f(t)\,dt\right\| < \infty$$

in the domain $\mathbb{C}^-_{\widehat{M}(x)}$, and therefore in the whole plane. Thus, $f \equiv 0$ and (4.112) yields $\varphi e_{k+1} = \widehat{\varphi}e_{k+1}$. □

4.1.4 Some generalizations

The considerations of Subsections 4.1.1–4.1.3 remain valid (or valid after slight modifications) for systems (4.1), where potentials ζ satisfy reductions of the form

$$\zeta(x) = -B\zeta(x)^*B, \quad B := \operatorname{diag}\{b_1, \dots, b_m\}, \quad b_i = \pm 1 \ (1 \le i \le m). \tag{4.113}$$

The reduction, which is given in (4.2), is obtained when $B = I_m$. For some fixed $B \neq I_m$, Definitions 4.1 and 4.7 remain valid after we change the condition $\zeta = -\zeta^*$ to (4.113).

The construction of the solution of the ISpP in the proof of Theorem 4.10 is modified in the following way. The operator identity (4.61) takes the form

$$AS - SA^* = \mathrm{i}\Pi B\Pi^*,$$

where Π is defined by (4.58). Hence, in formula (4.63), for $S = S_l$, we have

$$S_l f = BD^{-1}f(x) + \int_0^l \left\{s_{ij}(x,t)\right\}_{i,j=1}^m f(t)dt,$$

where the entries s_{ij} are defined by the modified formula (4.64) with

$$\breve{Y}(x,t) = \Pi'(x)B\Pi'(t)^*D^{-1}$$

and terms $d_i^{-1} b_j \Pi_{ij}'$ and $d_j^{-1} b_i \overline{\Pi}_{ji}'$ substituted in place of terms $d_i^{-1}\Pi_{ij}'$ and $d_j^{-1}\overline{\Pi}_{ji}'$, respectively. Modifying the formulas for S_l and $s = \{s_{ij}\}_{i,j=1}^m$ as described above and assuming in addition that the operators S_l are boundedly invertible for all l, we obtain the solution of the inverse problem from the formula (4.90), where

$$\Gamma(l,l) = BD^{-1}\left(S_l^{-1}s(x,l)\right)(l)D^{1/2}B.$$

It is important that, for $B = I_m$, the invertibility of S follows automatically from the conditions of Theorem 4.10.

Theorems 4.2 and 4.15 (and their proofs) are also valid for systems (4.1) with potentials considered without reductions (i.e. references to (4.2) can be deleted from these theorems).

4.2 Conditions on the potential and asymptotics of generalized Weyl functions

4.2.1 Preliminaries. Beals–Coifman asymptotics

It is essential to know under which conditions on ζ GW-functions satisfy requirements of Theorem 4.10. For this, we need a particular case of important results from [36] (see also [37, 38, 115]).

Recall that $[D, \mathcal{M}]$ stands for the commutator $D\mathcal{M} - \mathcal{M}D$. Consider system (4.1) on the whole axis, and let an $m \times m$ matrix function $\mathcal{W}$ satisfy (4.1). Then, the matrix function $\mathcal{M}(x,z) = \mathcal{W}(x,z)\exp\{-izxD\}$ satisfies equation

$$\mathcal{M}_x(x,z) = \mathrm{i}z\,[D,\, \mathcal{M}(x,z)] - \zeta(x)\mathcal{M}(x,z), \qquad -\infty < x < \infty \tag{4.114}$$

and vice versa. The function $\mathcal{M}$ is defined by (4.114) up to the right factor

$$\exp\{\mathrm{i}zxD\}a(z)\exp\{-\mathrm{i}zxD\}.$$

Normalization conditions

$$\lim_{x\to-\infty} \mathcal{M}(x,z) = I_m, \quad \overline{\lim}_{x\to\infty}\|\mathcal{M}(x,z)\| < \infty \tag{4.115}$$

are used to define $\mathcal{M}$ uniquely. Sufficient conditions on the potential ζ, under which relations (4.51) and (4.52) hold, follow from a particular case of [36, Theorem 6.1]:

Theorem 4.16. *Suppose that the $m \times m$ potential ζ is two times differentiable and $\zeta^{(i)}(x) = \frac{d^i\zeta}{dx^i} \in L^1(-\infty, \infty)$ for $i = 0, 1, 2$. Then, for some $M > 0$, the analytic in z matrix function $\mathcal{M}(x,z)$, which satisfies (4.114) and (4.115), is well-defined in the domain $\mathbb{C}_M^-$, the norm $\|\mathcal{M}\|$ is uniformly bounded:*

$$\sup_{x\in(-\infty,\infty),\, \Im z<-M} \|\mathcal{M}(x,z)\| < \infty, \tag{4.116}$$

and uniformly with respect to x, *we have*

$$\lim_{|z|\to\infty,\, \Im z<-M} \mathcal{M}(x,z) = I_m. \tag{4.117}$$

Finally, there is an $m \times m$ *matrix function* $\mathcal{M}_1(x)$ *which satisfies*

$$\left\| \mathcal{M}(x,z) - I_m - \frac{\mathcal{M}_1(x)}{z} \right\| = O(z^{-2}) \quad (\Im z < -M), \tag{4.118}$$

is two times differentiable and such that $\mathcal{M}_1^{(i)}(x) \in L^1(-\infty, \infty)$ *for* $i = 1, 2$.

Scheme of the proof. First, the case $\|\zeta\|_1 < 1$, where

$$\|f\|_1 := \int_a^b \left(\mathrm{Tr}\,(f(x)^* f(x))\right)^{\frac{1}{2}} dx,$$

is treated in [36, Theorem 3.8]. The operator $K_{z,\zeta}$ in the space of bounded matrix functions, where the norm is defined by the equality

$$\|f\|_\infty := \sup\,(\mathrm{Tr}\,(f(x)^* f(x)))^{\frac{1}{2}},$$

is introduced by the formula

$$\begin{aligned}\left(K_{z,\zeta} f\right)(x) &= \int_x^\infty \exp\{\mathrm{i}z(x-t)D\}\,(\zeta(t)f(t))_+ \exp\{-\mathrm{i}z(x-t)D\}\,dt \\ &\quad - \int_{-\infty}^x \exp\{\mathrm{i}z(x-t)D\}\,(\zeta(t)f(t))_- \exp\{-\mathrm{i}z(x-t)D\}\,dt.\end{aligned} \tag{4.119}$$

Here, $\mathcal{A}_+$ ($\mathcal{A}_-$) is the upper (lower) triangular part of the matrix $\mathcal{A}$, $\mathcal{A} = \mathcal{A}_+ + \mathcal{A}_-$, and the main diagonal is included in $\mathcal{A}_-$. One can see that $\|K_{z,\zeta}\| \le \|\zeta\|_1$ and so the operator $I - K_{z,\zeta}$ is invertible. Moreover, it is proved that

$$\mathcal{M}(x,z) = (I - K_{z,\zeta})^{-1} I_m$$

satisfies (4.114) and (4.115), and all the properties of $\mathcal{M}$ follow. The general case $\|\zeta\|_1 < 2^r$ ($r > 0$) follows by induction on r ([36, Theorem A]). In Theorem 6.1 from [36], matrix functions $\mathcal{M}_j(x)$ ($j = 1, 2$) are introduced by the equalities

$$\lim_{x\to-\infty} \mathcal{M}_j(x) = 0,$$

$$\frac{d}{dx}\mathcal{M}_{j-1} + \zeta\mathcal{M}_{j-1} = \mathrm{i}[D, \mathcal{M}_j^{od}], \quad \mathcal{M}_j^d(x) = -\int_{-\infty}^x \left(\zeta(t)\mathcal{M}_j^{od}(t)\right)^d dt, \tag{4.120}$$

where $\mathcal{M}_0 = I_m$, $\mathcal{M}_j = \mathcal{M}_j^d + \mathcal{M}_j^{od}$, and the entries of $\mathcal{M}_j$ and $\mathcal{M}_j^d$ coincide on the main diagonal, the entries of $\mathcal{M}_j$ and $\mathcal{M}_j^{od}$ coincide outside the main diagonal. Then, relation (4.118) follows from equality (6.14) in [36]:

$$\mathcal{M}(x,z)^{-1}\mathcal{M}^a(x,z) = I_m - \int_x^\infty \mathrm{e}^{\mathrm{i}z(x-t)D}(\hat{\nu}(t,z))_+\mathrm{e}^{-\mathrm{i}z(x-t)D}dt + \int_{-\infty}^x \mathrm{e}^{\mathrm{i}z(x-t)D}(\hat{\nu}(t,z))_-\mathrm{e}^{-\mathrm{i}z(x-t)D}dt, \tag{4.121}$$

where $\mathcal{M}^a(x,z) = \sum_{j=0}^2 z^{-j}\mathcal{M}_j(x)$ and

$$\hat{\nu}(x,z) = z^{-2}\mathcal{M}(x,z)^{-1}\left(\frac{d}{dx}\mathcal{M}_2(x) + \zeta(x)\mathcal{M}_2(x)\right).$$

4.2.2 Solvability of the inverse problem and Borg–Marchenko-type result

GW-functions are closely related to the functions $\mathcal{M}(x,z)$ from the previous section. Thus, Theorem 4.16 yields our next theorem.

Theorem 4.17. *Suppose that the $m \times m$ potential $\zeta = -\zeta^*$ is two times differentiable and $\zeta^{(i)}(x) = \frac{d^i\zeta}{dx^i} \in L^1(0, \infty)$ for $i = 0, 1, 2$. Then, for some $M > 0$, a* GW*-function of system* (4.1), (4.2) *satisfies conditions* (4.51) *and* (4.52).

Proof. Define $\zeta(x)$ on the semiaxis $x < 0$ so that the conditions of Theorem 4.16 hold. Then, we have

$$w(x,z) = \mathcal{M}(x,z)\exp\{\mathrm{i}zxD\}\mathcal{M}(0,z)^{-1} \quad (x \geq 0). \tag{4.122}$$

Hence, in view of Definition 4.7 and formulas (4.116) and (4.122), the function $\varphi(z) = \mathcal{M}(0,z)$ is a Weyl function. Now, it is immediate from (4.118) that the conditions (4.51) and (4.52) are fulfilled. □

Next, we consider a simple example where the conditions (4.51) and (4.52) are fulfilled, but the conditions of Theorem 4.17 are not. Namely, the condition $\zeta \in L^1(0, \infty)$ is not valid.

Example 4.18. Let $m = 2$ and

$$\zeta(x) \equiv \begin{bmatrix} 0 & -v \\ \overline{v} & 0 \end{bmatrix} = \mathrm{const} \quad (v \neq 0), \tag{4.123}$$

where const means a constant matrix. Calculating eigenvalues and eigenvectors of $\mathrm{i}zD - \zeta$, we obtain

$$\mathrm{i}zD - \zeta = T(z)\Lambda(z)T(z)^{-1}, \quad \Lambda(z) = \begin{bmatrix} \lambda_1(z) & 0 \\ 0 & \lambda_2(z) \end{bmatrix}, \tag{4.124}$$

$$T(z) = \begin{bmatrix} 1 & (\mathrm{i}zd_2 - \lambda_2(z))/\overline{v} \\ (\lambda_1(z) - \mathrm{i}zd_1)/v & 1 \end{bmatrix}, \tag{4.125}$$

where λ_k are the roots of equation

$$(\mathrm{i}zd_1 - \lambda)(\mathrm{i}zd_2 - \lambda) + |v|^2 = 0, \quad \text{i.e.} \tag{4.126}$$

$$\lambda_{1,2} = \frac{\mathrm{i}}{2}\left((d_1 + d_2)z \pm \sqrt{(d_1 - d_2)^2 z^2 + 4|v|^2}\right). \tag{4.127}$$

In particular, we have

$$\lambda_k(z) - \mathrm{i}zd_k = (-1)^{k+1} 2\mathrm{i}|v|^2 \left(\sqrt{(d_1 - d_2)^2 z^2 + 4|v|^2} + (d_1 - d_2)z\right)^{-1}. \tag{4.128}$$

In view of (4.124), we get the fundamental solution

$$w(x,z) = T(z)\exp\{x\Lambda(z)\}T(z)^{-1}. \tag{4.129}$$

From (4.128) and (4.129), we see that the matrix function $\varphi(z) = T(z)$ satisfies (4.14), and so this φ proves a GW-function of system (4.1), (4.2). (Recall that without normalization, a GW-function is not unique.) Moreover, from (4.125) and (4.128), it follows that

$$\varphi(z) = T(z) = I_2 + \frac{\mathrm{i}}{(d_1 - d_2)z}\begin{bmatrix} 0 & v \\ \overline{v} & 0 \end{bmatrix} + O\left(\frac{1}{z^3}\right). \tag{4.130}$$

Therefore, conditions (4.51)–(4.53) are fulfilled.

Under conditions (4.51)–(4.53), a Borg–Marchenko-type statement is valid.

Theorem 4.19. *Let the analytic $m \times m$ matrix functions φ_1 and φ_2 satisfy* (4.51)–(4.53). *Suppose that on some ray $c\Im z = \Re z$ ($c \in \mathbb{R}$, $\Im z < -M$), we have*

$$\varphi_1(z)^{-1} - \varphi_2(z)^{-1} = \mathrm{e}^{-\mathrm{i}zlD} O(z) \quad \text{for } |z| \to \infty. \tag{4.131}$$

Then, φ_1 and φ_2 are GW-functions of systems (4.1) *satisfying* (4.2), *where potentials ζ_1 and ζ_2, respectively, are such that*

$$\zeta_1(x) \equiv \zeta_2(x) \quad (0 < x \le l). \tag{4.132}$$

Proof. The fact that φ_1 and φ_2 are GW-functions follows from Theorem 4.10. From (4.51)–(4.53) follow relations (4.55) and (4.56). According to the classical results on the Fourier transform in the complex domain (see, e.g. Theorem E.11), the function $\frac{1}{z}\varphi(z)^{-1}$, where $\varphi(z)$ satisfies (4.55) and (4.56), admits Fourier representation. Moreover, also taking into account formula (4.54) and the Plancherel theorem, we obtain this representation for $z = \xi - \mathrm{i}\eta$ and fixed values of $\eta > M$ in terms of Π, that is,

$$\frac{1}{z}\varphi(z)^{-1} = \mathrm{i}D\int_0^\infty \exp\{-\mathrm{i}zxD\}\Pi(x)dx, \tag{4.133}$$

where $\exp\{-x(M+\varepsilon)D\}\Pi(x) \in L^2_{m\times m}(0,\infty)$ (for any $\varepsilon > 0$) and equalities (4.133) hold pointwise. We will need (4.133) in order to estimate $F(z)$ of the form

$$F(z) := \exp\{\mathrm{i}zlD\}\int_0^l \exp\{-\mathrm{i}zxD\}\,(\Pi_1(x) - \Pi_2(x))\,dx, \tag{4.134}$$

where Π_1 and Π_2 correspond via formula (4.58) to φ_1 and φ_2, respectively. From (4.133) and (4.134), we derive

$$\begin{aligned} F(z) = {} & -\frac{\mathrm{i}}{z}D^{-1}\exp\{\mathrm{i}zlD\}\left(\varphi_1(z)^{-1} - \varphi_2(z)^{-1}\right) \\ & - \int_l^\infty \exp\{\mathrm{i}z(l-x)D\}\,(\Pi_1(x) - \Pi_2(x))\,dx. \end{aligned} \tag{4.135}$$

In view of (4.131), the relation

$$\left\|\mathrm{i}D^{-1}\exp\{\mathrm{i}zlD\}\left(\varphi_1(z)^{-1} - \varphi_2(z)^{-1}\right)/z\right\| = O(1) \tag{4.136}$$

is valid. Since $\exp\{-x(M+\varepsilon)D\}\Pi_k(x) \in L^2_{m\times m}(0,\infty)$, for $z = \xi - \mathrm{i}\eta$, we obtain

$$\left\|\int_l^\infty \exp\{\mathrm{i}z(l-x)D\}\,(\Pi_1(x) - \Pi_2(x))\,dx\right\| = O(\frac{1}{\sqrt{\eta}}), \quad \eta \to \infty. \tag{4.137}$$

According to (4.135)–(4.137), $F(z)$ is bounded on the ray $c\Im z = \Re z$ ($\Im z < -M - \varepsilon$). It is immediate that F is also bounded on the axis $\Im z = -M - \varepsilon$. We proceed in the standard way. Namely, we use the fact that (in view of the said above) F given by (4.134) satisfies conditions of the Phragmen–Lindelöf theorem (Corollary E.7) in the angles with the boundaries $\Im z = -M - \varepsilon$ and $c\Im z = \Re z$ ($\Im z < -M - \varepsilon$) in the lower half-plane. Applying Corollary E.7, we derive that F is bounded for $\Im z \le -M - \varepsilon$. It easily follows from (4.134) that F is bounded for $\Im z > -M - \varepsilon$ too, and that $F(z) \to 0$ for $z = \overline{z}$ tending to infinity. Thus, we derive $F \equiv 0$. Hence, taking into account that Π is continuous, we have the pointwise identity

$$\Pi_1(x) \equiv \Pi_2(x) \quad (0 \le x \le l). \tag{4.138}$$

Finally, notice that according to (4.63), (4.64) and (4.91), the matrix function $\Gamma(x,x)$ on the interval $(0, l]$ is determined by $\Pi(x)$ on the same interval. Hence, formulas (4.90) and (4.138) imply (4.132). □

4.3 Direct and inverse problems: explicit solutions

As in Subsection 2.2.2, where the self-adjoint Dirac system is studied, the GBDT results from Subsection 1.1.3 can be successfully used in order to construct explicit solutions of the direct and inverse problems for system (4.1), (4.2), see also the corresponding results and bibliography in the review [264, Ch. 5]. We consider systems (4.1), where (4.2) holds and

$$\zeta = [D, \widetilde{\varrho}], \quad \widetilde{\varrho} = \widetilde{\varrho}^*, \tag{4.139}$$

so that (4.1) takes the form (1.67) ($B = \pm I_m$). We also put $B = I_m$, $\varrho \equiv 0$ and rewrite (1.72) as

$$\widetilde{\varrho}(x) = -\Pi(x)^* S(x)^{-1}\Pi(x). \tag{4.140}$$

Here, parameter matrices A, $S(0) = S(0)^*$ and $\Pi(0)$ satisfy the identity

$$AS(0) - S(0)A^* = \mathrm{i}\Pi(0)\Pi(0)^* \tag{4.141}$$

and determine matrix functions $\Pi(x)$ and $S(x)$ via equalities

$$\Pi' = -\mathrm{i}A\Pi D, \quad S' = \Pi D \Pi^*. \tag{4.142}$$

We require $S(0) > 0$ and so (4.142) yields

$$S(x) > 0 \quad (x \geq 0), \quad AS(x) - S(x)A^* = \mathrm{i}\Pi(x)\Pi(x)^*. \tag{4.143}$$

In particular, we see that $S(x)$ is invertible. The fundamental solution of system (4.1), where ζ is given by (4.139) and (4.140), are denoted by $\widetilde{w}$. Since relations (4.140)–(4.142) coincide with the corresponding relations in Subsection 1.1.3 (for the case that $B = I_m$, $\varrho \equiv 0$), we can use Proposition 1.8. This proposition implies

$$\widetilde{w}(x,z) = w_A(x,z)\mathrm{e}^{\mathrm{i}zxD} w_A(0,z)^{-1}, \tag{4.144}$$

where

$$w_A(x,z) = I_m - \mathrm{i}\Pi(x)^* S(x)^{-1}(A - zI_n)^{-1}\Pi(x). \tag{4.145}$$

Taking into account that $S(x)$ is invertible, we see that w_A is well-defined for all $x \geq 0$ and all $z \notin \sigma(A)$. In view of Corollary 1.15, w_A satisfies (1.88), where $J = I_m$. For $z \in \mathbb{C}_-$, the inequality $S(x) > 0$ and formula (1.88) imply

$$w_A(x,z)^* w_A(x,z) \leq I_m, \tag{4.146}$$

$$\mathrm{i}(z - \overline{z})\Pi(x)^*(A^* - zI_n)^{-1}S(x)^{-1}(A - zI_n)^{-1}\Pi(x) \leq I_m. \tag{4.147}$$

Proposition 4.20. *Let parameter matrices A, $S(0)$ and $\Pi(0)$ satisfy conditions* (4.141) *and $S(0) > 0$. Then, $\widetilde{\varrho}$ determined by the triple $\{A, S(0), \Pi(0)\}$ via* (4.140) *and* (4.142) *is bounded on $[0, \infty)$.*

Proof. It follows from (4.142) that Π has the form

$$\Pi(x) = [\exp\{-\mathrm{i}d_1 xA\}\, g_1 \;\dots\; \exp\{-\mathrm{i}d_m xA\}\, g_m], \qquad (4.148)$$
$$g_i \in \mathbb{C}^m \quad (1 \le i \le m).$$

One can easily see that $\mathrm{span}\cup_{z\in\mathcal{O}}(A - zI_n)^{-1}g_i \supseteq g_i$ for any open domain $\mathcal{O}$. Hence, from (4.147) and (4.148), we obtain

$$\sup_{x\in\mathbb{R}_+} \left\| S(x)^{-\frac{1}{2}} \mathrm{e}^{-\mathrm{i}d_i xA} g_i \right\| < \infty \quad (1 \le i \le m). \qquad (4.149)$$

Now, the boundedness of $\widetilde{\varrho}$ of the form (4.140) is immediate. □

Similar to Definition 2.28, we introduce below the class of generalized pseudoexponential potentials ζ of system (4.1).

Definition 4.21. The $m \times m$ potential $\zeta = [D, \widetilde{\varrho}]$, where $\widetilde{\varrho}$ is constructed in Proposition 4.20, is called a generalized pseudoexponential potential. The class of such potentials is denoted by PE. It is said that ζ is generated by the triple $\{A, S(0), \Pi(0)\}$.

According to Proposition 4.20, matrix functions $\zeta \in$ PE are bounded, and so Theorem 4.2 is applicable. That is, there is a unique (normalized) Weyl function of the corresponding system (4.1). However, it is more convenient to skip normalization and deal with GW-functions. Our next theorem is immediate from (4.144), (4.146) and Definition 4.7.

Theorem 4.22. *Let $\zeta \in$ PE be generated by the triple $\{A, S(0), \Pi(0)\}$. Then, a GW-function of the corresponding system* (4.1), (4.2) *is given by the formula*

$$\varphi(z) = w_A(0, z) = I_m - \mathrm{i}\Pi(0)^* S(0)^{-1}(A - zI_n)^{-1}\Pi(0). \qquad (4.150)$$

Notice that in view of (1.84) and (4.146), the matrix function $\varphi(z)$ given by (4.150) has the following properties:

$$\varphi(z)\varphi(\overline{z})^* = I_m, \quad \varphi(z)^*\varphi(z) \le I_m \quad (z \in \mathbb{C}_-), \quad \lim_{z\to\infty} \varphi(z) = I_m. \qquad (4.151)$$

This matrix function is rational and admits (Appendix B) a minimal realization

$$\varphi(z) = I_m + C(zI_n - \mathcal{A})^{-1}B. \qquad (4.152)$$

It follows from the second relation in (4.151) that

$$\sigma(\mathcal{A}) \in \mathbb{C}_+, \quad \sigma(\mathcal{A}) \cap \sigma(\mathcal{A}^*) = \emptyset. \qquad (4.153)$$

Hence, there is a unique and positive solution $S_0 > 0$ of the identity

$$\mathcal{A}S_0 - S_0\mathcal{A}^* = \mathrm{i}BB^*. \qquad (4.154)$$

Matrix functions admitting realization (4.152) satisfy conditions of Theorem 4.10 and so there is a unique solution of the corresponding ISpP. This solution is given below (see also [262, Theorem 5.6]).

Theorem 4.23. *Let an $m \times m$ rational matrix function φ satisfy conditions* (4.151). *Then, φ is a GW-function of the unique system* (4.1), *where $\zeta \in$ PE.*

To recover ζ, we take a minimal realization (4.152) *of φ, recover S_0 from* (4.154) *and put*

$$A = \mathcal{A}, \quad S(0) = S_0, \quad \Pi(0) = B. \tag{4.155}$$

Then, ζ is generated by the parameter matrices A, $S(0)$ and $\Pi(0)$ via formulas (4.139), (4.140) *and* (4.142).

Proof. Taking into account (4.152) and (4.154), we have

$$\begin{aligned}\varphi(z)\varphi(\overline{z})^* - I_m &= C(zI_n - \mathcal{A})^{-1}B + B^*(zI_n - \mathcal{A}^*)^{-1}C^* \\ &\quad - \mathrm{i}C(zI_n - \mathcal{A})^{-1}(\mathcal{A}S_0 - S_0\mathcal{A}^*)(zI_n - \mathcal{A}^*)^{-1}C^* \\ &= C(zI_n - \mathcal{A})^{-1}(B - \mathrm{i}S_0C^*) + (B^* + \mathrm{i}CS_0)(zI_n - \mathcal{A}^*)^{-1}C^*.\end{aligned} \tag{4.156}$$

From the first relation in (4.151), we see that the right-hand side of (4.156) equals zero. Thus, it has no poles and the second relation in (4.153) implies that both terms $C(zI_n - \mathcal{A})^{-1}(B - \mathrm{i}S_0C^*)$ and $(B^* + \mathrm{i}CS_0)(zI_n - \mathcal{A}^*)^{-1}C^*$ equal zero. Recall that (4.152) is a minimal realization, which yields the second equality in (B.2). It easily follows from the second equality in (B.2) that span $\cup_{z \in \mathcal{O}} (zI_n - A^*)^{-1}C^* = \mathbb{C}^m$. Therefore, using equality $(B^* + \mathrm{i}CS_0)(zI_n - \mathcal{A}^*)^{-1}C^* \equiv 0$, we derive

$$C = \mathrm{i}B^*S_0^{-1}. \tag{4.157}$$

Substitute (4.157) into (4.152) and use (4.155) to see that our φ admits representation (4.150). Hence, according to Theorem 4.22, our φ is a GW-function of the constructed system. □

5 Discrete systems

The approach presented in the previous chapters is easily modified for discrete systems. Discrete systems are of great interest and their study is sometimes more complicated than the study of the corresponding continuous systems (see, e.g. [1, 6, 16, 50, 102] and references therein). Here, we deal with discrete analogs of self-adjoint and skew-self-adjoint Dirac systems with the square matrix potentials following papers [105, 148, 259]. Similar results for discrete analogs of Dirac systems with rectangular potentials can be obtained in the same way (see, e.g. [109]).

5.1 Discrete self-adjoint Dirac system

In this section, we treat a discrete self-adjoint Dirac system (dDS) of the form:

$$y_{k+1}(z) = (I_m + \mathrm{i}zjC_k)y_k(z) \quad (k \in \mathbb{N}_0), \tag{5.1}$$

where $\mathbb{N}_0$ stands for the set of nonnegative integer numbers and the $m \times m$ matrices $\{C_k\}$ are positive and j-unitary:

$$C_k > 0, \quad C_k jC_k = j; \quad m = 2p, \quad j := \begin{bmatrix} I_p & 0 \\ 0 & -I_p \end{bmatrix}. \tag{5.2}$$

To show that the dDS is a discrete analog of Dirac system

$$y'(x,z) = \mathrm{i}(zj + jV(x))y(x,z), \tag{5.3}$$

we recall that, according to Subsection 1.1.1, Dirac systems form a subclass of canonical systems. Moreover, introducing $U(x)$ and $Y(x,z)$ by the equalities

$$Y(x,z) = U(x)y(x,z), \quad U' = -\mathrm{i}U(x)jV(x), \quad U(0) = I_m, \tag{5.4}$$

we see that

$$Y' = U'U^{-1}Y + \mathrm{i}U(zj + jV)U^{-1}Y = \mathrm{i}zjHY, \tag{5.5}$$

where $H = jUjU^{-1}$ and $U^*jU = j$ (i.e. U is j-unitary). Hence, we obtain

$$H = jUU^*j = H^* > 0, \quad HjH \equiv j. \tag{5.6}$$

Compare formulas (5.1) and (5.2) with formulas (5.5) and (5.6) to see that system (5.1) is, indeed, a discrete analog of (5.3).

It is essential that dDS (5.1), (5.2) is also equivalent to the very well-known (see, e.g. [92, 310]) Szegö recurrence. This connection is discussed in detail in Subsection 5.1.1. Direct and inverse problems for the subcase of the scalar Schur (or Verblunsky) coefficients were studied, for instance, in [16, 18, 310] (see also various references therein), and here we deal with the matrix Schur coefficients. Direct problems are considered in Subsection 5.1.2 and inverse problems are considered in Subsection 5.1.3.

5.1.1 Dirac system and Szegö recurrence

Remark 5.1. If $C \geq 0$, the matrix $C^{1/2} \geq 0$ is well-defined. Further, we always assume that the nonnegative square roots of matrices are chosen.

The next simple proposition is essential for the considerations in this subsection and could be of independent interest in the theory of functions of matrices, which is actively developed last years (see, e.g. [49] and references therein).

Proposition 5.2. *Let an $m \times m$ matrix C satisfy relations*

$$C > 0, \quad CjC = j \quad (j = j^* = j^{-1}). \tag{5.7}$$

Then, we have

$$C^{1/2} > 0, \quad C^{1/2} j C^{1/2} = j. \tag{5.8}$$

Proof. Since $C > 0$, it admits a representation

$$C = \mathcal{U}^* D \mathcal{U}, \tag{5.9}$$

where D is a diagonal matrix and

$$D > 0, \quad \mathcal{U}^* \mathcal{U} = \mathcal{U}\mathcal{U}^* = I_m. \tag{5.10}$$

We substitute (5.9) into the second relation in (5.7) to derive

$$\mathcal{U}^* D \mathcal{U} j \mathcal{U}^* D \mathcal{U} = j,$$

or, equivalently,

$$D\tilde{J}D = \tilde{J}, \quad \tilde{J} = \tilde{J}^* = \tilde{J}^{-1} := \mathcal{U} j \mathcal{U}^*. \tag{5.11}$$

Formula (5.11) yields $D^{-1} = \tilde{J} D \tilde{J}$ and, taking square roots of both parts of this equality, we obtain

$$D^{-1/2} = \tilde{J} D^{1/2} \tilde{J}, \qquad D^{1/2} \tilde{J} D^{1/2} = \tilde{J}. \tag{5.12}$$

Since $\mathcal{U}$ is unitary, using second relations in (5.11) and (5.12), we derive

$$\mathcal{U}^* D^{1/2} \mathcal{U} j \mathcal{U}^* D^{1/2} \mathcal{U} = j. \tag{5.13}$$

Because of (5.9), we have $C^{1/2} = \mathcal{U}^* D^{1/2} \mathcal{U}$ and so (5.13) proves the proposition. □

We apply Proposition 5.2 to matrices C_k in order to obtain the next proposition.

Proposition 5.3. *Let matrices C_k satisfy* (5.2). *Then, they admit representations*

$$C_k = 2\beta(k)^* \beta(k) - j, \quad \beta(k) j \beta(k)^* = I_p, \tag{5.14}$$

$$C_k = j + 2\gamma(k)^* \gamma(k), \quad \gamma(k) j \gamma(k)^* = -I_p, \tag{5.15}$$

where $\beta(k)$ and $\gamma(k)$ are $p \times m$ matrices.

Proof. We note that matrices C_k satisfy conditions of Proposition 5.2 and so (5.8) holds for $C = C_k$. Next, we put

$$\beta(k) := \begin{bmatrix} I_p & 0 \end{bmatrix} C_k^{1/2} \tag{5.16}$$

and take into account the equality

$$C_k = C_k^{1/2} \left(2 \begin{bmatrix} I_p & 0 \end{bmatrix}^* \begin{bmatrix} I_p & 0 \end{bmatrix} - j \right) C_k^{1/2}.$$

Now, representation (5.14) is apparent from (5.8). In a similar way, formula (5.8) and equality $I_m = j + 2[0 \; I_p]^*[0 \; I_p]$ imply representation (5.15) for

$$\gamma(k) := \begin{bmatrix} 0 & I_p \end{bmatrix} C_k^{1/2}. \tag{5.17}$$

□

Next, we will consider interrelations between Dirac system (5.1), (5.2) and the Szegö recurrence, which is given by the formula

$$X_{k+1}(\lambda) = \mathcal{D}_k H_k \begin{bmatrix} \lambda I_p & 0 \\ 0 & I_p \end{bmatrix} X_k(\lambda), \tag{5.18}$$

where

$$H_k = \begin{bmatrix} I_p & \rho_k \\ \rho_k^* & I_p \end{bmatrix}, \quad \mathcal{D}_k = \operatorname{diag}\left\{ \left(I_p - \rho_k \rho_k^* \right)^{-\frac{1}{2}}, \; \left(I_p - \rho_k^* \rho_k \right)^{-\frac{1}{2}} \right\}, \tag{5.19}$$

and the $p \times p$ matrices ρ_k are strictly contractive, that is,

$$\|\rho_k\| < 1. \tag{5.20}$$

Remark 5.4. When $p = 1$, one easily removes the factor $(1 - |\rho_k|^2)^{-1/2}$ in (5.18) to obtain systems as in [16, 18], where direct and inverse problems for the case of scalar strictly pseudoexponential potentials have been treated. The block matrix version (i.e. the version where $p > 1$) of Szegö recurrence, its connections with Schur coefficients ρ_k and applications are discussed in [89, 90] (see also references therein). We note that $\mathcal{D}_k H_k$ is the so called Halmos extension of ρ_k (see [92, p. 167]), and that the matrices $\mathcal{D}_k$ and H_k commute (which easily follows, e.g. from [92, Lemma 1.1.12]). The matrix $\mathcal{D}_k H_k$ is j-unitary and positive, that is,

$$\mathcal{D}_k H_k j H_k \mathcal{D}_k = H_k \mathcal{D}_k j \mathcal{D}_k H_k = j, \tag{5.21}$$

$$\mathcal{D}_k H_k > 0. \tag{5.22}$$

According to [94, Theorem 1.2], any j-unitary matrix C admits a representation, which is close to the Halmos extension:

$$C = \mathcal{D}(\rho)H(\rho)\begin{bmatrix} u_1 & 0 \\ 0 & u_2 \end{bmatrix}, \quad u_i^* u_i = u_i u_i^* = I_p; \quad H(\rho) = \begin{bmatrix} I_p & \rho \\ \rho^* & I_p \end{bmatrix}, \tag{5.23}$$

$$\mathcal{D}(\rho) = \operatorname{diag}\left\{ \left(I_p - \rho\rho^*\right)^{-\frac{1}{2}}, \left(I_p - \rho^*\rho\right)^{-\frac{1}{2}} \right\}, \quad \rho^*\rho < I_p. \tag{5.24}$$

The next statement, which is converse to (5.21), (5.22), follows from the more general [109, Proposition 2.4], where the case of rectangular Schur coefficients was also dealt with.

Proposition 5.5. *Let an $m \times m$ matrix C be j-unitary and positive. Then, it admits a representation*

$$C = \mathcal{D}(\rho)H(\rho), \tag{5.25}$$

where $H(\rho)$ and $\mathcal{D}(\rho)$ are of the form (5.23) and (5.24) (i.e. the last factor on the right-hand side of the first equality in (5.23) is removed).

Proposition 5.5 completes Propositions 5.2 and 5.3 on representations and properties of C_k. Taking into account (5.21), (5.22) and Proposition 5.5, we rewrite Szegö recurrence (5.18) in an equivalent form

$$X_{k+1}(\lambda) = \widetilde{C}_k \begin{bmatrix} \lambda I_p & 0 \\ 0 & I_p \end{bmatrix} X_k(\lambda), \quad k \in \mathbb{N}_0, \tag{5.26}$$

$$\widetilde{C}_k > 0, \quad \widetilde{C}_k j \widetilde{C}_k = j. \tag{5.27}$$

Using (5.27), we see that the matrix functions U_k, which are given by the equalities

$$U_0 := I_m, \quad U_{k+1} := \mathrm{i}U_k\widetilde{C}_k j = \prod_{r=0}^{k} (\mathrm{i}\widetilde{C}_r j) \quad (k \geq 0), \tag{5.28}$$

are also j-unitary. From (5.27) and (5.28), we have

$$\begin{aligned}(\mathrm{i}+z)U_{k+1}(I_m + \mathrm{i}zj)\widetilde{C}_k &\begin{bmatrix} \frac{z-\mathrm{i}}{z+\mathrm{i}} I_p & 0 \\ 0 & I_p \end{bmatrix} (I_m + \mathrm{i}zj)^{-1}U_k^{-1} \\ &= I_m + \mathrm{i}zU_{k+1}jU_{k+1}^{-1}. \end{aligned} \tag{5.29}$$

In view of (5.29), the function y_k of the form

$$y_k(z) = (\mathrm{i}+z)^k U_k (I_m + \mathrm{i}zj) X_k\left(\frac{z-\mathrm{i}}{z+\mathrm{i}}\right) \tag{5.30}$$

satisfies (5.1), where $y_0(z) = (I_m + \mathrm{i}zj)X_0(z)$ and $C_k = jU_{k+1}jU_{k+1}^{-1}$. Since U_{k+1} is j-unitary, we rewrite C_k as

$$C_k = jU_{k+1}U_{k+1}^* j, \tag{5.31}$$

and so (5.2) holds. Because of (5.28), (5.31) and j-unitarity of U_k, we have $jU_k^* C_k U_k j = \widetilde{C}_k^2$, that is,

$$\widetilde{C}_k = (jU_k^* C_k U_k j)^{1/2}. \tag{5.32}$$

The following theorem describes interconnections between systems (5.1) and (5.26).

Theorem 5.6. *Dirac systems* (5.1), (5.2) *and Szegö recurrences* (5.26), (5.27) *are equivalent. The transformation* $\widetilde{\mathfrak{M}} : \{\widetilde{C}_k\} \to \{C_k\}$ *of Szegö recurrence into the Dirac system and the transformation of their solutions are given, respectively, by formulas* (5.31) *and* (5.30), *where matrices* $\{U_k\}$ *are defined in* (5.28). *The mapping* $\widetilde{\mathfrak{M}}$ *is bijective, and the inverse mapping is obtained by applying* (5.32) *(and the substitution of the result into* (5.28)*) for the successive values of k.*

Proof. It is already proved above that the formulas (5.31) and (5.30) describe a mapping of Szegö recurrence and its solution into the Dirac system and its solution, respectively. Moreover, the mapping $\widetilde{\mathfrak{M}}$ is injective since we can successively and uniquely recover $\widetilde{C}_k$ and U_{k+1} from C_k and U_k using formulas (5.32) and (5.28), respectively.

Next, we prove that $\widetilde{\mathfrak{M}}$ is surjective. Indeed, given an arbitrary sequence $\{C_k\}$ satisfying (5.2), let us apply to the matrices from this sequence relation (5.32) (and substitute the result into (5.28)) for the successive values of k. In this way, we construct a sequence $\{\widetilde{C}_k\}$. Since the matrices $jU_k^* C_k U_k j$ are positive and j-unitary, we see, from (5.32) and Proposition 5.2, that the matrices $\widetilde{C}_k$ are also positive and j-unitary. Next, we apply to $\{\widetilde{C}_k\}$ the mapping $\widetilde{\mathfrak{M}}$. Taking into account (5.28) and (5.32), we derive

$$jU_{k+1}U_{k+1}^* j = jU_k \widetilde{C}_k^2 U_k^* j = jU_k(jU_k^* C_k U_k j)U_k^* j = C_k, \tag{5.33}$$

that is, $\widetilde{\mathfrak{M}}$ maps the constructed sequence $\{\widetilde{C}_k\}$ into the initial sequence $\{C_k\}$. Recall that we started from an arbitrary $\{C_k\}$ satisfying (5.2). Hence, $\widetilde{\mathfrak{M}}$ is surjective. □

5.1.2 Weyl theory: direct problems

In this subsection, we introduce Weyl functions for dDS (5.1). Next, we prove the Weyl function's existence and, moreover, give a procedure to construct it (direct problems). Finally, we construct the S-node, which corresponds to system (5.1), and the transfer matrix function representation of the fundamental solution W_k. The fundamental $m \times m$ solution $\{W_k\}$ of (5.1) is normalized by the condition

$$W_0(z) = I_m. \tag{5.34}$$

Similar to the continuous case, the Weyl functions of system (5.1) on the interval $[0, r]$ (i.e. system (5.1) considered for $0 \leq k \leq r$) are defined by the Möbius (linear-fractional) transformation:

$$\varphi_r(z, \mathcal{P}) = \begin{bmatrix} 0 & I_p \end{bmatrix} W_{r+1}(z)^{-1}\mathcal{P}(z) \left(\begin{bmatrix} I_p & 0 \end{bmatrix} W_{r+1}(z)^{-1}\mathcal{P}(z) \right)^{-1}, \tag{5.35}$$

where $\mathcal{P}(z)$ are nonsingular $m \times p$ matrix functions with property-j (in the sense of Definition 1.42, where we put $J = j$, $m = 2p$, $m_1 = p$ and $M = 0$, i.e. $\mathbb{C}_M = \mathbb{C}_+$). According to Definition 1.42, the inequalities

$$\mathcal{P}(z)^*\mathcal{P}(z) > 0, \quad \mathcal{P}(z)^* j\mathcal{P}(z) \geq 0 \tag{5.36}$$

hold in $\mathbb{C}_+$ (excluding, possibly, isolated points).

It is apparent from (5.1) and (5.34) that

$$W_{r+1}(z) = \prod_{k=0}^{r} (I_m + \mathrm{i}zjC_k). \tag{5.37}$$

In view of (5.15) and (5.37), we obtain

$$W_{r+1}(\mathrm{i}) = (-2)^{r+1} \prod_{k=0}^{r} \left(j\gamma(k)^*\gamma(k)\right). \tag{5.38}$$

Hence, $\det W_{r+1}(\mathrm{i}) = 0$, and we don't consider $z = \mathrm{i}$ in this subsection.

Remark 5.7. We note that the behavior of Weyl functions in the neighborhood of $z = \mathrm{i}$ is essential for the inverse problems that are dealt with in the next subsection. Therefore, unlike the Weyl disc case (see Notation 5.10), in the definition (5.35) of the Weyl functions on the interval, we assume that $\mathcal{P}$ is not only nonsingular with property-j, but also has an additional property. Namely, it is well-defined and nonsingular at $z = \mathrm{i}$. We don't use this additional property in this subsection, though, in important cases, it could be obtained via multiplication by a scalar function.

The lemma below shows that transformations $\varphi_r(z, \mathcal{P})$ are well-defined.

Lemma 5.8. *Fix any $z \in \mathbb{C}_+$ such that the inequalities* (5.36) *and* $\det W_r(z) \neq 0$ *hold. Then, we have the inequality*

$$\det \left(\begin{bmatrix} I_p & 0 \end{bmatrix} W_{r+1}(z)^{-1}\mathcal{P}(z) \right) \neq 0. \tag{5.39}$$

Proof. Using (5.2) and (5.15), we obtain

$$\begin{aligned} (I_m + \mathrm{i}zjC_k)^* j (I_m + \mathrm{i}zjC_k) &= (1 + \mathrm{i}(z - \overline{z}) + |z|^2)j + 2\mathrm{i}(z - \overline{z})\gamma(k)^*\gamma(k) \\ &\leq (1 - 2\Im(z) + |z|^2)j, \\ &\quad (1 - 2\Im(z) + |z|^2) > 0 \quad \text{for } z \neq \mathrm{i}. \end{aligned} \tag{5.40}$$

Since the equality (5.37) holds, formula (5.40) implies that

$$\left(W_{r+1}(z)^{-1}\right)^* jW_{r+1}(z)^{-1} \geq (1 - 2\Im(z) + |z|^2)^{-r-1} j \quad (z \in \mathbb{C}_+, \; z \neq \mathrm{i}). \tag{5.41}$$

Because of (1.170) and (5.41), we see that $\widetilde{\mathcal{P}} := W_{r+1}(z)^{-1}\mathcal{P}(z)$ satisfies the inequalities $\widetilde{\mathcal{P}}^*\widetilde{\mathcal{P}} > 0$ and $\widetilde{\mathcal{P}}^* j\widetilde{\mathcal{P}} \geq 0$. Now, (5.39) follows from Proposition 1.43. □

Corollary 5.9. *The following inequalities hold for the fundamental solution W_{r+1} of (5.1) (where $\{C_k\}$ satisfy (5.2)):*

$$\det W_{r+1}(z) \neq 0, \quad W_{r+1}(z)^{-1} = (1+z^2)^{-r-1} jW_{r+1}(\overline{z})^* j \qquad (z \neq \pm \mathrm{i}). \quad (5.42)$$

Proof. Relations (5.40) and (5.37) imply that

$$W_{r+1}(z)^* jW_{r+1}(z) = (1+z^2)^{r+1} j, \qquad z = \overline{z}.$$

Hence, using analyticity considerations, we obtain

$$W_{r+1}(\overline{z})^* jW_{r+1}(z) \equiv (1+z^2)^{r+1} j, \tag{5.43}$$

and (5.42) is apparent. □

Notation 5.10. *The set of values of matrices $\varphi_r(z, \mathcal{P})$, which are given by the transformation (5.35) where parameter matrices $\mathcal{P}(z)$ satisfy (5.36), is denoted by $\mathcal{N}(r, z)$ (or, sometimes, simply $\mathcal{N}(r)$).*

As already mentioned before, usually $\mathcal{N}(r, z)$ is called the Weyl disc.

Corollary 5.11. *The sets $\mathcal{N}(r, z)$ are embedded (i.e. $\mathcal{N}(r, z) \subseteq \mathcal{N}(r-1, z)$) for all $r > 0$ and $z \in \mathbb{C}_+$, $z \neq \mathrm{i}$. Moreover, for all $\varphi_k \in \mathcal{N}(k, z)$ $(k \geq 0)$, we have*

$$\varphi_k(z)^* \varphi_k(z) \leq I_p. \tag{5.44}$$

Proof. It follows from Corollary 5.9 that the matrices $(I_m + \mathrm{i}zjC_r)$, $W_r(z)$ and $W_{r+1}(z)$ are invertible. Hence, inequalities (5.36) (for $\mathcal{P}$) and (5.40) imply that

$$\widetilde{\mathcal{P}} := (I_m + \mathrm{i}zjC_r)^{-1}\mathcal{P}(z)$$

also satisfies (5.36). Therefore, we rewrite (5.35) in the form

$$\varphi_r(z, \mathcal{P}) = \begin{bmatrix} 0 & I_p \end{bmatrix} W_r(z)^{-1}\widetilde{\mathcal{P}}(z) \left(\begin{bmatrix} I_p & 0 \end{bmatrix} W_r(z)^{-1}\widetilde{\mathcal{P}}(z) \right)^{-1}, \tag{5.45}$$

and see that $\varphi_r(z) \in \mathcal{N}(r-1, z)$ $(r > 0)$. Inequality (5.44) is obtained for the matrices from $\mathcal{N}(0, z)$ via substitution of $r = 0$ into (5.45). □

Weyl functions of dDS (5.1) on the semiaxis $\mathbb{N}_0$ (of nonnegative integers) are defined in a different and more traditional way (in terms of summability), see the definition below. As in the previous chapters, the definitions of Weyl functions on the interval and semiaxis are interrelated.

Definition 5.12. The Weyl–Titchmarsh (or simply Weyl) function of dDS (5.1) (which is given on the semiaxis $0 \leq k < \infty$ and satisfies (5.2)) is a $p \times p$ matrix function $\varphi(z)$ $(z \in \mathbb{C}_+)$, such that the following inequality holds:

$$\sum_{k=0}^{\infty} q(z)^k \begin{bmatrix} I_p & \varphi(z)^* \end{bmatrix} W_k(z)^* C_k W_k(z) \begin{bmatrix} I_p \\ \varphi(z) \end{bmatrix} < \infty, \tag{5.46}$$

$$q(z) := (1 + |z|^2)^{-1}. \tag{5.47}$$

Lemma 5.13. *If $\varphi_r(z) \in \mathcal{N}(r,z)$, we have the inequality*

$$\sum_{k=0}^{r} q(z)^k \begin{bmatrix} I_p & \varphi_r(z)^* \end{bmatrix} W_k(z)^* C_k W_k(z) \begin{bmatrix} I_p \\ \varphi_r(z) \end{bmatrix} \le \frac{1+|z|^2}{\mathrm{i}(\overline{z}-z)} \left(I_p - \varphi_r(z)^* \varphi_r(z) \right). \tag{5.48}$$

Proof. Because of (5.1) and (5.2), we have

$$\begin{aligned} W_{k+1}(z)^* j W_{k+1}(z) &= W_k(z)^* \left(I_m - \mathrm{i}\overline{z} C_k j \right) j \left(I_m + \mathrm{i} z j C_k \right) W_k(z) \\ &= q(z)^{-1} W_k(z)^* j W_k(z) + \mathrm{i}(z-\overline{z}) W_k(z)^* C_k W_k(z). \end{aligned} \tag{5.49}$$

Using (5.34) and (5.49), we derive a summation formula (compare with [105, formula (4.2)])

$$\sum_{k=0}^{r} q(z)^k W_k(z)^* C_k W_k(z) = \frac{1+|z|^2}{\mathrm{i}(\overline{z}-z)} \left(j - q(z)^{r+1} W_{r+1}(z)^* j W_{r+1}(z) \right). \tag{5.50}$$

On the other hand, it follows from (5.35) that

$$\begin{bmatrix} I_p \\ \varphi_r(z) \end{bmatrix} = W_{r+1}(z)^{-1} \mathcal{P}(z) \left(\begin{bmatrix} I_p & 0 \end{bmatrix} W_{r+1}(z)^{-1} \mathcal{P}(z) \right)^{-1}, \tag{5.51}$$

and so formula (5.36) yields

$$\begin{bmatrix} I_p & \varphi_r(z)^* \end{bmatrix} W_{r+1}(z)^* j W_{r+1}(z) \begin{bmatrix} I_p \\ \varphi_r(z) \end{bmatrix} \ge 0. \tag{5.52}$$

Formulas (5.50) and (5.52) imply (5.48). □

Now, we are ready to prove the main direct theorem.

Theorem 5.14. *Let dDS* (5.1) *be given on the semiaxis $0 \le k < \infty$ and satisfy* (5.2). *Then, there is a unique Weyl function φ of this dDS. This φ is analytic and nonexpansive (i.e. $\varphi^*\varphi \le I_p$) in $\mathbb{C}_+$.*

Proof. The proof consists of three steps. First, we show that there is an analytic and nonexpansive function

$$\varphi_\infty(z) \in \bigcap_{r \ge 0} \mathcal{N}(r,z). \tag{5.53}$$

Next, we show that $\varphi_\infty(z)$ is a Weyl function. Finally, we prove the uniqueness.

Step 1. As in similar theorems for continuous cases before, this step is based on Montel's theorem (i.e. Theorem E.9). Indeed, from Corollary 5.11, we see that the set of functions $\varphi_r(z,\mathcal{P})$ of the form (5.35) is uniformly bounded in $\mathbb{C}_+$. So, Montel's theorem is applicable and there is an analytic matrix function which we denote by $\varphi_\infty(z)$ and which is a uniform limit of some sequence

$$\varphi_\infty(z) = \lim_{i\to\infty} \varphi_{r_i}(z, \mathcal{P}_i) \quad (i \in \mathbb{N}, \quad r_i \uparrow, \quad \lim_{i\to\infty} r_i = \infty) \tag{5.54}$$

on all the bounded and closed subsets of $\mathbb{C}_+$. Clearly, φ_∞ is nonexpansive. Since $r_i \uparrow$, the sets $\mathcal{N}(r, z)$ are embedded and equality (5.51) is valid, it follows that the matrix functions

$$\mathcal{P}_{ij}(z) := W_{r_i+1}(z) \begin{bmatrix} I_p \\ \varphi_{r_j}(z, \mathcal{P}_j) \end{bmatrix} \quad (j \geq i)$$

satisfy relations (5.36). Therefore, using (5.54), we derive that (5.36) holds for

$$\mathcal{P}_{i,\infty}(z) := W_{r_i+1}(z) \begin{bmatrix} I_p \\ \varphi_\infty(z) \end{bmatrix}, \tag{5.55}$$

implying that we can substitute $\mathcal{P} = \mathcal{P}_{i,\infty}$ and $r = r_i$ into (5.35) to obtain

$$\varphi_\infty(z) \in \mathcal{N}(r_i, z). \tag{5.56}$$

Since (5.56) holds for all $i \in \mathbb{N}$, we see that (5.53) is fulfilled.

Step 2. Because of (5.53), the function φ_∞ satisfies the condition of Lemma 5.13. Hence, after substitution $\varphi_r = \varphi_\infty$, the inequality (5.48) holds for any $r \geq 0$, which implies (5.46). Therefore, φ_∞ is a Weyl function.

Step 3. It is apparent from (5.14) that

$$W_k(z)^* C_k W_k(z) \geq W_k(z)^*(-j)W_k(z). \tag{5.57}$$

Using (5.49), we also derive

$$q(z)^k W_k(z)^*(-j)W_k(z) \geq q(z)^{k-1} W_{k-1}(z)^*(-j)W_{k-1}(z). \tag{5.58}$$

Formulas (5.34), (5.57) and (5.58) yield the basic for Step 3 inequality

$$q(z)^k W_k(z)^* C_k W_k(z) \geq -j. \tag{5.59}$$

Therefore, the following equality is immediate for any $g \in \mathbb{C}^p$:

$$\sum_{k=0}^{\infty} g^*[0 \quad I_p] q(z)^k W_k(z)^* C_k W_k(z) \begin{bmatrix} 0 \\ I_p \end{bmatrix} g = \infty. \tag{5.60}$$

It was shown in Step 2 that $\varphi = \varphi_\infty$ satisfies (5.46). Hence, in view of (5.60), the dimension of the subspace $L \in \mathbb{C}^m$ of vectors h such that

$$\sum_{k=0}^{\infty} h^* q(z)^k W_k(z)^* C_k W_k(z) h < \infty \tag{5.61}$$

equals p. Now, suppose that there is a Weyl function $\widetilde{\varphi} \neq \varphi_\infty$. Then, we have

$$\mathrm{Im}\begin{bmatrix} I_p \\ \varphi_\infty(z) \end{bmatrix} \subseteq L, \quad \mathrm{Im}\begin{bmatrix} I_p \\ \widetilde{\varphi}(z) \end{bmatrix} \subseteq L.$$

Thus, $\dim L > p$ (for those z, where $\widetilde{\varphi}(z) \neq \varphi_\infty(z)$) and we arrive at a contradiction. □

Finally, let us construct representations of W_{r+1} $(r \geq 0)$ via S-nodes. First, recall that matrices $\{C_k\}$ generate via formula (5.17) a set $\{\gamma(k)\}$ of the $p \times m$ matrices $\gamma(k)$. Using $\{\gamma(k)\}$, we introduce $p(r+1) \times m$ matrices Γ_r and $p(r+1) \times p(r+1)$ matrices K_r $(0 \leq r < \infty)$:

$$\Gamma_r := \begin{bmatrix} \gamma(0) \\ \gamma(1) \\ \dots \\ \gamma(r) \end{bmatrix}; \quad K_r := \begin{bmatrix} \kappa_r(0) \\ \kappa_r(1) \\ \dots \\ \kappa_r(r) \end{bmatrix}, \tag{5.62}$$

$$\kappa_r(i) := \mathrm{i}\gamma(i)j\begin{bmatrix} \gamma(0)^* & \dots & \gamma(i-1)^* & \gamma(i)^*/2 & 0 & \dots & 0 \end{bmatrix}. \tag{5.63}$$

It is apparent from (5.62) and (5.63) that the identity

$$K_r - K_r^* = \mathrm{i}\Gamma_r j \Gamma_r^* \tag{5.64}$$

holds. The $p(r+1) \times p(r+1)$ matrices A_r are introduced by the equalities

$$A_r = \{a_{k-i}\}_{i,k=0}^{r}, \quad a_n = -\begin{cases} 0 & \text{for } n > 0, \\ (\mathrm{i}/2)I_p & \text{for } n = 0, \\ \mathrm{i}I_p & \text{for } n < 0. \end{cases} \tag{5.65}$$

Proposition 5.15. *Matrices K_r and A_r are linear similar, that is,*

$$K_r = E_r A_r E_r^{-1}. \tag{5.66}$$

Moreover, the transformation operators E_r can be constructed so that

$$E_r = \begin{bmatrix} E_{r-1} & 0 \\ X_r & e_r^- \end{bmatrix} \quad (r > 0); \quad E_r^{-1}\Gamma_{r,2} = \Phi_{r,2}, \quad \Phi_{r,2} := \begin{bmatrix} I_p \\ \dots \\ I_p \end{bmatrix}; \tag{5.67}$$

$$E_0 = e_0^- = \gamma_2(0), \tag{5.68}$$

where $\Gamma_{r,i}$ $(i = 1, 2)$ are $p(r+1) \times p$ blocks of $\Gamma_r = [\Gamma_{r,1} \;\; \Gamma_{r,2}]$ and $\gamma_i(k)$ are $p \times p$ blocks of $\gamma(k) = [\gamma_1(k) \;\; \gamma_2(k)]$.

Proof. It follows from (5.15), (5.62), (5.63) and (5.65) that

$$K_0 = A_0 = -(\mathrm{i}/2)I_p, \quad \det \gamma_2(0) \neq 0, \tag{5.69}$$

$$\kappa_r(r) = \mathrm{i}\begin{bmatrix} \gamma(r)j\gamma(0)^* & \dots & \gamma(r)j\gamma(r-1)^* & -I_p/2 \end{bmatrix}. \tag{5.70}$$

We see that (5.68) and (5.69) imply (5.66) for $r = 0$. Next, we prove (5.66) by induction. Assume that $K_{r-1} = E_{r-1}A_{r-1}E_{r-1}^{-1}$ and let E_r have the form (5.67), where $\det e_r^- \neq 0$. Then, we obtain

$$E_r^{-1} = \begin{bmatrix} E_{r-1}^{-1} & 0 \\ -(e_r^-)^{-1}X_rE_{r-1}^{-1} & (e_r^-)^{-1} \end{bmatrix}, \tag{5.71}$$

and, in view of (5.62), (5.65), (5.67), (5.70), it is necessary and sufficient, for (5.66) to hold, that

$$\begin{aligned}&\left(\begin{bmatrix} X_rA_{r-1} & -(\mathrm{i}/2)e_r^- \end{bmatrix} - \mathrm{i}e_r^- \begin{bmatrix} I_p & \dots & I_p & 0 \end{bmatrix}\right) \begin{bmatrix} I_{rp} \\ -(e_r^-)^{-1}X_r \end{bmatrix} E_{r-1}^{-1} \\ &= \mathrm{i}\gamma(r)j\begin{bmatrix} \gamma(0)^* & \dots & \gamma(r-1)^* \end{bmatrix}. \end{aligned}\tag{5.72}$$

We can rewrite (5.72) in the form

$$\begin{aligned}X_r\left(A_{r-1} + (\mathrm{i}/2)I_{rp}\right) = \mathrm{i}\gamma(r)j&\begin{bmatrix} \gamma(0)^* & \dots & \gamma(r-1)^* \end{bmatrix} E_{r-1} \\ &+ \mathrm{i}e_r^-\begin{bmatrix} I_p & \dots & I_p \end{bmatrix}. \end{aligned}\tag{5.73}$$

We partition X_r $(r > 1)$ into two $p \times p$ and $p \times (r-1)p$, respectively, blocks

$$X_r = \begin{bmatrix} x_r^- & \widetilde{X}_r \end{bmatrix}, \tag{5.74}$$

and we will also need partitions of the matrices $A_{r-1} + (\mathrm{i}/2)I_{rp}$ and E_{r-1}, which follow (for $r > 1$) from (5.65) and (5.67):

$$A_{r-1} + (\mathrm{i}/2)I_{rp} = \begin{bmatrix} 0 & 0 \\ \left(A_{r-2} - (\mathrm{i}/2)I_{(r-1)p}\right) & 0 \end{bmatrix}, \quad E_{r-1}\begin{bmatrix} 0 \\ I_p \end{bmatrix} = \begin{bmatrix} 0 \\ e_{r-1}^- \end{bmatrix}. \tag{5.75}$$

Using (5.74) and (5.75), we see that (5.73) is equivalent to the relations

$$e_r^- = -\gamma(r)j\gamma(r-1)^*e_{r-1}^- \quad \text{for } r \geq 1; \tag{5.76}$$

$$\begin{aligned}\widetilde{X}_r = \mathrm{i}&\left(\gamma(r)j\begin{bmatrix} \gamma(0)^* & \dots & \gamma(r-1)^* \end{bmatrix} E_{r-1} + e_r^-\begin{bmatrix} I_p & \dots & I_p \end{bmatrix}\right) \\ &\times \begin{bmatrix} \left(A_{r-2} - (\mathrm{i}/2)I_{(r-1)p}\right)^{-1} \\ 0 \end{bmatrix} \quad \text{for } r > 1. \end{aligned}\tag{5.77}$$

Hence, if e_r^- and X_r satisfy (5.76) and (5.77), respectively, and $\det e_r^- \neq 0$, the similarity relation (5.66) holds. The inequalities $\det e_r^- \neq 0$ are apparent (by induction) from (5.68), (5.76) and the inequalities

$$\det(\gamma(r)j\gamma(r-1)^*) \neq 0, \tag{5.78}$$

so that (5.78) only remains to be proven. Indeed, substituting $J = -j$, $\vartheta = \gamma(r)^*$ and $\widetilde{\vartheta} = \gamma(r-1)^*$ into Proposition 1.43 and using the second equality in (5.15) (taken for $k = r$ and $k = r-1$), we derive (5.78). Thus, equality (5.66) is proven.

Formula (5.68) shows that the second equality in (5.67) holds for $r = 0$. Now, we choose X_r (for $r = 1$) and x_r^- (for $r > 1$) so that the second equality in (5.67) holds in the case that $r > 0$. Taking into account (5.71), (5.74) and using induction, we see that this equality is valid when

$$X_1 = y_2(1) - e_1^-, \qquad x_r^- = y_2(r) - e_r^- - \widetilde{X}_r \Phi_{r-2,2} \quad (r > 1). \tag{5.79}$$

□

Now, we substitute (5.66) into (5.64) to derive

$$E_r A_r E_r^{-1} - (E_r^*)^{-1} A_r^* E_r^* = \mathrm{i}\Gamma_r j \Gamma_r^*. \tag{5.80}$$

Multiplying both sides of (5.80) by E_r^{-1} and $(E_r^*)^{-1}$ from the left and right, respectively, we obtain the operator identity

$$A_r S_r - S_r A_r^* = \mathrm{i}\Pi_r j \Pi_r^* = \mathrm{i}(\Phi_{r,1}\Phi_{r,1}^* - \Phi_{r,2}\Phi_{r,2}^*), \tag{5.81}$$

where

$$S_r := E_r^{-1}(E_r^*)^{-1}, \quad \Pi_r := E_r^{-1}\Gamma_r = \begin{bmatrix} \Phi_{r,1} & \Phi_{r,2} \end{bmatrix}. \tag{5.82}$$

Hence, the triple $\{A_r, S_r, \Pi_r\}$ forms a symmetric S-node. The transfer matrix function, which corresponds to this S-node, is given by (1.85) after substitution $J = j$:

$$w_A(r,\lambda) = I_m - \mathrm{i}j\Pi_r^* S_r^{-1}\left(A_r - \lambda I_{(r+1)p}\right)^{-1}\Pi_r. \tag{5.83}$$

For $r > 0$, introduce $rp \times (r+1)p$ and $p \times (r+1)p$ matrices acting as projectors from $\mathbb{C}^{(r+1)p}$ into $\mathbb{C}^{rp}$ and $\mathbb{C}^p$, respectively,

$$P_1 := \begin{bmatrix} I_{rp} & 0 \end{bmatrix}, \quad P_2 = P := \begin{bmatrix} 0 & \dots & 0 & I_p \end{bmatrix}. \tag{5.84}$$

Since E_r^{-1} is a block lower triangular matrix, we easily derive from (5.71) and (5.82) that

$$P_1 S_r P_1^* = E_{r-1}^{-1}(E_{r-1}^*)^{-1} = S_{r-1}, \quad P_1\Pi_r = \Pi_{r-1}. \tag{5.85}$$

It is apparent that

$$\det S_{r-1} \neq 0, \quad P_1 A_r P_2^* = 0, \quad P_1 A_r P_1^* = A_{r-1}. \tag{5.86}$$

In view of (5.85) and (5.86), the factorization Theorem 1.16 yields

$$w_A(r,\lambda) = \left(I_m - \mathrm{i}j\Pi_r^* S_r^{-1} P^* \left(PA_r P^* - \lambda I_p\right)^{-1}\left(PS_r^{-1}P^*\right)^{-1} PS_r^{-1}\Pi_r\right) \times w_A(r-1,\lambda). \tag{5.87}$$

Proposition 5.16. *The fundamental solution W of the system* (5.1)*, where W is normalized by the condition* (5.34) *and the potential $\{C_k\}$ satisfies* (5.2)*, admits representation*

$$W_{r+1}(z) = (1+\mathrm{i}z)^{r+1} w_A\left(r, (2z)^{-1}\right). \tag{5.88}$$

Proof. Formulas (5.1) and (5.15) imply

$$W_{r+1}(z) = (1+\mathrm{i}z)\left(I_m + 2\mathrm{i}z(1+\mathrm{i}z)^{-1} j\gamma(r)^*\gamma(r)\right) W_r(z) \quad (r \geq 0). \tag{5.89}$$

On the other hand, we easily derive from (5.62), (5.65), (5.67) and (5.82) that

$$\left(PA_rP^* - \lambda I_p\right)^{-1} = -(\lambda + \mathrm{i}/2)^{-1} I_p, \quad S_r^{-1} = E_r^* E_r, \tag{5.90}$$

$$PS_r^{-1}P^* = (e_r^-)^* e_r^-, \quad PS_r^{-1}\Pi_r = PE_r^*\Gamma_r = (e_r^-)^*\gamma(r). \tag{5.91}$$

We substitute (5.90) and (5.91) into (5.87) to obtain

$$w_A(r,\lambda) = \left(I_m + \frac{2\mathrm{i}}{2\lambda + \mathrm{i}} j\gamma(r)^*\gamma(r)\right) w_A(r-1,\lambda) \quad (r \geq 1). \tag{5.92}$$

In a similar way, we rewrite (5.83) (for the case that $r = 0$) in the form

$$w_A(0,\lambda) = I_m + \frac{2\mathrm{i}}{2\lambda + \mathrm{i}} j\gamma(0)^*\gamma(0). \tag{5.93}$$

Finally, we compare (5.89) with (5.92) and (5.93) (and take into account (5.34)) to see that $W_1(z) = (1+\mathrm{i}z)w_A(0,(2z)^{-1})$ and the iterative relations for the left- and right-hand sides of (5.88) coincide. □

5.1.3 Weyl theory: inverse problems

The values of φ and its derivatives at $z = \mathrm{i}$ will be of interest in this subsection. Therefore, using (5.42), we rewrite (5.35) in the form

$$\varphi_r(z,\mathcal{P}) = -\begin{bmatrix} 0 & I_p \end{bmatrix} W_{r+1}(\overline{z})^*\mathcal{P}(z)\left(\begin{bmatrix} I_p & 0 \end{bmatrix} W_{r+1}(\overline{z})^*\mathcal{P}(z)\right)^{-1}, \tag{5.94}$$

where $\mathcal{P}$ in (5.94) differs from $\mathcal{P}$ in (5.35) by the factor j (and so this $\mathcal{P}$ is also a nonsingular matrix function with property-j).

Definition 5.17. Weyl functions of Dirac system (5.1) (which is given on the interval $0 \leq k \leq r$ and satisfies (5.2)) are $p \times p$ matrix functions $\varphi(z)$ of the form (5.94), where $\mathcal{P}$ are nonsingular matrix functions with property-j such that $\mathcal{P}(\mathrm{i})$ are well-defined and nonsingular.

It is apparent that (5.94) is equivalent to

$$\begin{bmatrix} I_p \\ \varphi_r(z,\mathcal{P}) \end{bmatrix} = jW_{r+1}(\overline{z})^*\mathcal{P}(z)\left(\begin{bmatrix} I_p & 0 \end{bmatrix} W_{r+1}(\overline{z})^*\mathcal{P}(z)\right)^{-1}. \tag{5.95}$$

Lemma 5.18. *Let $\mathcal{P}$ satisfy conditions from Definition 5.17. Then, we have the inequality*

$$\det\left(\begin{bmatrix} I_p & 0 \end{bmatrix} W_{r+1}(-\mathrm{i})^* \mathcal{P}(\mathrm{i})\right) \neq 0. \tag{5.96}$$

Proof. First, note that in view of (5.14), we obtain

$$I_m + C_k j = 2\beta(k)^* \beta(k) j. \tag{5.97}$$

Formulas (5.37) and (5.97) imply

$$\begin{aligned} \begin{bmatrix} I_p & 0 \end{bmatrix} W_{r+1}(-\mathrm{i})^* \mathcal{P}(\mathrm{i}) &= 2^{r+1} \left(\begin{bmatrix} I_p & 0 \end{bmatrix} \beta(0)^*\right) (\beta(0) j \beta(1)^*) \dots \\ &\quad \times (\beta(r-1) j \beta(r)^*)(\beta(r) j \mathcal{P}(\mathrm{i})). \end{aligned} \tag{5.98}$$

Substituting $J = j$, $\vartheta = \beta(k)^*$ and $\widetilde{\vartheta} = \beta(k+1)^*$ or $\widetilde{\vartheta} = \mathcal{P}(\mathrm{i})$ into Proposition 1.43 and using the second equality in (5.14), we derive inequalities

$$\det(\beta(k) j \beta(k+1)^*) \neq 0 \quad \text{and} \quad \det(\beta(r) j \mathcal{P}(\mathrm{i})) \neq 0, \tag{5.99}$$

respectively. In the same way, we obtain $\det([I_p \quad 0]\beta(0)^*) \neq 0$. Now, inequality (5.96) follows from (5.98). □

Our next proposition is proved similar to Corollary 5.11.

Proposition 5.19. *Let dDS (5.1), where the potential $\{C_k\}$ satisfies (5.2), be given on the interval $0 \leq k \leq r$. Assume that φ is its Weyl function. Then, φ is a Weyl function of the same system on all the intervals $0 \leq k \leq \widetilde{r}$ ($\widetilde{r} \leq r$).*

Proof. Clearly, it suffices to show that the statement of the proposition holds for $\widetilde{r} = r - 1$ (if $r > 0$). That is, in view of Definition 5.17, we should prove that $\widetilde{\mathcal{P}}(z) := (I_m - \mathrm{i}zC_r j)\mathcal{P}(z)$ has the property-j, that $\widetilde{\mathcal{P}}(\mathrm{i})$ is well-defined and that the first inequality in (5.36) written for $\widetilde{\mathcal{P}}$ at $z = \mathrm{i}$ always holds (i.e. $\widetilde{\mathcal{P}}(\mathrm{i})$ is nonsingular) if $\mathcal{P}$ has these properties.

Indeed, since we have

$$(I_m - \mathrm{i}zC_r j)^* j (I_m - \mathrm{i}zC_r j) = (1 + |z|^2) j + \mathrm{i}(\overline{z} - z) j C_r j \geq (1 + |z|^2) j, \tag{5.100}$$

the matrix function $\widetilde{\mathcal{P}}$ has the property-j. The nonsingularity of

$$\widetilde{\mathcal{P}}(\mathrm{i}) = (I_m + C_r j)\mathcal{P}(\mathrm{i})$$

is apparent from (5.97) and (5.99). □

Theorem 5.20. *Let dDS (5.1), where the potential $\{C_k\}$ satisfies (5.2), be given on the interval $0 \leq k \leq r$. Assume that φ is its Weyl function. Then, $\{C_k\}_{k=0}^r$ is uniquely recovered from the first $r + 1$ Taylor coefficients of $\varphi(\mathrm{i}\frac{1-z}{1+z})$ at $z = 0$.*

Namely, if $\varphi(\mathrm{i}\frac{1-z}{1+z}) = \sum_{k=0}^{r} \phi_k z^k + O(z^{r+1})$, *then matrices* $\Phi_{k,1}$ *are recovered via the formula*

$$\Phi_{k,1} = -\begin{bmatrix} \phi_0 \\ \phi_0 + \phi_1 \\ \cdots \\ \phi_0 + \phi_1 + \ldots + \phi_k \end{bmatrix}. \tag{5.101}$$

Using $\Phi_{k,1}$, *we easily consecutively recover* $\Pi_k = \begin{bmatrix} \Phi_{k,1} & \Phi_{k,2} \end{bmatrix}$ *(where* $\Phi_{k,2}$ *is given in (5.67)) and* S_k, *which is the unique solution of the matrix identity*

$$A_k S_k - S_k A_k^* = \mathrm{i}\Pi_k j \Pi_k^*.$$

Next, we construct

$$\gamma(k)^*\gamma(k) = \Pi_k^* S_k^{-1} P^* (P S_k^{-1} P^*)^{-1} P S_k^{-1} \Pi_k, \quad P = \begin{bmatrix} 0 & \ldots & 0 & I_p \end{bmatrix}. \tag{5.102}$$

Finally, we use $\gamma(k)^*\gamma(k)$ *to recover* C_k *via (5.15).*

Proof. Put

$$\daleth(z) = \daleth_r(z) := |1+z^2|^{-2(r+1)} \begin{bmatrix} I_p & \varphi(z)^* \end{bmatrix} W_{r+1}(z)^* j W_{r+1}(z) \begin{bmatrix} I_p \\ \varphi(z) \end{bmatrix}. \tag{5.103}$$

According to (5.42) and (5.95), we have

$$\daleth(z) = \left(\left(\begin{bmatrix} I_p & 0 \end{bmatrix} W_{r+1}(\overline{z})^* \mathcal{P}(z) \right)^{-1} \right)^* \mathcal{P}(z)^* j \mathcal{P}(z) \times \left(\begin{bmatrix} I_p & 0 \end{bmatrix} W_{r+1}(\overline{z})^* \mathcal{P}(z) \right)^{-1}. \tag{5.104}$$

From (5.96) and (5.104), we see that $\daleth$ is bounded in the neighborhood of $z = \mathrm{i}$:

$$\|\daleth(z)\| = O(1) \quad \text{for } z \to \mathrm{i}. \tag{5.105}$$

Let us include into considerations the S-node corresponding to dDS (i.e. the S-node given by equalities (5.65) and (5.82), where Γ_r is given in (5.62)). Substitute (5.88) into (5.103) and use (1.88) to obtain

$$\daleth(z) = ((1 - \mathrm{i}z)(1 + \mathrm{i}\overline{z}))^{-r-1} \begin{bmatrix} I_p & \varphi(z)^* \end{bmatrix} \times \left(j - \frac{\Im(z)}{|z|^2} \Pi_r^* \left(A_r^* - \frac{1}{2\overline{z}} I \right)^{-1} S_r^{-1} \left(A_r - \frac{1}{2z} I \right)^{-1} \Pi_r \right) \begin{bmatrix} I_p \\ \varphi(z) \end{bmatrix}, \tag{5.106}$$

where $I = I_{(r+1)p}$. Notice that $S_r > 0$. Hence, formulas (5.44), (5.105) and (5.106) imply that

$$\left\| \left(A_r - \frac{1}{2z} I \right)^{-1} \Pi_r \begin{bmatrix} I_p \\ \varphi(z) \end{bmatrix} \right\| = O(1) \quad \text{for } z \to \mathrm{i}. \tag{5.107}$$

Using the block representation $\Pi_r = \begin{bmatrix} \Phi_{r,1} & \Phi_{r,2} \end{bmatrix}$ from (5.82) and multiplying both sides of (5.107) by $\|(\Phi_{r,2}^*(A_r - \frac{1}{2z}I)^{-1}\Phi_{r,2})^{-1}\Phi_{r,2}^*\|$, we rewrite the result:

$$\left\| \varphi(z) + \left(\Phi_{r,2}^*\left(A_r - \frac{1}{2z}I\right)^{-1}\Phi_{r,2}\right)^{-1}\Phi_{r,2}^*\left(A_r - \frac{1}{2z}I\right)^{-1}\Phi_{r,1}\right\|$$
$$= O\left(\left\|\left(\Phi_{r,2}^*\left(A_r - \frac{1}{2z}I\right)^{-1}\Phi_{r,2}\right)^{-1}\right\|\right) \quad \text{for } z \to \mathrm{i}. \tag{5.108}$$

In order to obtain (5.108), we also applied the matrix (operator) norm inequality $\|X_1X_2\| \le \|X_1\|\|X_2\|$.

The resolvent $(A_r - \lambda I)^{-1}$ is easily constructed explicitly (see, e.g. [250, (1.10)]):

$$(A_r - \lambda I)^{-1} = \{R_{ik}(\lambda)\}_{i,k=0}^r, \tag{5.109}$$
$$R_{ik}(\lambda) = \begin{cases} 0 \quad \text{for } i < k, \\ -(\lambda + (\mathrm{i}/2))^{-1} I_p \quad \text{for } i = k, \\ \mathrm{i}(\lambda + (\mathrm{i}/2))^{k-i-1}(\lambda - (\mathrm{i}/2))^{i-k-1} I_p \quad \text{for } i > k. \end{cases}$$

In particular, we derive

$$\Phi_{r,2}^*\left(A_r - \frac{1}{2z}I\right)^{-1} = -\frac{2z}{1+\mathrm{i}z}\begin{bmatrix}\hat{q}(z)^r & \hat{q}(z)^{r-1} & \dots & I_p\end{bmatrix}, \quad \hat{q} := \frac{1-\mathrm{i}z}{1+\mathrm{i}z}I_p. \tag{5.110}$$

From (5.110), we see that

$$\Phi_{r,2}^*\left(A_r - \frac{1}{2z}I\right)^{-1}\Phi_{r,2} = \mathrm{i}\left(1 - \left(\frac{1-\mathrm{i}z}{1+\mathrm{i}z}\right)^{r+1}\right)I_p. \tag{5.111}$$

Partitioning $\Phi_{r,1}$ into the $p \times p$ blocks $\Phi_{r,1}(k)$ and using (5.108)–(5.111), we obtain

$$\varphi\left(\mathrm{i}\frac{1-z}{1+z}\right) + \frac{1-z}{1-z^{r+1}}\sum_{k=0}^r z^k\Phi_{r,1}(k) = O(z^{r+1}) \quad \text{for } z \to 0,$$

which can be easily transformed into

$$\varphi\left(\mathrm{i}\frac{1-z}{1+z}\right) + (1-z)\sum_{k=0}^r z^k\Phi_{r,1}(k) = O(z^{r+1}) \quad \text{for } z \to 0, \tag{5.112}$$

and (5.101) follows for $k = r$. Since $\sigma(A_r) \cap \sigma(A_r^*) = \emptyset$, the matrix S_r is uniquely recovered from the matrix identity (5.81). Finally, (5.102) (for the case where $k = r$) is apparent from (5.91). From Proposition 5.19, we see that φ is a Weyl function of our Dirac system on all the intervals $0 \le k \le \tilde{r}$ ($\tilde{r} \le r$) and so all $C_{\tilde{r}}$ are recovered in the same way as C_r. □

The next corollary is a discrete version of Borg–Marchenko-type uniqueness theorems. Recall that the active study of the Borg–Marchenko-type theorems was initiated by the seminal papers by F. Gesztesy and B. Simon [121, 122]. Interesting Borg–Marchenko-type results for discrete systems are contained, for instance, in [77, 125, 259, 329].

Corollary 5.21. *Suppose φ and $\widetilde{\varphi}$ are Weyl functions of two Dirac systems with potentials $\{C_k\}$ and $\{\widetilde{C}_k\}$, which are given on the intervals $0 \le k \le r$ and $0 \le k \le \widetilde{r}$, respectively. We suppose that matrices $\{C_k\}$ and $\{\widetilde{C}_k\}$ are positive and j-unitary. Moreover, we assume that*

$$\varphi\left(\mathrm{i}\frac{1-z}{1+z}\right) - \widetilde{\varphi}\left(\mathrm{i}\frac{1-z}{1+z}\right) = O(z^{r_0+1}), \quad z \to \infty, \quad r_0 \in \mathbb{N}_0, \quad r_0 \le \min(r, \widetilde{r}). \tag{5.113}$$

Then, we have $C_k = \widetilde{C}_k$ for all $0 \le k \le r_0$.

Proof. According to Proposition 5.19, both functions φ and $\widetilde{\varphi}$ are Weyl functions of the corresponding Dirac systems on the same interval $[0, r_0]$. From (5.113), we see that the first $r_0 + 1$ Taylor coefficients of $\varphi(\mathrm{i}\frac{1-z}{1+z})$ and $\widetilde{\varphi}(\mathrm{i}\frac{1-z}{1+z})$ coincide. Hence, the uniqueness of the recovery of the potential from Taylor coefficients in Theorem 5.20 yields $C_k = \widetilde{C}_k$ $(0 \le k \le r_0)$. □

Taking into account (5.101), we derive that the first $r + 1$ Taylor coefficients of $\varphi_r(\mathrm{i}\frac{1-z}{1+z})$ at $z = 0$ (for any Weyl function φ_r of a fixed dDS on the interval $0 \le k \le r$) can be uniquely recovered from the matrix $\Phi_{r,1}$. The matrix $\Phi_{r,1}$, in turn, can be constructed via (5.62) and (5.82). Therefore, the next theorem is apparent.

Theorem 5.22. *Let Dirac system* (5.1), *where matrices C_k satisfy* (5.2), *be given on the interval $0 \le k \le r$. Then, all the functions $\varphi_d(z) := \varphi_r(\mathrm{i}\frac{1-z}{1+z}, \mathcal{P})$, where φ_r are Weyl functions of this Dirac system, are nonexpansive in the unit disc and have the same first $r + 1$ Taylor coefficients $\{\phi_k\}_0^r$ at $z = 0$.*

Step 1 in the proof of Theorem 5.14 shows that the Weyl function φ_∞ of dDS on the semiaxis can be constructed as a uniform limit of Weyl functions φ_r on increasing intervals. Hence, using Theorem 5.22, we obtain the following corollary.

Corollary 5.23. *Let dDS* (5.1) *be given on the semiaxis $0 \le k \le \infty$ and satisfy* (5.2). *Assume that φ is the Weyl function of this dDS and φ_r is a Weyl function of the same system on the finite interval $0 \le k \le r$. Then, the first $r + 1$ Taylor coefficients of $\varphi(\mathrm{i}\frac{1-z}{1+z})$ and $\varphi_r(\mathrm{i}\frac{1-z}{1+z})$ coincide. Therefore, dDS on the semiaxis can be uniquely recovered from φ via the procedure from Theorem* 5.20.

5.2 Discrete skew-self-adjoint Dirac system

In this section, we consider the case of a discrete skew-self-adjoint Dirac system, that is,

$$y_{k+1}(z) = (I_m - \mathrm{i}C_k/z)\, y_k(z), \quad C_k = C_k^* = C_k^{-1}, \tag{5.114}$$

where C_k are $m \times m$ matrices, $m = 2p$ and $0 \le k \le n$. If $p = 1$, then either $C_k = \pm I_2$ or $C_k = \mathcal{U}(k)^* j\mathcal{U}(k)$, where $j = \operatorname{diag}\{1, -1\}$ and $\mathcal{U}(k)^*\mathcal{U}(k) = I_2$. If

$C_k = \mathcal{U}(k)^* j\mathcal{U}(k)$ $(k \geq 0)$, system (5.114) is an auxiliary system for isotropic Heisenberg magnet model [148] (see the next section for more details). Therefore, generalizing this important case, we assume

$$C_k = \mathcal{U}(k)^* j\mathcal{U}(k), \quad \mathcal{U}(k)^*\mathcal{U}(k) = I_{2p}, \quad j = \mathrm{diag}\,\{I_p, -I_p\}, \quad 0 \leq k \leq n \tag{5.115}$$

for $p \geq 1$. Clearly, (5.115) is equivalent to the representation

$$C_k = I_{2p} - 2\gamma(k)^*\gamma(k), \quad \gamma(k)\gamma(k)^* = I_p, \tag{5.116}$$

$$\gamma(k) = [\gamma_1(k) \quad \gamma_2(k)] := [0 \quad I_p]\mathcal{U}(k), \quad 0 \leq k \leq n, \tag{5.117}$$

where the blocks $\gamma_1(k)$ and $\gamma_2(k)$ are $p \times p$ blocks of $\gamma(k)$. Compare (5.116) with the representation (5.15) in the self-adjoint case. Now, introduce the simple additional conditions

$$\det \gamma_1(0) \neq 0, \quad \det \gamma(k-1)\gamma(k)^* \neq 0, \quad 0 < k \leq n. \tag{5.118}$$

In this section, we require that $\mathcal{P}_1$ and $\mathcal{P}_2$ are $p \times p$ matrix functions analytic in the neighborhood of $z = \mathrm{i}$ and such that

$$\det\left(\mathcal{W}_{21}(\mathrm{i})\mathcal{P}_1(\mathrm{i}) + \mathcal{W}_{22}(\mathrm{i})\mathcal{P}_2(\mathrm{i})\right) \neq 0, \tag{5.119}$$

where

$$\mathcal{W}(z) = \{\mathcal{W}_{ij}(z)\}_{i,j=1}^2 := W_{n+1}(\overline{z})^*, \tag{5.120}$$

$$W_{k+1}(z) = (I_m - \mathrm{i}C_k/z)\, W_k(z), \quad W_0(z) = I_m, \tag{5.121}$$

that is, W is a normalized fundamental solution of (5.114).

Remark 5.24. The pairs $\{\mathcal{P}_1, \mathcal{P}_2\}$ satisfying (5.119) always exist since the rows of $[\mathcal{W}_{21}(\mathrm{i}) \quad \mathcal{W}_{22}(\mathrm{i})]$ are linearly independent, that is,

$$\mathrm{rank}\,[\mathcal{W}_{21}(\mathrm{i}) \quad \mathcal{W}_{22}(\mathrm{i})] = p. \tag{5.122}$$

Indeed, first, using (5.121) and representation $C_k = \mathcal{U}(k)^* j\mathcal{U}(k)$, we derive

$$W(-\mathrm{i}) = 2^{n+1} \prod_{k=0}^{n} \beta(k)^*\beta(k), \quad \beta(k) = [\beta_1(k) \quad \beta_2(k)] := [I_p \quad 0]\mathcal{U}(k). \tag{5.123}$$

Notice that from $\mathcal{U}(k)^*\mathcal{U}(k) = I_{2p}$, we have

$$\beta(k)\gamma(k)^* = 0, \quad \beta(k)\beta(k)^* = I_p. \tag{5.124}$$

From (5.120) and (5.123), it follows that

$$[\mathcal{W}_{21}(\mathrm{i}) \quad \mathcal{W}_{22}(\mathrm{i})] = 2^{n+1}\beta_2(0)^* \left(\beta(0)\beta(1)^*\right) \ldots \left(\beta(n-1)\beta(n)^*\right) \beta(n). \tag{5.125}$$

It remains to show that inequalities (5.118) imply

$$\det(\beta_2(0)^*) \neq 0; \quad \det(\beta(k-1)\beta(k)^*) \neq 0, \quad 0 < k \leq n. \tag{5.126}$$

Suppose there is $g \neq 0$ such that $g^*\beta(k-1)\beta(k)^* = 0$. Since $\beta(k-1)^*g \perp \mathrm{Im}\,(\beta(k)^*)$, from (5.124), we see that $\beta(k-1)^*g \in \mathrm{Im}\,(\gamma(k)^*)$, and $\beta(k-1)^*g \neq 0$ (because $g \neq 0$). In other words, we have

$$\beta(k-1)^*g = \gamma(k)^*\widetilde{g}, \quad \widetilde{g} \neq 0.$$

Hence, taking into account the second inequality in (5.118), we obtain

$$\gamma(k-1)\beta(k-1)^*g = \gamma(k-1)\gamma(k)^*\widetilde{g} \neq 0.$$

Since the first equality in (5.124) yields $\gamma(k-1)\beta(k-1)^* = 0$, we arrive at a contradiction, that is, the second inequality in (5.126) is valid. The first inequality in (5.126) may be rewritten in the form $\det([0 \quad I_p]\beta(0)^*) \neq 0$ and proven in the same way as the second inequality.

Similar to the continuous case and discrete system (5.1), we define Weyl functions of the system (5.114) via Möbius (linear-fractional) transformation. Namely, we set

$$\varphi(z) = (\mathcal{W}_{11}(z)\mathcal{P}_1(z) + \mathcal{W}_{12}(z)\mathcal{P}_2(z))\,(\mathcal{W}_{21}(z)\mathcal{P}_1(z) + \mathcal{W}_{22}(z)\mathcal{P}_2(z))^{-1}. \tag{5.127}$$

Definition 5.25. Let system (5.114) be given on the interval $0 \leq k \leq n$ and satisfy (5.116), (5.118). Suppose $\mathcal{P}_1$ and $\mathcal{P}_2$ are $p \times p$ matrix functions analytic in the neighborhood of $z = \mathrm{i}$ and such that inequality (5.119) holds. Then, linear-fractional transformations φ of the form (5.127) are called Weyl functions of this system. The pair $\mathcal{P}_1$, $\mathcal{P}_2$ satisfying our conditions is called admissible.

Example 5.26. Put

$$n = 1, \quad C_0 = -j \quad \text{and} \quad C_1 = J, \tag{5.128}$$

where j, J and also Θ below coincide with j, J and Θ in (1.7), that is,

$$J = \begin{bmatrix} 0 & I_p \\ I_p & 0 \end{bmatrix} = \Theta j \Theta^*, \quad \Theta = \frac{1}{\sqrt{2}}\begin{bmatrix} I_p & -I_p \\ I_p & I_p \end{bmatrix}, \quad \Theta^* = \Theta^{-1}. \tag{5.129}$$

It is easy to recover matrices $\mathcal{U}(k)$ such that representation (5.115) holds. In view of (5.128), we have $C_0 = JjJ$, and so $\mathcal{U}(0) = J$. From (5.129), we obtain $\mathcal{U}(1) = \Theta$. Hence, we see that

$$\gamma(0) = [0 \quad I_p]\mathcal{U}(0) = [I_p \quad 0], \quad \gamma(1) = [0 \quad I_p]\mathcal{U}(1) = \frac{1}{\sqrt{2}}[-I_p \quad I_p]. \tag{5.130}$$

Thus, the conditions (5.116) and (5.118) are fulfilled. Moreover, we have

$$W_2(z) = (I_m - \mathrm{i}J/z)(I_m + \mathrm{i}j/z) = I_m + \mathrm{i}z^{-1}(j - J) + z^{-2}Jj, \quad \text{i.e.}$$
$$\mathcal{W}(z) = I_m - \mathrm{i}z^{-1}(j - J) - z^{-2}Jj. \tag{5.131}$$

In particular, we have $\mathcal{W}_{21}(\mathrm{i}) = \mathcal{W}_{22}(\mathrm{i}) = 2I_p$ and so (5.119) takes the form

$$\det\,(\mathcal{P}_1(\mathrm{i}) + \mathcal{P}_2(\mathrm{i})) \neq 0. \tag{5.132}$$

In view of (5.131), Möbius transformation (5.127), which describes Weyl functions, takes the form

$$\varphi(z) = \frac{z - \mathrm{i}}{z + \mathrm{i}}(z\mathcal{P}_1 + \mathrm{i}\mathcal{P}_2)(\mathrm{i}\mathcal{P}_1 + z\mathcal{P}_2)^{-1}, \tag{5.133}$$

where $\mathcal{P}_1$ and $\mathcal{P}_2$ satisfy (5.132).

Notation 5.27. *The set of Weyl functions of system* (5.114) *on the interval* $0 \le k \le n$ *is denoted by* $\mathcal{N}(n)$.

Remark 5.28. Like in the continuous case, the sets $\mathcal{N}(n)$ are nested, that is, given system (5.114), we have

$$\mathcal{N}(l) \supseteq \mathcal{N}(n) \quad \text{for } l < n. \tag{5.134}$$

Indeed, for the pair

$$\begin{bmatrix} \widetilde{\mathcal{P}}_1(z) \\ \widetilde{\mathcal{P}}_2(z) \end{bmatrix} := (I_m + \mathrm{i}C_{l+1}/z) \times \ldots \times (I_m + \mathrm{i}C_n/z) \begin{bmatrix} \mathcal{P}_1(z) \\ \mathcal{P}_2(z), \end{bmatrix} \tag{5.135}$$

we have

$$W_{l+1}(\overline{z})^* \begin{bmatrix} \widetilde{\mathcal{P}}_1(z) \\ \widetilde{\mathcal{P}}_2(z) \end{bmatrix} = W_{n+1}(\overline{z})^* \begin{bmatrix} \mathcal{P}_1(z) \\ \mathcal{P}_2(z) \end{bmatrix} = \mathcal{W}(z) \begin{bmatrix} \mathcal{P}_1(z) \\ \mathcal{P}_2(z) \end{bmatrix}. \tag{5.136}$$

So, if the pair $\mathcal{P}_1$, $\mathcal{P}_2$ is admissible for system (5.114) on the interval $0 \le k \le n$, then the pair $\widetilde{\mathcal{P}}_1$, $\widetilde{\mathcal{P}}_2$ is admissible for system (5.114) on the interval $0 \le k \le l$. Moreover, let φ be a Weyl function of system (5.114) on the interval $0 \le k \le n$ and be determined by some admissible pair $\mathcal{P}_1$, $\mathcal{P}_2$. Then, φ coincides with the Weyl function of the same system (5.114) on a smaller interval $0 \le k \le l$, which is determined by the admissible pair $\widetilde{\mathcal{P}}_1$, $\widetilde{\mathcal{P}}_2$. Therefore, (5.134) holds.

The next theorem solves the inverse problem to recover system (5.114) from its Weyl function.

Theorem 5.29. *Let system* (5.114) *satisfying conditions* (5.116) *and* (5.118) *be given on the interval* $0 \le k \le n$ *and suppose that* φ *is a Weyl function of this system. Then, system* (5.114) *is uniquely recovered from the first* $n + 1$ *Taylor coefficients* $\{\phi_k\}_{k=0}^n$ *of* $\varphi(\mathrm{i}\frac{1+\lambda}{1-\lambda})$ *at* $\lambda = 0$ *using the following procedure.*

First, introduce $(r+1)p \times p$ *matrices* $\Phi_{r,1}$ *and* $\Phi_{r,2}$ $(r \leq n)$:

$$\Phi_{r,1} = -\begin{bmatrix} \phi_0 \\ \phi_0 + \phi_1 \\ \cdots \\ \phi_0 + \phi_1 + \ldots + \phi_r \end{bmatrix}, \quad \Phi_{r,2} = \begin{bmatrix} I_p \\ I_p \\ \cdots \\ I_p \end{bmatrix}. \tag{5.137}$$

Then, introduce $(r+1)p \times 2p$ *matrices* Π_r *and* $(r+1)p \times (r+1)p$ *block lower triangular matrices* $A(r)$ *by their blocks:* $\Pi_r = [\Phi_{r,1} \quad \Phi_{r,2}]$,

$$A_r := \left\{a_{k-i}\right\}_{i,k=0}^{n}, \qquad a_l = \begin{cases} 0 & \text{for} \quad l > 0 \\ (\mathrm{i}/2)I_p & \text{for} \quad l = 0 \\ \mathrm{i}I_p & \text{for} \quad l < 0 \end{cases}. \tag{5.138}$$

Next, we recover $(r+1)p \times (r+1)p$ *matrices* S_r *as unique solutions of the matrix identities*

$$A_r S_r - S_r A_r^* = \mathrm{i}\Pi_r \Pi_r^* \quad (0 \leq r \leq n). \tag{5.139}$$

These solutions are invertible and positive, that is, $S_r > 0$. *Finally, matrices* $\gamma(k)^*\gamma(k)$ *are easily recovered from the formula*

$$\Pi_r^* S_r^{-1} \Pi_r = \Gamma_r^* \Gamma_r, \quad \Gamma_r := \begin{bmatrix} \gamma(0) \\ \gamma(1) \\ \cdots \\ \gamma(r) \end{bmatrix} J, \tag{5.140}$$

where J *is given in* (5.129), *that is, we have*

$$\gamma(0)^*\gamma(0) = J\Pi_0^* S_0^{-1} \Pi_0 J; \quad \gamma(k)^*\gamma(k) = J(\Pi_k^* S_k^{-1} \Pi_k - \Pi_{k-1}^* S_{k-1}^{-1} \Pi_{k-1})J \tag{5.141}$$

for $k > 0$. *Now, matrices* C_k *and system* (5.114) *are defined via* (5.116).

Proof. Step 1. The method of the proof coincides with the method of the proof of Theorem 5.20. As in (5.62), we introduce matrices K_r via blocks $\kappa_r(i)$, that is,

$$K_r = \begin{bmatrix} \kappa_r(0) \\ \kappa_r(1) \\ \cdots \\ \kappa_r(r) \end{bmatrix}, \quad \kappa_r(i) := \mathrm{i}\gamma(i)[\gamma(0)^* \ldots \gamma(i-1)^* \quad \gamma(i)^*/2 \quad 0 \ldots 0], \tag{5.142}$$

where the expression for the $p \times (r+1)p$ matrix $\kappa_r(i)$ only slightly differs from (5.63). From (5.140) and (5.142), we obtain

$$K_r - K_r^* = \mathrm{i}\Gamma_r \Gamma_r^*. \tag{5.143}$$

By induction, we show in the next step that K_r is similar to A_r:

$$K_r = E_r A_r E_r^{-1} \quad (0 \leq r \leq n), \tag{5.144}$$

where $E_r^{\pm 1}$ are block lower triangular matrices, which satisfy relations

$$E_n = \begin{bmatrix} E_r & 0 \\ \dots & \dots \end{bmatrix}, \quad E_n^{-1} = \begin{bmatrix} E_r^{-1} & 0 \\ \dots & \dots \end{bmatrix}. \tag{5.145}$$

Taking into account (5.144) (and (5.145)) and multiplying both sides of (5.143) by E_r^{-1} from the left and by $(E_r^*)^{-1}$ from the right, we derive

$$A_r\widetilde{S}_r - \widetilde{S}_r A_r^* = \mathrm{i}\widetilde{\Pi}_r\widetilde{\Pi}_r^*, \tag{5.146}$$

$$\widetilde{S}_r := E_r^{-1}\left(E_r^*\right)^{-1}, \quad \widetilde{\Pi}_r := E_r^{-1}\Gamma_r, \quad \widetilde{\Pi}_r = \begin{bmatrix} I_{(r+1)p} & 0 \end{bmatrix}\widetilde{\Pi}_n. \tag{5.147}$$

Moreover, Step 3 shows that the matrices E_r can be chosen so that $\widetilde{\Pi}_n = \Pi_n = [\Phi_{n,1} \ \ \Phi_{n,2}]$, where $\Phi_{n,k}$ $(k = 1,2)$ are given in (5.137). In view of the last equality in (5.147), the relation $\widetilde{\Pi}_n = \Pi_n$ implies that $\widetilde{\Pi}_r = \Pi_r$. Hence, identities (5.139) and (5.146) coincide and they also have unique solutions because the spectra of A and A^* don't intersect. That is, we have $S_r = \widetilde{S}_r > 0$. Using $\widetilde{\Pi}_r = \Pi_r$ and $S_r = \widetilde{S}_r$, we also derive from (5.147) that (5.140) holds. It remains only to prove that (5.144) holds and that $\widetilde{\Pi}_n = \Pi_n$.

Step 2. We introduce matrices E_r $(0 \le r \le n)$ via relations

$$E_0 = e_0^- = \gamma_1(0), \quad E_r = \begin{bmatrix} E_{r-1} & 0 \\ X_r & e_r^- \end{bmatrix} \quad (r > 0), \tag{5.148}$$

where e_r^- is a $p \times p$ matrix, X_r is a $p \times rp$ matrix and equalities (5.144) and

$$E_r^{-1}\Gamma_{r,2} = \Phi_{r,2}, \qquad \Gamma_{r,2} := \Gamma_r \begin{bmatrix} 0 \\ I_p \end{bmatrix} \tag{5.149}$$

are required. As in (5.74), we partition X_r into two blocks $X_r = \begin{bmatrix} x_r^- & \widetilde{X}_r \end{bmatrix}$. We set

$$\widetilde{X}_r = \mathrm{i}\left(\gamma(r)[\gamma(0)^* \ \dots \ \gamma(r-1)^*]E_{r-1}\begin{bmatrix} I_{(r-1)p} \\ 0 \end{bmatrix} - e_r^-[I_p \ \dots \ I_p]\right)$$
$$\times \left(A_{r-2} + \frac{\mathrm{i}}{2}I_{(r-1)p}\right)^{-1}, \quad e_r^- = \gamma(r)\gamma(r-1)^* e_{r-1}^-. \tag{5.150}$$

Let us show that (5.144) is valid. (The choice of x_r^- and proof of (5.149) follow in Step 3.) According to (5.138), we have $A_0 = (\mathrm{i}/2)I_p$. From the second relation in (5.116) and definition (5.142), it is immediate that $K_0 = (\mathrm{i}/2)I_p$, and so (5.144) is valid for $r = 0$.

Assume that (5.144) is valid for $r = k-1$, and let us show that (5.144) is valid for $r = k$ too. It is easy to see that

$$E_k^{-1} = \begin{bmatrix} E_{k-1}^{-1} & 0 \\ -(e_k^-)^{-1}X_k E_{k-1}^{-1} & (e_k^-)^{-1} \end{bmatrix}. \tag{5.151}$$

Then, in view of definitions (5.138) and (5.148), our assumption implies

$$E_k A_k E_k^{-1} = \begin{bmatrix} K_{k-1} & 0 \\ Y_k & \frac{\mathrm{i}}{2}I_p \end{bmatrix}, \tag{5.152}$$

where

$$Y_k := \left[\left(X_k A_{k-1} + \mathrm{i}e_k^-[I_p \ \dots \ I_p]\right) \quad \tfrac{\mathrm{i}}{2}e_k^-\right]\begin{bmatrix} E_{k-1}^{-1} \\ -(e_k^-)^{-1}X_k E_{k-1}^{-1}\end{bmatrix}.$$

Rewrite the product on the right-hand side of the last formula as

$$Y_k = \left(X_k\left(A_{k-1} - (\mathrm{i}/2)I_{kp}\right) + \mathrm{i}e_k^-[I_p \ \dots \ I_p]\right)E_{k-1}^{-1}. \tag{5.153}$$

From (5.74), (5.138) and (5.153), we derive

$$Y_k = \left[\left(\widetilde{X}_k\left(A_{k-2} + \tfrac{\mathrm{i}}{2}I_{(k-1)p}\right) + \mathrm{i}e_k^-[I_p \ \dots \ I_p]\right) \quad \mathrm{i}e_k^-\right]E_{k-1}^{-1}. \tag{5.154}$$

Notice that the row $[I_p \ \dots \ I_p]$ of identity matrices' blocks in (5.154) is one block shorter than in (5.153). According to (5.150) and (5.154), we have

$$Y_k = \mathrm{i}\gamma(k)\left[[\gamma(0)^* \ \dots \ \gamma(k-1)^*]E_{k-1}\begin{bmatrix} I_{(k-1)p} \\ 0\end{bmatrix} \quad \gamma(k-1)^*e_{k-1}^-\right]E_{k-1}^{-1}. \tag{5.155}$$

Finally, formulas (5.148) and (5.155) imply

$$Y_k = \mathrm{i}\gamma(k)[\gamma(0)^* \ \dots \ \gamma(k-1)^*]. \tag{5.156}$$

Taking into account the second relation in (5.116) and formulas (5.142) and (5.156), we obtain

$$\begin{bmatrix} Y_k & \tfrac{\mathrm{i}}{2}I_p\end{bmatrix} = \kappa_k(k). \tag{5.157}$$

Now, using (5.142) and (5.156), we see that the right-hand side of (5.152) equals K_k. Thus, (5.144) is valid for $r = k$ and, therefore, it holds for all $0 \le r \le n$.

Step 3. In order to obtain (5.149), we note that the equality (5.149) for $r = 0$ is immediate from the definitions of Γ_r and E_0 in (5.140) and (5.148), respectively. Assuming that (5.149) is valid for $r = k-1$ and taking into account (5.151), we see that (5.149) holds for $r = k$ if and only if

$$-(e_k^-)^{-1}X_k\begin{bmatrix} I_p \\ \cdots \\ I_p\end{bmatrix} + (e_k^-)^{-1}\gamma_1(k) = I_p. \tag{5.158}$$

Using the partitioning (5.74), we rewrite (5.158) as a definition of x_k^-:

$$x_k^- = \gamma_1(k) - e_k^- - \widetilde{X}(k)\begin{bmatrix} I_p \\ \cdots \\ I_p\end{bmatrix}. \tag{5.159}$$

In other words, we introduce x_k^- via (5.159) and obtain (5.149) for all $r \le n$.

Finally, in order to prove the equality $\widetilde{\Pi}_n = \Pi_n$, we show that

$$E_n^{-1}\Gamma_{n,1} = \Phi_{n,1}, \quad \text{where} \quad \Gamma_{n,1} := \Gamma_n \begin{bmatrix} I_p \\ 0 \end{bmatrix} = \begin{bmatrix} y_2(0) \\ \cdots \\ y_2(n) \end{bmatrix}. \tag{5.160}$$

For that purpose, we first prove the transfer matrix function representation of the fundamental solution W_k at $k = n+1$:

$$W_{n+1}(z) = z^{-n-1}(z-\mathrm{i})^{n+1} w_A(n, z/2). \tag{5.161}$$

Here, formula (5.161) is an analog (for the skew-self-adjoint case) of (5.88), $W_k(z)$ is introduced in (5.121) and w_A has the form

$$w_A(r,z) = I_{2p} - \mathrm{i}J\widetilde{\Pi}_r^* \widetilde{S}_r^{-1}\left(A_r - zI_{(r+1)p}\right)^{-1}\widetilde{\Pi}_r J, \tag{5.162}$$

that is, we substitute $\Pi = \widetilde{\Pi}J$ into the definition (1.64) of the transfer matrix function for the skew-self-adjoint Dirac system (and note that the triple $\{A_r, \widetilde{S}_r, \widetilde{\Pi}_r J\}$ forms a symmetric S-node). It is immediate from (5.138), (5.146) and the last equality in (5.147) that

$$P_1 A_r P_1^* = A_{r-1}, \quad P_1 \widetilde{S}_r P_1^* = \widetilde{S}_{r-1} \quad \text{for } r > 0, \quad P_1 := \begin{bmatrix} I_{rp} & 0 \end{bmatrix}. \tag{5.163}$$

Therefore, putting

$$P_2 = P = [0 \quad \dots \quad 0 \quad I_p] \tag{5.164}$$

and using factorization Theorem 1.16, we obtain

$$\begin{aligned} w_A(r,z) = {}& \left(I_{2p} - \mathrm{i}J\widetilde{\Pi}_r^* \widetilde{S}_r^{-1}P^*\left(PA_rP^* - zI_p\right)^{-1}\left(P\widetilde{S}_r^{-1}P^*\right)^{-1}P\widetilde{S}_r^{-1}\widetilde{\Pi}_r J\right) \\ & \times w_A(r-1,z). \end{aligned} \tag{5.165}$$

Taking into account (5.138), (5.147) and (5.148), we derive

$$\left(PA_rP^* - zI_p\right)^{-1} = ((\mathrm{i}/2) - z)^{-1} I_p, \quad P\widetilde{S}_r^{-1}P^* = (e_r^-)^* e_r^-, \tag{5.166}$$

$$P\widetilde{S}_r^{-1}\widetilde{\Pi}_r = (e_r^-)^* P\Gamma_r = (e_r^-)^* y(r) J. \tag{5.167}$$

Substituting (5.166) and (5.167) into (5.165), we have

$$w_A(r, z/2) = \left(I_{2p} - 2\mathrm{i}(\mathrm{i}-z)^{-1} y(r)^* y(r)\right) w_A(r-1, z/2), \quad r > 0. \tag{5.168}$$

From the definitions (5.138), (5.147), (5.148) and (5.162), we also easily derive

$$w_A(0, z/2) = I_{2p} - 2\mathrm{i}(\mathrm{i}-z)^{-1} y(0)^* y(0). \tag{5.169}$$

On the other hand, using (5.116), we rewrite (5.121) as

$$W(r+1,z) = (z-\mathrm{i})z^{-1}\left(I_m - 2\mathrm{i}(\mathrm{i}-z)^{-1}\gamma(r)^*\gamma(r)\right)W(r,z), \quad W(0,z) = I_m. \tag{5.170}$$

It is easy to see that formulas (5.168)–(5.170) imply (5.161).

Now, as in Chapter 3, we include Weyl functions into consideration. Introduce

$$\beth(z) := \left|\frac{z}{z-\mathrm{i}}\right|^{2n+2} [\varphi(z)^* \quad I_p] W(n+1,z)^* W(n+1,z) \begin{bmatrix} \varphi(z) \\ I_p \end{bmatrix}. \tag{5.171}$$

Notice that for w_A from this section, we have $J = I_m$ in (1.87), and so $w_A(n, z/2) w_A(n, \overline{z}/2)^* = I_m$. Hence, in view of (5.161), we obtain

$$W(n+1,z)W(n+1,\overline{z})^* = (z-\mathrm{i})^{n+1}(z+\mathrm{i})^{n+1}z^{-2n-2}I_m. \tag{5.172}$$

From (5.120), (5.127) and (5.172), we derive

$$W(n+1,z)\begin{bmatrix} \varphi(z) \\ I_p \end{bmatrix} = \left(\frac{z^2+1}{z^2}\right)^{n+1} \begin{bmatrix} \mathcal{P}_1(z) \\ \mathcal{P}_2(z) \end{bmatrix} (\mathcal{W}_{21}(z)\mathcal{P}_1(z) + \mathcal{W}_{22}(z)\mathcal{P}_2(z))^{-1}. \tag{5.173}$$

Therefore, (5.171) takes the form

$$\begin{aligned} \beth(z) = \left|\frac{z+\mathrm{i}}{z}\right|^{2n+2} & \left((\mathcal{W}_{21}(z)\mathcal{P}_1(z) + \mathcal{W}_{22}(z)\mathcal{P}_2(z))^*\right)^{-1} \\ & \times (\mathcal{P}_1(z)^*\mathcal{P}_1(z) + \mathcal{P}_2(z)^*\mathcal{P}_2(z)) (\mathcal{W}_{21}(z)\mathcal{P}_1(z) + \mathcal{W}_{22}(z)\mathcal{P}_2(z))^{-1}. \end{aligned} \tag{5.174}$$

According to (5.119) and (5.174), the matrix function $\beth$ is bounded in the neighborhood of $z = \mathrm{i}$:

$$\|\beth(z)\| = O(1) \quad \text{for } z \to \mathrm{i}. \tag{5.175}$$

Now, we rewrite (5.171) in a different way. For that purpose, we substitute (5.161) into (5.171) and use (1.88):

$$\begin{aligned} \beth(z) = [\varphi(z)^* \quad I_p] & \Big(I_{2p} + (\mathrm{i}/2)(\overline{z}-z)J\widetilde{\Pi}_n^* \left(A_n^* - (\overline{z}/2)I_{(n+1)p}\right)^{-1} \widetilde{S}_n^{-1} \\ & \times \left(A_n - (z/2)I_{(n+1)p}\right)^{-1} \widetilde{\Pi}_n J \Big) \begin{bmatrix} \varphi(z) \\ I_p \end{bmatrix}. \end{aligned} \tag{5.176}$$

Recall that $\widetilde{S}_n > 0$. Hence, formulas (5.175) and (5.176) imply that

$$\left\| \left(A_n - (z/2)I_{(n+1)p}\right)^{-1} \widetilde{\Pi}_n J \begin{bmatrix} \varphi(z) \\ I_p \end{bmatrix} \right\| = O(1) \quad \text{for } z \to \mathrm{i}. \tag{5.177}$$

Applying $\widetilde{\Phi}_{n,2}^*$ to the matrix function in the left-hand side of (5.177) and taking into account the block representation $\widetilde{\Pi}_n = \begin{bmatrix} \widetilde{\Phi}_{n,1} & \widetilde{\Phi}_{n,2} \end{bmatrix}$, we obtain

$$\begin{aligned} & \left\| \varphi(z) + \left(\widetilde{\Phi}_{n,2}^* \left(A_n - (z/2)I_{(n+1)p}\right)^{-1} \widetilde{\Phi}_{n,2} \right)^{-1} \widetilde{\Phi}_{n,2}^* \left(A_n - (z/2)I_{(n+1)p}\right)^{-1} \widetilde{\Phi}_{n,1} \right\| \\ & = O\left(\left\| \left(\widetilde{\Phi}_{n,2}^* \left(A_n - (z/2)I_{(n+1)p}\right)^{-1} \widetilde{\Phi}_{n,2} \right)^{-1} \right\| \right) \quad \text{for } z \to \mathrm{i}. \end{aligned} \tag{5.178}$$

It was already proved in Step 3 that $\widetilde{\Phi}_{n,2} = \Phi_{n,2}$, where $\Phi_{n,2}$ is defined in (5.137) (in the same way as in (5.67)). Thus, we can rewrite (5.110) and (5.111), taking into account that the matrix A_n is given in (5.138) (and coincides with $-A_n$ for A_n considered in (5.110) and (5.111)):

$$\widetilde{\Phi}_{n,2}^* \left(A_n - (z/2)I_{(n+1)p}\right)^{-1} = 2(\mathrm{i} - z)^{-1} \begin{bmatrix} \lambda(z)^{-n} & \lambda(z)^{1-n} & \dots & I_p \end{bmatrix},$$
$$\widetilde{\Phi}_{n,2}^* \left(A_n - (z/2)I_{(n+1)p}\right)^{-1} \widetilde{\Phi}_{n,2} = \mathrm{i}(\lambda(z)^{-n-1} - 1)I_p, \quad \lambda(z) := (z - \mathrm{i})(z + \mathrm{i})^{-1}. \tag{5.179}$$

We easily inverse the function $\lambda(z)$, which was introduced in (5.179), and derive the formula $z(\lambda) = \mathrm{i}(1 + \lambda)(1 - \lambda)^{-1}$. Hence, in view of (5.179), relation (5.178) yields

$$\left\| \varphi\left(\mathrm{i}\frac{1+\lambda}{1-\lambda}\right) + (1-\lambda)[I_p \quad \lambda I_p \quad \lambda^2 I_p \quad \dots]\widetilde{\Phi}_{n,1} \right\| = O(\lambda^{n+1}) \tag{5.180}$$

for $\lambda \to 0$. According to (5.137) and (5.180), the equality $\widetilde{\Phi}_{n,1} = \Phi_{n,1}$ holds. Thus, the equality $\widetilde{\Pi}_n = \Pi_n$ is also proved. As already explained in Step 1, the statement of the theorem follows from the relations (5.144) and $\widetilde{\Pi}_n = \Pi_n$. □

From Theorem 5.29 and Remark 5.28, we obtain a Borg–Marchenko-type result.

Theorem 5.30. *Let $\widetilde{\varphi}$ and $\widehat{\varphi}$ be Weyl functions of the systems* (5.114) *with the potentials $\widetilde{C}_k$ $(0 \leq k \leq \widetilde{n})$ and $\widehat{C}_k$ $(0 \leq k \leq \widehat{n})$, respectively. Suppose that these potentials satisfy conditions* (5.116) *and* (5.118). *Denote Taylor coefficients of $\widetilde{\varphi}(\mathrm{i}\frac{1+\lambda}{1-\lambda})$ and $\widehat{\varphi}(\mathrm{i}\frac{1+\lambda}{1-\lambda})$ at $\lambda = 0$ by $\{\widetilde{\phi}_k\}$ and $\{\widehat{\phi}_k\}$, respectively, and assume that $\widetilde{\phi}_k = \widehat{\phi}_k$ for $k \leq l$ $(l \leq \min(\widetilde{n}, \widehat{n}))$. Then, we have*

$$\widetilde{C}_k = \widehat{C}_k \quad \text{for } 0 \leq k \leq l. \tag{5.181}$$

Proof. Remark 5.28 implies that $\widetilde{\varphi}$ and $\widehat{\varphi}$ are Weyl functions of the first and second systems, respectively, on the interval $0 \leq k \leq l$. According to Theorem 5.29, these systems (i.e. their potentials) on the interval $0 \leq k \leq l$ are uniquely recovered from the first $l + 1$ Taylor coefficients of the Weyl functions. □

Step 3 of the proof of Theorem 5.29 leads us to the following corollary.

Corollary 5.31. *All Weyl functions of system* (5.114) *that satisfy conditions* (5.116) *and* (5.118) *admit a Taylor representation*

$$\varphi\left(\mathrm{i}\frac{1+\lambda}{1-\lambda}\right) = -\psi_0 + (\psi_0 - \psi_1)\lambda + \dots + (\psi_{n-1} - \psi_n)\lambda^n + O(\lambda^{n+1}) \tag{5.182}$$

for $\lambda \to 0$. Here, ψ_k are $p \times p$ blocks of $\widetilde{\Phi}_{n,1} = \{\psi_k\}_{k=0}^n$; $\widetilde{\Phi}_{n,1}$, being the left block of $\widetilde{\Pi}_n$, is given by the second equality in (5.147); *Γ_r in* (5.147) *is defined in* (5.140) *and E_n in* (5.147) *is defined via formulas* (5.148), (5.150) *and* (5.159). *In particular, the matrices ψ_k $(0 \leq k \leq n)$ do not depend on the choice of the Weyl function.*

Moreover, from the proof of Theorem 5.29 follows a complete description of the Weyl functions, given in terms of Taylor coefficients. We recall first that S_n is uniquely defined by the identity

$$A_nS_n - S_nA_n^* = \mathrm{i}\Pi_n\Pi_n^*, \tag{5.183}$$

if only Π_n is fixed.

Theorem 5.32.

(a) *Let system* (5.114) *be given on the interval* $0 \le k \le n$ *and satisfy* (5.116) *and* (5.118). *Then, the matrix function* φ, *analytic at* $z = \mathrm{i}$, *is a Weyl function of this system if and only if it admits expansion* (5.182), *where matrices* ψ_k *are defined in Corollary* 5.31.

(b) *Suppose* φ *is a* $p \times p$ *matrix function analytic at* $z = \mathrm{i}$. *Then,* φ *is a Weyl function of some system* (5.114), *which is given on the interval* $0 \le k \le n$ *and satisfies* (5.116) *and* (5.118), *if and only if the matrix* S_n, *determined by* (5.183), *is invertible. Here, the matrix* $\Pi_n = [\Phi_{n,1} \quad \Phi_{n,2}]$ *is given by* (5.137), *where* $\{\phi_k\}$ *are Taylor coefficients of* $\varphi(\mathrm{i}\frac{1+\lambda}{1-\lambda})$ *at* $\lambda = 0$.

Proof. Suppose φ is a $p \times p$ matrix function which is analytic at $z = \mathrm{i}$ and satisfies (5.182), and set

$$\begin{bmatrix} \mathcal{P}_1(z) \\ \mathcal{P}_2(z) \end{bmatrix} := \mathcal{W}(z)^{-1} \begin{bmatrix} \varphi(z) \\ I_p \end{bmatrix}. \tag{5.184}$$

Fix a Weyl function $\tilde{\varphi}$ of system (5.114), and denote by $\tilde{\mathcal{P}}_1, \tilde{\mathcal{P}}_2$ some admissible pair that provides representation (5.127) of $\tilde{\varphi}$. Rewrite (5.184) in the form

$$\begin{bmatrix} \mathcal{P}_1(z) \\ \mathcal{P}_2(z) \end{bmatrix} = \mathcal{W}(z)^{-1} \begin{bmatrix} \tilde{\varphi}(z) \\ I_p \end{bmatrix} + \mathcal{W}(z)^{-1} \begin{bmatrix} \varphi(z) - \tilde{\varphi}(z) \\ 0 \end{bmatrix}. \tag{5.185}$$

Since $\tilde{\varphi}$ is the Möbius transformation (5.127) of the admissible pair $\tilde{\mathcal{P}}_1, \tilde{\mathcal{P}}_2$, we have

$$\mathcal{W}(z)^{-1} \begin{bmatrix} \tilde{\varphi}(z) \\ I_p \end{bmatrix} = \begin{bmatrix} \tilde{\mathcal{P}}_1(z) \\ \tilde{\mathcal{P}}_2(z) \end{bmatrix} \left(\mathcal{W}_{21}(z)\tilde{\mathcal{P}}_1(z) + \mathcal{W}_{22}(z)\tilde{\mathcal{P}}_2(z)\right)^{-1}. \tag{5.186}$$

Thus, the first summand on the right-hand side of (5.185) is analytic at $z = \mathrm{i}$. Taking into account that expansion (5.182) is valid for φ and $\tilde{\varphi}$, we derive

$$\varphi(z) - \tilde{\varphi}(z) = O\left((z - \mathrm{i})^{n+1}\right) \quad \text{for } z \to \mathrm{i}.$$

From (5.172), it follows also that

$$\mathcal{W}(z)^{-1} = z^{2n+2}(z + \mathrm{i})^{-n-1}(z - \mathrm{i})^{-n-1}W(n + 1, z).$$

So, the second summand on the right-hand side of (5.185) is analytic at $z = \mathrm{i}$ too. Therefore, the pair $\mathcal{P}_1, \mathcal{P}_2$ is analytic at $z = \mathrm{i}$. Moreover, according to (5.184), we have $\mathcal{W}_{21}(z)\mathcal{P}_1(z) + \mathcal{W}_{22}(z)\mathcal{P}_2(z) = I_p$. Hence, inequality (5.119) holds and the pair $\mathcal{P}_1$,

$\mathcal{P}_2$ is admissible. It easily follows from (5.184) that φ admits representation (5.127) with this pair $\mathcal{P}_1, \mathcal{P}_2$, that is, φ is a Weyl function of our system.

Vice versa, if φ is a Weyl function of our system, then according to Corollary 5.31, the expansion (5.182) holds. The statement (a) is proven.

The proof of Theorem 5.29 shows that if φ is a Weyl function, then the matrix S_n uniquely defined by (5.183) coincides with $\widetilde{S}_n$ of the form (5.147) and so S_n is invertible.

It remains to be shown that if S_n is invertible, then φ is a Weyl function. Assume that $\det S_n \neq 0$. Since the identity (5.183) is equivalent to

$$\begin{aligned} &S_n(A_n^* - zI_{(n+1)p})^{-1} - (A_n - zI_{(n+1)p})^{-1}S_n \\ &= \mathrm{i}(A_n - zI_{(n+1)p})^{-1}\Pi_n\Pi_n^*(A_n^* - zI_{(n+1)p})^{-1}, \end{aligned}$$

using Cauchy's residue theorem, we derive

$$S_n = \frac{1}{2\pi}\int_{-\infty}^{\infty}(A_n - tI_{(n+1)p})^{-1}\Pi_n\Pi_n^*(A_n^* - tI_{(n+1)p})^{-1}dt. \tag{5.187}$$

Therefore, the inequality $\det S_n \neq 0$ leads us to $S_n > 0$. In addition to the notations P_1 and P (introduced in (5.163) and (5.164)), we set

$$\widetilde{P}_1 := \begin{bmatrix} I & 0 \end{bmatrix}, \quad \widetilde{P} = \begin{bmatrix} 0 & \dots & 0 & I_p \end{bmatrix} \tag{5.188}$$

where $\widetilde{P}_1$ and $\widetilde{P}$ are $(r+1)p \times (n+1)p$ and $p \times rp$, respectively, matrices and I stands for $I_{(r+1)p}$ in this chapter (see also formula (5.106) and further). It is easy to see that $\Pi_r = \widetilde{P}_1\Pi_n$ and $S_r = \widetilde{P}_1 S_n \widetilde{P}_1^*$. Hence, $S_n > 0$ implies $S_r > 0$ and (in particular) $\det S_r \neq 0$. Introduce

$$\gamma(r) := \left(PS_r^{-1}P^*\right)^{-\frac{1}{2}} PS_r^{-1}\Pi_r J \quad (0 \leq r \leq n). \tag{5.189}$$

Matrices $\gamma(r)$ satisfy conditions (5.116) and (5.118). Indeed, from (5.189), we obtain

$$\gamma(r)\gamma(r)^* = \left(PS_r^{-1}P^*\right)^{-\frac{1}{2}} PS_r^{-1}\Pi_r\Pi_r^*S_r^{-1}P^*\left(PS_r^{-1}P^*\right)^{-\frac{1}{2}}. \tag{5.190}$$

The identity

$$S_r^{-1}\Pi_r\Pi_r^*S_r^{-1} = -\mathrm{i}(S_r^{-1}A_r - A_r^*S_r^{-1})$$

immediately follows from (5.139). We also see that $PA_r^* = -(\mathrm{i}/2)P$ and $A_rP^* = (\mathrm{i}/2)P^*$. Thus, we rewrite (5.190) as

$$\gamma(r)\gamma(r)^* = -\mathrm{i}\left(PS_r^{-1}P^*\right)^{-\frac{1}{2}} P\left(S_r^{-1}A_r - A_r^*S_r^{-1}\right)P^*\left(PS_r^{-1}P\right)^{-\frac{1}{2}} = I_p, \tag{5.191}$$

and so the second equality from (5.116) is proven. Notice further that according to (5.137) and (5.189), we have

$$\gamma_1(0) = \left(S_0^{-1}\right)^{-\frac{1}{2}} S_0^{-1}I_p = S_0^{-\frac{1}{2}}, \tag{5.192}$$

and therefore the first inequality in (5.118) is also valid. In order to prove the second inequality in (5.118), we write down S_r^{-1} $(r > 0)$ in the block form

$$S_r^{-1} = \begin{bmatrix} S_{r-1}^{-1} + S_{r-1}^{-1}S_{12}T_{22}S_{21}S_{r-1}^{-1} & -S_{r-1}^{-1}S_{12}T_{22} \\ -T_{22}S_{21}S_{r-1}^{-1} & T_{22} \end{bmatrix}, \tag{5.193}$$

where $T_{22} = \left(s - S_{21}S_{r-1}^{-1}S_{12}\right)^{-1}$ and S_{12}, S_{21} and s are blocks of S_r, that is,

$$S_r = \begin{bmatrix} S_{r-1} & S_{12} \\ S_{21} & s \end{bmatrix}. \tag{5.194}$$

Here, conditions $\det S_r \neq 0$ and $\det S_{r-1} \neq 0$ imply (see, e.g. [290, p. 21]) the invertibility of $s - S_{21}S_{r-1}^{-1}S_{12}$. Since $s - S_{21}S_{r-1}^{-1}S_{12}$ is invertible, the right-hand side of (5.193) is well-defined and representation (5.193) is easily checked directly. Using (5.189) and (5.193), after easy transformations we rewrite the second inequality in (5.118) in an equivalent form

$$\det\left([-S_{21}S_{r-1}^{-1} \quad I_p]\Pi_r\Pi_{r-1}^*S_{r-1}^{-1}\widetilde{P}^*\right) \neq 0 \quad (0 < r \leq n), \tag{5.195}$$

where $\widetilde{P}$ contains one zero block less than P (see (5.188)). Recall that S_{r-1} satisfies the identity $A_{r-1}S_{r-1} - S_{r-1}A_{r-1}^* = \mathrm{i}\Pi_{r-1}\Pi_{r-1}^*$, that is,

$$S_{r-1}^{-1}\Pi_{r-1}\Pi_{r-1}^*S_{r-1}^{-1} = -\mathrm{i}(S_{r-1}^{-1}A_{r-1} - A_{r-1}^*S_{r-1}^{-1}).$$

Therefore, following some elements of the proof of (5.191), we obtain

$$\begin{aligned} &[-S_{21}S_{r-1}^{-1} \quad I_p]\Pi_r\Pi_{r-1}^*S_{r-1}^{-1}\widetilde{P}^* \\ &= -(S_{21}S_{r-1}^{-1}\widetilde{P}^*/2) - \mathrm{i}S_{21}A_{r-1}^*S_{r-1}^{-1}\widetilde{P}^* + [\psi_r \quad I_p]\Pi_{r-1}^*S_{r-1}^{-1}\widetilde{P}^*, \end{aligned} \tag{5.196}$$

where $\{\psi_k\}$ are the blocks of Φ_1. Consider now the first r blocks in the lowest block rows on both sides of the identity (5.139). In view of (5.194), we have

$$\begin{aligned} &\mathrm{i}[I_p \quad \dots \quad I_p]S_{r-1} + (\mathrm{i}/2)S_{21} - S_{21}A_{r-1}^* = \mathrm{i}[\psi_r \quad I_p]\Pi_{r-1}^*, \quad \text{and so} \\ &\mathrm{i}\left((\mathrm{i}/2)S_{21} - S_{21}A_{r-1}^* - \mathrm{i}[\psi_r \quad I_p]\Pi_{r-1}^*\right)S_{r-1}^{-1}\widetilde{P}^* = I_p. \end{aligned} \tag{5.197}$$

In view of (5.196) and (5.197), inequality (5.195) holds.

Thus, formula (5.189) determines via (5.116) a system (5.114) which satisfies conditions (5.116) and (5.118). Moreover, according to Theorem 1.16, the transfer matrix functions $w_A(r,z)$ corresponding to our matrices S_r admit factorizations of the form (5.165) (where Π_r and S_r should be substituted in place of $\widetilde{\Pi}_r$ and $\widetilde{S}_r$, respectively). These factorizations and definition (5.189) imply

$$w_A(n, z/2) = \prod_{r=0}^{n}\left(I_{2p} - 2\mathrm{i}(\mathrm{i} - z)^{-1}\gamma(r)^*\gamma(r)\right). \tag{5.198}$$

Compare (5.198) with (5.170) in order to derive (for the fundamental solution of the constructed system) the equality $W(n+1,z) = ((z-\mathrm{i})/z)^{n+1}w_A(n,z/2)$, that is,

$$\mathcal{W}(z)^{-1} = z^{n+1}(z+\mathrm{i})^{-n-1}w_A(n,z/2). \tag{5.199}$$

Now, let us consider $\varphi(z)$ and prove an analog of (5.177), namely,

$$\left\| \left(A_n - (z/2)I_{(n+1)p}\right)^{-1} \Pi_n J \begin{bmatrix} \varphi(z) \\ I_p \end{bmatrix} \right\| = O(1) \quad \text{for } z \to \mathrm{i}. \tag{5.200}$$

Besides equalities (5.179), which we used earlier, we need some other formulas for the resolvent of A_n. From (5.109) (after taking into account that A_r in (5.109) differs by sign from A_r in this section), we see that the i-th block row $\breve{R}_i$ of the resolvent $(A_n - (z/2)I_{(n+1)p})^{-1}$ is given by the formula

$$\begin{aligned} \breve{R}_i(z) &= \mathrm{i}\,(1-\lambda)\,\lambda^{-i-1} \\ &\times \begin{bmatrix} (1-\lambda)I_p & (1-\lambda)\lambda I_p & \dots & (1-\lambda)\lambda^{i-1}I_p & \lambda^i I_p & 0 & \dots & 0 \end{bmatrix}, \end{aligned} \tag{5.201}$$

where

$$\lambda = \lambda(z) = (z-\mathrm{i})/(z+\mathrm{i}), \quad z = \mathrm{i}(1+\lambda)(1-\lambda)^{-1}. \tag{5.202}$$

It is immediate from (5.201) that

$$\left(A - (z/2)I_{(n+1)p}\right)^{-1} \Phi_{n,2} = \mathrm{i}\frac{1-\lambda(z)}{\lambda(z)} \mathrm{col}\begin{bmatrix} I_p & \lambda(z)^{-1}I_p & \dots & \lambda(z)^{-n}I_p \end{bmatrix}, \tag{5.203}$$

where col means column. In view of (5.201)–(5.203), we have

$$\begin{aligned} &\left(A - (z/2)I_{(n+1)p}\right)^{-1} \Pi_n \begin{bmatrix} I_p \\ \varphi(z) \end{bmatrix} \\ &= \mathrm{i}\left(1 - \frac{z-\mathrm{i}}{z+\mathrm{i}}\right)\left(\frac{z-\mathrm{i}}{z+\mathrm{i}}\right)^{-k-1} \\ &\quad\times \left\{ \varphi(z) + \psi_0 + (\psi_1 - \psi_0)\frac{z-\mathrm{i}}{z+\mathrm{i}} + \dots + (\psi_k - \psi_{k-1})\left(\frac{z-\mathrm{i}}{z+\mathrm{i}}\right)^k \right\}_{k=0}^n . \end{aligned} \tag{5.204}$$

Recall that ψ_k are blocks of Π_n and, in view of (5.137), we have $\psi_0 = -\phi_0$ and $\psi_k - \psi_{k-1} = -\phi_k$ $(k > 0)$, where ϕ_k are Taylor coefficients of $\varphi(\mathrm{i}\frac{1+\lambda}{1-\lambda})$. Hence, formulas (5.202) and (5.204) yield (5.200).

Relations (5.162), (5.199) and (5.200) lead us to the equality

$$\left\| \mathcal{W}(z)^{-1} \begin{bmatrix} \varphi(z) \\ I_p \end{bmatrix} \right\| = O(1) \quad \text{for } z \to \mathrm{i}.$$

Thus, the pair $\mathcal{P}_1$, $\mathcal{P}_2$, given by (5.184), is admissible. Since φ admits representation (5.127) with $\mathcal{P}_1$, $\mathcal{P}_2$ given by (5.184), the matrix function φ is a Weyl function of the constructed system. □

Example 5.33. According to (5.133), for the system considered in Example 5.26, we have $\phi_0 = 0$ and $\phi_1 = I_p$. Hence, from Theorem 5.32, we see that the set of Weyl functions φ of this system is defined by the expansion

$$\varphi\left(\mathrm{i}\frac{1+\lambda}{1-\lambda}\right) = \lambda I_p + O(\lambda^2), \quad \lambda \to 0. \tag{5.205}$$

5.3 GBDT for the discrete skew-self-adjoint Dirac system

In this last section of the present chapter, we consider GBDT for the discrete skew-self-adjoint system (5.114) and those transformations of the Weyl functions, which correspond to this GBDT. Discrete GBDT is somewhat different from the GBDT for continuous systems, which was considered in Subsections 1.1.3, 2.2.2 and 4.3. However, the discrete GBDT is based on the same ideas.

We note that, in the case $p = 1$, the system (5.114) is an auxiliary linear system for the integrable isotropic Heisenberg magnet (IHM) model [311] (see also the historical remarks in [101]). More precisely, the IHM equation is equivalent to the compatibility condition of the systems

$$y_{k+1}(t,z) = G_k(t,z)y_k(t,z), \quad G_k(t,z) := I_2 - (\mathrm{i}/z)C_k(t); \tag{5.206}$$

$$\frac{d}{dt}y_k(t,z) = F_k(t,z)y_k(t,z), \tag{5.207}$$

where F_k is introduced several paragraphs below. It is apparent that system (5.206) has, for each fixed t, the form (5.114). The compatibility condition of systems (5.206) and (5.207) is usually derived via differentiation of y_{k+1}. Indeed, by first using (5.206) and then (5.207), we obtain $(d/dt)y_{k+1} = (d/dt)(G_k y_k) = ((d/dt)G_k + G_k F_k)y_k$. On the other hand, using first (5.207) and then (5.206), we obtain $(d/dt)y_{k+1} = F_{k+1}G_k y_k$. Thus, the compatibility condition is given by the equality

$$\frac{d}{dt}G_k(t,z) = F_{k+1}(t,z)G_k(t,z) - G_k(t,z)F_k(t,z). \tag{5.208}$$

Equality (5.208) is also called (for our choice of G_k and F_k) a zero curvature representation of the IHM equation. (See a more detailed discussion on the zero curvature representation in Chapter 6 and in the book [101].) We will use the GBDT to construct explicit solutions of the IHM model and study the evolution of the corresponding Weyl functions.

The IHM equation mentioned above has the form [311]:

$$\frac{d\vec{S}_k}{dt} = 2\vec{S}_k \bigwedge \left(\frac{\vec{S}_{k+1}}{1+\vec{S}_k\cdot\vec{S}_{k+1}} + \frac{\vec{S}_{k-1}}{1+\vec{S}_{k-1}\cdot\vec{S}_k}\right), \tag{5.209}$$

where the vectors $\vec{S}_k = [S_k^1 \ S_k^2 \ S_k^3]$ belong to $\mathbb{R}^3$, the dot "$\cdot$" denotes the scalar product in $\mathbb{R}^3$ and $\bigwedge$ stands for the vector product in $\mathbb{R}^3$. The vector $\vec{S}_k$ is interpreted in physics as a *spin vector* and the correspondence between $\vec{S}_k$ and the so called spin matrix C_k is given by the equality

$$C_k = \begin{bmatrix} S_k^3 & S_k^1 - \mathrm{i}S_k^2 \\ S_k^1 + \mathrm{i}S_k^2 & -S_k^3 \end{bmatrix}. \tag{5.210}$$

In (5.209), it is required that $\|\vec{S}_r\| = 1$ for $r = k,\ k \pm 1$, and it is easy to see that the representation (5.210), where $\vec{S}_k \cdot \vec{S}_k = 1$, is equivalent to the relations

$$C_k = C_k^* = C_k^{-1}, \quad C_k \neq \pm I_2. \tag{5.211}$$

Recall that the relations in (5.211) coincide with the conditions on C_k in Section 5.2.

Finally, the matrix function $F_k(t,z)$ in (5.207) is also expressed via the spin vectors and matrices:

$$F_k(t,z) = (z-\mathrm{i})^{-1}V_k^+(t) + (z+\mathrm{i})^{-1}V_k^-(t), \tag{5.212}$$

$$V_k^{\pm}(t) := (1 + \vec{S}_{k-1}(t) \cdot \vec{S}_k(t))^{-1}(I_2 \pm C_k(t))(I_2 \pm C_{k-1}(t)). \tag{5.213}$$

Then, the equivalence of (5.208) and (5.209) is easily checked directly. We also note that the spin matrices $C_{k-1}(t)$ and $C_k(t)$ corresponding to the IHM equation (5.208) on the semiaxis $k \geq 1$ coincide with the potentials $C_k(t)$ of systems (5.206) (or, equivalently, systems (5.114)) on the semiaxis $k \geq 0$. We will deal with the systems (5.114) on the semiaxis $k \geq 0$.

5.3.1 Main results

In order to apply GBDT, we fix, like in Subsection 1.1.3, an integer $n > 0$ and three parameter matrices with the following properties: we consider an $n \times n$ matrix A with $\det A \neq 0$, an $n \times n$ matrix $S(0)$ such that $S(0) = S(0)^*$ and an $n \times m$ ($m = 2p$) matrix $\Pi(0)$. These matrices should satisfy the matrix identity

$$AS(0) - S(0)A^* = \mathrm{i}\Pi(0)\Pi(0)^*. \tag{5.214}$$

Notice that in this section (like in the majority of other works on GBDT), n denotes the order of the matrices A and $S(k)$.

As in Subsections 2.2.2 and 4.3, we transform the initial system with the trivial (identical zero) potential and obtain explicit expressions for the transformed potentials and fundamental solutions. Namely, given the fixed three matrices A, $S(0)$ and $\Pi(0)$ mentioned above, we define, for $k = 1, 2, \ldots$, the $n \times 2p$ matrix function $\Pi(k)$ and the $n \times n$ matrix function $S(k)$ via recursions:

$$\Pi(k+1) = \Pi(k) + \mathrm{i}A^{-1}\Pi(k)j, \quad j = \begin{bmatrix} I_p & 0 \\ 0 & -I_p \end{bmatrix}, \tag{5.215}$$

$$S(k+1) = S(k) + A^{-1}S(k)(A^*)^{-1} + A^{-1}\Pi(k)j\Pi(k)^*(A^*)^{-1}. \tag{5.216}$$

Definition 5.34. If the matrices $S(k)$ $(k = 0, 1, 2, \ldots)$ are invertible, we say that the sequence of matrices $\{C_k\}$ defined by

$$C_k = j + \Pi(k)^* S(k)^{-1}\Pi(k) - \Pi(k+1)^* S(k+1)^{-1}\Pi(k+1), \quad k \geq 0, \tag{5.217}$$

is the potential (or spin sequence) *determined* by the parameter matrices A, $S(0)$ and $\Pi(0)$. (Notice the requirement of the invertibility of the matrices $S(k)$.)

For potentials defined in this way, our first theorem presents a formula for the fundamental solution $W_k(z)$ of (5.114).

Theorem 5.35. *Let* A $(\det A \neq 0)$, $S(0)$ $(S(0) = S(0)^*)$ *and* $\Pi(0)$ *satisfy* (5.214), *and assume that* $\det S(k) \neq 0$ *for* $0 \leq k \leq r$, *where* $S(k)$ *is given by* (5.216). *For* $0 \leq k \leq r-1$, *let* C_k *be the matrices determined by* A, $S(0)$ *and* $\Pi(0)$ *via* (5.215)–(5.217). *Then,* $C_k = C_k^* = C_k^{-1}$ *for* $0 \leq k \leq r-1$ *and, for* $0 \leq k \leq r$, *the fundamental solution* $W_k(z)$ *of the discrete system* (5.114) *can be represented in the form*

$$W_k(z) = w_A(k,z)\left(I_m - (\mathrm{i}/z)j\right)^k w_A(0,z)^{-1}, \tag{5.218}$$

where $w_A(k,z)$ *is defined by*

$$w_A(k,z) = I_m - \mathrm{i}\Pi(k)^* S(k)^{-1}(A - zI_n)^{-1}\Pi(k). \tag{5.219}$$

When $S(0) > 0$, there exist simple conditions on A and $\Pi(0)$ to guarantee that $\det S(k) \neq 0$. First, if $S(0) > 0$, then without loss of generality, we can assume that $S(0) = I_n$. Indeed, it is easy to see that the matrices $S(k)$ $(k > 0)$ and matrices $S_{tr}(k)$ corresponding (via (5.215) and (5.216)) to the transformed parameter set $S(0)^{-\frac{1}{2}}AS(0)^{\frac{1}{2}}$, I_n and $S(0)^{-\frac{1}{2}}\Pi(0)$ are invertible (or singular) simultaneously. Moreover, the sequence of matrices $\{C_k\}$ defined by (5.215)–(5.217) does not change if we substitute A, $S(0)$ and $\Pi(0)$ by $S(0)^{-\frac{1}{2}}AS(0)^{\frac{1}{2}}$, I_n and $S(0)^{-\frac{1}{2}}\Pi(0)$. Therefore, let us assume that $S(0) = I_n$.

Next, we partition the matrix $\Pi(0)$ into two $n \times p$ blocks $\varkappa_1$ and $\varkappa_2$ as follows: $\Pi(0) = [\varkappa_1 \quad \varkappa_2]$. This, together with $S(0) = I_n$, allows us to rewrite (5.214) in the form

$$A - A^* = \mathrm{i}(\varkappa_1\varkappa_1^* + \varkappa_2\varkappa_2^*). \tag{5.220}$$

If $S(0) = I_n$, we also say that the corresponding potential is determined by the triple A, $\varkappa_1$ and $\varkappa_2$. Furthermore, in this case, $\Pi(k)$ is given by

$$\Pi(k) = [(I_n + \mathrm{i}A^{-1})^k\varkappa_1 \quad (I_n - \mathrm{i}A^{-1})^k\varkappa_2]. \tag{5.221}$$

Finally, we shall assume that the pair $\{A, \varkappa_1\}$ is *controllable*, meaning that

$$\mathbb{C}^n = \operatorname{span}\{A^k\varkappa_1\mathbb{C}^p \mid k = 0, 1, 2, \ldots, n-1\}$$

(see Appendix B). The following proposition shows that, under these conditions, $\det A \neq 0$ and $\det S(k) \neq 0$ automatically for $k = 0, 1, 2, \ldots$.

Proposition 5.36. *Let A be a square matrix of order n, and $\varkappa_1$ and $\varkappa_2$ be $n \times p$ matrices satisfying* (5.220). *Assume that the pair $\{A, \varkappa_1\}$ is controllable. Then, all the eigenvalues of A are in the open upper half-plane $\mathbb{C}_+$, and, for $k = 1, 2, \ldots$, the matrices $S(k)$ defined by* (5.216), *with $S(0) = I_n$ and $\Pi(k)$ given by* (5.221), *are positive-definite and satisfy the identity*

$$AS(k) - S(k)A^* = \mathrm{i}\Pi(k)\Pi(k)^*. \qquad (5.222)$$

Definition 5.37. A triple of matrices A, $\varkappa_1$ and $\varkappa_2$, with A square of order n and $\varkappa_1$ and $\varkappa_2$ of size $n \times p$, is called admissible if the pairs $\{A, \varkappa_1\}$ and $\{A, \varkappa_2\}$ are controllable and the identity (5.220) holds.

Notation 5.38. *We denote by the acronym* EG (explicitly generated) *the class of potentials $\{C_k\}$ determined by the matrices A, $S(0) = I_n$ and $\Pi(0) = [\varkappa_1 \quad \varkappa_2]$, where A, $\varkappa_1$ and $\varkappa_2$ form an admissible triple. In this case, we also say that these potentials are* determined *by the corresponding admissible triples.*

In this section, we introduce and study Weyl functions $\varphi(z)$ for systems of the form (5.114) given on the semiaxis $k \geq 0$. As in Chapter 4, we consider here Weyl functions in the lower semiplane. Similar to the continuous systems, where Weyl functions were introduced in terms of the solutions which belong to $L^2(0, \infty)$, we introduce Weyl functions of systems (5.114) in terms of the solutions from l^2.

Definition 5.39. A $p \times p$ matrix function φ, which is holomorphic in $\mathbb{C}_M^-$ (for some $M > 0$) and satisfies the inequality

$$\sum_{k=0}^{\infty} \begin{bmatrix} \varphi(z)^* & I_p \end{bmatrix} W_k(z)^* W_k(z) \begin{bmatrix} \varphi(z) \\ I_p \end{bmatrix} < \infty, \quad z \in \mathbb{C}_M^-, \qquad (5.223)$$

where $W_k(z)$ is the fundamental solution of system (5.114) given on the semiaxis $k \geq 0$, is called a Weyl function of system (5.114).

The next two theorems present the solutions of the direct and inverse problem in terms of the Weyl function.

Theorem 5.40. *Assume that the potential $\{C_k\}_{k=0}^{\infty}$ of system* (5.114) *belongs to the class* EG *and is determined by the admissible triple A, $\varkappa_1$ and $\varkappa_2$. Then, the system* (5.114) *has a unique Weyl function φ which satisfies* (5.223) *on the half-plane $\mathbb{C}_{1/2}^- = \{z : \Im z < -1/2\}$, a finite number of poles excluded, and this function is given by the formula*

$$\varphi(z) = -\mathrm{i}\varkappa_1^*(\alpha - zI_n)^{-1}\varkappa_2, \qquad (5.224)$$

where $\alpha = A - \mathrm{i}\varkappa_2\varkappa_2^*$.

Notice that the Weyl function φ in (5.224) is a strictly proper $p \times p$ rational matrix function and the expression on the right-hand side of (5.224) is its realization. (See Appendix B, where these terms are explained.) Conversely, if φ is a strictly proper

$p \times p$ rational matrix function, then it admits a realization of the form

$$\varphi(z) = -\mathrm{i}C(\mathcal{A} - zI_n)^{-1}B, \tag{5.225}$$

where $\mathcal{A}$ is a square matrix and C and B, respectively, are $p \times n$ and $n \times p$ matrices. (Recall that we refer to the right-hand side of (5.225) as a *minimal realization* of φ if among all possible representations (5.225) of φ, the order n of the matrix $\mathcal{A}$ is as small as possible.) We can now state the solution of the inverse problem.

Theorem 5.41. *Let φ be a strictly proper rational $p \times p$ matrix function given by the minimal realization (5.225). There is a unique positive-definite $n \times n$ matrix solution X of the algebraic Riccati equation*

$$\mathcal{A}X - X\mathcal{A}^* = \mathrm{i}(XC^*CX - BB^*). \tag{5.226}$$

Using X, define matrices $\varkappa_1, \varkappa_2$ and $A = \alpha + \mathrm{i}\varkappa_2\varkappa_2^*$ by

$$\varkappa_1 = X^{\frac{1}{2}}C^*, \quad \varkappa_2 = X^{-\frac{1}{2}}B, \quad \alpha = X^{-\frac{1}{2}}\mathcal{A}X^{\frac{1}{2}}. \tag{5.227}$$

Then, A, $\varkappa_1$ and $\varkappa_2$ form an admissible triple, and the given matrix function φ is the Weyl function of a system (5.114) *of which the potential $\{C_k\}$ ($\{C_k\} \in \mathrm{EG}$) is uniquely determined by the admissible triple A, $\varkappa_1$ and $\varkappa_2$.*

5.3.2 The fundamental solution

In this subsection, we prove Theorem 5.35 and present some results that will be used in Subsection 5.3.4 (a result on the invertibility of matrices $S(k)$, in particular).

Proof of Theorem 5.35. First, we shall show that equalities (5.214)–(5.216) yield the identity (5.222) for all $k \geq 0$. The statement is proved by induction. Indeed, for $k = 0$, it is valid by assumption. Suppose (5.222) is true for $k = i$. Then, using the expression for $S(i+1)$ from (5.216) and identity (5.222) for $k = i$, we obtain

$$\begin{aligned} AS(i+1) - S(i+1)A^* = {} & \mathrm{i}\Pi(i)\Pi(i)^* + \mathrm{i}A^{-1}\Pi(i)\Pi(i)^*(A^*)^{-1} \\ & + \Pi(i)j\Pi(i)^*(A^*)^{-1} - A^{-1}\Pi(i)j\Pi(i)^*. \end{aligned} \tag{5.228}$$

Formulas (5.215) and (5.228) yield (5.222) for $k = i+1$ and thus for all $k \geq 0$.

The next equality will be crucial for our proof. Namely, we shall show that on the interval $0 \leq k \leq r-1$ (where $\det S(k) \neq 0$ and $\det S(k+1) \neq 0$), we have

$$w_A(k+1,z)(I_m - (\mathrm{i}/z)j) = (I_m - (\mathrm{i}/z)C_k)w_A(k,z). \tag{5.229}$$

By virtue of (5.219), formula (5.229) is equivalent to the formula

$$\begin{aligned} z^{-1}(C_k - j) = {} & -(I_m - \mathrm{i}z^{-1}C_k)\Pi(k)^*S(k)^{-1}(A - zI_n)^{-1}\Pi(k) \\ & + \Pi(k+1)^*S(k+1)^{-1}(A - zI_n)^{-1}\Pi(k+1)(I_m - \mathrm{i}z^{-1}j). \end{aligned} \tag{5.230}$$

Using the Taylor expansion of $(A - zI_n)^{-1}$ at infinity, one shows that (5.230) is, in turn, equivalent to the set of equalities:

$$C_k - j = \Pi(k)^* S(k)^{-1}\Pi(k) - \Pi(k+1)^* S(k+1)^{-1}\Pi(k+1), \tag{5.231}$$

$$\Pi(k+1)^* S(k+1)^{-1} A^l \Pi(k+1) - \mathrm{i}\Pi(k+1)^* S(k+1)^{-1} A^{l-1}\Pi(k+1) j$$
$$= \Pi(k)^* S(k)^{-1} A^l \Pi(k) - \mathrm{i}C_k \Pi(k)^* S(k)^{-1} A^{l-1}\Pi(k) \quad (l > 0). \tag{5.232}$$

Equality (5.231) is equivalent to (5.217). In order to prove (5.232), we note that (5.215) yields

$$A\Pi(k+1) - \mathrm{i}\Pi(k+1)j = A\Pi(k) + A^{-1}\Pi(k).$$

Thus, the equalities in (5.232) can be rewritten in the form $M_k A^{l-2}\Pi(k) = 0$, where

$$M_k := \Pi(k+1)^* S(k+1)^{-1}(A^2 + I_n) - \Pi(k)^* S(k)^{-1} A^2 + \mathrm{i}C_k \Pi(k)^* S(k)^{-1} A. \tag{5.233}$$

Therefore, if we prove that $M_k = 0$, then equalities (5.232) will be proved, and so formula (5.229) will be proved too. Substitute (5.217) into (5.233), and again use (5.215) to obtain

$$M_k = \Pi(k+1)^* S(k+1)^{-1}(A^2 + I_n) - \Pi(k)^* S(k)^{-1} A^2 + \mathrm{i}j\Pi(k)^* S(k)^{-1} A$$
$$+ \mathrm{i}\Pi(k)^* S(k)^{-1}\Pi(k)\Pi(k)^* S(k)^{-1} A \tag{5.234}$$
$$- \mathrm{i}\Pi(k+1)^* S(k+1)^{-1}\left(\Pi(k) + \mathrm{i}A^{-1}\Pi(k)j\right)\Pi(k)^* S(k)^{-1} A.$$

Now, using (5.222), we obtain $\mathrm{i}\Pi(k)\Pi(k)^* S(k)^{-1} = A - S(k)A^* S(k)^{-1}$ and substitute this relation into (5.234). After easy transformations, it follows that

$$M_k = \Pi(k+1)^* S(k+1)^{-1}\left(A^{-1}S(k)(A^*)^{-1} + S(k) + A^{-1}\Pi(k)j\Pi(k)^*(A^*)^{-1}\right)$$
$$\times A^* S(k)^{-1} A + \mathrm{i}j\Pi(k)^* S(k)^{-1} A - \Pi(k)^* A^* S(k)^{-1} A. \tag{5.235}$$

In view of (5.216), the first term on the right-hand side of the equality (5.235) can be rewritten as $\Pi(k+1)^* A^* S(k)^{-1} A$ and we have

$$M_k = (\Pi(k+1)^* + \mathrm{i}j\Pi(k)^*(A^*)^{-1} - \Pi(k)^*)A^* S(k)^{-1} A. \tag{5.236}$$

The equality $M_k = 0$ is now immediate from (5.215), that is, (5.229) holds.

Notice that equality (5.218) is valid for $k = 0$. Suppose that it is valid for $k = i$. Then, after taking into account that W_{i+1} satisfies (5.114), formula (5.218) yields

$$W_{i+1}(z) = (I_m - \mathrm{i}z^{-1}C_i)w_A(i,z)\left(I_m - \mathrm{i}z^{-1}j\right)^i w_A(0,z)^{-1}. \tag{5.237}$$

By virtue of (5.229) and (5.237), the validity of (5.218) for $k = i + 1$ easily follows, that is, (5.218) is proved by induction.

Consider now the matrices C_k given by (5.217). It is easy to see that $C_k = C_k^*$. Notice also that in view of (1.87), we have

$$w_A(k,z)w_A(k,\overline{z})^* = I_m \quad (k \geq 0), \tag{5.238}$$

where $\overline{z}$ stands for complex conjugate for z. From (5.229) and (5.238), it follows that

$$(I_m - \mathrm{i}z^{-1}C_k)(I_m + \mathrm{i}z^{-1}C_k) = z^{-2}(z^2+1)I_m.$$

Thus, the equality $C_k = C_k^{-1}$ holds, and so the proof of the theorem is finished. □

The case when $\pm\mathrm{i} \notin \sigma(A)$ (σ means spectrum) is important for the study of the IHM model. The following proposition will be useful for formulating the conditions of invertibility of $S(k)$ in a somewhat different form then those in Proposition 5.36.

Proposition 5.42. *Let the matrices A ($\det A \neq 0$), $S(0) = S(0)^*$ and $\Pi(0)$ satisfy (5.214), and let the matrices $S(k)$ be given by (5.216). If $\mathrm{i} \notin \sigma(A)$, then the sequence of matrices*

$$Q^+(k) = (I_n - \mathrm{i}A^{-1})^{-k}S(k)(I_n + \mathrm{i}(A^*)^{-1})^{-k} \tag{5.239}$$

is well-defined and nondecreasing. If $-\mathrm{i} \notin \sigma(A)$, then the sequence of matrices

$$Q^-(k) = (I_n + \mathrm{i}A^{-1})^{-k}S(k)(I_n - \mathrm{i}(A^*)^{-1})^{-k} \tag{5.240}$$

is well-defined and nonincreasing.

Proof. To prove that the sequence $\{Q^+(k)\}$ is nondecreasing it will suffice to show that

$$S(k+1) - (I_n - \mathrm{i}A^{-1})S(k)(I_n + \mathrm{i}(A^*)^{-1}) \geq 0. \tag{5.241}$$

For this purpose, notice that

$$\begin{aligned} &S(k+1) - (I_n - \mathrm{i}A^{-1})S(k)(I_n + \mathrm{i}(A^*)^{-1}) \\ &\quad = S(k+1) - S(k) - A^{-1}S(k)(A^*)^{-1} - \mathrm{i}A^{-1}\left(AS(k) - S(k)A^*\right)(A^*)^{-1}. \end{aligned}$$

Hence, in view of (5.216) and (5.222), we derive

$$\begin{aligned} &S(k+1) - (I_n - \mathrm{i}A^{-1})S(k)(I_n + \mathrm{i}(A^*)^{-1}) \\ &\quad = A^{-1}\left(\Pi(k)j\Pi(k)^* + \Pi(k)\Pi(k)^*\right)(A^*)^{-1}. \end{aligned} \tag{5.242}$$

Since $j + I_m \geq 0$, the inequality (5.241) is immediate from (5.242).

Similarly, from (5.216) and (5.222), we obtain

$$\begin{aligned} &S(k+1) - (I_n + \mathrm{i}A^{-1})S(k)(I_n - \mathrm{i}(A^*)^{-1}) \\ &\quad = A^{-1}(\Pi(k)j\Pi(k)^* - \Pi(k)\Pi(k)^*)(A^*)^{-1} \leq 0, \end{aligned} \tag{5.243}$$

and so the sequence of matrices $\{Q^-(k)\}$ is nonincreasing. □

According to Proposition 5.42, when $\mathrm{i} \notin \sigma(A)$ and $S(0) > 0$, we have $Q^+(k) > 0$.

Corollary 5.43. *Let the matrices A ($\det A \neq 0$), $S(0) = S(0)^*$ and $\Pi(0)$ satisfy (5.214), and let the matrices $S(k)$ be given by (5.216). Assume that $\mathrm{i} \notin \sigma(A)$ and $S(0) > 0$. Then, we have $S(k) > 0$ for all $k > 0$.*

Partition the matrices $w_A(k,z)$ and $\Pi(k)$ into two p-column blocks each, that is,

$$w_A(k,z) = [(w_A(k,z))_1 \quad (w_A(k,z))_2], \quad \Pi(k) = [\Lambda_1(k) \quad \Lambda_2(k)]. \tag{5.244}$$

The next lemma will be used in Subsection 5.3.4.

Lemma 5.44. *Let the matrices A $(0, \pm\mathrm{i} \notin \sigma(A))$, $S(0)$ $(S(0) = S(0)^*)$ and $\Pi(0)$ satisfy (5.214), and let the matrices $S(k)$ be given by (5.216). Then, for $k \geq 0$, the following relations hold:*

$$\begin{aligned}(w_A(k,\mathrm{i}))_1 & \qquad\qquad (5.245)\\ &= (w_A(k+1,-\mathrm{i}))_1 \left(I_p + 2\,(w_A(k,\mathrm{i}))_1^*\,\Pi(k)^*S(k)^{-1}(A^2+I_n)^{-1}\Lambda_1(k)\right),\\ (w_A(k,-\mathrm{i}))_2 & \qquad\qquad (5.246)\\ &= (w_A(k+1,\mathrm{i}))_2 \left(I_p - 2\,(w_A(k,-\mathrm{i}))_2^*\,\Pi(k)^*S(k)^{-1}(A^2+I_n)^{-1}\Lambda_2(k)\right).\end{aligned}$$

Proof. From the proof of Theorem 5.35, we know that $M_k = 0$, where M_k is given by (5.233). In particular, we have

$$M_kA^{-1}(A^2+I_n)^{-1}\Lambda_1(k) = 0, \quad M_kA^{-1}(A^2+I_n)^{-1}\Lambda_2(k) = 0. \tag{5.247}$$

In order to prove (5.245), notice that $\Lambda_1(k+1) = A^{-1}(A+\mathrm{i}I_n)\Lambda_1(k)$ and rewrite the first equality in (5.247) as

$$\begin{aligned}&\Pi(k+1)^*S(k+1)^{-1}(A+\mathrm{i}I_n)^{-1}\Lambda_1(k+1) - \Pi(k)^*S(k)^{-1}(A-\mathrm{i}I_n)^{-1}\Lambda_1(k)\\ &+\mathrm{i}(I_m+C_k)\Pi(k)^*S(k)^{-1}(A^2+I_n)^{-1}\Lambda_1(k) = 0.\end{aligned} \tag{5.248}$$

Put $z = -\mathrm{i}$ in (5.229) and take into account (5.238) to derive

$$I_m + C_k = 2\,(w_A(k+1,-\mathrm{i}))_1\,(w_A(k,\mathrm{i}))_1^*. \tag{5.249}$$

In view of definition (5.219) of w_A, equality (5.245) follows from (5.248) and (5.249).

Putting $z = \mathrm{i}$ in (5.229), we obtain

$$I_m - C_k = 2\,(w_A(k+1,\mathrm{i}))_2\,(w_A(k,-\mathrm{i}))_2^*. \tag{5.250}$$

Similar to the proof of (5.245), we derive (5.246) from (5.250) and the second equality in (5.247). □

Remark 5.45. According to (5.249) and (5.250), the rank of the matrices $I_m \pm C_k$ is less than or equal to p. Together with the equalities $C_k = C_k^* = C_k^{-1}$, this implies that, under the conditions of Lemma 5.44, we have $C_k = \mathcal{U}(k)^*j\mathcal{U}(k)$, where $\mathcal{U}(k)$ are unitary matrices.

5.3.3 Weyl functions: direct and inverse problems

In this subsection, we prove Theorems 5.40 and 5.41, and Proposition 5.36. At the end of the section, a lemma on the case $\mathrm{i} \notin \sigma(A)$ is dealt with too.

Proof of Proposition 5.36. Suppose g is an eigenvector of A, that is, $Ag = cg$, $g \neq 0$. Then, formula (5.220) yields the equality

$$\mathrm{i}(\overline{c} - c)g^*g = g^*(\varkappa_1\varkappa_1^* + \varkappa_2\varkappa_2^*)g \geq 0. \tag{5.251}$$

Thus, $c \in \overline{\mathbb{C}_+}$. Moreover, if $c \in \mathbb{R}$, then, according to (5.251), we have $\varkappa_1^* g = \varkappa_2^* g = 0$, and therefore $Ag = A^*g$. It follows that

$$g^*\varkappa_1 = 0, \quad g^*(A - cI_n) = 0 \quad (g \neq 0). \tag{5.252}$$

As $\{A, \varkappa_1\}$ is a controllable pair, so the pair $\{A - cI_n, \varkappa_1\}$ is also full range, which contradicts (5.252). This implies that $c \in \mathbb{C}_+$, that is, $\sigma(A) \subset \mathbb{C}_+$.

Recall identity (5.222), which was deduced in the proof of Theorem 5.35. In view of $\sigma(A) \subset \mathbb{C}_+$, identity (5.222) yields

$$S(k) = \frac{1}{2\pi}\int_{-\infty}^{\infty}(A - tI_n)^{-1}\Pi(k)\Pi(k)^*(A^* - tI_n)^{-1}dt, \tag{5.253}$$

of which the proof is similar to the proof of (5.187). Notice now that the pair $\{A, (I_n + \mathrm{i}A^{-1})^k\varkappa_1\}$ is controllable and use (5.221) and(5.253) to obtain $S(k) > 0$ for all $k \geq 0$. □

Remark 5.46. In the same way as in the proof of Proposition 5.36 above, the inclusion $\sigma(A) \subset \mathbb{C}_+$ follows from the weaker (than in Proposition 5.36) condition that the pair $\{A, \Pi(0)\}$ is controllable. However, the example $n = 1$, $A = \mathrm{i}$, $\varkappa_1 = 0$, $\varkappa_2\varkappa_2^* = 2$, which yields $S(k) \equiv 0$ for $k > 0$, shows that we have to require that the pair $\{A, \varkappa_1\}$ is controllable in order to obtain $S(k) > 0$. The controllability condition on the pair $\{A, \Pi(0)\}$ is not enough for this conclusion.

Recall now Definition 5.37 of the admissible triple. Proposition 5.36 implies, in particular, that $\det A \neq 0$ for the admissible triple and the spin sequences $\{C_k\}$ determined by it are well-defined for all $k \geq 0$. In other words, the class EG is well-defined.

Proof of Theorem 5.40. Let $w_A(k, z)$ be given by (5.219) and partition $w_A(0, z)$ into $p \times p$ blocks

$$w_A(0, z) = \{w_{ij}(z)\}_{i,j=1}^2. \tag{5.254}$$

Then, the $w_{i2}(z)$ blocks are given by the equalities

$$w_{12}(z) = -\mathrm{i}\varkappa_1^*(A - zI_n)^{-1}\varkappa_2, \quad w_{22}(z) = I_p - \mathrm{i}\varkappa_2^*(A - zI_n)^{-1}\varkappa_2. \tag{5.255}$$

We first prove that

$$w_{12}(z)w_{22}(z)^{-1} = -\mathrm{i}\varkappa_1^*(\alpha - zI_n)^{-1}\varkappa_2. \tag{5.256}$$

Using (B.5), from $\alpha = A - \mathrm{i}\varkappa_2\varkappa_2^*$ and (5.255), we obtain

$$w_{22}(z)^{-1} = I_p + \mathrm{i}\varkappa_2^*(\alpha - zI_n)^{-1}\varkappa_2. \tag{5.257}$$

From $\alpha = A - \mathrm{i}\varkappa_2\varkappa_2^*$ and the equalities (5.255) and (5.257), we see that

$$\begin{aligned} &w_{12}(z)w_{22}(z)^{-1} \\ &= -\mathrm{i}\varkappa_1^*(A - zI_n)^{-1}\varkappa_2 + \mathrm{i}\varkappa_1^*(A - zI_n)^{-1}(\alpha - A)(\alpha - zI_n)^{-1}\varkappa_2. \end{aligned} \tag{5.258}$$

Finally, from (5.258), formula (5.256) follows.

Letting φ be defined via (5.224), by virtue of (5.256), we have

$$\varphi(z) = w_{12}(z)w_{22}(z)^{-1}. \tag{5.259}$$

From (5.254), (5.259) and the representation (5.218) of the fundamental solution, we derive

$$W_k(z)\begin{bmatrix}\varphi(z)\\ I_p\end{bmatrix} = (z+\mathrm{i})^k z^{-k} w_A(k,z)\begin{bmatrix}0\\ w_{22}(z)^{-1}\end{bmatrix}. \tag{5.260}$$

Substituting $w_A(\cdot) = w_A(k,\cdot)$, $J = I_m$ and $\zeta = \overline{z}$ into (1.88), we have

$$\begin{aligned} &w_A(k,z)^* w_A(k,z) \\ &= I_m - \mathrm{i}(z-\overline{z})\Pi(k)^*(A^* - \overline{z}I_n)^{-1}S(k)^{-1}(A - zI_n)^{-1}\Pi(k). \end{aligned} \tag{5.261}$$

Since, according to Proposition 5.36, $S(k) > 0$ for $k \geq 0$, the right-hand side of (5.261) is contractive in $\mathbb{C}_-$, that is,

$$w_A(k,z)^* w_A(k,z) \leq I_m \quad (z \in \mathbb{C}_-). \tag{5.262}$$

Formulas (5.260) and (5.262) lead us to (5.223), that is, φ is a Weyl function.

The uniqueness of the Weyl function remains to be proven. Let us first show that, for some $M > 0$ and all $k \geq 0$, we have the inequality

$$g^*(I_n - \mathrm{i}(A^*)^{-1})^k S(k)^{-1}(I_n + \mathrm{i}A^{-1})^k g \leq M g^* g, \quad g \in L, \tag{5.263}$$

where

$$L := \operatorname{span} \bigcup_{z \notin \sigma(A)} \operatorname{Im}\,(A - zI_n)^{-1}\varkappa_1.$$

In view of (5.221), formula (5.261) yields

$$\begin{aligned} &\varkappa_1^*(A^* - \overline{z}I_n)^{-1}(I_n - \mathrm{i}(A^*)^{-1})^k S(k)^{-1} \\ &\times (I_n + \mathrm{i}A^{-1})^k (A - zI_n)^{-1}\varkappa_1 \leq \mathrm{i}(\overline{z} - z)^{-1} I_p. \end{aligned} \tag{5.264}$$

In order to derive (5.263) from (5.264), we only need to note that

$$\operatorname{span} \bigcup_{z \notin \sigma(A)} \operatorname{Im} (A - zI_n)^{-1} \varkappa_1$$

coincides with the same span when z runs over an ε-neighborhood O_ε of any point $z_0 \notin \sigma_A$, for any sufficiently small $\varepsilon > 0$. By virtue of (5.261) and (5.263), we can choose $M_1 > 0$ such that we have

$$[I_p \quad 0] w_A(k,z)^* w_A(k,z) \begin{bmatrix} I_p \\ 0 \end{bmatrix} \geq \frac{1}{2} \quad \text{for all } |z| > M_1. \tag{5.265}$$

Without loss of generality, we may assume that M_1 is large enough in order that $w_{11}(z)$ is invertible (for $|z| > M_1$) and $M_1 > ||A||$. Then, taking into account (5.218) and (5.265), we obtain

$$\sum_{k=0}^{r} \left[I_p \quad (w_{21}(z) w_{11}(z)^{-1})^* \right] W_k(z)^* W_k(z) \begin{bmatrix} I_p \\ w_{21}(z) w_{11}(z)^{-1} \end{bmatrix}$$
$$\geq (r/2) \left(w_{11}(z)^{-1} \right)^* w_{11}(z)^{-1} \quad \text{for } z \in \Omega = \{z : |z| > M_1, \quad \Im z < -1/2\}.$$

In other words, for all $z \in \Omega$, we have

$$\sum_{k=0}^{\infty} g^* W_k(z)^* W_k(z) g = \infty \quad \left(g \in \operatorname{Im} \begin{bmatrix} I_p \\ w_{21}(z) w_{11}(z)^{-1} \end{bmatrix} \right). \tag{5.266}$$

In view of (5.266), the uniqueness of $\varphi(z)$ satisfying (5.223) is proved (for any $z \in \Omega$) quite similar to the proof of the corresponding uniqueness in Corollary 2.21. □

For the proof of Theorem 5.41, we shall need the following lemma which is of independent interest.

Lemma 5.47. *A strictly proper rational $p \times p$ matrix function φ admits a minimal realization of the form*

$$\varphi(z) = -\mathrm{i} \varkappa_1^* (\alpha - zI_n)^{-1} \varkappa_2, \tag{5.267}$$

such that $\alpha - \alpha^* = \mathrm{i}(\varkappa_1 \varkappa_1^* - \varkappa_2 \varkappa_2^*)$.

Proof. Without loss of generality, we assume that φ is given by the minimal realization (5.225).

First, we derive that equation (5.226) has a unique solution $X > 0$. For that, we use Theorem B.2. Indeed, the minimality of the realization (5.225) means that the pair $\{C, \mathcal{A}\}$ is observable and the pair $\{\mathcal{A}, B\}$ is controllable. Notice that, for any matrix ω, we have $\operatorname{Im} \omega \supseteq \operatorname{Im} \omega\omega^*$ and $g^*\omega\omega^* = 0$ yields $g^*\omega = 0$, that is, $\operatorname{Im} \omega = \operatorname{Im} \omega\omega^*$. Hence, the pair $\{\mathcal{A}, BB^*\}$ is controllable too. Therefore, the pair $\{BB^*, \mathrm{i}\mathcal{A}^*\}$ is observable. Since $\{C, \mathcal{A}\}$ is observable, the pair $\{\mathrm{i}\mathcal{A}^*, C^*\}$ is controllable and hence c-stabilizable (Appendix B). Thus, we can apply Theorem B.2 to show that the equation (5.226) has a unique nonnegative solution X and that this solution X is positive-definite.

Next, let $\varkappa_1$, $\varkappa_2$ and α be defined by (5.227). From (5.226) and (5.227), we see that $\alpha - \alpha^* = \mathrm{i}(\varkappa_1\varkappa_1^* - \varkappa_2\varkappa_2^*)$.

According to (5.225) and (5.227), the function φ is also given by the realization (5.267). Moreover, since the realization (5.225) is minimal, the same holds for the realization (5.267). □

Proof of Theorem 5.41. Let φ be a strictly proper rational $p \times p$ matrix function. Let $\varkappa_1$, $\varkappa_2$, α be as in Lemma 5.47, and put $A = \alpha + \mathrm{i}\varkappa_2\varkappa_2^*$. Then, the triple A, $\varkappa_1$, and $\varkappa_2$ satisfies (5.220). Furthermore, the pairs $\{\alpha, \varkappa_2\}$ and $\{\alpha^*, \varkappa_1\}$ are controllable. Since $A = \alpha + \mathrm{i}\varkappa_2\varkappa_2^*$, it is immediate that the pair $\{A, \varkappa_2\}$ is controllable (Appendix B). From (5.220), we have $\alpha^* = A^* + \mathrm{i}\varkappa_2\varkappa_2^* = A - \mathrm{i}\varkappa_1\varkappa_1^*$. Hence, as $\{\alpha^*, \varkappa_1\}$ is a controllable pair, so the pair $\{A, \varkappa_1\}$ is controllable as well. Therefore, the triple A, $\varkappa_1$ and $\varkappa_2$ is admissible. From Theorem 5.40, it follows now that the potential $\{C_k\}$ determined by the admissible triple A, $\varkappa_1$ and $\varkappa_2$ is indeed a solution of the inverse problem.

Let us prove now the uniqueness of the solution of the inverse problem. Suppose that there is system (5.114) with another spin sequence $\{\widetilde{C}_k\}$, given by the admissible triple $\widetilde{A}$, $\widetilde{\varkappa}_1$, $\widetilde{\varkappa}_2$, and with the same Weyl function φ. According to Theorem 5.40, we have another realization for φ, namely,

$$\varphi(z) = -\mathrm{i}\widetilde{\varkappa}_1^*(\widetilde{\alpha} - zI_{\widetilde{n}})^{-1}\widetilde{\varkappa}_2, \quad \widetilde{\alpha} = \widetilde{A} - \mathrm{i}\widetilde{\varkappa}_2\widetilde{\varkappa}_2^*. \tag{5.268}$$

Since the pairs $\{\widetilde{A}, \widetilde{\varkappa}_1\}$ and $\{\widetilde{A}, \widetilde{\varkappa}_2\}$ are controllable, and

$$\widetilde{\alpha} = \widetilde{A} - \mathrm{i}\widetilde{\varkappa}_2\widetilde{\varkappa}_2^*, \quad \widetilde{\alpha}^* = \widetilde{A} - \mathrm{i}\widetilde{\varkappa}_1\widetilde{\varkappa}_1^*,$$

it follows that the pairs $\{\widetilde{\alpha}, \widetilde{\varkappa}_2\}$ and $\{\widetilde{\alpha}^*, \widetilde{\varkappa}_1\}$ are also controllable. Thus the realization (5.268) is minimal and $\widetilde{n} = n$. Moreover, there is (see Appendix B) a similarity transformation, which transforms the realization (5.224) into (5.268), that is, there exists an invertible $n \times n$ matrix $\mathcal{S}$ such that

$$\widetilde{\alpha} = \mathcal{S}\alpha\mathcal{S}^{-1}, \quad \widetilde{\varkappa}_2 = \mathcal{S}\varkappa_2, \quad \widetilde{\varkappa}_1^* = \varkappa_1^*\mathcal{S}^{-1}. \tag{5.269}$$

Recall that $\widetilde{A}$, $\widetilde{\varkappa}_1$, $\widetilde{\varkappa}_2$ is an admissible triple, and therefore we have

$$\widetilde{\alpha} - \widetilde{\alpha}^* = \mathrm{i}(\widetilde{\varkappa}_1\widetilde{\varkappa}_1^* - \widetilde{\varkappa}_2\widetilde{\varkappa}_2^*). \tag{5.270}$$

From (5.269) and (5.270), it follows that $Z = \mathcal{S}^{-1}(\mathcal{S}^*)^{-1}$ satisfies

$$\alpha Z - Z\alpha^* = \mathrm{i}(Z\varkappa_1\varkappa_1^*Z - \varkappa_2\varkappa_2^*), \quad Z > 0. \tag{5.271}$$

Completely similar to the uniqueness of $X > 0$ in (5.226), one obtains the uniqueness of the solution $Z > 0$ of (5.271). By comparing the identity $\alpha - \alpha^* = \mathrm{i}(\varkappa_1\varkappa_1^* - \varkappa_2\varkappa_2^*)$

with (5.271), we see that $Z = I_n$ and thus $\mathcal{S}$ is unitary. In view of this, we have $\widetilde{A} = \mathcal{S}A\mathcal{S}^*$, $\widetilde{\varkappa}_2 = \mathcal{S}\varkappa_2$, and $\widetilde{\varkappa}_1 = \mathcal{S}\varkappa_1$. This unitary equivalence transformation does not change the potential, that is, $\{C_k\} = \{\widetilde{C}_k\}$. □

We already mentioned that the case when $\mathrm{i} \notin \sigma(A)$ is important for the study of the IHM model.

Notation 5.48. *We use the acronym* $\widetilde{\mathrm{EG}}$ *to denote the class of potentials* $\{C_k\}$ *determined by the triples* A, $\varkappa_1$, $\varkappa_2$, *with* A *an* $n \times n$ *nonsingular matrix and* $\varkappa_1$, $\varkappa_2$ *of size* $n \times p$, *satisfying the identity* (5.220) *and the additional special condition* $\mathrm{i} \notin \sigma(A)$.

Notice (see the beginning of the proof of Proposition 5.36) that (5.220) implies that $\sigma(A) \subset \overline{\mathbb{C}_+}$. The next lemma shows that without loss of generality, we can also require that $\{A, \varkappa_1\}$ and $\{A, \varkappa_2\}$ are controllable pairs, that is, $\widetilde{\mathrm{EG}} \subseteq \mathrm{EG}$.

Lemma 5.49. *Assume that the potential* $\{C_k\}$ *belongs* $\widetilde{\mathrm{EG}}$. *Then, it can be determined by a triple* A, $\varkappa_1$, $\varkappa_2$ *such that* $\det A \neq 0$, *(5.220) holds,* $\mathrm{i} \notin \sigma(A)$ *and the pairs* $\{A, \varkappa_1\}$ *and* $\{A, \varkappa_2\}$ *are controllable.*

Proof. Let n denote the minimal order of A $(0, \mathrm{i} \notin \sigma(A))$ in the set of triples that satisfy (5.220) and determine the given potential $\{C_k\}$. Suppose the $n \times n$ matrix $\widehat{A}$ and the $n \times p$ matrices $\widehat{\varkappa}_1$ and $\widehat{\varkappa}_2$ form such a triple, but the pair $\{\widehat{A}, \widehat{\varkappa}_2\}$ is not controllable. Put

$$\widehat{L} := \operatorname{span}\left(\bigcup_{k=0}^{\infty} \operatorname{Im} \widehat{A}^k \widehat{\varkappa}_2, \right), \qquad \widetilde{n} := \dim \widehat{L},$$

and choose a unitary matrix ω which maps $\widehat{L}$ onto the $\operatorname{Im} \left[\begin{smallmatrix} I_{\widetilde{n}} \\ 0 \end{smallmatrix}\right]$. Then, we have

$$A := \omega \widehat{A} \omega^* = \begin{bmatrix} \widetilde{A} & A_{12} \\ 0 & A_{22} \end{bmatrix}, \quad \varkappa_1 := \omega \widehat{\varkappa}_1 = \begin{bmatrix} \widetilde{\varkappa}_1 \\ \widetilde{\varkappa}_0 \end{bmatrix}, \quad \varkappa_2 := \omega \widehat{\varkappa}_2 = \begin{bmatrix} \widetilde{\varkappa}_2 \\ 0 \end{bmatrix}, \tag{5.272}$$

where the $\widetilde{n} \times \widetilde{n}$ matrix $\widetilde{A}$ $(0, \mathrm{i} \notin \sigma(\widetilde{A}))$ and the $\widetilde{n} \times p$ matrices $\widetilde{\varkappa}_1$, $\widetilde{\varkappa}_2$ form a triple which satisfies (5.220) and determines $\{C_k\}$.

To show this, we need to make some preparations. As $\widehat{L}$ is an invariant subspace of $\widehat{A}$, $\omega\widehat{L}$ is an invariant subspace of A, and thus A has the block triangular form given in (5.272). Moreover, in view of the inclusion $\operatorname{Im} \widehat{\varkappa}_2 \subseteq \widehat{L}$, we have $\operatorname{Im} \varkappa_2 \subseteq \omega\widehat{L}$, that is, $\varkappa_2$ has the block form given in (5.272). Taking into account that ω is unitary, $0, \mathrm{i} \notin \sigma(\widehat{A})$ and that $\widehat{A}$, $\widehat{\varkappa}_1$, $\widehat{\varkappa}_2$ satisfy (5.220) and determine $\{C_k\}$, we see that $0, \mathrm{i} \notin \sigma(A)$ and that A, $\varkappa_1$, $\varkappa_2$ satisfy (5.220) and determine $\{C_k\}$ too. Therefore, in view of (5.272), we have $0, \mathrm{i} \notin \sigma(\widetilde{A})$ and the triple $\widetilde{A}$, $\widetilde{\varkappa}_1$, $\widetilde{\varkappa}_2$ satisfies (5.220) as well.

Now, using the partitioning of $\Pi(k)$ in (5.244), we partition $\Pi(k)^* S(k)^{-1} \Pi(k)$ into the blocks of size $p \times p$, that is, $\Pi(k)^* S(k)^{-1} \Pi(k) = \{\Lambda_i(k)^* S(k)^{-1} \Lambda_j(k)\}_{i,j=1}^2$.

We express these blocks via the matrices $Q^-(k)$ which are given in (5.240):

$$\begin{aligned}
\Lambda_1(k)^*S(k)^{-1}\Lambda_1(k) &= \varkappa_1^*Q^-(k)^{-1}\varkappa_1,\\
\Lambda_1(k)^*S(k)^{-1}\Lambda_2(k) &= \varkappa_1^*Q^-(k)^{-1}(I_n+\mathrm{i}A^{-1})^{-k}(I_n-\mathrm{i}A^{-1})^k\varkappa_2,\\
\Lambda_2(k)^*S(k)^{-1}\Lambda_1(k) &= \varkappa_2^*(I_n+\mathrm{i}(A^*)^{-1})^k(I_n-\mathrm{i}(A^*)^{-1})^{-k}Q^-(k)^{-1}\varkappa_1,\\
\Lambda_2(k)^*S(k)^{-1}\Lambda_2(k) &= \varkappa_2^*(I_n+\mathrm{i}(A^*)^{-1})^k(I_n-\mathrm{i}(A^*)^{-1})^{-k}Q^-(k)^{-1}\\
&\quad\times(I_n+\mathrm{i}A^{-1})^{-k}(I_n-\mathrm{i}A^{-1})^k\varkappa_2.
\end{aligned}\tag{5.273}$$

Partition $Q^-(k+1)-Q^-(k)$ into four blocks:

$$Q^-(k+1)-Q^-(k)=\{q_{ij}\}_{i,j=1}^2,\tag{5.274}$$

where q_{11} is an $\tilde{n}\times\tilde{n}$ block. In view of (5.221), (5.243) and (5.272), we obtain

$$\begin{aligned}
q_{11} &= -2\tilde{A}^{-1}(I_{\tilde{n}}+\mathrm{i}\tilde{A}^{-1})^{-k-1}(I_{\tilde{n}}-\mathrm{i}\tilde{A}^{-1})^k\tilde{\varkappa}_2\tilde{\varkappa}_2^*(I_{\tilde{n}}+\mathrm{i}(\tilde{A}^*)^{-1})^k\\
&\quad\times(I_{\tilde{n}}-\mathrm{i}(\tilde{A}^*)^{-1})^{-k-1}(\tilde{A}^*)^{-1},\quad q_{21}=0,\quad q_{12}=0,\quad q_{22}=0.
\end{aligned}\tag{5.275}$$

Denote by $\tilde{\Pi}(k)$, $\tilde{S}(k)$, $\tilde{Q}^-(k)$, $\tilde{C}_k$, etc. the matrices generated by the triple $\tilde{A}$, $\tilde{\varkappa}_1$, $\tilde{\varkappa}_2$. We see that $q_{11}=\tilde{Q}^-(k+1)-\tilde{Q}^-(k)$. Taking into account the equalities (5.274), (5.275) and $Q^-(0)=I_n$, we obtain that the matrices $Q^-(k)$ are block diagonal:

$$Q^-(k)=\mathrm{diag}\,\{\tilde{Q}^-(k),\,I_{n-\tilde{n}}\}.\tag{5.276}$$

Now, according to (5.272), (5.273) and (5.276), it follows that

$$\Pi(k)^*S(k)^{-1}\Pi(k)-\tilde{\Pi}(k)^*\tilde{S}(k)^{-1}\tilde{\Pi}(k)=\begin{bmatrix}\tilde{\varkappa}_0^*\tilde{\varkappa}_0 & 0\\ 0 & 0\end{bmatrix}.\tag{5.277}$$

From (5.217) and (5.277), we derive $C_k=\tilde{C}_k$, that is, the triple $\tilde{A}$, $\tilde{\varkappa}_1$, $\tilde{\varkappa}_2$ determines the potential $\{C_k\}$. This contradicts the assumption that n is minimal and therefore the pair $\{\hat{A},\hat{\varkappa}_2\}$ should be controllable.

In the same way, we shall show that the pair $\{\hat{A},\hat{\varkappa}_1\}$ is controllable too. Indeed, suppose $\{\hat{A},\hat{\varkappa}_1\}$ is not controllable. Now, put

$$\hat{L}:=\mathrm{span}\left(\bigcup_{k=0}^{\infty}\mathrm{Im}\,\hat{A}^k\hat{\varkappa}_1,\right),\qquad \tilde{n}:=\dim\hat{L},$$

and choose a unitary matrix ω which maps $\hat{L}$ onto the Im $\begin{bmatrix}I_{\tilde{n}}\\0\end{bmatrix}$. Then, similar to the previous case, we obtain

$$A:=\omega\hat{A}\omega^*=\begin{bmatrix}\tilde{A} & A_{12}\\ 0 & A_{22}\end{bmatrix},\quad \varkappa_1:=\omega\hat{\varkappa}_1=\begin{bmatrix}\tilde{\varkappa}_1\\0\end{bmatrix},\quad \varkappa_2:=\omega\hat{\varkappa}_2=\begin{bmatrix}\tilde{\varkappa}_2\\ \tilde{\varkappa}_0\end{bmatrix},\tag{5.278}$$

where the $\widetilde{n} \times \widetilde{n}$ matrix $\widetilde{A}$ $(0, \mathrm{i} \notin \sigma(\widetilde{A}))$ and the $\widetilde{n} \times p$ matrices $\widetilde{\varkappa}_1$, $\widetilde{\varkappa}_2$ form a triple which satisfies (5.220). To show that the triple $\widetilde{A}$, $\widetilde{\varkappa}_1$, $\widetilde{\varkappa}_2$ determines C_k, we shall use the fact that A, $\varkappa_1$, $\varkappa_2$ determines C_k and rewrite the blocks of $\Pi(k)^*S(k)^{-1}\Pi(k)$ in terms of $Q^+(k)$, which is given in (5.239), that is,

$$\begin{aligned}
\Lambda_2(k)^*S(k)^{-1}\Lambda_2(k) &= \varkappa_2^*Q^+(k)^{-1}\varkappa_2, \\
\Lambda_2(k)^*S(k)^{-1}\Lambda_1(k) &= \varkappa_2^*Q^+(k)^{-1}(I_n - \mathrm{i}A^{-1})^{-k}(I_n + \mathrm{i}A^{-1})^k\varkappa_1, \\
\Lambda_1(k)^*S(k)^{-1}\Lambda_2(k) &= \varkappa_1^*(I_n - \mathrm{i}(A^*)^{-1})^k(I_n + \mathrm{i}(A^*)^{-1})^{-k}Q^+(k)^{-1}\varkappa_2, \\
\Lambda_1(k)^*S(k)^{-1}\Lambda_1(k) &= \varkappa_1^*(I_n - \mathrm{i}(A^*)^{-1})^k(I_n + \mathrm{i}(A^*)^{-1})^{-k}Q^+(k)^{-1} \\
&\quad \times (I_n - \mathrm{i}A^{-1})^{-k}(I_n + \mathrm{i}A^{-1})^k\varkappa_1.
\end{aligned} \tag{5.279}$$

Partition now $Q^+(k+1) - Q^+(k)$ into four blocks, namely,

$$Q^+(k+1) - Q^+(k) = \{q_{ij}\}_{i,j=1}^2. \tag{5.280}$$

In view of (5.242) and (5.278), we obtain an analog of (5.275), that is,

$$\begin{aligned}
q_{11} &= 2\widetilde{A}^{-1}(I_{\widetilde{n}} - \mathrm{i}\widetilde{A}^{-1})^{-k-1}(I_{\widetilde{n}} + \mathrm{i}\widetilde{A}^{-1})^k\widetilde{\varkappa}_1\widetilde{\varkappa}_1^*(I_{\widetilde{n}} - \mathrm{i}(\widetilde{A}^*)^{-1})^k \\
&\quad \times (I_{\widetilde{n}} + \mathrm{i}(\widetilde{A}^*)^{-1})^{-k-1}(\widetilde{A}^*)^{-1}, \quad q_{21} = 0, \quad q_{12} = 0, \quad q_{22} = 0.
\end{aligned} \tag{5.281}$$

Denote by $\widetilde{\Pi}(k)$, $\widetilde{S}(k)$, $\widetilde{Q}^+(k)$, $\widetilde{C}_k$, etc. the matrices generated by the triple $\widetilde{A}$, $\widetilde{\varkappa}_1$, $\widetilde{\varkappa}_2$. We see that $q_{11} = \widetilde{Q}^+(k+1) - \widetilde{Q}^+(k)$. Taking into account (5.280), (5.281) and $Q^+(0) = I_n$, we obtain that the matrices $Q^+(k)$ are block diagonal: $Q^+(k) = \operatorname{diag}\{\widetilde{Q}^+(k), I_{n-\widetilde{n}}\}$. Now, according to (5.278) and (5.279), it follows that

$$\Pi(k)^*S(k)^{-1}\Pi(k) - \widetilde{\Pi}(k)^*\widetilde{S}(k)^{-1}\widetilde{\Pi}(k) = \begin{bmatrix} 0 & 0 \\ 0 & \widetilde{\varkappa}_0^*\widetilde{\varkappa}_0 \end{bmatrix}. \tag{5.282}$$

From (5.217) and (5.282), we derive $C_k = \widetilde{C}_k$, that is, the triple $\widetilde{A}$, $\widetilde{\varkappa}_1$, $\widetilde{\varkappa}_2$ determines the potential $\{C_k\}$. Thus, the pair $\{\widehat{A}, \widehat{\varkappa}_1\}$ should also be controllable. □

Finally, recalling Corollary 5.43, from the proof of Theorem 5.40, we obtain the following corollary.

Corollary 5.50. *Let the parameter matrices A $(0, \mathrm{i} \notin \sigma(A))$, $S(0) > 0$ and $\Pi(0)$ satisfy the identity (5.214). Then the Weyl function φ of the system determined by these matrices is given by the formula*

$$\varphi(z) = -\mathrm{i}\varkappa_1^*S(0)^{-1}(\widetilde{\alpha} - zI_n)^{-1}\varkappa_2, \quad \widetilde{\alpha} = A - \mathrm{i}\varkappa_2\varkappa_2^*S(0)^{-1}.$$

This corollary is proved by transforming the matrices A, $S(0)$ and $\Pi(0)$ into the equivalent set $S(0)^{-\frac{1}{2}}AS(0)^{\frac{1}{2}}$, I_n, $S(0)^{-\frac{1}{2}}\Pi(0)$.

5.3.4 Isotropic Heisenberg magnet

Explicit solutions of the discrete integrable nonlinear equations form an interesting and actively studied domain (see, e.g. [7, 97, 101, 119, 164, 184, 192, 209, 321], and see also [119] for further references). In order to study the IHM model, we insert an additional variable t in our notations: $\Pi(k,t)$, $\Lambda_i(k,t)$, $S(k,t)$, $C_k(t)$, $w_A(k,t,z)$, $\varphi(t,z)$ and so on. Notice that the order n and the parameter matrix A do not depend on t, and we preserve the notation $\varkappa_i$ for $\Lambda_i(0,0)$. The dependence on t of the other matrix functions is defined by

$$\frac{d}{dt}\Lambda_1(0,t) = -2(A - \mathrm{i}I_n)^{-1}\Lambda_1(0,t), \quad \frac{d}{dt}\Lambda_2(0,t) = -2(A + \mathrm{i}I_n)^{-1}\Lambda_2(0,t), \tag{5.283}$$

where $\Pi(k,t) =: \begin{bmatrix}\Lambda_1(k,t) & \Lambda_2(k,t)\end{bmatrix}$, and

$$\begin{aligned}\frac{d}{dt}S(0,t) = &-\Big((A - \mathrm{i}I_n)^{-1}S(0,t) + (A + \mathrm{i}I_n)^{-1}S(0,t) + S(0,t)(A^* + \mathrm{i}I_n)^{-1} \\ &+S(0,t)(A^* - \mathrm{i}I_n)^{-1} + 2(A^2 + I_n)^{-1} \\ &\times (A\Pi(0,t)j\Pi(0,t)^* + \Pi(0,t)j\Pi(0,t)^*A^*)\left((A^*)^2 + I_n\right)^{-1}\Big).\end{aligned} \tag{5.284}$$

We assume that the parameter matrices A, $S(0,0)$ and $\Pi(0,0)$ satisfy the identity

$$AS(0,t) - S(0,t)A^* = \mathrm{i}\Pi(0,t)\Pi(0,t)^* \tag{5.285}$$

at $t = 0$.

According to equalities (5.283), (5.284) and identity (5.285) at $t = 0$, the identity (5.285) holds for all t, which can be derived by differentiating both sides of (5.285).

Remark 5.51. We already mentioned (in the proof of Proposition 5.36) that (5.220) implies $\sigma(A) \subset (\mathbb{C}_+ \cup \mathbb{R})$. Clearly, the same property $\sigma(A) \subset (\mathbb{C}_+ \cup \mathbb{R})$ follows from the inequality $S(0,0) > 0$ and identity

$$AS(0,0) - S(0,0)A^* = \mathrm{i}\Pi(0,0)\Pi(0,0)^*. \tag{5.286}$$

In particular, the relation $-\mathrm{i} \notin \sigma(A)$ is immediate.

Theorem 5.52. *Assume the parameter matrices A $(0, \mathrm{i} \notin \sigma(A))$, $S(0,0) > 0$ and $\Pi(0,0)$ satisfy the identity (5.286). Define $S(0,t)$ and $\Pi(0,t)$ by equations (5.283) and (5.284). Then, the inequality $S(0,t) > 0$ holds on some interval $-\varepsilon < t < \varepsilon$, and the potentials $\{C_k(t)\}$ given by (5.215)–(5.217) belong to $\widetilde{\mathrm{EG}}$ for each t from this interval. Moreover, $\{C_k(t)\}$ $(-\varepsilon < t < \varepsilon)$ generate via (5.210) the spins $\{\vec{S}_k(t)\}$ $(k \geq 0)$ which satisfy the IHM equation (5.209).*

Proof. Step 1. Since A does not depend on t and (5.285) is valid, it is immediate that $\{C_k(t)\} \in \widetilde{\mathrm{EG}}$. In order to prove other statements of the theorem, we shall show that systems (5.206) and (5.207) are compatible and derive in this way the zero curvature equation (5.208), which is equivalent to (5.209), (5.210) (see, e.g. [101] and references therein). For that, we first successively obtain the derivatives $\frac{d}{dt}\Pi(k,t)$, $\frac{d}{dt}S(k,t)$, $\frac{d}{dt}(\Pi(k,t)^*S(k,t)^{-1})$ and $\frac{d}{dt}w_A(k,t,z)$.

Recall that $\varkappa_i$ are the $n \times p$ blocks of $\Pi(0,0)$. From (5.221) and (5.283), we have

$$\Pi(k,t) = \left[(I_n + \mathrm{i}A^{-1})^k \mathrm{e}^{-2t(A-\mathrm{i}I_n)^{-1}}\varkappa_1 \quad (I_n - \mathrm{i}A^{-1})^k \mathrm{e}^{-2t(A+\mathrm{i}I_n)^{-1}}\varkappa_2\right]. \tag{5.287}$$

Hence, $\frac{d}{dt}\Pi(k,t)$ is easily expressed in terms of its $n \times p$ blocks $\Lambda_1(k,t)$, $\Lambda_2(k,t)$:

$$\frac{d}{dt}\Pi(k,t) = -2\left[(A-\mathrm{i}I_n)^{-1}\Lambda_1(k,t) \quad (A+\mathrm{i}I_n)^{-1}\Lambda_2(k,t)\right]. \tag{5.288}$$

Step 2. Now, we shall show by induction that

$$\begin{aligned}\frac{d}{dt}S(k,t) = &-\Big((A-\mathrm{i}I_n)^{-1}S(k,t) + (A+\mathrm{i}I_n)^{-1}S(k,t) \\ &+S(k,t)(A^*+\mathrm{i}I_n)^{-1} + S(k,t)(A^*-\mathrm{i}I_n)^{-1} + 2(A^2+I_n)^{-1} \\ &\times (A\Pi(k,t)j\Pi(k,t)^* + \Pi(k,t)j\Pi(k,t)^*A^*)\left((A^*)^2+I_n\right)^{-1}\Big).\end{aligned} \tag{5.289}$$

In view of (5.284), formula (5.289) holds for $k=0$. Suppose it is valid for $k=i$. Then, using (5.216), for $\Sigma(t) := \frac{d}{dt}S(i+1,t)$, we obtain

$$\begin{aligned}\Sigma(t) = &-\Big((A-\mathrm{i}I_n)^{-1}S(i+1,t) - (A-\mathrm{i}I_n)^{-1}A^{-1}\Pi(i,t)j\Pi(i,t)^*(A^*)^{-1} \\ &+(A+\mathrm{i}I_n)^{-1}S(i+1,t) - (A+\mathrm{i}I_n)^{-1}A^{-1}\Pi(i,t)j\Pi(i,t)^*(A^*)^{-1} \\ &+S(i+1,t)(A^*+\mathrm{i}I_n)^{-1} - A^{-1}\Pi(i,t)j\Pi(i,t)^*(A^*)^{-1}(A^*+\mathrm{i}I_n)^{-1} \\ &+S(i+1,t)(A^*-\mathrm{i}I_n)^{-1} - A^{-1}\Pi(i,t)j\Pi(i,t)^*(A^*)^{-1}(A^*-\mathrm{i}I_n)^{-1} \\ &+2(A^2+I_n)^{-1}\Big((A+A^{-1})\Pi(i,t)j\Pi(i,t)^* \\ &+\Pi(i,t)j\Pi(i,t)^*(A^*+(A^*)^{-1})\Big)\left((A^*)^2+I_n\right)^{-1}\Big) \\ &+\frac{d}{dt}\left(A^{-1}\Pi(i,t)j\Pi(i,t)^*(A^*)^{-1}\right).\end{aligned} \tag{5.290}$$

Taking into account (5.288), we easily calculate that

$$\begin{aligned}\Sigma_1(t) := &\frac{d}{dt}\left(A^{-1}\Pi(i,t)j\Pi(i,t)^*(A^*)^{-1}\right) + (A-\mathrm{i}I_n)^{-1}A^{-1}\Pi(i,t)j\Pi(i,t)^*(A^*)^{-1} \\ &+(A+\mathrm{i}I_n)^{-1}A^{-1}\Pi(i,t)j\Pi(i,t)^*(A^*)^{-1} \\ &+A^{-1}\Pi(i,t)j\Pi(i,t)^*(A^*)^{-1}(A^*+\mathrm{i}I_n)^{-1} \\ &+A^{-1}\Pi(i,t)j\Pi(i,t)^*(A^*)^{-1}(A^*-\mathrm{i}I_n)^{-1} \\ = &-(A-\mathrm{i}I_n)^{-1}A^{-1}\Pi(i,t)\Pi(i,t)^*(A^*)^{-1} \\ &+(A+\mathrm{i}I_n)^{-1}A^{-1}\Pi(i,t)\Pi(i,t)^*(A^*)^{-1} \\ &-A^{-1}\Pi(i,t)\Pi(i,t)^*(A^*)^{-1}(A^*+\mathrm{i}I_n)^{-1} \\ &+A^{-1}\Pi(i,t)\Pi(i,t)^*(A^*)^{-1}(A^*-\mathrm{i}I_n)^{-1}.\end{aligned} \tag{5.291}$$

Notice that $(A - \mathrm{i}I_n)^{-1} - (A + \mathrm{i}I_n)^{-1} = 2\mathrm{i}(A^2 + I_n)^{-1}$ and so we rewrite (5.291) as

$$\Sigma_1(t) = 2\mathrm{i}\Big(A^{-1}\Pi(i,t)\Pi(i,t)^*(A^*)^{-1}((A^*)^2 + I_n)^{-1} \\ -(A^2 + I_n)^{-1}A^{-1}\Pi(i,t)\Pi(i,t)^*(A^*)^{-1}\Big). \tag{5.292}$$

Notice also that

$$\Sigma_2(t) := 2(A^2 + I_n)^{-1}\Big((A + A^{-1})\Pi(i,t)j\Pi(i,t)^* \\ +\Pi(i,t)j\Pi(i,t)^*(A^* + (A^*)^{-1})\Big)((A^*)^2 + I_n)^{-1} \\ = 2\Big(A^{-1}\Pi(i,t)j\Pi(i,t)^*((A^*)^2 + I_n)^{-1} \\ +(A^2 + I_n)^{-1}\Pi(i,t)j\Pi(i,t)^*(A^*)^{-1}\Big). \tag{5.293}$$

Taking into account (5.215), (5.292) and (5.293), we see that

$$\Sigma_2(t) - \Sigma_1(t) = 2\Big(A^{-1}\Pi(i,t)j\Pi(i+1,t)^*((A^*)^2 + I_n)^{-1} \\ +(A^2 + I_n)^{-1}\Pi(i+1,t)j\Pi(i,t)^*(A^*)^{-1}\Big) \\ = 2(A^2 + I_n)^{-1}(A\Pi(i+1,t)j\Pi(i+1,t)^* \\ +\Pi(i+1,t)j\Pi(i+1,t)^*A^*)((A^*)^2 + I_n)^{-1}. \tag{5.294}$$

In view of (5.291), (5.293) and (5.294), we rewrite (5.290) in the form (5.289), with $i+1$ in place of k. It follows that (5.289) is valid for $k = i+1$ and thus for all $k \ge 0$.

Step 3. Using (5.288) and (5.289), we shall obtain the equation

$$\frac{d}{dt}\Big(\Pi(k,t)^*S(k,t)^{-1}\Big) = \mathfrak{H}^+(k,t)\Pi(k,t)^*S(k,t)^{-1}(A - \mathrm{i}I_n)^{-1} \\ + \mathfrak{H}^-(k,t)\Pi(k,t)^*S(k,t)^{-1}(A + \mathrm{i}I_n)^{-1}, \tag{5.295}$$

where

$$\mathfrak{H}^\pm(k,t) = 2w_A(k,t,\pm\mathrm{i})P_\pm w_A(k,t,\mp\mathrm{i})^*, \quad P_\pm := \frac{1}{2}(I_m \pm j). \tag{5.296}$$

Indeed, taking into account (5.288), we have

$$\frac{d}{dt}\Big(\Pi(k,t)^*S(k,t)^{-1}\Big) = -2\Big(P_+\Pi(k,t)^*(A^* + \mathrm{i}I_n)^{-1} + P_-\Pi(k,t)^*(A^* - \mathrm{i}I_n)^{-1} \\ +\frac{1}{2}\Pi(k,t)^*S(k,t)^{-1}\frac{d}{dt}S(k,t)\Big)S(k,t)^{-1}. \tag{5.297}$$

Identity (5.222) yields

$$(A^* \pm \mathrm{i}I_n)^{-1}S(k,t)^{-1} = S(k,t)^{-1}(A \pm \mathrm{i}I_n)^{-1} + \mathrm{i}(A^* \pm \mathrm{i}I_n)^{-1} \\ \times S(k,t)^{-1}\Pi(k,t)\Pi(k,t)^*S(k,t)^{-1}(A \pm \mathrm{i}I_n)^{-1}. \tag{5.298}$$

Finally, notice that

$$2(A^2 + I_n)^{-1}A = (A - \mathrm{i}I_n)^{-1} + (A + \mathrm{i}I_n)^{-1}. \tag{5.299}$$

By using (5.289), (5.298) and (5.299), and after some calculations, we rewrite (5.297) in the form (5.295), where

$$\begin{aligned}\mathfrak{H}^+(k,t) = {} & I_m + j - \mathrm{i}\Pi(k,t)^* S(k,t)^{-1}(A - \mathrm{i}I_n)^{-1}\Pi(k,t)j \\ & + \mathrm{i}j\Pi(k,t)^*(A^* - \mathrm{i}I_n)^{-1}S(k,t)^{-1}\Pi(k,t) + \Pi(k,t)^* S(k,t)^{-1} \\ & \times (A - \mathrm{i}I_n)^{-1}\Pi(k,t)j\Pi(k,t)^*(A^* - \mathrm{i}I_n)^{-1}S(k,t)^{-1}\Pi(k,t),\end{aligned} \tag{5.300}$$

$$\begin{aligned}\mathfrak{H}^-(k,t) = {} & I_m - j + \mathrm{i}\Pi(k,t)^* S(k,t)^{-1}(A + \mathrm{i}I_n)^{-1}\Pi(k,t)j \\ & - \mathrm{i}j\Pi(k,t)^*(A^* + \mathrm{i}I_n)^{-1}S(k,t)^{-1}\Pi(k,t) - \Pi(k,t)^* S(k,t)^{-1} \\ & \times (A + \mathrm{i}I_n)^{-1}\Pi(k,t)j\Pi(k,t)^*(A^* + \mathrm{i}I_n)^{-1}S(k,t)^{-1}\Pi(k,t).\end{aligned} \tag{5.301}$$

From (5.219) and (5.300), it follows that

$$\mathfrak{H}^+(k,t) = I_m + w_A(k,t,\mathrm{i})jw_A(k,t,-\mathrm{i})^*,$$

and so, taking into account (5.238), we derive (5.296) for the case of $\mathfrak{H}^+(k,t)$. Equality (5.296) for $\mathfrak{H}^-(k,t)$ follows from (5.301) in a similar way. Thus, (5.295) (for $\mathfrak{H}^\pm(k,t)$ given by (5.296)) is proved.

Step 4. Recall now that $m = 2$ and (5.238) holds. Then, according to (5.245), (5.249) and (5.296), we have

$$\mathfrak{H}^+(k,t) = c_k^+(t)(I_2 + C_k(t))(I_2 + C_{k-1}(t)), \quad k \geq 1, \tag{5.302}$$

and according to (5.246), (5.250) and (5.296), we have

$$\mathfrak{H}^-(k,t) = c_k^-(t)(I_2 - C_k(t))(I_2 - C_{k-1}(t)), \tag{5.303}$$

where $c_k^\pm(t)$ are scalar functions. In view of (5.296), we easily calculate $\operatorname{Tr}\mathfrak{H}^\pm$, where Tr denotes the trace:

$$\operatorname{Tr}\mathfrak{H}^\pm(k,t) \equiv 2. \tag{5.304}$$

From Remark 5.45 and formula (5.210), we derive

$$\begin{aligned}\operatorname{Tr}(I_2 \pm C_k(t))(I_2 \pm C_{k-1}(t)) &= \operatorname{Tr}(I_2 + C_k(t)C_{k-1}(t)) \\ &= 2\left(1 + \vec{\mathcal{S}}_{k-1}(t) \cdot \vec{\mathcal{S}}_k(t)\right).\end{aligned} \tag{5.305}$$

Formulas (5.302)–(5.305) imply that $1 + \vec{\mathcal{S}}_{k-1}(t) \cdot \vec{\mathcal{S}}_k(t) \neq 0$. Now, relations (5.213), (5.304) and (5.305) yield that

$$\operatorname{Tr}V_k^\pm(t) \equiv 2 \equiv \operatorname{Tr}\mathfrak{H}^\pm(k,t). \tag{5.306}$$

Taking into account (5.213), (5.302) and (5.303), we see that

$$V_k^{\pm}(t) = \hat{c}_k^{\pm}(t)\mathfrak{H}^{\pm}(k,t),$$

where $c_k^{\pm}$ are scalar functions, and so (5.306) yields equalities $V_k^{\pm}(t) \equiv \mathfrak{H}^{\pm}(k,t)$. Hence, according to (5.295), we have

$$\frac{d}{dt}\left(\Pi(k,t)^*S(k,t)^{-1}\right) = V_k^{+}(t)\Pi(k,t)^*S(k,t)^{-1}(A - \mathrm{i}I_n)^{-1} + V_k^{-}(t)\Pi(k,t)^*S(k,t)^{-1}(A + \mathrm{i}I_n)^{-1}. \tag{5.307}$$

From $V_k^{\pm}(t) \equiv \mathfrak{H}^{\pm}(k,t)$, (5.238) and (5.296), we also derive

$$V_k^{\pm}(t)w_A(k,t,\pm\mathrm{i}) = 2w_A(k,t,\pm\mathrm{i})P_{\pm}. \tag{5.308}$$

Let us differentiate now w_A. For this purpose, notice that

$$(A \pm \mathrm{i}I_n)^{-1}(A - zI_n)^{-1} = (z \pm \mathrm{i})^{-1}\left((A - zI_n)^{-1} - (A \pm \mathrm{i}I_n)^{-1}\right). \tag{5.309}$$

Using (5.288), (5.307) and (5.309), we derive

$$\begin{aligned}\frac{d}{dt}w_A(k,t,z) = {}& (z-\mathrm{i})^{-1}V_k^{+}(t)(w_A(k,t,z) - w_A(k,t,\mathrm{i})) \\ &+ (z+\mathrm{i})^{-1}V_k^{-}(t)(w_A(k,t,z) - w_A(k,t,-\mathrm{i})) \\ &- 2(z-\mathrm{i})^{-1}(w_A(k,t,z) - w_A(k,t,\mathrm{i}))P_+ \\ &- 2(z+\mathrm{i})^{-1}(w_A(k,t,z) - w_A(k,t,-\mathrm{i}))P_-.\end{aligned} \tag{5.310}$$

Step 5. In view of (5.308), we can rewrite (5.310) as

$$\frac{d}{dt}w_A(k,t,z) = F_k(t,z)w_A(k,t,z) - w_A(k,t,z)\hat{F}(z), \quad k \geq 1, \tag{5.311}$$

where F_k is given by (5.212) and

$$\hat{F}(z) = 2((z-\mathrm{i})^{-1}P_+ + (z+\mathrm{i})^{-1}P_-).$$

Thus, according to Theorem 5.35 and formula (5.311), the nondegenerate matrix functions $\{\widehat{W}_k\}$ given by

$$\widehat{W}_k(t,z) = z^{-k}w_A(k,t,z)\begin{bmatrix}(z-\mathrm{i})^k \mathrm{e}^{2(z-\mathrm{i})^{-1}t} & 0 \\ 0 & (z+\mathrm{i})^k \mathrm{e}^{2(z+i)^{-1}t}\end{bmatrix}$$

satisfy equations (5.206) and (5.207), that is, the compatibility condition (5.208) is valid for $k \geq 1$. Since equation (5.208) is equivalent to (5.209), (5.210), the theorem is proved. □

We note that the IHM equation (5.209) is considered in Theorem 5.52 on the semiaxis $k \geq 1$. Theorem 5.52 together with Corollary 5.50 yields the following result.

Corollary 5.53. *Under the conditions of Theorem 5.52, the evolution of the Weyl function φ of the system $y_{k+1}(t,z) = G_k(t,z)y_k(t,z)$ is given by the formula*

$$\varphi(t,z) = -\mathrm{i}\varkappa_1^* \mathrm{e}^{-2t(A^*+\mathrm{i}I_n)^{-1}} S(0,t)^{-1}(\alpha(t) - zI_n)^{-1}\mathrm{e}^{-2t(A+\mathrm{i}I_n)^{-1}}\varkappa_2, \tag{5.312}$$
$$\alpha(t) = A - \mathrm{i}\mathrm{e}^{-2t(A+\mathrm{i}I_n)^{-1}}\varkappa_2\varkappa_2^*\mathrm{e}^{-2t(A^*-\mathrm{i}I_n)^{-1}}S(0,t)^{-1}.$$

As an illustration, let us consider a simple example.

Example 5.54. Put $p = n = 1$ and $A = \mathrm{i}\varepsilon$ $(\varepsilon > 0,\ \varepsilon \neq 1)$, and choose scalars $\varkappa_1$, $\varkappa_2$ such that $|\varkappa_1|^2 + |\varkappa_2|^2 = 2\varepsilon$. Then, A, $\varkappa_1$, $\varkappa_2$ form an admissible triple and A, $S(0,0) = 1$, $\Pi(0,0) = [\varkappa_1 \quad \varkappa_2]$ satisfy the conditions of Theorem 5.52 and Corollary 5.53. Therefore, according to (5.287), we have

$$\Pi(k,t) = \varepsilon^{-k}\Big[(1+\varepsilon)^k\varkappa_1 \exp\{-2\mathrm{i}t/(1-\varepsilon)\} \quad (\varepsilon-1)^k\varkappa_2\exp\{2\mathrm{i}t/(1+\varepsilon)\}\Big]. \tag{5.313}$$

From (5.222) and (5.313), we obtain

$$S(k,t) \equiv c_k(\varepsilon)\varepsilon^{-2k-1}/2, \quad c_k(\varepsilon) := (1+\varepsilon)^{2k}|\varkappa_1|^2 + (1-\varepsilon)^{2k}|\varkappa_2|^2. \tag{5.314}$$

In view of (5.217), (5.313) and (5.314), we now derive

$$(C_k(t))_{11} = 1 - 8\varepsilon^2|\varkappa_1\varkappa_2|^2(1-\varepsilon^2)^{2k}\left(c_k(\varepsilon)c_{k+1}(\varepsilon)\right)^{-1}, \quad (C_k(t))_{22} = -(C_k(t))_{11},$$
$$(C_k(t))_{12} = \overline{(C_k(t))_{21}} = 4\varepsilon\overline{\varkappa_1}\varkappa_2\left(c_k(\varepsilon)c_{k+1}(\varepsilon)\right)^{-1}\exp\left\{4\mathrm{i}t(1-\varepsilon^2)^{-1}\right\}$$
$$\times(\varepsilon^2-1)^k\left((1+\varepsilon)^{2k+1}|\varkappa_1|^2 + (1-\varepsilon)^{2k+1}|\varkappa_2|^2\right).$$

Finally, Corollary 5.53 yields

$$\varphi(t,z) = \mathrm{i}\overline{\varkappa_1}\varkappa_2\left(z + \mathrm{i}(|\varkappa_2|^2 - \varepsilon)\right)^{-1}\exp\left\{4\mathrm{i}t(1-\varepsilon^2)^{-1}\right\}.$$

6 Integrable nonlinear equations

The zero curvature representation of the integrable nonlinear equations is a well-known approach which was developed soon after seminal Lax pairs appeared in [188] (see [5, 101, 216, 336] and references in [101]). Namely, many integrable nonlinear equations admit the representation (zero curvature representation)

$$\frac{\partial}{\partial t}G(x,t,z) - \frac{\partial}{\partial x}F(x,t,z) + [G(x,t,z), F(x,t,z)] = 0, \tag{6.1}$$

which is the compatibility condition of the auxiliary linear systems

$$\frac{\partial}{\partial x}w(x,t,z) = G(x,t,z)w(x,t,z), \quad \frac{\partial}{\partial t}w(x,t,z) = F(x,t,z)w(x,t,z). \tag{6.2}$$

Here, G and F are $m \times m$ matrix functions, and z is the spectral parameter which will be omitted sometimes in our notations. We already discussed zero curvature representation (compatibility condition) for the case of one discrete (and one continuous) variable in Section 5.3, where IHM model was dealt with. Note that formula (6.1) for the case of continuous variables x and t differs from the compatibility condition (5.208) for the discrete case.

The solution of the integrable nonlinear equations is closely related to the Lax pairs and zero curvature representations mentioned above. This solution has been a great breakthrough in the second half of the twentieth century. Following successes in the theory of integrable nonlinear equations, an active study of the cases, which are close to integrable in a certain sense, was undertaken (see, e.g. some references in [43, 163]). Initial-boundary value problems for the integrable nonlinear equations can be considered as an important example, where integrability is "spoiled" by the boundary conditions.

The breakthrough in the initial value problems for integrable nonlinear equations was achieved using the Inverse Scattering Transform method. The initial-boundary value problems are more complicated. The inverse scattering transform and several other methods help to obtain some interesting results on the initial-boundary value problems (in particular, see more on the *global relation method* [52, 103] in Remark 6.24), but there are still many open problems of great current interest. The Inverse Spectral Transformation (ISpT) method [44, 46, 150, 281, 286] is one of the fruitful approaches in this domain. In particular, some further developments of the results and methods from [281, 286] are given in [238, 242, 256, 262, 289, 290].

The basic factorization theorem for fundamental solutions and applications to evolution of the Weyl functions are studied in Section 6.1. Somewhat more complicated results are obtained in Section 6.2 on sine-Gordon theory in a semistrip. In particular, we obtain results on the evolution of GW-functions, uniqueness of the solution of the *complex sine-Gordon equation* and the existence of the solution of the *sine-Gordon equation*.

In this chapter, we assume that $M > 0$ in the notation $\mathbb{C}_M$, and that the variables x and t are such that the points (x, t) belong to a semistrip

$$\Omega_a = \{(x, t) : 0 \le x < \infty, \ 0 \le t < a\}. \tag{6.3}$$

6.1 Compatibility condition and factorization formula

6.1.1 Main results

We normalize fundamental solutions of the auxiliary systems by the initial conditions

$$\frac{d}{dx}W(x,t,z) = G(x,t,z)W(x,t,z), \quad W(0,t,z) = I_m; \tag{6.4}$$

$$\frac{d}{dt}R(x,t,z) = F(x,t,z)R(x,t,z), \quad R(x,0,z) = I_m. \tag{6.5}$$

If condition (6.1) holds, the fundamental solution of (6.4) admits factorization

$$W(x,t,z) = R(x,t,z)W(x,0,z)R(t,z)^{-1}, \quad R(t,z) := R(0,t,z). \tag{6.6}$$

Formula (6.6) is one of the basic and actively used formulas in the ISpT method (see [238, 242, 256, 262, 281, 286, 289, 290] and references therein). It was derived in [281, 286] under some smoothness conditions (continuous differentiability of G and F, in particular): see formulas (1.6) in [281, p.22] and in [286, p.39].

Here, we prove (6.6) under weaker conditions and in much greater detail, which is important for applications. Namely, we prove the following theorem.

Theorem 6.1. *Let $m \times m$ matrix functions G and F and their derivatives $\frac{\partial}{\partial t}G$ and $\frac{\partial}{\partial x}F$ exist on the semistrip Ω_a, let G, $\frac{\partial}{\partial t}G$, and F be continuous with respect to x and t on Ω_a, and let* (6.1) *hold. Then, we have the equality*

$$W(x,t,z)R(t,z) = R(x,t,z)W(x,0,z), \quad R(t,z) := R(0,t,z). \tag{6.7}$$

Constructions similar to (6.6) also appear in the theory of the Knizhnik–Zamolodchikov equation (see Theorem 3.1 in [298] and see also [297]).

Remark 6.2. Formula (6.7) implies that the condition (6.1) is, indeed, the compatibility condition and

$$w(x,t,z) = W(x,t,z)R(t,z) = R(x,t,z)W(x,0,z)$$

satisfies (6.2).

We note that the solvability of system (6.2) in the domain Ω_a is of independent interest as one of the well-posedness and compatibility problems in domains with a boundary. The well-posedness of initial and initial-boundary value problems is a difficult area which is actively studied (see, e.g. recent works [52, 58, 183] and references therein).

Theorem 6.1 is proved in Subsection 6.1.2. Subsection 6.1.3 is dedicated to applications to initial-boundary value problems, and Theorem 6.6 on the evolution of the Weyl function for the "focusing" modified Korteweg–de Vries (mKdV) equation is proved there as an example. Another interesting example, namely, the second harmonic generation model, is dealt with in Subsection 6.1.4.

Recall that by $C^N(\Omega_a)$, we denote the functions and matrix functions which are N times continuously differentiable on Ω_a.

6.1.2 Proof of Theorem 6.1

The spectral parameter z is nonessential for the formulation of Theorem 6.1 and for its proof, and we shall omit it in this section. We shall need the proposition below.

Proposition 6.3. *Let the $m \times m$ matrix function W be given on Ω_a by equation (6.4), where $G(x,t)$ and $\frac{\partial}{\partial t}G(x,t)$ are continuous matrix functions in x and t.*

(a) *Then, the derivative $\frac{\partial}{\partial t}W$ exists and matrix functions W and $\frac{\partial}{\partial t}W$ are continuous with respect to x and t on the semistrip Ω_a.*

(b) *Moreover, the mixed derivative $\frac{\partial}{\partial x}\frac{\partial}{\partial t}W$ exists and the equality*

$$\frac{\partial}{\partial x}\frac{\partial}{\partial t}W = \frac{\partial}{\partial t}\frac{\partial}{\partial x}W \quad \text{holds on} \quad \Omega_a.$$

Proof. Consider system

$$\frac{d}{dx}y = \widehat{G}(x,y)y, \quad \widehat{G}(x,y) = \widehat{G}(x,y_{m+1}) := \begin{bmatrix} G(x,y_{m+1}) & 0 \\ 0 & 0 \end{bmatrix}, \tag{6.8}$$

where $\widehat{G}$ is an $(m+1)\times(m+1)$ matrix function and y_{m+1} is the last entry of the column vector $y \in \mathbb{C}^{m+1}$. Denote by W_i and e_i the i-th columns of W and I_m, respectively $(1 \le i \le m)$. It easily follows from (6.4) that the solution of (6.8) with the initial condition

$$y(0) = g = \begin{bmatrix} e_i \\ t \end{bmatrix} \tag{6.9}$$

has the form

$$y(x,g) = \begin{bmatrix} W_i(x,t) \\ t \end{bmatrix}. \tag{6.10}$$

Putting $G(x,t) = G(0,t)$ for $-\varepsilon \le x \le 0$ whereas $t \ge 0$, and putting

$$G(x,t) = G(x,0) + t\left(\frac{\partial}{\partial t}G\right)(x,0) \quad \text{for} \ -\varepsilon \le t \le 0 \quad (\varepsilon > 0),$$

we extend G so that G and $\frac{\partial}{\partial t}G$ remain continuous on the rectangles

$$\Omega(a_1, a_2) := \{(x, t) : -\varepsilon \le x \le a_1, \ -\varepsilon \le t \le a_2 < a\}, \quad a_1, a_2 \in \mathbb{R}_+. \tag{6.11}$$

Hence, it follows from the definition of $\widehat{G}$ in (6.8) that $\widehat{G}(x, y)$ and, as a consequence, the vector function $\widehat{G}(x, y)y$, are continuous on $\Omega(a_1, a_2)$ together with their derivatives with respect to the entries of y. Thus, according to the classical theory of ordinary differential equations (see, e.g. theorem on pp. 305–306 in [312]), the partial first derivatives of $y(x, g)$ with respect to the entries of g exist in the interior $\widetilde{\Omega}(a_1, a_2)$ of $\Omega(a_1, a_2)$. Moreover, y and its partial derivatives with respect to the entries of g are continuous. In particular, since from (6.9) we have $g_{m+1} = t$, the functions y and $\frac{\partial}{\partial t}y$ are continuous in all rectangles $\widetilde{\Omega}(a_1, a_2)$. Taking into account (6.10), we see that W and $\frac{\partial}{\partial t}W$ are continuous in the rectangles $\widetilde{\Omega}(a_1, a_2)$, and statement (a) is proven.

In view of (6.4) and the considerations above, the derivatives $\frac{\partial}{\partial x}W$, $\frac{\partial}{\partial t}\frac{\partial}{\partial x}W$ and $\frac{\partial}{\partial t}W$ exist and are continuous in the rectangles $\widetilde{\Omega}(a_1, a_2)$. Hence, by a stronger formulation (see, e.g. the notes [13, 306] or the book [207, p.201]) of the well-known (so called Clairaut's, or Schwarz's) theorem on mixed derivatives, $\frac{\partial}{\partial x}\frac{\partial}{\partial t}W$ exists in $\widetilde{\Omega}(a_1, a_2)$ and $\frac{\partial}{\partial x}\frac{\partial}{\partial t}W = \frac{\partial}{\partial t}\frac{\partial}{\partial x}W$. Thus, the statement (b) is valid. □

Now, we can follow the scheme from [286, Chapter 3] (see also [290, Chapter 12]).

Proof of Theorem 6.1. According to statement (a) in Proposition 6.3, the matrix function $\frac{\partial}{\partial t}W$ exists and is continuous. Introduce $U(x, t)$ by the equality

$$U := \frac{\partial W}{\partial t} - FW. \tag{6.12}$$

By virtue of (6.4), (6.12) and statement (b) in Proposition 6.3, we have

$$\begin{aligned} \frac{\partial U}{\partial x} &= \frac{\partial^2 W}{\partial x \partial t} - \frac{\partial F}{\partial x}W - F\frac{\partial W}{\partial x} = \frac{\partial^2 W}{\partial t \partial x} - \frac{\partial F}{\partial x}W - FGW, \\ \frac{\partial^2 W}{\partial x \partial t} &:= \frac{\partial}{\partial x}\frac{\partial}{\partial t}W = \frac{\partial}{\partial x}\left(\frac{\partial}{\partial t}W\right). \end{aligned} \tag{6.13}$$

It is also immediate from (6.4) that

$$\frac{\partial^2 W}{\partial t \partial x} = \frac{\partial}{\partial t}(GW) = \frac{\partial G}{\partial t}W + G\frac{\partial W}{\partial t}. \tag{6.14}$$

Formulas (6.13) and (6.14) imply

$$\begin{aligned} \frac{\partial U}{\partial x} &= \frac{\partial G}{\partial t}W + G\frac{\partial W}{\partial t} - \frac{\partial F}{\partial x}W - FGW \\ &= \left(\frac{\partial G}{\partial t} - \frac{\partial F}{\partial x} + GF - FG\right)W + G\frac{\partial W}{\partial t} - GFW. \end{aligned} \tag{6.15}$$

From (6.1), (6.15), and definition (6.12), we see that $\frac{\partial}{\partial x}U = GU$, that is, U and W satisfy the same equation. Since $W(0,t) = I_m$, we derive $(\frac{\partial}{\partial t}W)(0,t) = 0$ and so, in view of (6.12), we have $U(0,t) = -F(0,t)$. Finally, since

$$\frac{\partial}{\partial x}U = GU, \quad \frac{\partial}{\partial x}W = GW, \quad U(0,t) = -F(0,t), \quad W(0,t) = I_m,$$

we have $U(x,t) = -W(x,t)F(0,t)$ or, equivalently,

$$\left(\frac{\partial}{\partial t}W\right)(x,t) - F(x,t)W(x,t) = -W(x,t)F(0,t). \tag{6.16}$$

Put

$$Y(x,t) = W(x,t)R(t), \quad Z(x,t) = R(x,t)W(x,0). \tag{6.17}$$

Recall that $R(t) := R(0,t)$. Therefore, (6.5), (6.16) and (6.17) imply that

$$\begin{aligned}\left(\frac{\partial}{\partial t}Y\right)(x,t) &= (F(x,t)W(x,t) - W(x,t)F(0,t))\,R(t) + W(x,t)F(0,t)R(t)\\ &= F(x,t)Y(x,t), \quad Y(x,0) = W(x,0).\end{aligned} \tag{6.18}$$

Formulas (6.5) and (6.17) yield

$$\left(\frac{\partial}{\partial t}Z\right)(x,t) = F(x,t)Z(x,t), \quad Z(x,0) = W(x,0). \tag{6.19}$$

From (6.18) and (6.19), we see that $Y = Z$, that is, (6.7) holds. □

Remark 6.4. Though the case of continuous F is more convenient for applications, it is immediate from the proof that the statement of Theorem 6.1 is valid when F remains differentiable with respect to x, but is only measurable and summable with respect to t on all finite intervals belonging to $\mathbb{R}_+$.

From the proof of Theorem 6.1, we also obtain the following remark.

Remark 6.5. Theorem 6.1 holds on the domains more general than Ω_a. In particular, it holds if we consider $(x,\, t) \in \mathcal{I}_1 \times \mathcal{I}_2$, where $\mathcal{I}_k$ $(k = 1,2)$ is the interval $[0,\, a_k)$ $(0 < a_k \le \infty)$.

Other interesting cases of matrix factorizations related to boundary value problems are treated in [60, 155].

6.1.3 Application to the matrix "focusing" modified Korteweg-de Vries

The *matrix "focusing" mKdV equation* has the form

$$4\frac{\partial v}{\partial t} = \frac{\partial^3 v}{\partial x^3} + 3\left(\frac{\partial v}{\partial x}v^* v + vv^*\frac{\partial v}{\partial x}\right), \tag{6.20}$$

where $v(x,t)$ is a $p \times p$ matrix function. Equation (6.20) is equivalent (see [61, 101, 327] and references therein) to zero curvature equation (6.1), where the $m \times m$ ($m = 2p$) matrix functions $G(x,t,z)$ and $F(x,t,z)$ are given by the formulas

$$G = \mathrm{i}zj + jV, \quad j = \begin{bmatrix} I_p & 0 \\ 0 & -I_p \end{bmatrix}, \quad V = \begin{bmatrix} 0 & v \\ v^* & 0 \end{bmatrix}, \tag{6.21}$$

$$F = -\mathrm{i}z^3 j - z^2 jV + \frac{\mathrm{i}z}{2}\left(V^2 + \frac{\partial V}{\partial x}\right) + \frac{1}{4}j\left(\frac{\partial^2 V}{\partial x^2} + 2V^3 + j\frac{\partial V}{\partial x}V - jV\frac{\partial V}{\partial x}\right). \tag{6.22}$$

Clearly, in the case that G is given by (6.21), the first auxiliary system from (6.2) coincides (for each fixed t) with the skew-self-adjoint Dirac system

$$\frac{d}{dx}y(x,z) = \left(\mathrm{i}zj + jV(x)\right)y(x,z). \tag{6.23}$$

The Weyl theory of system (6.23) (also called Zakharov–Shabat or AKNS system) was dealt with in Chapter 3 (and some useful references were given there). Our next theorem on the evolution of the Weyl function in the case of the focusing mKdV follows from Theorem 6.1 and Corollary 3.8. The case of the defocusing mKdV was studied earlier in [281, 286, 290].

Theorem 6.6. *Let a $p \times p$ matrix function $v \in C^1(\Omega_a)$ have a continuous partial second derivative $\frac{\partial^2}{\partial x^2}v$ and let $\frac{\partial^3}{\partial x^3}v$ exist. Assume that v satisfies mKdV (6.20) and that the following inequalities are valid:*

$$\sup_{(x,t)\in\Omega_a} \|v(x,t)\| \le M, \qquad \sup_{(x,t)\in\Omega_a} \left(\left\| \left(\frac{\partial}{\partial x}v\right)(x,t) \right\| + \left\| \left(\frac{\partial^2}{\partial x^2}v\right)(x,t) \right\| \right) < \infty. \tag{6.24}$$

Then, the evolution $\varphi(t,z)$ of the Weyl function of the skew-self-adjoint Dirac system (6.23), where the system is considered on the semiaxis $[0, \infty)$ and its potential $V = V(x,t)$ has the form (6.21), is given by the equality

$$\varphi(t,z) = \left(R_{21}(t,z) + R_{22}(t,z)\varphi(0,z)\right)\left(R_{11}(t,z) + R_{12}(t,z)\varphi(0,z)\right)^{-1} \tag{6.25}$$

in the half-plane $\mathbb{C}_M$. Here, the block matrix function

$$R(t,z) = \{R_{ik}(t,z)\}_{i,k=1}^2 = R(0,t,z) \tag{6.26}$$

is defined by the boundary values $v(0,t)$, $(\frac{\partial}{\partial x}v)(0,t)$ and $(\frac{\partial^2}{\partial x^2}v)(0,t)$ via formulas (6.5) and (6.22).

Proof. Because of the smoothness conditions on v, we see that G and F given by (6.21) and (6.22), respectively, satisfy the requirements of Theorem 6.1, that is, (6.7) holds. Let us rewrite (6.7) in the form

$$W(x,t,z)^{-1}R(x,t,z) = R(t,z)W(x,0,z)^{-1}. \tag{6.27}$$

Recall that the class of matrix functions, which are nonsingular with property-j (Definition 1.42) in $\mathbb{C}_M$, is denoted by $\mathbf{P}_M(j)$ and let $\mathcal{P}(x,z) \in \mathbf{P}_M(j)$. Putting

$$\widetilde{\mathcal{P}}(x,t,z) := R(x,t,z)\mathcal{P}(x,z), \tag{6.28}$$

from (6.27), we have

$$W(x,t,z)^{-1}\widetilde{\mathcal{P}}(x,t,z) = R(t,z)W(x,0,z)^{-1}\mathcal{P}(x,z). \tag{6.29}$$

Now, we shall show that

$$\widetilde{\mathcal{P}}(x,t,z)^*\widetilde{\mathcal{P}}(x,t,z) > 0, \quad \widetilde{\mathcal{P}}(x,t,z)^* j\widetilde{\mathcal{P}}(x,t,z) \geq 0 \quad (z \in \Omega_1), \tag{6.30}$$

$$\Omega_1 := \{z : z \in \mathbb{C}, \quad \Im z > M_1, \quad 0 < \arg z < \pi/4\} \tag{6.31}$$

for some $M_1 > M$. Indeed, using (6.5), (6.22) and (6.24), we obtain

$$\frac{\partial}{\partial t}\left(R(x,t,z)^* jR(x,t,z)\right) = R(x,t,z)^*\left(\mathrm{i}(\overline{z}^3 - z^3)I_m + O(z^2)\right)R(x,t,z) \tag{6.32}$$

for $z \to \infty$. Formula (6.32) implies that for some $M_1 > M$ and all $z \in \Omega_1$, we have $\frac{\partial}{\partial t}(R(x,t,z)^* jR(x,t,z)) > 0$, and so

$$R(x,t,z)^* jR(x,t,z) \geq j. \tag{6.33}$$

Relations $\mathcal{P}(x,z) \in \mathbf{P}_M(j)$, (6.28) and (6.33) imply (6.30).

Clearly, it suffices to prove (6.25) for values of z from Ω_1. (According to (6.31), the domain Ω_1 belongs to the half-plane $\mathbb{C}_M = \{z : \Im z > M > 0\}$.)

Substitute $u = W$ into relations (3.10) and (3.11) and rewrite them in the form

$$\left(W(x,t,z)^{-1}\right)^* jW(x,t,z)^{-1} > j \quad (x > 0, \quad z \in \mathbb{C}_M); \tag{6.34}$$

$$\det\left(\begin{bmatrix} I_p & 0 \end{bmatrix} W(x,0,z)^{-1}\mathcal{P}(x,z)\right) \neq 0 \quad (z \in \mathbb{C}_M), \tag{6.35}$$

$$\det\left(\begin{bmatrix} I_p & 0 \end{bmatrix} W(x,t,z)^{-1}\widetilde{\mathcal{P}}(x,t,z)\right) \neq 0 \quad (z \in \Omega_1). \tag{6.36}$$

In view of (6.35), we rewrite (6.29) as

$$\begin{aligned} &W(x,t,z)^{-1}\widetilde{\mathcal{P}}(x,t,z) \\ &= R(t,z)\begin{bmatrix} I_p \\ \phi(x,0,z) \end{bmatrix}\left(\begin{bmatrix} I_p \, 0 \end{bmatrix} W(x,0,z)^{-1}\mathcal{P}(x,z)\right), \end{aligned} \tag{6.37}$$

$$\phi(x,0,z) := \begin{bmatrix} 0 & I_p \end{bmatrix} W(x,0,z)^{-1}\mathcal{P}(x,z)\left(\begin{bmatrix} I_p & 0 \end{bmatrix} W(x,0,z)^{-1}\mathcal{P}(x,z)\right)^{-1}. \tag{6.38}$$

According to (6.35)–(6.37), we have

$$\begin{aligned} &\begin{bmatrix} 0 & I_p \end{bmatrix} W(x,t,z)^{-1}\widetilde{\mathcal{P}}(x,t,z)\left(\begin{bmatrix} I_p & 0 \end{bmatrix} W(x,t,z)^{-1}\widetilde{\mathcal{P}}(x,t,z)\right)^{-1} \\ &= \left(R_{21}(t,z) + R_{22}(t,z)\phi(x,0,z)\right)\left(R_{11}(t,z) + R_{12}(t,z)\phi(x,0,z)\right)^{-1}. \end{aligned} \tag{6.39}$$

Since $\widetilde{\mathcal{P}}(x,t,z)$ satisfies (6.30), using Remark 3.9, we derive

$$\varphi(t,z) = \lim_{x\to\infty} \begin{bmatrix} 0 & I_p \end{bmatrix} W(x,t,z)^{-1}\widetilde{\mathcal{P}}(x,t,z) \times \left(\begin{bmatrix} I_p & 0 \end{bmatrix} W(x,t,z)^{-1}\widetilde{\mathcal{P}}(x,t,z)\right)^{-1} \quad (z \in \Omega_1). \tag{6.40}$$

In a similar way, we derive from (3.27) and (6.38) that

$$\varphi(0,z) = \lim_{x\to\infty} \phi(x,0,z) \quad (z \in \mathbb{C}_M). \tag{6.41}$$

In order to show that

$$\det\left(R_{11}(t,z) + R_{12}(t,z)\varphi(0,z)\right) \neq 0 \quad (z \in \Omega_1), \tag{6.42}$$

we recall (3.17), which yields the inequality

$$[I_p \quad \varphi(0,z)^*]j\begin{bmatrix} I_p \\ \varphi(0,z) \end{bmatrix} \geq 0. \tag{6.43}$$

According to formula (6.33) and Corollary E.3, we have $RjR^* \geq j$, which implies

$$\vartheta^* j \vartheta \geq I_p \quad \text{for } \vartheta^* = \begin{bmatrix} R_{11}(t,z) & R_{12}(t,z) \end{bmatrix} \quad (z \in \Omega_1). \tag{6.44}$$

Finally, inequality (6.42) follows from relations (6.43), (6.44) and Proposition 1.43. In view of analyticity of both parts of (6.25), relations (6.39)–(6.42) imply (6.25) for z in the whole half-plane $\mathbb{C}_M$, that is, not only in Ω_1. □

Remark 6.7. One can apply Theorem 6.6 and the solution of the inverse problem from Theorem 3.21 in order to recover solutions of mKdV. (Here, we take into account Corollary 3.8 stating that the Weyl function of system (6.23) on the semiaxis is a Weyl function of the same system on all the intervals $[0, l]$.) Theorems on the evolution of the Weyl functions also constitute the first step in proofs of uniqueness and existence of the solutions of nonlinear equations via the ISpT method, see Section 6.2 of this chapter (see also [268]).

Remark 6.8. The Möbius transformation-type formula for the evolution $\varphi(t,z)$ of the Weyl function yields the corresponding Riccati equation on φ ([242, 281, 290]). In particular, formula (6.25), which we shall also prove for other cases considered in this chapter, implies the Riccati equation

$$\frac{d}{dt}\varphi(t,z) = F_{21}(t,z) + F_{22}(t,z)\varphi(t,z) - \varphi(t,z)F_{11}(t,z) - \varphi(t,z)F_{12}(t,z)\varphi(t,z), \tag{6.45}$$

where $F_{ik}(t,z)$ are $p \times p$ blocks of $F(0,t,z)$. Indeed, differentiating both sides of (6.25) and using formula (6.5), we derive

$$\begin{aligned} \frac{d}{dt}\varphi(t,z) = &\begin{bmatrix} 0 & I_p \end{bmatrix} F(0,t,z)R(t,z) \begin{bmatrix} I_p \\ \varphi(0,z) \end{bmatrix} \left(R_{11}(t,z) + R_{12}(t,z)\varphi(0,z)\right)^{-1} \\ &- \left(R_{21}(t,z) + R_{22}(t,z)\varphi(0,z)\right) \left(R_{11}(t,z) + R_{12}(t,z)\varphi(0,z)\right)^{-1} \\ &\times \begin{bmatrix} I_p & 0 \end{bmatrix} F(0,t,z)R(t,z) \begin{bmatrix} I_p \\ \varphi(0,z) \end{bmatrix} \left(R_{11}(t,z)+R_{12}(t,z)\varphi(0,z)\right)^{-1}. \end{aligned} \tag{6.46}$$

After we substitute the equality

$$F(0,t,z)R(t,z) = F(0,t,z)\left(\begin{bmatrix} I_p \\ 0 \end{bmatrix}\begin{bmatrix} I_p & 0 \end{bmatrix} + \begin{bmatrix} 0 \\ I_p \end{bmatrix}\begin{bmatrix} 0 & I_p \end{bmatrix}\right) R(t,z)$$

into (6.46) and simplify the result, formula (6.45) follows.

Another case of Riccati equations for Weyl functions can be seen, for instance, in [126].

6.1.4 Second harmonic generation: Goursat problem

The second harmonic generation is one of the simplest nonlinear interactions, which is described [127] by the system

$$\frac{\partial}{\partial x}\psi_1 = -2\overline{\psi_1}\psi_2, \quad \frac{\partial}{\partial t}\psi_2 = \psi_1^2, \tag{6.47}$$

where $\overline{\psi_1}$ is the complex conjugate of ψ_1. The second harmonic generation model (SHG), that is, system (6.47), is essential in the study of impulse propagation. The results related to the case of a purely amplitude-modulated fundamental wave go back to Liouville [35, 198, 314]. The SHG integrability was proven and the Lax pair was constructed in [162]. As mentioned in Chapter 0, the Cauchy problem is physically meaningless for the SHG system, and only the Goursat problem has physical meaning. The case of the Goursat problem for small and intermediate values of x and t has been studied in [166]. In spite of many important results (see [12, 166, 168, 314] and references therein), the SHG remained unsolved until 2005. It was finally solved in [256] using the ISpT method.

It is easy to check that the SHG system is equivalent to the compatibility condition (6.1) of the auxiliary systems (6.2), where

$$G(x,t,z) = \mathrm{i}\,(zj + jV(x,t)), \quad j = \mathrm{diag}\,\{1,\,-1\}, \quad v(x,t) = -2\mathrm{i}\psi_2(x,t); \tag{6.48}$$

$$F(x,t,z) = \frac{\mathrm{i}}{z} jH(x,t), \quad H(x,t) = \vartheta(x,t)\vartheta(x,t)^*, \quad \vartheta(x,t) = \begin{bmatrix} \psi_1(x,t) \\ \overline{\psi_1}(x,t) \end{bmatrix}. \tag{6.49}$$

(More precisely, in the case that G and F are given by (6.48) and (6.49), equation (6.1) follows from the SHG system, and the inverse holds under some simple additional conditions.) Clearly, the auxiliary system $y'(x,t,z) = G(x,t,z)y$ is the self-adjoint Dirac system (2.18), which was studied in Chapter 2.

Theorem 6.9. *Let continuous functions $\psi_1(x,t)$ and $\psi_2(x,t)$ and their partial derivatives $(\frac{\partial}{\partial x}\psi_1)(x,t)$ and $(\frac{\partial}{\partial t}\psi_2)(x,t)$ exist on the semistrip*

$$\Omega_a = \{(x,t) : 0 \le x < \infty,\, 0 \le t < a\}.$$

Suppose that ψ_1 and ψ_2 satisfy the SHG system (6.47). Then, the evolution $\varphi(t,z)$ of the Weyl function of the self-adjoint Dirac system $y'(x,t,z) = G(x,t,z)y$, where G has the form (6.48), is given by the formula

$$\varphi(t,z) = \left(R_{21}(t,z) + R_{22}(t,z)\varphi(0,z)\right) / \left(R_{11}(t,z) + R_{12}(t,z)\varphi(0,z)\right). \tag{6.50}$$

Moreover, ψ_1 and ψ_2 in the semistrip Ω_a are uniquely recovered from the initial-boundary values $\psi_2(x,0)$ and $\psi_1(0,t)$.

Here, $\varphi(0,z)$ is defined via $\psi_2(x,0)$ using formula (2.26) and Corollary 2.21, and $R(t,z) = R(0,t,z)$ is defined via $\psi_1(0,t)$ using formulas (6.5) and (6.49). Evolution $\varphi(t,z)$ of the Weyl function then follows from (6.50).

The function $\psi_2(x,t)$ is recovered from $\varphi(t,z)$ in several steps. First, we recover $\beta(x,t)$ using Proposition 2.50. Next, we recover $y(x,t)$:

$$y(x,t) = \begin{bmatrix} \overline{\beta_2}(x,t) & \overline{\beta_1}(x,t) \end{bmatrix}. \tag{6.51}$$

Finally, we obtain $\psi_2(x,t)$ from (2.140) and the third equality in (6.48).

After ψ_2 is recovered, we recover $\vartheta(x,t)$ of the form (6.49) using the fact that ϑ is the unique solution of the equation

$$\frac{\partial}{\partial x}\vartheta(x,t) = -2\begin{bmatrix} 0 & \psi_2(x,t) \\ \overline{\psi_2}(x,t) & 0 \end{bmatrix}\vartheta(x,t), \quad \vartheta(0,t) = \begin{bmatrix} \psi_1(0,t) \\ \overline{\psi_1}(0,t) \end{bmatrix}. \tag{6.52}$$

The function ψ_1 is immediately recovered from ϑ.

Proof. Formula (6.52) follows from the first equation in system (6.47). We also note that ψ_2 is continuous and, in particular, locally bounded with respect to x on $[0, \infty)$. Hence, we may apply Corollary 2.48 and Proposition 2.50. Thus, in order to show that the procedure to recover ψ_1 and ψ_2 from $\varphi(t, z)$ is valid, it remains only to prove that γ is given by (6.51). Indeed, in view of (2.87), (2.90), (2.113) and (2.137), we see that β satisfies (2.162) and γ satisfies (2.163), where $m_1 = m_2 = 1$. Furthermore, according to Proposition 2.53, there is a unique matrix function satisfying (2.163). Taking into account (2.162), we see that the matrix function on the right-hand side of (6.51) satisfies (2.163), therefore (6.51) holds.

Now, we prove the evolution formula (6.50). Since $\psi_1(x, t)$ and $\psi_2(x, t)$ are continuous, the system (6.47) implies that the partial derivatives $(\frac{\partial}{\partial x}\psi_1)(x, t)$, $(\frac{\partial}{\partial t}\psi_2)(x, t)$ are also continuous. In particular, it means that the conditions of Theorem 6.1 are fulfilled and so (6.27) holds in the SHG case. It is immediate from (6.49) that

$$\frac{\partial}{\partial t}\left(R(x, t, z)^* jR(x, t, z)\right) \geq 0, \tag{6.53}$$

which yields inequality (6.33) in $\mathbb{C}_+$. Inequality (2.24) takes the form (6.34), where we put $\mathbb{C}_M = \mathbb{C}_+$. Using relations (2.44), (6.27), (6.33), (6.34) and Corollary 2.18, the evolution formula (6.50) is proved precisely like a similar evolution formula in the proof of Theorem 6.6. □

The evolution of the somewhat differently defined Weyl functions (Definition 1.51), which belong to the Nevanlinna class, was given in [256]. In that case, the evolution of the Weyl functions describes the evolution of the corresponding spectral functions of Dirac systems.

Similar to the Goursat problem on Ω_a, the Goursat problem on the semistrip

$$\Omega_{-a} = \{(x, t) : 0 \leq x < \infty, \ -a < t \leq 0\}$$

can be solved using the ISpT method. Certain modifications are only required in the proof of the evolution formula (6.50).

Theorem 6.10. *Let continuous functions $\psi_1(x, t)$ and $\psi_2(x, t)$ and their partial derivatives $(\frac{\partial}{\partial x}\psi_1)(x, t)$ and $(\frac{\partial}{\partial t}\psi_2)(x, t)$ exist on the semistrip Ω_{-a}. Suppose that ψ_1 and ψ_2 satisfy the SHG system (6.47). Then, the Weyl functions $\varphi(t, z)$ $(-a < t \leq 0)$ of the corresponding Dirac systems are given by the formula*

$$\varphi(t, z) = \left(R_{21}(t, z) + R_{22}(t, z)\varphi(0, z)\right) / \left(R_{11}(t, z) + R_{12}(t, z)\varphi(0, z)\right). \tag{6.54}$$

Proof. First, note that Theorem 6.1 is easily reformulated for the domain Ω_{-a}, and so (6.7) holds under our conditions on ψ_1 and ψ_2. Next, we take into account that (6.53) yields $R(x, t, z)^* jR(x, t, z) \leq j$ for $t \leq 0$, that is, in place of (6.33), we now have the inequality

$$\left(R(x, t, z)^{-1}\right)^* jR(x, t, z)^{-1} \geq j \quad (z \in \mathbb{C}_+, \quad t \leq 0). \tag{6.55}$$

Because of (6.55), it is convenient to rewrite (6.7) in the form

$$W(x,t,z)^{-1} = R(t,z)W(x,0,z)^{-1}R(x,t,z)^{-1}. \tag{6.56}$$

Now, we follow the scheme of the proof of Theorem 6.6, using (6.56) instead of (6.27), matrix function $\mathcal{P}(x,z)$ instead of $\widetilde{\mathcal{P}}(x,t,z)$ and matrix function $\widetilde{\mathcal{P}}(x,t,z){:=}R(x,t,z)^{-1}\mathcal{P}(x,z)$ instead of $\mathcal{P}(x,z)$. □

6.2 Sine-Gordon theory in a semistrip

The well-known sine-Gordon and complex sine-Gordon equations are actively used in the study of various physical models and processes such as self-induced transparency and coherent optical pulse propagation, relativistic vortices in a superfluid, nonlinear sigma models, the motion of rigid pendulum, dislocations in crystals and so on (see, e.g. references in [7, 30, 33, 144, 221, 222, 305]).

The *sine-Gordon equation* (SGE) in the light cone coordinates has the form

$$\frac{\partial^2}{\partial t \partial x}\psi = 2\sin(2\psi). \tag{6.57}$$

It is the first equation to which the so-called auto-Bäcklund transformation was applied. It is also one of the first equations for which a Lax pair was found and which was consequently solved by the inverse scattering transform method [4]. The more general *complex sine-Gordon equation* (CSGE) was introduced (and its integrability was treated) only several years later [204, 222]. For further developments of the theory of CSGE and various applications, see, for instance, [30, 33, 57, 91, 220, 221] and references therein. The CSGE has the form

$$\frac{\partial^2\psi}{\partial t\partial x} + \frac{4\cos\psi}{(\sin\psi)^3}\frac{\partial\chi}{\partial x}\frac{\partial\chi}{\partial t} = 2\sin(2\psi), \quad \frac{\partial^2\chi}{\partial t\partial x} = \frac{2}{\sin(2\psi)}\left(\frac{\partial\psi}{\partial x}\frac{\partial\chi}{\partial t} + \frac{\partial\psi}{\partial t}\frac{\partial\chi}{\partial x}\right), \tag{6.58}$$

where $\psi = \overline{\psi}$ and $\chi = \overline{\chi}$. There are also two constraint equations

$$2(\cos\psi)^2\frac{\partial\chi}{\partial x} - (\sin\psi)^2\frac{\partial\omega}{\partial x} = 2c(\sin\psi)^2, \quad 2(\cos\psi)^2\frac{\partial\chi}{\partial t} + (\sin\psi)^2\frac{\partial\omega}{\partial t} = 0, \tag{6.59}$$

where c is a constant ($c = \overline{c} \equiv \mathrm{const}$) and $\omega = \overline{\omega}$. For the particular case $\chi \equiv 0$ and $\omega = -2cx$, the CSGE turns into the sine-Gordon equation (6.57) and the constraint equations hold automatically.

In spite of many interesting developments, including results obtained by variational and topological methods, the rigorous uniqueness and existence results on integrable equations in a semistrip or quarter-plane are comparatively rare. Nevertheless, one could mention, for instance, important results in [48, 53, 54, 63, 65, 67, 99, 143, 325, 328] (see also references therein). Here, we use the ISpT method in order to obtain a uniqueness result for CSGE and a global existence result for SGE in the semistrip $\Omega_a = \{(x,t) : 0 \le x < \infty, \ 0 \le t < a\}$.

Notice that the initial-boundary problem for SGE, where the values of ψ are given on the characteristics $x = -\infty$ and $t = 0$, is treated in [165] (see [338] for the related Cauchy problem for SGE in laboratory coordinates). A local solution of the Goursat problem for SGE, where ψ is given on the characteristics $x = 0$ and $t = 0$, is described in [182] (see also [197]). See Remark 6.24 for some other related results. The results for SGE, which we present below, originate from [268] and are based on a global existence theorem from [242] and on important results from [36–38] (see a discussion of those results in Subsection 4.2.1).

The study of the wave collapse, including unbounded and blow-up solutions, is of interest (see [315] and references therein). The approach presented here regarding the uniqueness and unboundedness of the solutions via evolution of the Weyl functions could be applied (see, e.g. [269]) to other integrable nonlinear equations.

Evolution of the Weyl function and the uniqueness result for the CSGE are given in Subsection 6.2.1, and the existence and recovery of the solution of the initial-boundary value problem for the SGE are dealt with in Subsection 6.2.2. Furthermore, in Subsection 6.2.2, we construct a class of unbounded in the quarter-plane solutions of the initial-boundary value problem.

6.2.1 Complex sine-Gordon equation: evolution of the Weyl function and uniqueness of the solution

If $\sin(2\psi) \neq 0$, the compatibility condition for constraint equations (6.59) is equivalent to the second equation in (6.58). If $\sin(2\psi) \neq 0$ and (6.59) holds, then CSGE (i.e. equations (6.58)) is equivalent to the compatibility condition (6.1), where

$$G(x,t,z) := \mathrm{i}zj + jV(x,t), \quad F(x,t,z) := -\frac{\mathrm{i}}{z+c}\vartheta(x,t)^* j\vartheta(x,t). \tag{6.60}$$

Here,

$$j = \begin{bmatrix} 1 & 0 \\ 0 & -1 \end{bmatrix}, \quad \vartheta(x,t) := D_1(x,t)\begin{bmatrix} \cos\psi & \mathrm{i}\sin\psi \\ \mathrm{i}\sin\psi & \cos\psi \end{bmatrix} D_2(x,t), \tag{6.61}$$

$$D_1 = \exp\{\mathrm{i}(\chi + (\omega/2))j\}, \quad D_2 = \exp\{\mathrm{i}(\chi - (\omega/2))j\}, \tag{6.62}$$

$$V = -j\left(\vartheta^*\frac{\partial\vartheta}{\partial x} + \mathrm{i}c(\vartheta^* j\vartheta - j)\right). \tag{6.63}$$

We shall consider the CSGE in the semistrip Ω_a. The following statement is valid according to [221, 222] (and can be checked directly as well).

Proposition 6.11. *Let $\{\psi(x,t),\ \chi(x,t),\ \omega(x,t)\}$ be a triple of real-valued and twice continuously differentiable functions on Ω_a. Assume that $\sin(2\psi) \neq 0$ and that equations (6.58) and (6.59) hold. Then, the zero curvature equation (6.1) holds too.*

Moreover, ϑ given by (6.61) belongs to $SU(2)$ (i.e. $\vartheta^*\vartheta = I_2$, $\det\vartheta = 1$) and satisfies relations

$$\begin{aligned}\vartheta^*\frac{\partial\vartheta}{\partial x} &+ \mathrm{i}c\vartheta^* j\vartheta \\ &= \mathrm{i}\left(\frac{\partial\chi}{\partial x}+\frac{\partial\omega}{2\partial x}+c\right)\vartheta^* j\vartheta + \mathrm{i}\left(\frac{\partial\chi}{\partial x}-\frac{\partial\omega}{2\partial x}\right)j + \mathrm{i}\frac{\partial\psi}{\partial x}D_2^* J D_2, \qquad (6.64)\end{aligned}$$

$$\vartheta^* j\vartheta = D_2^*\begin{bmatrix}\cos(2\psi) & \mathrm{i}\sin(2\psi)\\ -\mathrm{i}\sin(2\psi) & -\cos(2\psi)\end{bmatrix}D_2, \quad J = \begin{bmatrix}0 & 1\\ 1 & 0\end{bmatrix} \quad (\text{i.e. } J = \sigma_1). \qquad (6.65)$$

Recall that $\cos(2\psi) = (\cos\psi)^2 - (\sin\psi)^2$. Hence, in view of the first constraint in (6.59), we have

$$\left(\frac{\partial\chi}{\partial x}+\frac{\partial\omega}{2\partial x}+c\right)\cos(2\psi) + \frac{\partial\chi}{\partial x}-\frac{\partial\omega}{2\partial x} = c. \qquad (6.66)$$

Taking into account equalities (6.64)–(6.66), we rewrite the matrix function V introduced by (6.63) in the form

$$V = \begin{bmatrix}0 & v\\ \overline{v} & 0\end{bmatrix}, \quad v = \left(-\mathrm{i}\frac{\partial\psi}{\partial x} + 2(\cot\psi)\frac{\partial\chi}{\partial x}\right)\mathrm{e}^{\mathrm{i}(\omega-2\chi)}. \qquad (6.67)$$

According to (6.60) and (6.67), the auxiliary system $\frac{\partial}{\partial x}y = Gy$ coincides with the skew-self-adjoint Dirac system (6.23), where $p = 1$ and v is defined by the second relation in (6.67). This fact will be used in the study of the initial-boundary value problem for CSGE, where initial-boundary conditions are given by the equalities

$$v(x,0) = h_1(x), \quad \psi(0,t) = h_2(t), \quad \chi(0,t) = h_3(t), \quad \omega(0,0) = h_4. \qquad (6.68)$$

Remark 6.12. In fact, the requirements on smoothness of the functions ψ, χ and ω could be reduced in Proposition 6.11 and the theorem below. However, for convenience, we assume that the conditions of Proposition 6.11 are fulfilled.

Theorem 6.13. *Let $\{\psi(x,t),\ \chi(x,t),\ \omega(x,t)\}$ be a triple of real-valued and twice continuously differentiable functions on Ω_a. Assume that $\sin(2\psi) \neq 0$, that v is bounded, that is,*

$$\sup_{(x,t)\in\Omega_a}\left|\frac{\partial\psi}{\partial x} + 2\mathrm{i}(\cot\psi)\frac{\partial\chi}{\partial x}\right| \leq M, \qquad (6.69)$$

and that relations (6.58), (6.59) and (6.68) hold. Then, the Weyl functions $\varphi(t,z)$ of the auxiliary skew-self-adjoint Dirac systems $\frac{\partial}{\partial x}y = Gy$, where $G(x,t,z)$ is defined via (6.60) and (6.67), exist in $\mathbb{C}_M$ and have the form

$$\varphi(t,z) = \frac{R_{21}(t,z) + R_{22}(t,z)\varphi(0,z)}{R_{11}(t,z) + R_{12}(t,z)\varphi(0,z)}. \qquad (6.70)$$

Here, $R =: \{R_{ik}\}_{i,k=1}^{2}$ is defined by the equalities

$$\frac{d}{dt}R(t,z) = \frac{1}{\mathrm{i}(z+c)}\mathrm{e}^{-\mathrm{i}d(t)j}\begin{bmatrix} \cos(2h_2(t)) & \mathrm{i}\sin(2h_2(t)) \\ -\mathrm{i}\sin(2h_2(t)) & -\cos(2h_2(t)) \end{bmatrix}\mathrm{e}^{\mathrm{i}d(t)j}R(t,z), \tag{6.71}$$

$$R(0,z) = I_2, \quad d(t) := h_3(0) - \frac{1}{2}h_4 + \int_0^t h_3'(\xi)\,(\sin h_2(\xi))^{-2}\,d\xi, \tag{6.72}$$

and $\varphi(0,z)$ is the Weyl function of the system

$$\frac{d}{dx}y(x,z) = (\mathrm{i}zj + jV(x))\,y(x,z) \quad (x \ge 0), \quad V(x) := \begin{bmatrix} 0 & h_1(x) \\ \overline{h_1(x)} & 0 \end{bmatrix}. \tag{6.73}$$

Proof. In view of (6.69), the existence of the Weyl functions $\varphi(t,z)$ is immediate from Corollary 3.8.

Since G and F are continuously differentiable, we can use again the Factorization Theorem 6.1, that is, we have

$$W(x,t,z)^{-1}R(x,t,z) = R(t,z)W(x,0,z)^{-1}. \tag{6.74}$$

Next, we set $\mathcal{P}(x,z) = \begin{bmatrix} 1 \\ 0 \end{bmatrix}$ and our proof of (6.70) will coincide with the proof of Theorem 6.6 once we show that the matrix function

$$\widetilde{\mathcal{P}}(x,t,z) = R(x,t,z)\mathcal{P}(x,z) = R(x,t,z)\begin{bmatrix} 1 \\ 0 \end{bmatrix} \tag{6.75}$$

is nonsingular, with property-j. For that (as well as for some further applications in this section), we derive the inequality

$$\|R(x,t,z) - I_2 - \int_0^t F(x,\xi,z)d\xi\| \le M_1|z+c|^{-2}, \tag{6.76}$$

where t is fixed, $z \in \mathbb{C}_\delta$ $(\delta > 0)$ and M_1 does not depend on x and z. Indeed, using (6.5), (6.60) and the *multiplicative integral* representation of the fundamental solutions of the first order systems, we obtain

$$R(x,t,z) = \lim_{N\to\infty}\prod_{k=1}^{N}\mathrm{e}^{F_k(N)} = \lim_{N\to\infty}\prod_{k=1}^{N}\left(I_2 + \frac{t\theta_k(N)}{(z+c)N} + \frac{t^2\widetilde{\theta}_k(N)}{(z+c)^2N^2}\right), \tag{6.77}$$

$$F_k(N) := tF(x,kt/N,z)/N, \quad \theta_k(N) := -\mathrm{i}\vartheta(x,kt/N)^*j\vartheta(x,kt/N), \tag{6.78}$$

$$\widetilde{\theta}_k(N) := \theta_k(N)^2\sum_{i=0}^{\infty}\frac{1}{(i+2)!}\left(\frac{t\theta_k(N)}{(z+c)N}\right)^i, \tag{6.79}$$

where we sometimes omit the variables x, t and z in the notations. From the inequality $\|\theta_k(x,t,N)\| \le 1$, we have $\|\widetilde{\theta}_k(x,t,z,N)\| \le 1$ for sufficiently large values of N (uniformly with respect to k, x and $z \in \mathbb{C}_\delta$). Hence, the inequality

$$\left\| \prod_{k=1}^{N} \left(I_2 + \frac{t\theta_k(N)}{(z+c)N} + \frac{t^2\widetilde{\theta}_k(N)}{(z+c)^2N^2} \right) - I_2 - \sum_{k=1}^{N} \frac{t\theta_k(N)}{(z+c)N} \right\| \le M_1|z+c|^{-2} \quad (6.80)$$

holds for some M_1 and sufficiently large values of N. Relations (6.77) and (6.80) yield (6.76).

According to (6.76) and the definition of F in (6.60), there is a value $\widetilde{M} > M$ such that the vector functions $\widetilde{\mathcal{P}}(x,t,z)$ given by (6.75) are nonsingular, with property-j for $z \in \mathbb{C}_{\widetilde{M}}$ and $x \in \mathbb{R}_+$. Following the proof of Theorem 6.6, we see that (6.70) is valid in $\mathbb{C}_{\widetilde{M}}$. Now, using analyticity, we obtain that (6.70) is valid in $\mathbb{C}_M$. □

Corollary 6.14. *There is at most one triple* $\{\psi(x,t),\ \chi(x,t),\ \theta(x,t)\}$ *of real-valued and twice continuously differentiable functions on* Ω_a *such that* $\sin(2\psi) \ne 0$, *that* v *is bounded, and that CSGE* (6.58), *constraints* (6.59) *and initial-boundary conditions* (6.68) *are satisfied.*

Proof. Suppose the triple $\{\psi(x,t),\ \chi(x,t),\ \theta(x,t)\}$ satisfies conditions of the corollary. Then, the conditions of Theorem 6.13 are fulfilled and the evolution of the Weyl function is given by formula (6.70). In view of Corollary 3.8, $\varphi(t,z)$ is also a Weyl function of the system $\frac{\partial}{\partial x} = Gy$ considered on intervals. Hence, according to Theorem 3.21, the function $v(x,t)$ is uniquely recovered.

Next, we recover ψ from v. Using (6.58), (6.59) and (6.67), we derive

$$\begin{aligned}
\frac{\partial v}{\partial t} &= e^{i(\omega-2\chi)} \left(-i\frac{\partial^2\psi}{\partial t\partial x} + 2(\cot\psi)\frac{\partial^2\chi}{\partial t\partial x} - 2(\sin\psi)^{-2}\frac{\partial\chi}{\partial x}\frac{\partial\psi}{\partial t} \right. \\
&\quad \left. + \left(\frac{\partial\psi}{\partial x} + 2i(\cot\psi)\frac{\partial\chi}{\partial x}\right)\left(\frac{\partial\omega}{\partial t} - 2\frac{\partial\chi}{\partial t}\right)\right) \\
&= e^{i(\omega-2\chi)} \left(-i\frac{\partial^2\psi}{\partial t\partial x} + 2(\cot\psi)\frac{\partial^2\chi}{\partial t\partial x} - 2(\sin\psi)^{-2}\frac{\partial\chi}{\partial x}\frac{\partial\psi}{\partial t} \right. \\
&\quad \left. -2\left(\frac{\partial\psi}{\partial x} + 2i(\cot\psi)\frac{\partial\chi}{\partial x}\right)(\sin\psi)^{-2}\frac{\partial\chi}{\partial t}\right) \\
&= 2e^{i(\omega-2\chi)} \left(-i\sin(2\psi) + \left(\frac{\partial^2\chi}{\partial t\partial x} - \frac{1}{\sin\psi\cos\psi}\left(\frac{\partial\psi}{\partial x}\frac{\partial\chi}{\partial t} + \frac{\partial\psi}{\partial t}\frac{\partial\chi}{\partial x}\right)\right)\cot\psi \right) \\
&= -2ie^{i(\omega-2\chi)}\sin(2\psi). \qquad (6.81)
\end{aligned}$$

It follows from (6.81) that $|\sin 2\psi| = \frac{1}{2}|\frac{\partial}{\partial t}v|$. Since $\sin(2\psi) \ne 0$ and ψ is continuous, the function $\sin(2\psi)$ is uniquely recovered from the values of $|\sin(2\psi(x,t))|$ and the sign of $\sin(2\psi(0,t)) = \sin(2h_2(t))$. Using again (6.67) and (6.81), we obtain

$$\frac{\partial\psi}{\partial x} = 2(\sin(2\psi))\Re\left(v/\frac{\partial v}{\partial t}\right),$$

and thus we also recover $\frac{\partial}{\partial x}\psi$. Therefore, the function

$$\psi(x,t) = h_2(t) + \int_0^x \frac{\partial \psi}{\partial \xi}(\xi,t)d\xi$$

is uniquely recovered too. Moreover, we have

$$\chi(x,t) = h_3(t) + \int_0^x \frac{\partial \chi}{\partial \xi}(\xi,t)d\xi, \quad \frac{\partial}{\partial x}\chi = 2(\sin\psi)^2 \Im\left(v\Big/\frac{\partial v}{\partial t}\right). \tag{6.82}$$

Using (6.82), we recover χ from v and ψ. Finally, ω is uniquely recovered from the value $\omega(0,0)$ and constraint equations (6.59). □

Remark 6.15. Note that the derivatives of ω appear in CSGE (more precisely, in constraint equations) instead of ω itself. Therefore, the condition $\omega(0,0) = h_4$ in (6.68) is only used for the sake of uniqueness of ω and can be chosen arbitrarily.

6.2.2 Sine-Gordon equation in a semistrip

In the case of the sine-Gordon equation (6.57), the matrix functions G and F, which we use in the zero curvature representation (6.1) of SGE, have the form [4], that is,

$$G(x,t,z) = \mathrm{i}zj + jV(x,t), \quad V = \begin{bmatrix} 0 & v \\ v & 0 \end{bmatrix}, \quad v = -\frac{\partial}{\partial x}\psi, \quad \psi = \overline{\psi}, \tag{6.83}$$

$$F(x,t,z) = \frac{1}{\mathrm{i}z}\begin{bmatrix} \cos(2\psi) & \sin(2\psi) \\ \sin(2\psi) & -\cos(2\psi) \end{bmatrix}, \tag{6.84}$$

though an equivalent pair $\hat{G} = \hat{D}G\hat{D}^{-1}$, $\hat{F} = \hat{D}F\hat{D}^{-1}$, where G and F are defined in (6.83) and (6.84), and $\hat{D} := \mathrm{diag}\,\{\mathrm{i}, 1\}$ would be closer to G and F from (6.60).

Theorem 6.16. *Let $\psi(0,t)$ and $(\frac{\partial}{\partial x}\psi)(x,t)$ be continuous functions on $[0, a)$ and Ω_a, respectively, and let $(\frac{\partial^2}{\partial t \partial x}\psi)(x,t)$ exist. Let also (6.57) hold on Ω_a. Assume that $\frac{\partial}{\partial x}\psi$ is bounded, that is,*

$$\sup\left|\left(\frac{\partial}{\partial x}\psi\right)(x,t)\right| < \infty \quad ((x,t) \in \Omega_a), \tag{6.85}$$

and that $\varphi(0,z)$ is the Weyl function of the system $\frac{\partial}{\partial x}y = G(x,0,z)y$, where G is given by (6.83).

Then, the function $\varphi(t,z)$ of the form (6.70), where $R(t,z) = R(0,t,z)$ is given by (6.5) and (6.84), is the Weyl function of the system $\frac{\partial}{\partial x}y = G(x,t,z)y$.

Proof. Since $\psi(0,t)$ and $(\frac{\partial}{\partial x}\psi)(x,t)$ are continuous, the function $\psi(x,t)$ is continuous on Ω_a. Hence, (6.57) implies that $(\frac{\partial^2}{\partial t \partial x}\psi)(x,t)$ is continuous. Taking into account that ψ, $\frac{\partial}{\partial x}\psi$ and $\frac{\partial^2}{\partial t \partial x}\psi$ are continuous, we see that the conditions of Theorem 6.1 are satisfied. From the equality

$$\left\| \begin{bmatrix} \cos(2\psi) & \sin(2\psi) \\ \sin(2\psi) & -\cos(2\psi) \end{bmatrix} \right\| = 1,$$

it follows that, quite similar to the CSGE case, formula (6.76) (where $c = 0$) is also derived for SGE. Thus, the remaining part of the proof coincides with the corresponding arguments from the proof of Theorem 6.13. □

Remark 6.17. We note that from the proof of formula (6.76) (and of its subcase $c = 0$, which was mentioned above), the fact that M_1 does not depend on t is immediate. In particular, for $R(x,t,z)$ given by (6.5) and (6.84), and for any fixed values $\delta > 0$ and $0 < \tilde{a} < a$, we have

$$\left\| R(x,t,z) - I_2 - \int_0^t F(x,\xi,z)d\xi \right\| \le M_1 |z|^{-2} \quad \text{for } z \in \mathbb{C}_\delta, \quad 0 \le t \le \tilde{a}. \tag{6.86}$$

The slightly stronger version (6.86) of the formula (6.76) (with $c = 0$) will be used in the proof of the Existence Theorem 6.19.

Clearly, we can omit condition (6.85) in Theorem 6.16 and consider the evolution of GW-functions.

Theorem 6.18. *Let $\psi(0,t)$ and $(\frac{\partial}{\partial x}\psi)(x,t)$ be continuous functions on $[0,a)$ and Ω_a, respectively, and let $(\frac{\partial^2}{\partial t \partial x}\psi)(x,t)$ exist. Let also (6.57) hold on Ω_a, and assume that $\varphi(0,z)$ is a GW-function of the system $\frac{\partial}{\partial x}y = G(x,0,z)y$, where G is given by (6.83). Then, the function*

$$\varphi(t,z) = \frac{R_{21}(t,z) + R_{22}(t,z)\varphi(0,z)}{R_{11}(t,z) + R_{12}(t,z)\varphi(0,z)} \tag{6.87}$$

is a GW-function of the system $\frac{\partial}{\partial x}y = G(x,t,z)y$, where G is again given by (6.83).

Proof. According to Definition 3.27 of the GW-function, the inequality (3.104), where $W(x,0,z)$ is substituted in place of $u(x,z)$ and $\varphi(0,z)$ is substituted in place of $\varphi(z)$, is valid for some $M > 0$. Hence, using (6.86), we immediately derive

$$\sup_{x \le l,\, z \in \mathbb{C}_M} \left\| \mathrm{e}^{-izx} R(x,t,z) W(x,0,z) \begin{bmatrix} 1 \\ \varphi(0,z) \end{bmatrix} \right\| < \infty. \tag{6.88}$$

Like in the previous theorem, we see that the conditions of Theorem 6.1 are satisfied, and so we have $W(x,t,z)R(t,z) = R(x,t,z)W(x,0,z)$. Therefore, we rewrite (6.88) in the form

$$\sup_{x \le l,\, z \in \mathbb{C}_M} \left\| \mathrm{e}^{-izx} W(x,t,z) R(t,z) \begin{bmatrix} 1 \\ \varphi(0,z) \end{bmatrix} \right\| < \infty. \tag{6.89}$$

Using the arguments from the proof of Proposition 3.28, we easily derive that

$$(\mathrm{e}^{-2\mathrm{i}zl} + o(\mathrm{e}^{2\eta l}))\varphi(0,z) + o(\mathrm{e}^{2\eta l}), \quad \text{where} \quad z = \xi + \mathrm{i}\eta,$$

is uniformly bounded for any fixed $l > 0$ and sufficiently large values of η. In particular, for any $\delta > 0$, we can choose M so that $|\varphi(0,z)| \leq \delta$ in $\mathbb{C}_M$. Since, for sufficiently large values of η, $R_{11}(t,z)$ is close to 1 and $R_{22}(t,z)$ is close to zero uniformly with respect to $t \in [0, a)$, we can also assume that

$$|R_{11}(t,z) + R_{12}(t,z)\varphi(0,z)| > \varepsilon > 0 \quad \text{for } z \in \mathbb{C}_M. \tag{6.90}$$

Inequalities (6.89) and (6.90) yield

$$\sup_{x \leq l,\, z \in \mathbb{C}_M} \left\| \mathrm{e}^{-\mathrm{i}zx} W(x,t,z) \begin{bmatrix} 1 \\ \varphi(t,z) \end{bmatrix} \right\| < \infty, \tag{6.91}$$

where $\varphi(t,z)$ is given by (6.87). □

When a simple inequality (3.106) holds for $\varphi(t,z)$, Theorem 3.30 provides a procedure to recover ψ from $\varphi(t,z)$. However, we note that the existence of ψ satisfying SGE is required in Theorem 6.18 in order to prove the formula (6.87) giving $\varphi(t,z)$. Below, we show that $\varphi(t,z)$ satisfies (3.106) and the corresponding solution of SGE exists, once $\varphi(0,z)$ satisfies (3.106) ([242, Theorem 3]), which is an essentially more complicated task. Recall that according to Definition 3.29, $\mathfrak{M}(\varphi)$ stands for the potential v of system (3.1) recovered from the GW-function φ of this system.

Theorem 6.19. *Let the initial-boundary conditions*

$$\psi(x,0) = h_1(x), \quad \psi(0,t) = h_2(t), \quad h_k = \overline{h_k} \quad (k = 1,2), \quad h_1(0) = h_2(0) \tag{6.92}$$

be given. Assume that h_2 is continuous on $[0, a)$ and that h_1 is boundedly differentiable on all the finite intervals on $[0, \infty)$. Moreover, assume that the GW-function $\varphi(0,z)$ of the system

$$\frac{d}{dx}W = GW, \quad G(x,z) = \mathrm{i}zj + jV(x), \quad V(x) = V(x,0) = -\begin{bmatrix} 0 & h_1'(x) \\ h_1'(x) & 0 \end{bmatrix} \tag{6.93}$$

exists and satisfies (3.106). *Then, a solution of the initial-boundary value problem* (6.57), (6.92) *exists and is given by the equality*

$$\psi(x,t) = h_2(t) - \int_0^x \left(\mathfrak{M}\left(\varphi(t,z)\right)\right)(\xi)d\xi, \tag{6.94}$$

where $\varphi(t,z)$ is given by (6.87) *and R in* (6.87) *is defined by the relations*

$$\frac{d}{dt}R(t,z) = \frac{1}{\mathrm{i}z}\begin{bmatrix} \cos(2h_2(t)) & \sin(2h_2(t)) \\ \sin(2h_2(t)) & -\cos(2h_2(t)) \end{bmatrix} R(t,z), \quad R(0,z) = I_2. \tag{6.95}$$

Here, $\varphi(t,z)$ satisfies inequality (3.106), *where*

$$\phi_0(t) = \phi_0(0) - \mathrm{i}\int_0^t \sin(2h_2(\xi))\,d\xi \tag{6.96}$$

and on each interval $0 \le t \le \tilde{a}$, such that $0 < \tilde{a} < a$, we can fix $M(t) \equiv M(\tilde{a})$ (with some $M(\tilde{a}) > 0$).

Proof. Step 1. In order to derive

$$\sup \left\| z^2(\varphi(t,z) - \phi_0(t)/z) \right\| < \infty \quad (z \in \mathbb{C}_{M(\tilde{a})}, \quad 0 \le t \le \tilde{a}), \tag{6.97}$$

where φ is given by (6.87) and ϕ_0 is given by (6.96), we use (6.86). In view of (6.86), formula (6.97) follows from equality (6.87) and the validity of (3.106) for $\varphi(0,z)$.

We shall also need the equality

$$\varphi(t,z) = \overline{\varphi(t,-\overline{z})} \tag{6.98}$$

in order to show that $\psi = \overline{\psi}$. Indeed, we easily see that the fundamental solution $W(x,0,z)$ of (6.93) and the fundamental solution $R(t,z)$ of (6.95) have the properties

$$W(x,0,z) = \overline{W(x,0,-\overline{z})}, \quad R(t,z) = \overline{R(t,-\overline{z})}. \tag{6.99}$$

It follows from Definition 3.27 and formula (6.99) that $\overline{\varphi(0,-\overline{z})}$ is a GW-function of system (6.93) simultaneously with $\varphi(0,z)$. Hence, from Proposition 3.28 on uniqueness of GW-functions, we obtain $\varphi(0,z) = \overline{\varphi(0,-\overline{z})}$. Therefore, equality (6.98) is immediate from (6.87) and the second equality in (6.99).

Since (6.97) holds, the mapping $\mathfrak{M}(\varphi)$ is well-defined and the procedure to construct it is given in Theorem 3.30. In particular, we can recover Φ_1 using formula (3.107), which follows from (3.67). Formulas (3.107) and (6.98) yield $\Phi_1 = \overline{\Phi_1}$. Then we derive successively that $\beta = \overline{\beta}$,

$$\gamma = \overline{\gamma} = \begin{bmatrix} -\beta_2 & \beta_1 \end{bmatrix}, \tag{6.100}$$

and, finally, that $v = \overline{v}$, where $v = \mathfrak{M}(\varphi)$. Hence, $\psi = \overline{\psi}$ holds for ψ given by (6.94).

Next, we show that $\Phi_1(x,t)$ and some of its derivatives exist and are continuous with respect to x and t, that is,

$$\Phi_1, \ \frac{\partial}{\partial x}\Phi_1, \ \frac{\partial}{\partial t}\Phi_1, \ \frac{\partial^2}{\partial x \partial t}\Phi_1, \ \frac{\partial^3}{\partial x^2 \partial t}\Phi_1 \in C(\Omega_a). \tag{6.101}$$

In view of (6.97), we can rewrite representation (3.67) in the form (3.108). According to (3.108) and (6.97), the functions Φ_1 and $\frac{\partial}{\partial x}\Phi_1$ are continuous with respect to x. They are also continuous with respect to t since $\varphi(t,z) - \phi_0(t)/z$ is continuous with respect to t. Taking into account (6.84) and (6.96), we rewrite (6.45) in the form

$$\frac{d}{dt}\left(\varphi(t,z) - \frac{\phi_0(t)}{z}\right) = \frac{\mathrm{i}}{z}\varphi(t,z)\left(\varphi(t,z)\sin(2h_2(t)) + 2\cos(2h_2(t))\right). \tag{6.102}$$

From (3.108) and (6.102), we see that there is the derivative $\frac{\partial}{\partial t}\Phi_1$ given by

$$\left(\frac{\partial}{\partial t}\Phi_1\right)(x,t) = \frac{1}{2\pi\mathrm{i}}\mathrm{e}^{2\eta x}\int_{-\infty}^{\infty}(\xi+\mathrm{i}\eta)^{-1}\mathrm{e}^{-2\mathrm{i}\xi x}\frac{d}{dt}\left(\varphi(\xi+\mathrm{i}\eta,t)-\frac{\phi_0(t)}{(\xi+\mathrm{i}\eta)}\right)d\xi, \tag{6.103}$$

that the derivative $\frac{\partial^2}{\partial x\partial t}\Phi_1$ exists and that $\frac{\partial}{\partial t}\Phi_1$, $\frac{\partial^2}{\partial x\partial t}\Phi_1 \in C(\Omega_a)$. Rewriting (6.102), we obtain

$$\begin{aligned}
&\frac{d}{dt}\left(\varphi(t,z)-\frac{\phi_0(t)}{z}\right)\\
&=\frac{\mathrm{i}}{z}\left(\varphi(t,z)-\frac{\phi_0(t)}{z}\right)(\varphi(t,z)\sin(2h_2(t))+2\cos(2h_2(t)))\\
&\quad+\mathrm{i}\varphi(t,z)\sin(2h_2(t))\frac{\phi_0(t)}{z^2}+2\mathrm{i}\cos(2h_2(t))\frac{\phi_0(t)}{z^2}.
\end{aligned} \tag{6.104}$$

Substituting (6.104) into (6.103), we see that the integral corresponding to the last term on the right-hand side of (6.104) is easily calculated explicitly (and other terms on the right-hand side of (6.104) behave asymptotically as $O(z^{-3})$). Hence, $\frac{\partial^3}{\partial x^2\partial t}\Phi_1$ exists and belongs to $C(\Omega_a)$.

Using (6.101), we derive the smoothness of β and the inequality

$$\sup_{t\le\tilde a}\int_0^l\left(\left\|\left(\frac{\partial^2}{\partial x^2}\beta\right)(x,t)\right\|^2+\left\|\left(\frac{\partial^3}{\partial x\partial t\partial x}\beta\right)(x,t)\right\|^2\right)dx<\infty \tag{6.105}$$

for any $0<\tilde a<a$ and $0<l<\infty$. Indeed, formula (3.93) implies that for continuous functions f_1 and f_2, we have

$$\begin{aligned}
\frac{d}{d\xi}\int_0^\xi\left(S_\xi^{-1}f_1\right)(r)\overline{f_2(r)}dr &= \left(f_1(\xi)-\int_0^\xi\left(S_\xi^{-1}f_1\right)(r)\overline{s(r,\xi)}dr\right)\\
&\quad\times\left(\overline{f_2(\xi)}-\int_0^\xi\left(S_\xi^{-1}s(r,\xi)\right)(r)\overline{f_2(r)}dr\right).
\end{aligned} \tag{6.106}$$

In view of (3.93) and (6.106), we can differentiate the right-hand side of (3.73) and so $\frac{\partial}{\partial x}\beta$ exists. Moreover, $\frac{\partial}{\partial x}\beta$ is continuous on Ω_a and is given by the equality

$$\begin{aligned}
\left(\frac{\partial}{\partial x}\beta\right)(x,t) &= -\left(\left(\frac{\partial}{\partial x}\Phi_1\right)(x,t)-\int_0^x\left(S_x^{-1}s(r,x)\right)(r)\frac{\partial}{\partial r}\Phi_1(r,t)dr\right)\\
&\quad\times\left(\begin{bmatrix}\Phi_1(x,t) & 1\end{bmatrix}-\int_0^x\begin{bmatrix}(S_x^{-1}\Phi_1(r,t))(r) & (S_x^{-1}1)(r)\end{bmatrix}\right.\\
&\quad\left.\times\, s(r,x)dr\right).
\end{aligned} \tag{6.107}$$

From (3.73) and (6.101), we see that β and $\frac{\partial}{\partial t}\beta$ are also continuous on Ω_a. Finally, taking into account (3.108), (6.97), (6.104), (6.106), (6.107), and applying the same arguments as earlier in this step, we derive that $\frac{\partial^2}{\partial t\partial x}\beta$, $\frac{\partial^2}{\partial x^2}\beta$ and $\frac{\partial^3}{\partial x\partial t\partial x}\beta$ exist, $\frac{\partial^2}{\partial t\partial x}\beta \in C(\Omega_a)$ and (6.105) holds. (Clearly, the equality $\frac{\partial^2}{\partial t\partial x}\beta = \frac{\partial^2}{\partial x\partial t}\beta$ is also valid.)

Since $v(x,t) = (\frac{\partial}{\partial x}\beta)(x,t)\gamma(x,t)^*$ and γ is given by (6.100), we have

$$v, \ \frac{\partial}{\partial t}v \in C(\Omega_a). \tag{6.108}$$

Furthermore, the inequality (6.105) yields

$$\sup_{t\le\tilde{a}} \int_0^l \left(\left| \left(\frac{\partial}{\partial x}v\right)(x,t)\right|^2 + \left| \left(\frac{\partial}{\partial x\partial t}v\right)(x,t)\right|^2 \right) dx < \infty. \tag{6.109}$$

Step 2. Let $W(x,t,z)$ be the fundamental solution of (6.4), where G is given by the first two equalities in (6.83) and equality $v = \mathfrak{M}(\varphi(t,z))$. Note that representation (3.58) (with $u = W$ and additional parameter t) has the form

$$W(x,t,z) = \mathrm{e}^{\mathrm{i}zxj} + \int_{-x}^{x} \mathrm{e}^{\mathrm{i}zr} N(x,t,r)dr, \tag{6.110}$$

$$\sup \|N(x,t,r)\| < \infty \quad (0 < |r| < x < l). \tag{6.111}$$

Recall that $R(t,z)$ is given by (6.95) and introduce matrix function $\widehat{w}$ by the formula

$$\widehat{w}(x,t,z) := W(x,t,z)R(t,z)\begin{bmatrix} 1 & -\varphi(0,z) \\ \varphi(0,z) & 1 \end{bmatrix} \mathrm{e}^{-\mathrm{i}zxj}. \tag{6.112}$$

Then, using inequality (3.104), where we substitute $u = W$ and $\varphi = \varphi(t,z)$, and taking into account formula (6.87), we obtain the boundedness of the first column of $\widehat{w}(x,t,z)$ for $x \le l$, $z \in \mathbb{C}_{M(\tilde{a})}$, whereas the representation (6.110) implies the boundedness of the second column of $\widehat{w}$. Thus, for the fixed values of $t \le \tilde{a} < a$, we have

$$\sup_{x\le l,\, z\in\mathbb{C}_{M(\tilde{a})}} \|\widehat{w}(x,t,z)\| < \infty. \tag{6.113}$$

According to (6.76), (6.97) and (6.110), the relation

$$\widehat{w}(x,t,\xi+\mathrm{i}\eta) - I_2 \in L^2_{2\times 2}(-\infty,\infty)$$

holds for $\eta > M(\tilde{a})$. This relation and inequality (6.113) show that the conditions of Theorem E.11 are fulfilled. Hence, if $\Im(z) = \eta > M(\tilde{a})$, $\widehat{w}$ admits representation

$$\widehat{w}(x,t,z) = I_2 + \int_0^\infty \mathrm{e}^{\mathrm{i}zr}\Phi(x,t,r)dr, \quad \mathrm{e}^{-\eta r}\Phi(x,t,r) \in L^2_{2\times 2}(0,\infty). \tag{6.114}$$

In particular, setting $\Phi(t,r) = \Phi(0,t,r)$ and putting $x = 0$ in (6.114), we derive

$$\widehat{R}(t,z) := R(t,z)\begin{bmatrix} 1 & -\varphi(0,z) \\ \varphi(0,z) & 1 \end{bmatrix} = I_2 + \int_0^\infty e^{izr}\Phi(t,r)dr. \tag{6.115}$$

In view of the equality (6.112), there is a simple connection between $\Phi(x,t,r)$, $N(x,t,r)$ and $\Phi(t,r)$ from the representations (6.114), (6.110) and (6.115), respectively:

$$\Phi(x,t,r) = \begin{bmatrix} \Phi_{11}(t,r) & \Phi_{12}(t,r-2x) \\ \Phi_{21}(t,r+2x) & \Phi_{22}(t,r) \end{bmatrix} + \begin{bmatrix} 0 & N^-(x,t,r-x) \end{bmatrix} \tag{6.116}$$
$$+ \int_{-x}^{x} N(x,t,\xi)\begin{bmatrix} \Phi^+(t,r+x-\xi) & \Phi^-(t,r-x-\xi) \end{bmatrix} d\xi \quad (r \geq 0);$$

$$\begin{bmatrix} 0 \\ \Phi_{21}(t,r+2x) \end{bmatrix} + \int_{-x}^{r+x} N(x,t,\xi)\Phi^+(t,r+x-\xi)d\xi + N^+(x,t,r+x)$$
$$= 0 \quad (r < 0), \tag{6.117}$$

where Φ_{ik} are the entries of Φ, $\Phi = \begin{bmatrix} \Phi^+ & \Phi^- \end{bmatrix}$, $N = \begin{bmatrix} N^+ & N^- \end{bmatrix}$ and we set

$$\Phi(x,t,r) = \Phi(t,r) = 0 \quad \text{for } r < 0; \quad N(x,t,r) = 0 \quad \text{for } |r| > x. \tag{6.118}$$

(The symbols N^+ and N^- denote the columns of N because the notation N_i was already used differently in the proof of (3.58).)

Step 3. We derive the proof of our theorem from the equality

$$\left(\frac{\partial}{\partial t}\widehat{w}\right)(x,t,z)\widehat{w}(x,t,z)^{-1} = \frac{1}{iz}\left(\frac{\partial}{\partial t}\Phi\right)(x,t,0) + o\left(\frac{1}{z}\right). \tag{6.119}$$

In order to obtain (6.119), we first show that Φ and N are sufficiently smooth. From (6.115), we have

$$\Phi(t,r) = e^{\eta r}\,\mathrm{l.i.m.}_{c\to\infty}\frac{1}{2\pi}\int_{-c}^{c} e^{-i\xi r}(\widehat{R}(t,\xi+i\eta) - I_2)d\xi. \tag{6.120}$$

In view of (6.84) and (6.92), we represent $F(0,t,z)$ as

$$F(0,t,z) = \frac{\zeta(t)}{iz}, \quad \zeta(t) := \begin{bmatrix} \cos(2h_2(t)) & \sin(2h_2(t)) \\ \sin(2h_2(t)) & -\cos(2h_2(t)) \end{bmatrix}. \tag{6.121}$$

Now, using the definition of $\widehat{R}$ from (6.115) and formulas (6.86) and (6.95), we differentiate the right-hand side of (6.120). It easily follows that

$$\int_0^r \left\|\left(\frac{\partial}{\partial t}\Phi\right)(t,r)\right\|^2 dr < \infty,$$

and, moreover, $(\frac{\partial}{\partial t}\Phi)(t,r)$ is differentiable with respect to r and

$$\left(\frac{\partial}{\partial t}\Phi\right)(t,0) = -\zeta(t), \quad \left(\frac{\partial^2}{\partial r \partial t}\Phi\right)(t,r) = -\zeta(t)\Phi(t,r). \tag{6.122}$$

Next, we consider $N(x,t,r)$ from the representation (6.110). This representation was constructed using formulas (3.59)–(3.61) and setting there $u = W$. From (3.59)–(3.61), we derive that

$$N(x,t,r) = \sum_{i=1}^{\infty} N_i(x,t,r), \tag{6.123}$$

where

$$N_1(x,t,r) = \frac{1}{2}\begin{bmatrix} 0 & v\left(\frac{x-r}{2},t\right) \\ -v\left(\frac{x+r}{2},t\right) & 0 \end{bmatrix}, \tag{6.124}$$

$$\begin{aligned} N_i(x,t,r) = \frac{1}{2}\Bigg(&\int_{-x}^{r}\begin{bmatrix} 0 & v\left(\frac{x-s}{2},t\right) \\ 0 & 0 \end{bmatrix} N_{i-1}\left(\frac{x-s}{2},t,\frac{2r-x-s}{2}\right) ds \\ &-\int_{r}^{x}\begin{bmatrix} 0 & 0 \\ v\left(\frac{x+s}{2},t\right) & 0 \end{bmatrix} N_{i-1}\left(\frac{x+s}{2},t,\frac{2r+x-s}{2}\right) ds\Bigg), \quad i>1. \end{aligned} \tag{6.125}$$

In view of (6.108), (6.109) and (6.123)–(6.125), the matrix function N satisfies the inequality

$$\sup_{|r|\le x<l,\, t\le\tilde{a}} \left(\left\|\left(\frac{\partial}{\partial t}N\right)(x,t,r)\right\| + \left\|\left(\frac{\partial^2}{\partial r\partial t}N\right)(x,t,r)\right\|\right) < \infty \tag{6.126}$$

for all $l < \infty$ and $\tilde{a} < a$.

It is useful to note that in view of (6.120) the matrix function $\Phi(t,r)$ is continuous with respect to $r \ge 0$. Since $v(x,t)$ is continuous, the matrix functions N_i given by (6.124) and (6.125) are continuous with respect to r for the case that $|r| \le x$. Therefore, formulas (3.61) and (6.123) imply that N is continuous. Now, using (6.116) pointwise, we see that $\Phi(x,t,r)$ is continuous (with respect to r) on $[0,2x]$. Thus, $\Phi(x,t,0) = \Phi(x,t,+0)$ is well-defined and, taking into account (6.116)– (6.118), we obtain

$$\Phi(x,t,0) = \begin{bmatrix} \Phi_{11}(t,0) - N_{11}(x,t,x) & N_{12}(x,t,-x) \\ -N_{21}(x,t,x) & \Phi_{22}(t,0) + N_{22}(x,t,-x) \end{bmatrix}. \tag{6.127}$$

According to (6.116), (6.120), (6.122) and (6.126), the matrix function $\Phi(x,t,r)$ is differentiable with respect to t, its derivative $\frac{\partial}{\partial t}\Phi$ is continuous with respect to r on

$[0, \infty)$ (excluding, possibly, the point $r = 2x$), and $\frac{\partial^2}{\partial r \partial t}\Phi$ exists. Furthermore, we have

$$\lim_{\delta \to +0} \int_0^\infty \left\| e^{-\eta r} \left(\left(\frac{\partial}{\partial t}\Phi \right)(x,t,r) \mp (\Phi(x,t \pm \delta, r) - \Phi(x,t,r))/\delta \right) \right\|^2 dr = 0, \tag{6.128}$$

$$\int_0^\infty \left\| e^{-\eta r} \left(\frac{\partial^2}{\partial r \partial t}\Phi \right)(x,t,r) \right\|^2 dr < \infty \quad (t \le \tilde{a}, \eta > M(\tilde{a})), \tag{6.129}$$

where both cases (with $t+\delta$ and with $t-\delta$) are valid in (6.128) for $0 < t < \tilde{a}$, but only the case with $t + \delta$ is considered for $t = 0$ and only the case with $t - \delta$ is considered for $t = \tilde{a}$.

Formulas (6.114) and (6.128) yield the representation

$$\left(\frac{\partial}{\partial t}\widehat{w} \right)(x,t,z) = \int_0^\infty e^{izr} \left(\frac{\partial}{\partial t}\Phi \right)(x,t,r)dr. \tag{6.130}$$

Taking into account (6.129) and integrating the right-hand side of (6.130) by parts, we derive

$$\left(\frac{\partial}{\partial t}\widehat{w} \right)(x,t,z) = \frac{1}{iz}\left(\frac{\partial}{\partial t}\Phi \right)(x,t,0) + o\left(\frac{1}{z}\right) \tag{6.131}$$

$$\text{for } |z| \to \infty, \ \varepsilon < \arg(z) < \pi - \varepsilon \text{ and any } \varepsilon > 0. \tag{6.132}$$

According to (6.114), we have $\widehat{w}(x,t,z)^{-1} = I_2 + o(1)$ in the same domain (6.132). Thus, from (6.131), we see that (6.119) holds in this domain. Clearly, the result does not change if we remove the last two factors (which do not depend on t) in the definition (6.112) of $\widehat{w}$. Namely, we have

$$\left(\frac{\partial}{\partial t}w \right)(x,t,z)w(x,t,z)^{-1} = \frac{1}{iz}\left(\frac{\partial}{\partial t}\Phi \right)(x,t,0) + o\left(\frac{1}{z}\right), \tag{6.133}$$

$$|z| \to \infty, \quad \varepsilon < \arg(z) < \pi - \varepsilon; \quad w(x,t,z) := W(x,t,z)R(t,z). \tag{6.134}$$

From (6.83) and (6.84), it is apparent that

$$J_0 G(x,t,-z)J_0^* = G(x,t,z), \quad J_0 F(x,t,-z)J_0^* = F(x,t,z), \quad J_0 = \begin{bmatrix} 0 & 1 \\ -1 & 0 \end{bmatrix}. \tag{6.135}$$

Since $J_0^* = J_0^{-1}$, the definition of w (see (6.134)) and formulas (6.4), (6.5) and (6.135) yield

$$J_0^* w(x,t,z)J_0 = w(x,t,-z). \tag{6.136}$$

Moreover, we shall show that

$$J_0^*\left(\frac{\partial}{\partial t}\Phi\right)(x,t,0)J_0 = -\left(\frac{\partial}{\partial t}\Phi\right)(x,t,0). \tag{6.137}$$

For this purpose, we shall simplify the expression for $\Phi(x,t,0)$ given in (6.127), which simplification will also be used further in the proof. If $i > 1$, we derive from (6.125) that

$$N_i(x,t,x) = \frac{1}{2}\int_{-x}^{x}\begin{bmatrix} 0 & v\left(\frac{x-s}{2},t\right) \\ 0 & 0 \end{bmatrix} N_{i-1}\left(\frac{x-s}{2},t,\frac{x-s}{2}\right)ds, \tag{6.138}$$

$$N_i(x,t,-x) = -\frac{1}{2}\int_{-x}^{x}\begin{bmatrix} 0 & 0 \\ v\left(\frac{x+s}{2},t\right) & 0 \end{bmatrix} N_{i-1}\left(\frac{x+s}{2},t,\frac{-x-s}{2}\right)ds. \tag{6.139}$$

In particular, using (6.124), (6.138) and (6.139), we have

$$N_2(x,t,x) = -\frac{1}{4}\int_{-x}^{x}\begin{bmatrix} v\left(\frac{x-s}{2},t\right)^2 & 0 \\ 0 & 0 \end{bmatrix}ds, \tag{6.140}$$

$$N_2(x,t,-x) = -\frac{1}{4}\int_{-x}^{x}\begin{bmatrix} 0 & 0 \\ 0 & v\left(\frac{x+s}{2},t\right)^2 \end{bmatrix}ds. \tag{6.141}$$

Taking into account (6.138)–(6.141), we easily show by induction that

$$N_i(x,t,x) = N_i(x,t,-x) = 0 \quad \text{for} \quad i > 2. \tag{6.142}$$

In view of (6.123), (6.124), (6.140)–(6.142), we rewrite (6.127) in the form

$$\begin{aligned}\Phi(x,t,0) &= \operatorname{diag}\{\Phi_{11}(t,0),\ \Phi_{22}(t,0)\} \\ &\quad + \frac{1}{4}\begin{bmatrix} \int_{-x}^{x} v\left(\frac{x-s}{2},t\right)^2 ds & 2v(x,t) \\ 2v(x,t) & -\int_{-x}^{x} v\left(\frac{x+s}{2},t\right)^2 ds \end{bmatrix}.\end{aligned} \tag{6.143}$$

Since

$$\int_{-x}^{x} v\left(\frac{x-s}{2},t\right)^2 ds = \int_{-x}^{x} v\left(\frac{x+s}{2},t\right)^2 ds = 2\int_{0}^{x} v(s,t)^2 ds$$

and (6.122) holds, by differentiating (6.143), we have

$$\begin{aligned}\frac{\partial\Phi}{\partial t}(x,t,0) &= \frac{1}{2}\begin{bmatrix} 2\int_0^x v(s,t)\left(\frac{\partial}{\partial t}v\right)(s,t)\,ds & \left(\frac{\partial}{\partial t}v\right)(x,t) \\ \left(\frac{\partial}{\partial t}v\right)(x,t) & -2\int_0^x v(s,t)\left(\frac{\partial}{\partial t}v\right)(s,t)\,ds \end{bmatrix} \\ &\quad - \operatorname{diag}\{\zeta_{11}(t),\ \zeta_{22}(t)\},\end{aligned} \tag{6.144}$$

where ζ is given in (6.121). Equality (6.137) is immediate from (6.144).

Because of (6.136) and (6.137), formula (6.133) is also valid in the lower half-plane (for $\varepsilon - \pi < \arg(z) < -\varepsilon$), and so we obtain

$$z\left(\frac{\partial}{\partial t}w\right)(x,t,z)w(x,t,z)^{-1} = -\mathrm{i}\left(\frac{\partial}{\partial t}\Phi\right)(x,t,0) + o(1) \qquad (6.145)$$
$$\text{for } \varepsilon < |\arg(z)| < \pi - \varepsilon, \quad z \to \infty.$$

Recalling results from Step 1, we see that W is differentiable with respect to t. Thus, by virtue of (3.31), where we may put $u = W$, and of (6.134), the equality

$$z\left(\frac{\partial}{\partial t}w\right)(x,t,z)w(x,t,z)^{-1} = z\left(\left(\frac{\partial}{\partial t}W\right)(x,t,z) + W(x,t,z)F(0,t,z)\right) \times W(x,t,\overline{z})^{*} \qquad (6.146)$$

holds. Hence, $z(\frac{\partial}{\partial t}w)(x,t,z)w(x,t,z)^{-1}$ is an entire matrix function, the entries of which have growth of order 1. Therefore, using the Phragmen–Lindelöf and first Liouvile theorems, we can rewrite (6.145) in a more precise form

$$\left(\frac{\partial}{\partial t}w\right)(x,t,z)w(x,t,z)^{-1} = \frac{1}{\mathrm{i}z}\left(\frac{\partial}{\partial t}\Phi\right)(x,t,0). \qquad (6.147)$$

Step 4. It is also convenient to rewrite (6.146) as

$$\left(\frac{\partial}{\partial t}w\right)w^{-1} = \left(\frac{\partial}{\partial t}W\right)W^{-1} + W(\zeta/(\mathrm{i}z))W^{-1}. \qquad (6.148)$$

Formulas (6.147) and (6.148) yield

$$\left(\frac{\partial}{\partial t}W\right)(x,t,z) = \frac{1}{\mathrm{i}z}\left(\left(\frac{\partial}{\partial t}\Phi\right)(x,t,0)W(x,t,z) - W(x,t,z)\zeta(t)\right). \qquad (6.149)$$

Differentiating both sides of (6.149) with respect to x and taking into account that $\frac{\partial^2}{\partial x\partial t}W = \frac{\partial^2}{\partial t\partial x}W = (\frac{\partial}{\partial t}G)W + G\frac{\partial}{\partial t}W$, after easy transformations, we obtain

$$\mathrm{i}z\left(\frac{\partial}{\partial t}G\right)(x,t,z) - \left(\frac{\partial^2}{\partial x\partial t}\Phi\right)(x,t,0) = \left(\frac{\partial}{\partial t}\Phi\right)(x,t,0)G(x,t,z) - G(x,t,z)$$
$$\times W(x,t,z)\zeta(t)W(x,t,z)^{-1} - \mathrm{i}zG(x,t,z)\left(\frac{\partial}{\partial t}W\right)(x,t,z)W(x,t,z)^{-1}.$$

Next, we substitute (6.149) into the formula above and derive the result

$$\mathrm{i}z\left(\frac{\partial}{\partial t}G\right)(x,t,z) - \left(\frac{\partial^2}{\partial x\partial t}\Phi\right)(x,t,0) = \left[\left(\frac{\partial}{\partial t}\Phi\right)(x,t,0), G(x,t,z)\right]. \qquad (6.150)$$

Putting $z = 0$, for the entry in the first row and second column in (6.150), we have

$$\frac{\partial^2\Phi_{12}}{\partial x\partial t}(x,t,0) = v(x,t)\left(\frac{\partial\Phi_{22}}{\partial t}(x,t,0) - \frac{\partial\Phi_{11}}{\partial t}(x,t,0)\right). \qquad (6.151)$$

Taking into account (6.121) and (6.144), we rewrite (6.151) in the form

$$\frac{\partial^2 v}{\partial x \partial t}(x,t) = 4v(x,t)\left(\cos(2h_2(t)) - \int_0^x v(s,t)\frac{\partial v}{\partial t}(s,t)ds\right). \tag{6.152}$$

Recall that $\Phi_1 = \overline{\Phi_1}$. In order to find $v(0,t)$, we also recall some results on the inverse problem. Namely, from (3.73) and (3.93), we derive

$$\beta'(0) = -\overline{\Phi_1}'(0)\begin{bmatrix}\Phi_1(0) & 1\end{bmatrix} = -\Phi_1'(0)\begin{bmatrix}\Phi_1(0) & 1\end{bmatrix}. \tag{6.153}$$

Since $\gamma(0) = \begin{bmatrix}0 & 1\end{bmatrix}$, relations (3.92) and (6.153) yield $v(0) = -\Phi_1'(0)$. Using (3.108) and (6.97), we easily obtain $\Phi_1'(0) = 2\mathrm{i}\phi_0$. Thus, formulas (6.96) and (6.97) imply that

$$v(0,t) = -2\mathrm{i}\phi_0(t) = -2\left(\mathrm{i}\phi_0(0) + \int_0^t \sin(2h_2(\xi))\,d\xi\right). \tag{6.154}$$

Finally, we put

$$f(x,t) := \cos(2h_2(t)) - \int_0^x v(s,t)\frac{\partial v}{\partial t}(s,t)ds + \frac{\mathrm{i}}{2}\frac{\partial v}{\partial t}(x,t). \tag{6.155}$$

It follows from (6.152) that

$$\begin{aligned}\frac{\partial f}{\partial x}(x,t) &= 2\mathrm{i}v(x,t)\left(\cos(2h_2(t)) - \int_0^x v(s,t)\frac{\partial v}{\partial t}(s,t)ds + \frac{\mathrm{i}}{2}\frac{\partial v}{\partial t}(x,t)\right)\\ &= 2\mathrm{i}v(x,t)f(x,t).\end{aligned} \tag{6.156}$$

Moreover, relations (6.154) and (6.155) yield

$$f(0,t) = \cos(2h_2(t)) + \frac{\mathrm{i}}{2}\frac{\partial v}{\partial t}(0,t) = \cos(2h_2(t)) - \mathrm{i}\sin(2h_2(t)) = \mathrm{e}^{-2\mathrm{i}h_2(t)}. \tag{6.157}$$

It is easy to see that $\mathrm{e}^{-2\mathrm{i}\psi(x,t)}$, where ψ is given by (6.94), that is,

$$\psi(x,t) = h_2(t) - \int_0^x v(s,t)ds, \tag{6.158}$$

satisfies equations (6.156) and (6.157). Thus, $\mathrm{e}^{-2\mathrm{i}\psi(x,t)}$ coincides with f of the form (6.155):

$$\mathrm{e}^{-2\mathrm{i}\psi(x,t)} = \cos(2h_2(t)) - \int_0^x v(s,t)\frac{\partial v}{\partial t}(s,t)ds + \frac{\mathrm{i}}{2}\frac{\partial v}{\partial t}(x,t). \tag{6.159}$$

Taking imaginary parts of the expressions on both sides of (6.159), we obtain

$$-2\mathrm{i}\sin(\psi(x,t)) = \frac{\mathrm{i}}{2}\frac{\partial v}{\partial t}(x,t) = -\frac{\mathrm{i}}{2}\frac{\partial^2 \psi}{\partial t \partial x}(x,t), \tag{6.160}$$

that is, the function ψ, which is constructed via procedure described in our theorem, satisfies SGE (6.57). From (6.158), it is immediate that ψ satisfies the boundary condition $\psi(0,t) = h_2(t)$. Moreover, since the Dirac system is uniquely recovered from its GW-function, in view of (6.93), we have $v(x,0) = -h_1'(x)$. Therefore, formula (6.158) implies that $\psi(x,0) = h_2(0) - h_1(0) + h_1(x) = h_1(x)$, and so ψ also satisfies the initial condition. □

The next proposition is a particular case of Theorem 6.1 in [36] (see also [36, Theorem A] and see Subsection 4.2.1 and Theorem 4.16 in our book for more explanations).

Proposition 6.20. *Suppose $v(x) \in L^1(\mathbb{R}_+)$. Then, there is a fundamental solution $\widehat{W}$ of the system (3.1) which is analytic with respect to z everywhere in $\mathbb{C}_M$ (for some $M > 0$) and satisfies the equality*

$$\lim_{z\to\infty} \widehat{W}(x,z)\mathrm{e}^{-\mathrm{i}zxj} = I_2, \quad z \in \mathbb{C}_M \tag{6.161}$$

uniformly with respect to x. If, in addition, v is two times differentiable and $v'(x)$, $v''(x) \in L^1(\mathbb{R}_+)$, then there is a matrix C_0 such that

$$\widehat{W}(0,z) = I_2 + \frac{1}{z}C_0 + O\left(\frac{1}{z^2}\right), \quad z \to \infty, \quad z \in \mathbb{C}_M. \tag{6.162}$$

We note that, differently from W and other fundamental solutions that we deal with, it is not required that $\widehat{W}(0,z) = I_2$ since $\widehat{W}$ is normalized in another way, namely, via (6.161).

Corollary 6.21.

(a) *If $v \in L^1(\mathbb{R}_+)$, then there is a GW-function of the system (3.1) and this GW-function is given by the formula*

$$\varphi(z) = \widehat{W}_{21}(0,z)/\widehat{W}_{11}(0,z), \quad \widehat{W}(0,z) =: \{\widehat{W}_{ik}(0,z)\}_{i,k=1}^2 \tag{6.163}$$

for all z in the half-plane $\mathbb{C}_M$ for some $M > 0$.

(b) *Moreover, if v is two times differentiable and $v, v', v'' \in L^1(\mathbb{R}_+)$, then this GW-function φ satisfies the asymptotic condition* (3.106).

Proof. According to Proposition 6.20, there is $M > 0$ such that (6.162) holds and

$$\sup_{x\geq 0,\, \Im(z)>M} \|\widehat{W}(x,z)\mathrm{e}^{-\mathrm{i}zxj}\| < \infty, \tag{6.164}$$

$$\det \widehat{W}(0,z) \neq 0, \quad \widehat{W}_{11}(0,z) \neq 0, \quad \sup_{\Im(z)>M} |1/\widehat{W}_{11}(0,z)| < \infty. \tag{6.165}$$

Since W and $\widehat{W}$ are fundamental solutions of the same system, it is immediate that

$$W(x,z) = \widehat{W}(x,z)\widehat{W}(0,z)^{-1}.$$

Hence, we derive

$$\begin{aligned} e^{-izx}W(x,z)\begin{bmatrix} 1 \\ \frac{\widehat{W}_{21}(0,z)}{\widehat{W}_{11}(0,z)} \end{bmatrix} &= \frac{e^{-izx}}{\widehat{W}_{11}(0,z)}\widehat{W}(x,z)\widehat{W}(0,z)^{-1}\widehat{W}(0,z)\begin{bmatrix} 1 \\ 0 \end{bmatrix} \\ &= \frac{1}{\widehat{W}_{11}(0,z)}\widehat{W}(x,z)e^{-izxj}\begin{bmatrix} 1 \\ 0 \end{bmatrix}. \end{aligned} \tag{6.166}$$

In view of (6.164)–(6.166), the function φ given by (6.163) satisfies (3.104), where $u = W$. Recall also that $\widehat{W}$ is analytic, and so φ given by (6.163) is analytic. Thus, the statement (a) is proved.

If $v, v', v'' \in L^1(\mathbb{R}_+)$, then it follows from (6.162) that

$$W_{21}(0,z) = \frac{1}{z}(C_0)_{21} + O\left(\frac{1}{z^2}\right), \quad W_{11}(0,z) = 1 + \frac{1}{z}(C_0)_{11} + O\left(\frac{1}{z^2}\right).$$

Therefore, we derive for $z \in \mathbb{C}_M, \quad z \to \infty$ that

$$1/W_{11}(0,z) = 1 - \frac{1}{z}(C_0)_{11} + O\left(\frac{1}{z^2}\right), \; W_{21}(0,z)/W_{11}(0,z) = \frac{1}{z}(C_0)_{12} + O\left(\frac{1}{z^2}\right). \tag{6.167}$$

The statement (b) is immediate from (6.163) and (6.167). □

The next theorem easily follows from Theorem 6.19 and Corollary 6.21.

Theorem 6.22. *Assume that $h_1(x) = \overline{h_1(x)}$ is three times differentiable for $x \geq 0$, that*

$$h_1', h_1'', h_1''' \in L^1(\mathbb{R}_+), \tag{6.168}$$

and that $h_2 = \overline{h_2}$ is continuous on $[0, a)$ $(h_1(0) = h_2(0))$. Then, the GW-function $\varphi(0,z)$ of the system (6.93) exists and satisfies (3.106). A solution of the initial-boundary value problem (6.92) for the sine-Gordon equation (6.57) exists and is given by the equality (6.94), where $\varphi(t,z)$ is given by (6.87) and $R = \{R_{ik}\}_{i,k=1}^2$ in (6.87) is defined by the relations (6.95).

Remark 6.23. If the conditions of Theorem 6.22 hold, then formulas (6.94) and (6.108) imply that the functions ψ, $\frac{\partial}{\partial x}\psi$ and $\frac{\partial}{\partial t \partial x}\psi$ are continuous.

Remark 6.24. In view of Theorem 6.22, the continuity of h_2 and condition (6.168) for h_1 are sufficient for the existence and construction of the solution of the sine-Gordon equation (and (6.97) holds automatically), whereas a rapid decay of h_1' was required in [326]. The a priori existence of the solution, which vanishes with all derivatives as $x \to \infty$, was required in an interesting paper [189] where the evolution of the spectral data was used. Several related problems were studied [103, 104] using the "global relation" approach introduced by A. S. Fokas (see [52] and references therein on this approach and some remaining open problems). However, the existence of the solution was assumed a priori in the paper [104] on Raman scattering and initial conditions from the Schwartz class were required in the paper [103] on sine-Gordon in laboratory coordinates. Weyl functions, which are used in the present book, allow us to essentially weaken such restrictions as well as to rigorously prove interesting uniqueness and existence results.

6.2.3 Unbounded solutions in the quarter-plane

The behavior of the solutions of initial-boundary value problems is of interest. Notice also that it is difficult to treat unbounded solutions using the inverse scattering transform method. Here, we describe a family of unbounded solutions.

Recall definition (6.3) of the semistrip Ω_a. Clearly, when $a = \infty$, we obtain a quarter-plane Ω_∞ (instead of the semistrip). It is immediate that we may put $a = \infty$ (i.e. we may consider SGE in the quarter-plane Ω_∞) in Theorem 6.16. Then, the next corollary easily follows.

Corollary 6.25. *Assume that the conditions of Theorem 6.16 are valid in Ω_∞. Then, for values of z, such that the inequalities*

$$(\cos(2h_2(t)) - \varepsilon(z))\,\Im(z) \geq |\Re(z)\sin(2h_2(t))| \quad (z \in \mathbb{C}_M) \tag{6.169}$$

hold for some $\varepsilon(z) > 0$ and for all $t \geq 0$, we have

$$\varphi(0,z) = -\lim_{t\to\infty} R_{21}(t,z)/R_{22}(t,z). \tag{6.170}$$

Proof. Recalling Proposition 3.3 and Corollary 3.8, we see that the inequality

$$|\varphi(t,z)| \leq 1$$

is valid. Hence, in view of (6.70), we have

$$[1 \quad \varphi(0,z)^*]R(t,z)^* jR(t,z)\begin{bmatrix} 1 \\ \varphi(0,z)\end{bmatrix} \geq 0. \tag{6.171}$$

From (6.95) and (6.169), we derive

$$\frac{d}{dt}\left(-R(t,z)^* jR(t,z)\right) \geq \frac{2\varepsilon(z)}{|z|^2}\,(\Im(z))\,R(t,z)^*R(t,z). \tag{6.172}$$

Taking into account (6.171) and (6.172), we easily obtain

$$\int_0^\infty [1 \quad \varphi(0,z)^*]R(t,z)^*R(t,z)\begin{bmatrix} 1 \\ \varphi(0,z)\end{bmatrix} dt < \infty. \tag{6.173}$$

Since according to (6.172) we have

$$\frac{d}{dt}\left(R(t,z)^* jR(t,z)\right) \le 0,$$

it follows that $R(t,z)^* jR(t,z) \le j$. In particular, we see that

$$|R_{22}(t,z)| \ge 1. \tag{6.174}$$

The equation $\frac{d}{dt}R = (\zeta/\mathrm{i}z)R$, where $\zeta(t)$ is unitary and so $\|\zeta\| = 1$, yields the inequality $\frac{d}{dt}(\exp\{2t/|z|\}R(t,z)^*R(t,z)) \ge 0$, that is,

$$R(t,z)^*R(t,z) \ge \exp\{2(t_0 - t)/|z|\}R(t_0,z)^*R(t_0,z) \quad \text{for } t \ge t_0. \tag{6.175}$$

Inequalities (6.173) and (6.175) imply that

$$\lim_{t\to\infty}\left\| R(t,z)\begin{bmatrix} 1 \\ \varphi(0,z)\end{bmatrix}\right\| = 0. \tag{6.176}$$

Finally, (6.170) follows from (6.174) and (6.176). □

Example 6.26. Let $h_2(t) \equiv 0$ $(0 \le t < \infty)$. Putting $\varepsilon = 1/2$, we see that (6.169) holds for all z in the half-plane $\mathbb{C}_M$. In view of (6.95), we have

$$R(t,z) = \exp\left\{\frac{t}{\mathrm{i}z}j\right\}. \tag{6.177}$$

Thus, if the conditions of Theorem 6.16 hold, using Corollary 6.25, we easily derive $\varphi(0,z) \equiv 0$. Therefore, Theorem 6.16 and formula (6.177) lead us to the equality $\varphi(t,z) \equiv 0$. From Proposition 3.26 and Theorem 3.30, it follows that the Weyl function $\varphi(t,z)$ is also the GW-function of the same system and this system is recovered uniquely. Hence, since (3.104) holds for the case that $\varphi = 0$ and $v = 0$, the potential $v = 0$ is this unique solution of the inverse problem, that is,

$$\left(\frac{\partial}{\partial x}\psi\right)(x,t) = -\mathfrak{M}(\varphi(t,z)) \equiv 0. \tag{6.178}$$

From Theorem 6.22 and Example 6.26, we derive the next proposition.

Proposition 6.27. *Assume that $h_1(x) = \overline{h_1(x)} \not\equiv 0$ is three times differentiable for $x \ge 0$,*

$$h_1', h_1'', h_1''' \in L^1(\mathbb{R}_+), \quad h_1(0) = 0,$$

and that $h_2 \equiv 0$. Then, one can use the procedure given in Theorem 6.22 in order to construct a solution ψ of the initial-boundary value problem (6.92) for the sine-Gordon equation (6.57), and the absolute value of the derivative $\frac{\partial}{\partial x}\psi$ is always unbounded in the quarter-plane.

Proof. Since the conditions of Theorem 6.22 hold, we can construct a solution ψ of (6.57), (6.92). Moreover, according to Remark 6.23, the functions ψ, $\frac{\partial}{\partial x}\psi$ and $\frac{\partial^2}{\partial t \partial x}\psi$ are continuous.

Thus, if (6.85) is valid, then the conditions of Theorem 6.16 and Example 6.26 are satisfied. In particular, we obtain (6.178). Equalities (6.178) and $h_2(0) = 0$ yield $\psi(x,0) \equiv 0$ which contradicts our assumption

$$\psi(x,0) = h_1(x) \not\equiv 0.$$

So, (6.85) does not hold and the proposition is proved by contradiction. □

7 General GBDT theorems and explicit solutions of nonlinear equations

Bäcklund–Darboux transformations (BDTs) are well known as a versatile tool in spectral theory as well as for integrable nonlinear equations (see, e.g. some explanations and references in the Introduction). Several versions of BDT and their comparison are discussed in the interesting recent survey [73]. In this book, we present the GBDT version of the BDT. The GBDT for self-adjoint and skew-self-adjoint Dirac systems, as well as for the system auxiliary to the N-wave equation, was studied in Subsection 1.1.3. Further, in Chapters 2 and 4, we used these transformations to explicitly solve the corresponding direct and inverse problems. A discrete case (discrete Dirac system) was studied in Section 5.3. That result was used in Subsection 5.3.4 in order to construct explicit solutions of an integrable nonlinear equation, namely, isotropic Heisenberg magnet model, and to derive the evolution of the corresponding Weyl function. Here, we mostly present results from the paper [251] and survey [264]. Various other applications of GBDT (and explicit solutions of various physically important integrable equations, in particular) are given, for instance, in [111, 112, 131–133, 244–248, 253, 254, 256–258, 261–263, 266, 269, 271]. We derive results on the GBDT for systems depending on one variable (and spectral parameter) and apply them for the construction of explicit solutions of integrable systems depending on two variables.

Explicit solutions of the nonlinear optics (or N-wave) equation are constructed in Section 7.1. That section can be considered as an introductory example of application of the GBDT to a continuous integrable system. In Section 7.2 of this chapter, we formulate and prove general theorems on the GBDT for a system depending rationally on the spectral parameter. The last Section 7.3 is dedicated to the construction of explicit solutions in several somewhat more complicated cases, namely, the cases of the main chiral field, elliptic sine-Gordon and elliptic sinh-Gordon equations.

As in Chapter 6, by $\mathcal{I}$ and $\mathcal{I}_k$, we denote intervals on the real axis. By the neighborhood of zero, we mean the neighborhood of the form $(0, \varepsilon)$ or $[0, \varepsilon)$. We always assume that $\sum_{i=k}^{r} d_i = 0$ when $k > r$. The spectrum of an operator A is denoted by $\sigma(A)$.

7.1 Explicit solutions of the nonlinear optics equation

Consider the nonlinear optics (N-wave) equation

$$\left[D, \frac{\partial \varrho}{\partial t}\right] - \left[\hat{D}, \frac{\partial \varrho}{\partial x}\right] = \left[[D, \varrho], [\hat{D}, \varrho]\right], \quad (x, t) \in \mathcal{I}_1 \times \mathcal{I}_2, \tag{7.1}$$

on the product $\mathcal{I}_1 \times \mathcal{I}_2$ of intervals $\mathcal{I}_1$ and $\mathcal{I}_2$, where D and B have the form (1.68), $\varrho(x, t)$ are $m \times m$ matrix functions and

$$\varrho^* = B\varrho B, \quad (0,0) \in \mathcal{I}_1 \times \mathcal{I}_2, \quad \hat{D} = \operatorname{diag}\{\hat{d}_1, \dots, \hat{d}_m\} = \hat{D}^*. \tag{7.2}$$

Nonlinear integrable equation (7.1) is the compatibility condition of two auxiliary linear systems [101, 334] (see also [3] for the case $N > 3$):

$$\frac{\partial}{\partial x}w = (\mathrm{i}zD - [D,\varrho])\,w, \quad \frac{\partial}{\partial t}w = \left(\mathrm{i}z\hat{D} - [\hat{D},\varrho]\right)w. \tag{7.3}$$

Indeed, in view of (7.3), we have

$$\frac{\partial^2}{\partial t\partial x}w = -\left[D, \frac{\partial}{\partial t}\varrho\right]w + (\mathrm{i}zD - [D,\varrho])\left(\mathrm{i}z\hat{D} - [\hat{D},\varrho]\right)w, \tag{7.4}$$

$$\frac{\partial^2}{\partial x\partial t}w = -\left[\hat{D}, \frac{\partial}{\partial x}\varrho\right]w + \left(\mathrm{i}z\hat{D} - [\hat{D},\varrho]\right)(\mathrm{i}zD - [D,\varrho])\,w. \tag{7.5}$$

We note that systems (7.3) represent a particular case of systems (6.2), namely, the case where $G = \mathrm{i}zD - [D,\varrho]$ and $F = \mathrm{i}z\hat{D} - [\hat{D},\varrho]$. Clearly, according to relations (7.4) and (7.5), condition (6.1), that is, $\frac{\partial}{\partial t}G - \frac{\partial}{\partial x}F + [G,F] = 0$, coincides with the compatibility condition $\frac{\partial^2}{\partial t\partial x}w = \frac{\partial^2}{\partial x\partial t}w$, if only w is a fundamental solution. We see that condition (6.1) is also equivalent to (7.1).

In order to construct GBDT, we fix $n > 0$ and three parameter matrices, that is, two $n\times n$ matrices A and $S(0,0) = S(0,0)^*$, and an $n\times m$ matrix $\Pi(0,0)$ such that

$$AS(0,0) - S(0,0)A^* = \mathrm{i}\Pi(0,0)B\Pi(0,0)^*. \tag{7.6}$$

Now, introduce matrix functions $\Pi(x,t)$ and $S(x,t)$ by the equations

$$\frac{\partial}{\partial x}\Pi = -\mathrm{i}A\Pi D + \Pi[D,\varrho], \quad \frac{\partial}{\partial t}\Pi = -\mathrm{i}A\Pi\hat{D} + \Pi[\hat{D},\varrho], \tag{7.7}$$

$$\frac{\partial}{\partial x}S = \Pi DB\Pi^*, \quad \frac{\partial}{\partial t}S = \Pi\hat{D}B\Pi^*. \tag{7.8}$$

Quite similar to the equality $\frac{\partial^2}{\partial t\partial x}w = \frac{\partial^2}{\partial x\partial t}w$, one can show that, according to (7.1), equations (7.7) are compatible, that is, $\frac{\partial^2}{\partial t\partial x}\Pi = \frac{\partial^2}{\partial x\partial t}\Pi$. By virtue of (7.7) and (7.8), it is immediate that $\frac{\partial^2}{\partial t\partial x}S = \frac{\partial^2}{\partial x\partial t}S$. Proposition 1.8 implies the following.

Proposition 7.1. *Let an $m\times m$ matrix function ϱ ($\varrho^* = B\varrho B$) be continuously differentiable and satisfy nonlinear optics equation* (7.1). *Let matrix functions Π and $S = S^*$ satisfy* (7.6)–(7.8). *Then, in the points of invertibility of S, the matrix function*

$$\tilde{\varrho}(x,t) := \varrho(x,t) - B\Pi(x,t)^*S(x,t)^{-1}\Pi(x,t) \tag{7.9}$$

satisfies equation (7.1) *and an additional condition $\tilde{\varrho}^* = B\tilde{\varrho}B$.*

Proof. Equality $\tilde{\varrho}^* = B\tilde{\varrho}B$ is immediate. Now, let w be the fundamental solution of the systems (7.3), that is, let w satisfy (7.3) and equality $w(0,0,z) = I_m$. Since ϱ satisfies (7.1), the compatibility condition (6.1) for the systems (7.3) is fulfilled, and such a matrix function w exists (see Remarks 6.2 and 6.5). Put

$$w_A(x,t,z) = I_m - \mathrm{i}B\Pi(x,t)^*S(x,t)^{-1}(A - zI_n)^{-1}\Pi(x,t), \tag{7.10}$$

and calculate derivatives of w_A with respect to x and t using Proposition 1.8 in both cases. Then, for $\widetilde{w}(x,t,z) = w_A(x,t,z)w(x,t,z)$, we have

$$\frac{\partial}{\partial x}\widetilde{w} = \widetilde{G}\widetilde{w}, \quad \frac{\partial}{\partial t}\widetilde{w} = \widetilde{F}\widetilde{w}, \tag{7.11}$$

where $\widetilde{G}$ and $\widetilde{F}$ have the same form as G and F, namely,

$$\widetilde{G} = \mathrm{i}zD - [D, \widetilde{\varrho}], \quad \widetilde{F} = \mathrm{i}z\hat{D} - [\hat{D}, \widetilde{\varrho}]. \tag{7.12}$$

From the continuous differentiability of ϱ follows the continuous differentiability of $\widetilde{\varrho}$ in the points of invertibility of S. Hence, in view of (7.11), we obtain $\frac{\partial^2}{\partial t \partial x}\widetilde{w} = \frac{\partial^2}{\partial x \partial t}\widetilde{w}$ or equivalently $\frac{\partial}{\partial t}\widetilde{G} - \frac{\partial}{\partial x}\widetilde{F} + [\widetilde{G}, \widetilde{F}] = 0$ (see also the proof of Theorem 7.11). As we have already discussed here (for the similar matrix functions G and F), the last equality is in turn equivalent to the nonlinear optics equation

$$\left[D, \frac{\partial}{\partial t}\widetilde{\varrho}\right] - \left[\hat{D}, \frac{\partial}{\partial x}\widetilde{\varrho}\right] = \left[[D, \widetilde{\varrho}], [\hat{D}, \widetilde{\varrho}]\right]. \qquad \square$$

Remark 7.2. In the case of the trivial initial solution $\varrho \equiv 0$, we easily construct explicit solutions of the nonlinear optics equation. Namely, putting

$$\Pi(0,0) = [f_1 \ f_2 \ \dots \ f_m]$$

and using (7.7), we recover Π:

$$\Pi(x,t) = \left[\exp\left\{-\mathrm{i}(d_1 x + \hat{d}_1 t)A\right\} f_1 \quad \exp\left\{-\mathrm{i}(d_2 x + \hat{d}_2 t)A\right\} f_2 \quad \dots\right]. \tag{7.13}$$

Next, we recover S, and explicit formulas for solutions of the nonlinear optics equation are immediate from (7.9).

We note that GBDT-type formulas (1.61), (1.66) and (7.9) for the representation of explicit solutions differ from the well-known determinant formulas. Usually, it is not difficult to derive determinant-type representation for the scalar multisoliton solutions from the GBDT-type representation (see, e.g. [244, Ch.4] for this procedure for sine-Gordon and sinh-Gordon equations).

7.2 GBDT for linear system depending rationally on z

We start with an $m \times m$ first order system

$$u'(x,z) = G(x,z)u(x,z), \quad G(x,z) = -\sum_{k=-r}^{r} z^k q_k(x) \quad \left(u' = \frac{d}{dx}u, \ x \in \mathcal{I}\right), \tag{7.14}$$

where G has only one pole (at $z = 0$), $\mathcal{I}$ is an interval such that $0 \in \mathcal{I}$, and the coefficients $q_k(x)$ and $q_{-k}(x)$ are $m \times m$ locally summable matrix functions. The function u in (7.14) is an absolutely continuous matrix function (it may be either fundamental solution or vector function, in particular). As in the previous special cases, the GBDT of system (7.14) is determined by the three square matrices A_1, A_2 and $S(0)$ ($\det S(0) \neq 0$) of order n and by the two $n \times m$ matrices $\Pi_1(0)$ and $\Pi_2(0)$, satisfying the operator identity

$$A_1 S(0) - S(0) A_2 = \Pi_1(0)\Pi_2(0)^*. \tag{7.15}$$

Suppose that such parameter matrices are fixed. Then, we introduce matrix functions $S(x)$, $\Pi_1(x)$ and $\Pi_2(x)$ with the values $\Pi_1(0)$, $\Pi_2(0)$ and $S(0)$ at $x = 0$ as the solutions of the linear differential equations

$$\Pi_1'(x) = \sum_{k=-r}^{r} A_1^k \Pi_1(x) q_k(x), \quad \Pi_2'(x) = -\sum_{k=-r}^{r} (A_2^*)^k \Pi_2(x) q_k(x)^*, \tag{7.16}$$

$$\begin{aligned} S'(x) &= \sum_{k=1}^{r} \sum_{j=1}^{k} A_1^{k-j} \Pi_1(x) q_k(x) \Pi_2(x)^* A_2^{j-1} \\ &\quad - \sum_{k=-r}^{-1} \sum_{j=k+1}^{0} A_1^{k-j} \Pi_1(x) q_k(x) \Pi_2(x)^* A_2^{j-1}. \end{aligned} \tag{7.17}$$

Notice that equations (7.16) and (7.17) are chosen in such a way that the identity

$$A_1 S(x) - S(x) A_2 = \Pi_1(x)\Pi_2(x)^* \tag{7.18}$$

follows from (7.15)–(7.17) for all x in the domain $\mathcal{I}$, where the coefficients q_k are defined. (Indeed, direct differentiation of the both sides of (7.18) shows that these derivatives coincide, and so (7.18) is immediate from (7.15).) Since A_1 appears in the equation for Π_1 and A_2 appears in the equation for Π_2 (see (7.16)), it is convenient to use here notations A_1, A_2 instead of the notations A, B, which were used in the definition of the general-type S-node in the Introduction.

Assuming that the inequality $\det S(x) \not\equiv 0$ holds, we define the transfer matrix function w_A in the standard way:

$$w_A(x,z) = I_m - \Pi_2(x)^* S(x)^{-1} (A_1 - zI_n)^{-1} \Pi_1(x). \tag{7.19}$$

Theorem 7.3. *Let the parameter matrices A_1 and A_2 be invertible:*

$$\det A_1 \neq 0, \quad \det A_2 \neq 0. \tag{7.20}$$

Suppose that matrix functions w, Π_1, Π_2 and S satisfy equations (7.14)–(7.17). *Then, in the points of invertibility of S, the matrix function*

$$\tilde{u}(x,z) = w_A(x,z) u(x,z) \tag{7.21}$$

satisfies the system

$$\tilde{u}'(x,z) = \tilde{G}(x,z)\tilde{u}(x,z), \quad \tilde{G}(x,z) = -\sum_{k=-r}^{r} z^k \tilde{q}_k(x) \tag{7.22}$$

with coefficients

$$\tilde{q}_k(x) = q_k(x) - \sum_{j=k+1}^{r} \Big(q_j(x) Y_{j-k-1}(x) - X_{j-k-1}(x) q_j(x) + \sum_{i=k+2}^{j} X_{j-i}(x) q_j(x) Y_{i-k-2}(x) \Big) \quad \text{for } k \geq 0, \tag{7.23}$$

$$\tilde{q}_k(x) = q_k(x) + \sum_{j=-r}^{k} \Big(q_j(x) Y_{j-k-1}(x) - X_{j-k-1}(x) q_j(x) - \sum_{i=j+1}^{k+1} X_{j-i}(x) q_j(x) Y_{i-k-2}(x) \Big) \quad \text{for } k < 0, \tag{7.24}$$

where

$$X_k(x) = \Pi_2(x)^* S(x)^{-1} A_1^k \Pi_1(x), \qquad Y_k(x) = \Pi_2(x)^* A_2^k S(x)^{-1} \Pi_1(x). \tag{7.25}$$

Proof. Step 1. From (7.18), it follows that

$$S(x)^{-1} A_1 - A_2 S(x)^{-1} = S(x)^{-1} \Pi_1(x) \Pi_2(x)^* S(x)^{-1}, \tag{7.26}$$

$$A_2^{-1} S(x)^{-1} - S(x)^{-1} A_1^{-1} = A_2^{-1} S(x)^{-1} \Pi_1(x) \Pi_2(x)^* S(x)^{-1} A_1^{-1}. \tag{7.27}$$

Taking into account (7.26) and (7.27) and using induction, we obtain

$$A_2^k S(x)^{-1} = S(x)^{-1} A_1^k - \sum_{j=0}^{k-1} A_2^{k-j-1} S(x)^{-1} \Pi_1(x) \Pi_2(x)^* S(x)^{-1} A_1^j \quad (k \geq 0), \tag{7.28}$$

$$A_2^k S(x)^{-1} = S(x)^{-1} A_1^k - \sum_{j=k}^{-1} A_2^{k-j-1} S(x)^{-1} \Pi_1(x) \Pi_2(x)^* S(x)^{-1} A_1^j \quad (k < 0). \tag{7.29}$$

Next, we obtain an expression for $(\Pi_2^* S^{-1})'$. In view of (7.16), (7.17) and (7.25), we have

$$\begin{aligned}(\Pi_2^* S^{-1})'(x) &= -\Pi_2(x)^* S^{-1}(x) S'(x) S(x)^{-1} - \sum_{k=-r}^{r} q_k(x) \Pi_2(x)^* A_2^k S(x)^{-1} \\ &= f_1(x) + f_2(x), \end{aligned} \tag{7.30}$$

where

$$f_1(x) = -\sum_{k=1}^{r}\sum_{j=1}^{k} X_{k-j}(x)q_k(x)\Pi_2(x)^* A_2^{j-1} S(x)^{-1}$$
$$-\sum_{k=0}^{r} q_k(x)\Pi_2(x)^* A_2^k S(x)^{-1}, \tag{7.31}$$
$$f_2(x) = \sum_{k=-r}^{-1}\sum_{j=k+1}^{0} X_{k-j}(x)q_k(x)\Pi_2(x)^* A_2^{j-1} S(x)^{-1}$$
$$-\sum_{k=-r}^{-1} q_k(x)\Pi_2(x)^* A_2^k S(x)^{-1}. \tag{7.32}$$

Using (7.25) and (7.28), we rewrite (7.31) in the form

$$f_1(x) = -\sum_{k=1}^{r}\sum_{j=1}^{k} X_{k-j}(x)q_k(x)\Pi_2(x)^* S(x)^{-1} A_1^{j-1}$$
$$+\sum_{k=1}^{r}\sum_{j=1}^{k}\sum_{i=0}^{j-2} X_{k-j}(x)q_k(x)Y_{j-i-2}(x)\Pi_2(x)^* S(x)^{-1} A_1^{i}$$
$$-\sum_{k=0}^{r} q_k(x)\Pi_2(x)^* S(x)^{-1} A_1^k \tag{7.33}$$
$$+\sum_{k=0}^{r}\sum_{i=0}^{k-1} q_k(x)Y_{k-i-1}(x)\Pi_2(x)^* S(x)^{-1} A_1^i.$$

In a similar way (using (7.29)), we also rewrite (7.32):

$$f_2(x) = \sum_{k=-r}^{-1}\sum_{j=k+1}^{0} X_{k-j}(x)q_k(x)\Pi_2(x)^* S(x)^{-1} A_1^{j-1}$$
$$+\sum_{k=-r}^{-1}\sum_{j=k+1}^{0}\sum_{i=j-1}^{-1} X_{k-j}(x)q_k(x)Y_{j-i-2}(x)\Pi_2(x)^* S(x)^{-1} A_1^{i}$$
$$-\sum_{k=-r}^{-1} q_k(x)\Pi_2(x)^* S(x)^{-1} A_1^k \tag{7.34}$$
$$-\sum_{k=-r}^{-1}\sum_{i=k}^{-1} q_k(x)Y_{k-i-1}(x)\Pi_2(x)^* S(x)^{-1} A_1^i.$$

Let us now calculate the coefficients before the terms $\Pi_2^* S^{-1} A_1^k$ $(-r \le k \le r)$ in (7.33) and (7.34):

$$f_1(x) = -\sum_{k=0}^{r}\Bigg(q_k(x) - \sum_{j=k+1}^{r}\Bigg(q_j(x)Y_{j-k-1}(x) - X_{j-k-1}(x)q_j(x)$$

$$+\sum_{i=k+2}^{j} X_{j-i}(x)q_j(x)Y_{i-k-2}(x)\Bigg)\Bigg)\Pi_2(x)^*S(x)^{-1}A_1^k,$$

$$f_2(x) = -\sum_{k=-r}^{-1}\Bigg(q_k(x) + \sum_{j=-r}^{k}\Bigg(q_j(x)Y_{j-k-1}(x) - X_{j-k-1}(x)q_j(x)$$

$$-\sum_{i=j+1}^{k+1} X_{j-i}(x)q_j(x)Y_{i-k-2}(x)\Bigg)\Bigg)\Pi_2(x)^*S(x)^{-1}A_1^k.$$

Because of (7.23), (7.24) and (7.30), the last relations yield the result

$$(\Pi_2^* S^{-1})'(x) = -\sum_{k=-r}^{r}\tilde{q}_k(x)\Pi_2(x)^*S(x)^{-1}A_1^k. \tag{7.35}$$

Step 2. Taking into account (7.16) and (7.35), and using the equality

$$A_1^k = A_1^k - z^k I_n + z^k I_n,$$

we differentiate the right-hand side of (7.19):

$$\begin{aligned} w_A'(x,z) = \sum_{k=-r}^{r}\Big(&z^k\tilde{q}_k(x)\Pi_2(x)^*S(x)^{-1}(A_1 - zI_n)^{-1}\Pi_1(x) \\ &+\tilde{q}_k(x)\Pi_2(x)^*S(x)^{-1}(A_1^k - z^kI_n)(A_1 - zI_n)^{-1}\Pi_1(x) \\ &-z^k\Pi_2(x)^*S(x)^{-1}(A_1 - zI_n)^{-1}\Pi_1(x)q_k(x) \\ &-\Pi_2(x)^*S(x)^{-1}(A_1^k - z^kI_n)(A_1 - zI_n)^{-1}\Pi_1(x)q_k(x)\Big). \end{aligned} \tag{7.36}$$

Recall now that

$$G(x,z) = -\sum_{k=-r}^{r} z^k q_k(x), \quad \tilde{G}(x,z) = -\sum_{k=-r}^{r} z^k\tilde{q}_k(x). \tag{7.37}$$

From (7.25), (7.36) and (7.37), we derive

$$\begin{aligned} w_A'(x,z) &= \tilde{G}(x,z)(w_A(x,z) - I_m) - (w_A(x,z) - I_m)G(x,z) \\ &\quad + \sum_{k=0}^{r}\Bigg(\tilde{q}_k(x)\sum_{j=0}^{k-1} z^jX_{k-j-1}(x) - \sum_{j=0}^{k-1} z^jX_{k-j-1}(x)q_k(x)\Bigg) \\ &\quad - \sum_{k=-r}^{-1}\Bigg(\tilde{q}_k(x)\sum_{j=k}^{-1} z^jX_{k-j-1}(x) - \sum_{j=k}^{-1} z^jX_{k-j-1}(x)q_k(x)\Bigg) \\ &= \tilde{G}(x,z)w_A(x,z) - w_A(x,z)G(x,z) + \sum_{i=-r}^{r} z^i h_i(x), \end{aligned} \tag{7.38}$$

where

$$h_i(x) = \tilde{q}_i(x) - q_i(x) + \sum_{k=i+1}^{r} \left(\tilde{q}_k(x) X_{k-i-1}(x) - X_{k-i-1}(x) q_k(x)\right), \quad i \geq 0; \tag{7.39}$$

$$h_i(x) = \tilde{q}_i(x) - q_i(x) - \sum_{k=-r}^{i} \left(\tilde{q}_k(x) X_{k-i-1}(x) - X_{k-i-1}(x) q_k(x)\right), \quad i < 0. \tag{7.40}$$

We shall show that $h_i(x) \equiv 0$ $(-r \leq i \leq r)$. In order to shorten and simplify the formulas, we omit the variable x during these calculations. For $i \geq 0$, according to (7.23) and (7.39), we have

$$\begin{aligned} h_i = & -\sum_{k=i+1}^{r} \left(q_k Y_{k-i-1} - X_{k-i-1} q_i + \sum_{j=i+2}^{k} X_{k-j} q_k Y_{j-i-2} \right) \\ & + \sum_{k=i+1}^{r} (\tilde{q}_k X_{k-i-1} - X_{k-i-1} q_k) \\ = & \sum_{k=i+1}^{r} \left(\left(q_k - \sum_{j=k+1}^{r} \left(q_j Y_{j-k-1} - X_{j-k-1} q_j + \sum_{l=k+2}^{j} X_{j-l} q_j Y_{l-k-2} \right) \right) X_{k-i-1} \right. \\ & \left. - q_k Y_{k-i-1} - \sum_{j=i+2}^{k} X_{k-j} q_k Y_{j-i-2} \right). \end{aligned} \tag{7.41}$$

Notice now that formulas (7.25) and (7.26) lead us to the equality

$$\begin{aligned} Y_{N-R} X_R &= \Pi_2^* A_2^{N-R} S^{-1} \Pi_1 \Pi_2^* S^{-1} A_1^R \Pi_1 \\ &= \Pi_2^* A_2^{N-R} S^{-1} A_1^{R+1} \Pi_1 - \Pi_2^* A_2^{N-R+1} S^{-1} A_1^R \Pi_1 . \end{aligned} \tag{7.42}$$

Taking into account (7.42), we easily obtain

$$\sum_{R=0}^{N} Y_{N-R} X_R = \Pi_2^* S^{-1} A_1^{N+1} \Pi_1 - \Pi_2^* A_2^{N+1} S^{-1} \Pi_1 = X_{N+1} - Y_{N+1} \quad (N \geq 0). \tag{7.43}$$

In view of (7.43), changing the order of the summation below, we derive

$$\sum_{k=i+1}^{r} \sum_{j=k+1}^{r} q_j Y_{j-k+1} X_{k-i-1} = \sum_{j=i+2}^{r} q_j (X_{j-i-1} - Y_{j-i-1}),$$

$$\sum_{k=i+1}^{r} \sum_{j=k+1}^{r} \sum_{l=k+2}^{j} X_{j-l} q_j Y_{l-k-2} X_{k-i-1} = \sum_{j=i+2}^{r} \sum_{l=i+3}^{j} X_{j-l} q_j (X_{l-i-2} - Y_{l-i-2}).$$

Using $X_0 = Y_0$ and two equalities above, we transform (7.41) into the equality

$$h_i = \sum_{k=i+1}^{r} \left(\sum_{j=k+1}^{r} X_{j-k-1} q_j X_{k-i-1} - \sum_{j=i+2}^{k} X_{k-j} q_k Y_{j-i-2} \right) - \sum_{j=i+2}^{r} \sum_{l=i+2}^{j} X_{j-l} q_j (X_{l-i-2} - Y_{l-i-2}). \tag{7.44}$$

The change of the order of the summation on the left-hand side below implies the equality

$$\sum_{k=i+1}^{r} \sum_{j=k+1}^{r} X_{j-k-1} q_j X_{k-i-1} = \sum_{j=i+2}^{r} \sum_{k=i+1}^{j-1} X_{j-k-1} q_j X_{k-i-1} = \sum_{j=i+2}^{r} \sum_{l=i+2}^{j} X_{j-l} q_j X_{l-i-2}. \tag{7.45}$$

Finally, substituting (7.45) into (7.44) and collecting similar terms, we obtain

$$h_i \equiv 0 \quad (i \geq 0). \tag{7.46}$$

In a similar way, we deal with the case $i < 0$. Namely, formulas (7.24) and (7.40) lead us to the equality

$$h_i = - \sum_{k=-r}^{i} \left(\left(q_k + \sum_{j=-r}^{k} \left(q_j Y_{j-k-1} - X_{j-k-1} q_j - \sum_{l=j+1}^{k+1} X_{j-l} q_j Y_{l-k-2} \right) \right) \times X_{k-i-1} - q_k Y_{k-i-1} + \sum_{j=k+1}^{i+1} X_{k-j} q_k Y_{j-i-2} \right) \quad (i < 0). \tag{7.47}$$

We also note that according to (7.25) and (7.42), we have

$$\sum_{R=N+1}^{-1} Y_{N-R} X_R = Y_{N+1} - X_{N+1} \quad (N < -1). \tag{7.48}$$

Using (7.48), after some calculations, we derive

$$\sum_{k=-r}^{i} \sum_{j=-r}^{k} q_j Y_{j-k+1} X_{k-i-1} = \sum_{j=-r}^{i} q_j \sum_{k=j}^{i} Y_{j-k-1} X_{k-i-1} = - \sum_{j=-r}^{i} q_j (X_{j-i-1} - Y_{j-i-1}),$$

$$\sum_{k=-r}^{i} \sum_{j=-r}^{k} \sum_{l=j+1}^{k+1} X_{j-l} q_j Y_{l-k-2} X_{k-i-1} = - \sum_{j=-r}^{i} \sum_{l=j+1}^{i+1} X_{j-l} q_j (X_{l-i-2} - Y_{l-i-2}).$$

Taking into account the equalities above, we rewrite (7.47) in the form

$$h_i = \sum_{k=-r}^{i} \sum_{j=-r}^{k} X_{j-k-1} q_j X_{k-i-1} - \sum_{j=-r}^{i} \sum_{l=j+1}^{i+1} X_{j-l} q_j X_{l-i-2} = 0 \quad (i < 0). \tag{7.49}$$

By virtue of (7.38), (7.46) and (7.49), we have the final expression for the derivative of the transfer matrix function:

$$w_A'(x,z) = \widetilde{G}(x,z) w_A(x,z) - w_A(x,z) G(x,z). \tag{7.50}$$

Formulas (7.14), (7.21) and (7.50) immediately yield (7.22). □

Next, we consider a general case of the first order system depending rationally on the spectral parameter z, that is, the case of multiple poles with respect to z:

$$u' = Gu, \quad G(x,z) = -\left(\sum_{k=0}^{r} z^k q_k(x) + \sum_{s=1}^{l} \sum_{k=1}^{r_s} (z - c_s)^{-k} q_{sk}(x) \right). \tag{7.51}$$

The case of several poles c_s is dealt with quite similar to the "one pole case" before since each pole can be dealt with separately. Like before, we assume that $x \in \mathcal{I}$, where $\mathcal{I}$ is an interval such that $0 \in \mathcal{I}$, and that the coefficients $q_k(x)$ and $q_{sk}(x)$ are $m \times m$ locally summable matrix functions. The GBDT is again determined by an integer $n > 0$, by $n \times n$ matrices A_k $(k = 1, 2)$ and $S(0)$, and by $n \times m$ matrices $\Pi_k(0)$ $(k = 1, 2)$. It is required that these matrices form an S-node, that is, the identity

$$A_1 S(0) - S(0) A_2 = \Pi_1(0) \Pi_2(0)^* \tag{7.52}$$

holds. In the "multiple pole case," matrix functions $\Pi_k(x)$ are introduced via their values at $x = 0$ and equations

$$(\Pi_1)' = \sum_{k=0}^{r} A_1^k \Pi_1 q_k + \sum_{s=1}^{l} \sum_{k=1}^{r_s} (A_1 - c_s I_n)^{-k} \Pi_1 q_{sk}, \tag{7.53}$$

$$(\Pi_2^*)' = -\left(\sum_{k=0}^{r} q_k \Pi_2^* A_2^k + \sum_{s=1}^{l} \sum_{k=1}^{r_s} q_{sk} \Pi_2^* (A_2 - c_s I_n)^{-k} \right). \tag{7.54}$$

Compare (7.51) with (7.54) to see that Π_2^* can be viewed as a generalized eigenfunction of the system $u' = Gu$.

Matrix function $S(x)$ is introduced via S' by the equality

$$S' = \sum_{k=1}^{r} \sum_{j=1}^{k} A_1^{k-j} \Pi_1 q_k \Pi_2^* A_2^{j-1} - \sum_{s=1}^{l} \sum_{k=1}^{r_s} \sum_{j=1}^{k} (A_1 - c_s I_n)^{j-k-1} \times \Pi_1 q_{sk} \Pi_2^* (A_2 - c_s I_n)^{-j}. \tag{7.55}$$

Like in the "one pole case," equations on Π_k and S (i.e. equations (7.53)–(7.55)) yield the identity $(A_1S - SA_2)' = (\Pi_1\Pi_2^*)'$, and so, taking into account (7.52), we have

$$A_1S(x) - S(x)A_2 = \Pi_1(x)\Pi_2(x)^*, \quad x \in \mathcal{I}. \tag{7.56}$$

The matrix function $\widetilde{G}$ has the same structure as G from (7.51). More precisely, we set

$$\widetilde{G}(x,z) = -\left(\sum_{k=0}^{r} z^k\widetilde{q}_k(x) + \sum_{s=1}^{l}\sum_{k=1}^{r_s}(z-c_s)^{-k}\widetilde{q}_{sk}(x)\right) \tag{7.57}$$

where the transformed coefficients $\widetilde{q}_k$ and $\widetilde{q}_{sk}$ are given by the formulas

$$\widetilde{q}_k = q_k - \sum_{j=k+1}^{r}\left(q_jY_{j-k-1} - X_{j-k-1}q_j + \sum_{i=k+2}^{j} X_{j-i}q_jY_{i-k-2}\right), \tag{7.58}$$

$$\widetilde{q}_{sk} = q_{sk} + \sum_{j=k}^{r_s}\left(q_{sj}Y_{s,k-j-1} - X_{s,k-j-1}q_{sj} - \sum_{i=k}^{j} X_{s,i-j-1}q_{sj}Y_{s,k-i-1}\right), \tag{7.59}$$

the matrix functions $X_k(x)$ and $Y_k(x)$ are given by (7.25), and $X_{sk}(x)$ and $Y_{sk}(x)$ have the form

$$X_{sk} = \Pi_2^*S^{-1}(A_1 - c_sI_n)^k\Pi_1, \quad Y_{sk} = \Pi_2^*(A_2 - c_sI_n)^kS^{-1}\Pi_1. \tag{7.60}$$

The following result from [264] (see also [246]) can be proved precisely in the same way as Theorem 7.3.

Theorem 7.4. *Let the first order system (7.51) and five matrices $S(0)$, A_1, A_2 and $\Pi_1(0), \Pi_2(0)$ be given. Assume that the identity (7.52) holds and that $\{c_s\} \cap \sigma(A_k) = \emptyset$ $(k = 1, 2)$. Then, in the points of invertibility of S, the transfer matrix function w_A given by (7.19), where S and Π_k are determined by (7.53)–(7.55), satisfies equation (7.50), where $\widetilde{G}$ is determined by the formulas (7.25) and (7.57)–(7.60).*

The important formula (7.35) has the following "multipole" analog:

$$\left(\Pi_2^*S^{-1}\right)' = -\left(\sum_{k=0}^{r}\widetilde{q}_k\Pi_2^*S^{-1}A_1^k + \sum_{s=1}^{l}\sum_{k=1}^{r_s}\widetilde{q}_{sk}\Pi_2^*S^{-1}(A_1 - c_sI_n)^{-k}\right). \tag{7.61}$$

Formula (7.61) means that by multiplying Π_2^* by S^{-1} from the right, we transform a generalized eigenfunction of system (7.51) into a generalized eigenfunction of the transformed system $\widetilde{u}' = \widetilde{G}\widetilde{u}$. (Compare formula (7.61) with formula (7.54).)

Remark 7.5. Theorem 7.4 is also valid for the cases $l = 0$ and $l = 1$.

Remark 7.6. It is immediate from (7.58) and (7.25), respectively, that

$$\widetilde{q}_r = q_r \quad \text{and} \quad X_0 = Y_0.$$

Remark 7.7. If $\sigma(A_1) \cap \sigma(A_2) = \emptyset$, the matrix function $S(x)$ is uniquely defined by the matrix identity (7.56).

Remark 7.8. In the points of the invertibility of S and for $z \notin (\sigma(A_1) \cup \sigma(A_2))$, the matrix function $w_A(x, z)$ is invertible (see Theorem 1.10) and we have

$$w_A(x,z)^{-1} = I_m + \Pi_2(x)^*(A_2 - zI_n)^{-1}S(x)^{-1}\Pi_1(x). \tag{7.62}$$

7.3 Explicit solutions of nonlinear equations

We apply Theorem 7.4 in order to construct solutions of nonlinear integrable equations and corresponding wave functions similar to the way in which it was done in Subsection 5.3.4 and Section 7.1. For this purpose, we use auxiliary linear systems for the integrable nonlinear equation

$$\frac{\partial}{\partial x}w = Gw, \quad \frac{\partial}{\partial t}w = Fw; \tag{7.63}$$

$$G(x,t,z) = -\sum_{k=0}^{r} z^k q_k(x,t) - \sum_{s=1}^{l}\sum_{k=1}^{r_s}(z-c_s)^{-k}q_{sk}(x,t), \tag{7.64}$$

$$F(x,t,z) = -\sum_{k=0}^{R} z^k Q_k(x,t) - \sum_{s=1}^{L}\sum_{k=1}^{R_s}(z-C_s)^{-k}Q_{sk}(x,t), \tag{7.65}$$

and the zero curvature (compatibility condition) representation (6.1) of the integrable nonlinear equation itself. We consider nonlinear equations in the domain $(x,t) \in \mathcal{I}_1 \times \mathcal{I}_2$ and assume $(0,0) \in \mathcal{I}_1 \times \mathcal{I}_2$. Recall that, according to Theorem 6.1 and Remark 6.5, a fundamental solution of (7.63) exists if only G, F and $\frac{\partial}{\partial t}G$ are continuous, $\frac{\partial}{\partial x}F$ exists and (6.1) holds. We assume that w is normalized by the condition

$$w(0,0) = I_m. \tag{7.66}$$

For simplicity, we require that G and F are continuously differentiable in this section.

When we deal with two auxiliary linear systems, the $n \times n$ matrix function S and the $n \times m$ matrix functions Π_k depend on two variables x and t, and the matrix identity (7.52) for parameter matrices A_k, $\Pi_k(0)$ and $S(0)$ is substituted by the identity

$$A_1S(0,0) - S(0,0)A_2 = \Pi_1(0,0)\Pi_2(0,0)^* \tag{7.67}$$

for parameter matrices A_k, $\Pi_k(0,0)$ and $S(0,0)$. Equations (7.53)–(7.55) should be completed by the similar equations with respect to derivatives in t:

$$\frac{\partial}{\partial t}\Pi_1 = \sum_{k=0}^{R} A_1^k \Pi_1 Q_k + \sum_{s=1}^{L}\sum_{k=1}^{R_s}(A_1 - C_s I_n)^{-k}\Pi_1 Q_{sk}, \tag{7.68}$$

$$\frac{\partial}{\partial t}\Pi_2^* = -\left(\sum_{k=0}^{R} Q_k \Pi_2^* A_2^k + \sum_{s=1}^{L}\sum_{k=1}^{R_s} Q_{sk}\Pi_2^*(A_2 - C_s I_n)^{-k}\right), \tag{7.69}$$

$$\frac{\partial}{\partial t}S = \sum_{k=1}^{R}\sum_{j=1}^{k} A_1^{k-j}\Pi_1 Q_k \Pi_2^* A_2^{j-1} - \sum_{s=1}^{L}\sum_{k=1}^{R_s}\sum_{j=1}^{k}(A_1 - C_s I_n)^{j-k-1} \times \Pi_1 Q_{sk}\Pi_2^*(A_2 - C_s I_n)^{-j}. \tag{7.70}$$

Clearly, we require

$$\{c_s\} \cap \sigma(A_k) = \emptyset, \quad \{C_s\} \cap \sigma(A_k) = \emptyset \quad (k = 1, 2). \tag{7.71}$$

Thus, we can apply the transformation of G into $\widetilde{G}$ and a similar transformation of F into $\widetilde{F}$.

Notation 7.9. *The notation $\widetilde{\mathfrak{W}}(G,x)$ stands for the mapping of G into $\widetilde{G}$ via formulas (7.25) and (7.57)–(7.60).*

Remark 7.10. Since $\widetilde{G}$ in this section depends on x and t (and matrices S and Π_k, which determine $\widetilde{G}$, depend on x and t), the variable x in $\widetilde{\mathfrak{W}}(G,x)$ only shows that the dependence of S and Π_k on x is determined (in (7.53)–(7.55)) by G. On the other hand, the equality $\widetilde{F} = \widetilde{\mathfrak{W}}(F,t)$ shows that the dependence of S and Π_k on t is determined (in (7.68)–(7.70)) by F.

Theorem 7.4 provides expressions for derivatives $\frac{\partial}{\partial x}w_A$ and $\frac{\partial}{\partial t}w_A$, where

$$w_A(x,t,z) = I_m - \Pi_2(x,t)^* S(x,t)^{-1}(A_1 - zI_n)^{-1}\Pi_1(x,t), \tag{7.72}$$

in terms of $\widetilde{G}$ and $\widetilde{F}$, respectively. Hence, equations

$$\frac{\partial}{\partial x}\widetilde{w} = \widetilde{G}\widetilde{w}, \quad \frac{\partial}{\partial t}\widetilde{w} = \widetilde{F}\widetilde{w}, \quad \widetilde{w}(x,t,z) := w_A(x,t,z)w(x,t,z) \tag{7.73}$$

hold. Finally, in a way which is similar to the proof of (7.56), we can show that

$$A_1 S(x,t) - S(x,t)A_2 = \Pi_1(x,t)\Pi_2(x,t)^*, \quad (x,t) \in \mathcal{I}_1 \times \mathcal{I}_2. \tag{7.74}$$

It follows from (7.73) that

$$\frac{\partial^2}{\partial t \partial x}\widetilde{w} = \left(\frac{\partial}{\partial t}\widetilde{G} + \widetilde{G}\widetilde{F}\right)\widetilde{w}, \quad \frac{\partial^2}{\partial x \partial t}\widetilde{w} = \left(\frac{\partial}{\partial x}\widetilde{F} + \widetilde{F}\widetilde{G}\right)\widetilde{w}. \tag{7.75}$$

If G and F are continuously differentiable, then $\widetilde{G}$ and $\widetilde{F}$ are continuously differentiable too. Therefore, we have $\frac{\partial^2}{\partial x \partial t}\widetilde{w} = \frac{\partial^2}{\partial t \partial x}\widetilde{w}$, and so formula (7.75) implies that

$$\left(\frac{\partial}{\partial t}\widetilde{G} - \frac{\partial}{\partial x}\widetilde{F} + [\widetilde{G}, \widetilde{F}]\right)\widetilde{w} = 0. \tag{7.76}$$

According to Remark 7.8, the matrix function $w_A(x,t,z)$ is invertible, and, according to (7.66), w is invertible. Thus, $\widetilde{w}$ is also invertible. Now, it is immediate from (7.76) that

$$\frac{\partial}{\partial t}\widetilde{G} - \frac{\partial}{\partial x}\widetilde{F} + [\widetilde{G}, \widetilde{F}] = 0. \tag{7.77}$$

We proved the following general theorem.

Theorem 7.11. *Let G and F of the form* (7.64) *and* (7.65)*, respectively, be continuously differentiable and satisfy zero curvature equation* (6.1)*. Assume that relations* (7.67)) *and* (7.71) *hold, and let the matrix functions Π_k and S satisfy equations* (7.53)–(7.55) *and* (7.68)–(7.70)*. Then, in the points of the invertibility of S, the zero curvature equation* (7.77) *holds. Here, $\widetilde{G} = \widetilde{\mathfrak{M}}(G,x)$, $\widetilde{F} = \widetilde{\mathfrak{M}}(F,t)$ (see Notation 7.9 explaining $\widetilde{\mathfrak{M}}$).*

Example 7.12. The main chiral field equation for an $m \times m$ invertible matrix function $\psi(x,t)$ has the form

$$2\frac{\partial^2}{\partial t \partial x}\psi = \left(\frac{\partial}{\partial x}\psi\right)\psi^{-1}\left(\frac{\partial}{\partial t}\psi\right) + \left(\frac{\partial}{\partial t}\psi\right)\psi^{-1}\left(\frac{\partial}{\partial x}\psi\right), \tag{7.78}$$

and is equivalent [222, 336] to the compatibility condition (6.1) of the auxiliary systems (7.63) where

$$G(x,t,z) = -(z-1)^{-1}q_{11}(x,t), \quad q_{11} = \left(\frac{\partial}{\partial x}\psi\right)\psi^{-1}; \tag{7.79}$$

$$F(x,t,z) = -(z+1)^{-1}Q_{11}(x,t), \quad Q_{11} = -\left(\frac{\partial}{\partial t}\psi\right)\psi^{-1}. \tag{7.80}$$

Let ψ satisfy (7.78). Then, for the case of G and F given by (7.79) and (7.80), respectively, (6.1) holds, equations (7.53)–(7.55) take the form

$$\frac{\partial}{\partial x}\Pi_1 = (A_1 - I_n)^{-1}\Pi_1\left(\frac{\partial}{\partial x}\psi\right)\psi^{-1}, \quad \frac{\partial}{\partial x}\Pi_2^* = -\left(\frac{\partial}{\partial x}\psi\right)\psi^{-1}\Pi_2^*(A_2 - I_n)^{-1}, \tag{7.81}$$

$$\frac{\partial}{\partial x}S = -(A_1 - I_n)^{-1}\Pi_1\left(\frac{\partial}{\partial x}\psi\right)\psi^{-1}\Pi_2^*(A_2 - I_n)^{-1}, \tag{7.82}$$

and equations (7.68)–(7.70) take the form

$$\frac{\partial}{\partial t}\Pi_1 = -(A_1 + I_n)^{-1}\Pi_1\left(\frac{\partial}{\partial t}\psi\right)\psi^{-1}, \quad \frac{\partial}{\partial t}\Pi_2^* = \left(\frac{\partial}{\partial t}\psi\right)\psi^{-1}\Pi_2^*(A_2 + I_n)^{-1}, \tag{7.83}$$

$$\frac{\partial}{\partial t}S = (A_1 + I_n)^{-1}\Pi_1\left(\frac{\partial}{\partial t}\psi\right)\psi^{-1}\Pi_2^*(A_2 + I_n)^{-1}. \tag{7.84}$$

Now, let matrices A_k, $S(0,0)$ and $\Pi_k(0,0)$ be fixed. Assume that (7.67) holds and that $\pm 1 \notin \sigma(A_k)$ $(k = 1,2)$. Let matrix functions S and Π_k satisfy (7.81)–(7.84). Taking into account (7.57) and the first equalities in (7.79) and (7.80), we have

$$\tilde{G}(x,t,z) = -(z-1)^{-1}\tilde{q}_{11}(x,t), \quad \tilde{F}(x,t,z) = -(z+1)^{-1}\tilde{Q}_{11}(x,t). \tag{7.85}$$

From Theorem 7.4, we obtain

$$\frac{\partial}{\partial x} w_A = \tilde{G} w_A - w_A G, \quad \frac{\partial}{\partial t} w_A = \tilde{F} w_A - w_A F. \tag{7.86}$$

Assume additionally that $\det A_k \neq 0$ $(k = 1,2)$. Then, according to Remark 7.8, the matrix function $w_A(x,t,0)^{-1}$ is well-defined in the points of invertibility of S. It follows from (7.85) and (7.86) that

$$\frac{\partial}{\partial x} w_A(x,t,0) = \tilde{q}_{11}(x,t) w_A(x,t,0) - w_A(x,t,0) q_{11}(x,t), \tag{7.87}$$

$$\frac{\partial}{\partial t} w_A(x,t,0) = -\tilde{Q}_{11}(x,t) w_A(x,t,0) + w_A(x,t,0) Q_{11}(x,t). \tag{7.88}$$

Rewrite the second relations in (7.79) and (7.80):

$$\frac{\partial}{\partial x}\psi = q_{11}\psi, \quad \frac{\partial}{\partial t}\psi = -Q_{11}\psi. \tag{7.89}$$

Putting

$$\tilde{\psi}(x,t) := w_A(x,t,0)\psi(x,t), \tag{7.90}$$

from formulas (7.87)–(7.90), we derive

$$\frac{\partial}{\partial x}\tilde{\psi} = \tilde{q}_{11}\tilde{\psi}, \quad \frac{\partial}{\partial t}\tilde{\psi} = -\tilde{Q}_{11}\tilde{\psi}.$$

Since $w_A(x,t,0)$ and $\psi(x,t)$ are invertible, the matrix function $\tilde{\psi}$ is also invertible, and we obtain

$$\tilde{q}_{11} = \left(\frac{\partial}{\partial x}\tilde{\psi}\right)\tilde{\psi}^{-1}, \quad \tilde{Q}_{11} = -\left(\frac{\partial}{\partial t}\tilde{\psi}\right)\tilde{\psi}^{-1}. \tag{7.91}$$

Recall that if formulas (7.79) and (7.80) hold, then the compatibility condition (6.1) is equivalent to (7.78) (and so (6.1) is valid). Moreover, according to Theorem 7.11, (6.1) yields (7.77). The only difference between equalities in (6.1), (7.79), (7.80) and equalities in (7.77), (7.85), (7.91) is the "tilde" in the notations. Hence (like in the case without the "tilde"), formulas (7.77), (7.85) and (7.91) imply (7.78), where ψ is substituted by $\tilde{\psi}$. In other words, $\tilde{\psi}(x,t)$ satisfies the main chiral field equation:

$$2\frac{\partial^2}{\partial t \partial x}\tilde{\psi} = \left(\frac{\partial}{\partial x}\tilde{\psi}\right)\tilde{\psi}^{-1}\left(\frac{\partial}{\partial t}\tilde{\psi}\right) + \left(\frac{\partial}{\partial t}\tilde{\psi}\right)\tilde{\psi}^{-1}\left(\frac{\partial}{\partial x}\tilde{\psi}\right).$$

Corollary 7.13. *Assume that the parameter matrices satisfy identity* (7.67) *and that*

$$\{0, 1, -1\} \cap \sigma(A_k) = \emptyset \quad (k = 1, 2).$$

Let an invertible matrix function ψ satisfy the main chiral field equation (7.78) *and be two times continuously differentiable. Then, the matrix function $\widetilde{\psi}$ given by* (7.90) *in the points of invertibility of S also satisfies the main chiral field equation.*

Our next examples deal with the construction of new (local) solutions of integrable elliptic sine-Gordon and sinh-Gordon equations from the initial solutions. See [55, 147] for auxiliary systems (see also the references therein for some related literature). We note that the matrices j and J in this section have the form

$$j = \begin{bmatrix} 1 & 0 \\ 0 & -1 \end{bmatrix}, \quad J = \begin{bmatrix} 0 & 1 \\ 1 & 0 \end{bmatrix}. \tag{7.92}$$

Example 7.14. Elliptic sine-Gordon equation

$$\frac{\partial^2 \psi}{\partial t^2} + \frac{\partial^2 \psi}{\partial x^2} = \sin \psi \quad (\psi = \overline{\psi}) \tag{7.93}$$

is equivalent to the compatibility condition (6.1) of the auxiliary systems (7.63) where

$$G = \frac{1}{4}\left(\mathrm{i}z\zeta + \left(\frac{\partial}{\partial t}\psi\right)j - \frac{\mathrm{i}}{z}J\zeta J\right), \quad \zeta = \begin{bmatrix} 0 & \mathrm{e}^{-\mathrm{i}\psi/2} \\ \mathrm{e}^{\mathrm{i}\psi/2} & 0 \end{bmatrix}, \tag{7.94}$$

$$F = -\frac{1}{4}\left(z\zeta + \left(\frac{\partial}{\partial x}\psi\right)j + \frac{1}{z}J\zeta J\right), \tag{7.95}$$

and matrices j and J are defined in (7.92). We also put

$$A_1 = A, \quad A_2 = -(A^*)^{-1}, \quad \Pi_1 \equiv \Pi, \quad \Pi_2(0,0) = A^{-1}\Pi(0,0)J. \tag{7.96}$$

Thus, we have three parameter matrices, namely, $n \times n$ matrices A and $S(0,0)$ and an $n \times m$ matrix $\Pi(0,0)$. We assume that ψ satisfies (7.93), that $\det A \neq 0$ and $S(0,0) = S(0,0)^*$, and that there is a matrix $\mathcal{V}$ such that equalities

$$\overline{A} = \mathcal{V}A^{-1}\mathcal{V}^{-1}, \quad \overline{\Pi}(0,0) = \mathcal{V}\Pi(0,0), \quad \overline{S}(0,0) = \mathcal{V}AS(0,0)A^*\mathcal{V}^* \tag{7.97}$$

hold. Here, $\overline{A}$ is the matrix with the entries which are complex conjugate to the corresponding entries of A. In view of (7.96), the identity (7.67), which should be satisfied by the parameter matrices, takes the form

$$AS(0,0)A^* + S(0,0) = \Pi(0,0)J\Pi(0,0)^*. \tag{7.98}$$

Compare (7.64) and (7.65) with (7.94) and (7.95), respectively, to see that

$$r = 1, \; q_1 = -(\mathrm{i}/4)\zeta, \; q_0 = -\frac{\partial}{\partial t}(\psi/4)j; \; l = r_1 = 1, \; c_1 = 0, \; q_{11} = (\mathrm{i}/4)J\zeta J;$$

$$R = 1,\ Q_1 = (1/4)\zeta,\ Q_0 = \frac{\partial}{\partial x}(\psi/4)j;\ L = R_1 = 1,\ C_1 = 0,\ Q_{11} = (1/4)J\zeta J.$$

Thus, taking into account (7.96), we rewrite equations (7.53) and (7.68) in the form

$$\frac{\partial}{\partial x}\Pi = \frac{1}{4}\left(-\mathrm{i}A\Pi\zeta - \left(\frac{\partial}{\partial t}\psi\right)\Pi j + \mathrm{i}A^{-1}\Pi J\zeta J\right), \tag{7.99}$$

$$\frac{\partial}{\partial t}\Pi = \frac{1}{4}\left(A\Pi\zeta + \left(\frac{\partial}{\partial x}\psi\right)\Pi j + A^{-1}\Pi J\zeta J\right). \tag{7.100}$$

Since $\overline{\zeta} = J\zeta J$ and $\overline{A} = \mathcal{V}A^{-1}\mathcal{V}^{-1}$, we see that both Π and $\mathcal{V}\overline{\Pi}$ satisfy relations (7.99) and (7.100). From the second equality in (7.97), we also have $\Pi(0,0) = \mathcal{V}\overline{\Pi}(0,0))$. Hence, we derive

$$\Pi(x,t) \equiv \mathcal{V}\overline{\Pi}(x,t). \tag{7.101}$$

Equations (7.54) and (7.69), which define Π_2^*, take the form

$$\frac{\partial}{\partial x}\Pi_2^* = \frac{1}{4}\left(\mathrm{i}\zeta\Pi_2^* A_2 + \left(\frac{\partial}{\partial t}\psi\right) j\Pi_2^* - \mathrm{i}J\zeta J\Pi_2^* A_2^{-1}\right), \tag{7.102}$$

$$\frac{\partial}{\partial t}\Pi_2^* = \frac{1}{4}\left(-\zeta\Pi_2^* A_2 - \left(\frac{\partial}{\partial x}\psi\right) j\Pi_2^* - J\zeta J\Pi_2^* A_2^{-1}\right). \tag{7.103}$$

Since $\zeta = \zeta^*$ and $A_2 = -(A^*)^{-1}$, it follows from (7.99) and (7.100) that the matrix function $A^{-1}\Pi(x,t)J$ satisfies equations (7.102) and (7.103) for Π_2. Moreover, we have $\Pi_2(0,0) = A^{-1}\Pi(0,0)J$. In other words, we have

$$\Pi_2(x,t) \equiv A^{-1}\Pi(x,t)J. \tag{7.104}$$

Therefore, identity (7.74) takes the form

$$AS(x,t)A^* + S(x,t) = \Pi(x,t)J\Pi(x,t)^*. \tag{7.105}$$

Finally, using the second equality in (7.96) (i.e. $A_2 = -(A^*)^{-1}$) and (7.104), we rewrite (7.55) and (7.70) in the form

$$\begin{aligned}\frac{\partial S}{\partial x} &= \frac{\mathrm{i}}{4}\left(A^{-1}\Pi J\zeta\Pi^* - \Pi\zeta J\Pi^*(A^*)^{-1}\right),\\ \frac{\partial S}{\partial t} &= \frac{1}{4}\left(A^{-1}\Pi J\zeta\Pi^* + \Pi\zeta J\Pi^*(A^*)^{-1}\right).\end{aligned} \tag{7.106}$$

Formulas (7.97), (7.101) and (7.106) imply that

$$\frac{\partial}{\partial x}\overline{S} = \mathcal{V}A\left(\frac{\partial}{\partial x}S\right)A^*\mathcal{V}^*, \quad \frac{\partial}{\partial t}\overline{S} = \mathcal{V}A\left(\frac{\partial}{\partial t}S\right)A^*\mathcal{V}^*,$$

and, taking into account equality $\overline{S}(0,0) = \mathcal{V}AS(0,0)A^*\mathcal{V}^*$ from (7.97), we obtain

$$\overline{S} \equiv \mathcal{V}ASA^*\mathcal{V}^*. \tag{7.107}$$

We note that (in view of the equality $c_1 = C_1 = 0$) the matrix functions X_{11} and Y_{11} do not depend on whether we construct the transformation of G into $\widetilde{G}$ or F into $\widetilde{F}$ and coincide with X_{-1} and Y_{-1} from (7.25), respectively. According to (7.25), (7.101), (7.104) and (7.107), we have

$$X_{-1} = J\Pi^*(A^*)^{-1}S^{-1}A^{-1}\Pi = J\Pi^*\mathcal{V}^*(\mathcal{V}ASA^*\mathcal{V}^*)^{-1}\mathcal{V}\Pi = J\overline{(\Pi^*S^{-1}\Pi)}. \quad (7.108)$$

Formulas (7.25), (7.62) and (7.72) lead us to

$$w_A(x,t,0) = I_2 - X_{-1}, \quad w_A(x,t,0)^{-1} = I_2 + Y_{-1}. \quad (7.109)$$

Moreover, using (7.96), (7.104) and equality $\Pi^*S^{-1}\Pi = (\Pi^*S^{-1}\Pi)^*$, we derive

$$Y_{-1} = -J\Pi^*S^{-1}\Pi, \quad -\Pi^*S^{-1}\Pi = \begin{bmatrix} a & b \\ \overline{b} & d \end{bmatrix}, \quad a = \overline{a}, \quad d = \overline{d}. \quad (7.110)$$

From (7.108)–(7.110), we see that

$$\begin{bmatrix} 1+b & d \\ a & 1+\overline{b} \end{bmatrix}\begin{bmatrix} 1+\overline{b} & d \\ a & 1+b \end{bmatrix} = (I_2 - X_{-1})(I_2 + Y_{-1}) = I_2. \quad (7.111)$$

If $1 + b \neq 0$, formula (7.111) implies that

$$a = d = 0, \quad |1+b| = 1, \quad I_2 - X_{-1} = \operatorname{diag}\{1+b,\ 1+\overline{b}\}. \quad (7.112)$$

Put

$$\widehat{w} = (I_2 - X_{-1})^{-\frac{1}{2}}\widetilde{w}, \quad \widehat{G} = \left(\frac{\partial}{\partial x}\widehat{w}\right)\widehat{w}^{-1}, \quad \widehat{F} = \left(\frac{\partial}{\partial t}\widehat{w}\right)\widehat{w}^{-1}, \quad (7.113)$$

where $\widetilde{w}(x,t,z) = w_A(x,t,z)w(x,t,z)$. In a way, which is similar to the proof of (7.77) in Theorem 7.11, we derive from (7.113) that

$$\frac{\partial}{\partial t}\widehat{G} - \frac{\partial}{\partial x}\widehat{F} + [\widehat{G},\widehat{F}] = 0. \quad (7.114)$$

Moreover, it is easy to see that $\widehat{G}$ and $\widehat{F}$ have the form (7.94) and (7.95), respectively, after the substitution of

$$\widehat{\psi} = \psi + 2\arg(1+b) \quad (7.115)$$

instead of ψ into the right-hand sides of (7.94) and (7.95). Therefore, formula (7.114) implies that $\widehat{\psi}$ satisfies (7.93).

Corollary 7.15. *Let an integer $n > 0$ and matrices A $(\det A \neq 0)$, $\Pi(0,0)$ and $S(0,0) = S(0,0)^*$ be fixed and satisfy conditions (7.97) and (7.98). Let ψ satisfy elliptic sine-Gordon equation (7.93) and be two times continuously differentiable. Then, in the points where $\det S \neq 0$ and $1 + b \neq 0$, the function $\widehat{\psi}$ given by (7.115) satisfies the elliptic sine-Gordon equation too.*

If $\sigma(A) \cap \sigma((-A^*)^{-1}) = \emptyset$, then the last equality in (7.97) follows from the first two equalities in (7.97) and formula (7.98).

Remark 7.16. Considerations from the proof of (7.109) show that the equality

$$(I_m - X_{-1})(I_m + Y_{-1}) = I_m,$$

where X_{-1} and Y_{-1} are given in (7.25), is valid in the general case.

Our next example is dealt with in the same way as Example 7.14.

Example 7.17. Elliptic sinh-Gordon equation

$$\frac{\partial^2 \psi}{\partial t^2} + \frac{\partial^2 \psi}{\partial x^2} = \sinh \psi \quad (\psi = \overline{\psi}) \tag{7.116}$$

is equivalent to the compatibility condition (6.1) of the auxiliary systems (7.63) where

$$G = -\frac{1}{4}\left(z\zeta - \mathrm{i}\frac{\partial \psi}{\partial t} j + \frac{1}{z}\zeta^*\right), \quad \zeta = \begin{bmatrix} 0 & \mathrm{e}^{-\psi/2} \\ \mathrm{e}^{\psi/2} & 0 \end{bmatrix}, \tag{7.117}$$

$$F = \frac{\mathrm{i}}{4}\left(z\zeta - \frac{\partial \psi}{\partial x} j - \frac{1}{z}\zeta^*\right). \tag{7.118}$$

We put

$$A_1 = A, \quad A_2 = -(A^*)^{-1}, \quad \Pi_1 \equiv \Pi, \quad \Pi_2(0,0) = A^{-1}\Pi(0,0), \tag{7.119}$$

and assume that

$$\det A \neq 0, \quad S(0,0) = S(0,0)^*, \tag{7.120}$$

and that there is a matrix $\mathcal{V}$ such that

$$\overline{A} = \mathcal{V}A^{-1}\mathcal{V}^{-1}, \quad \overline{\Pi}(0,0) = \mathcal{V}\Pi(0,0)J, \quad \overline{S}(0,0) = \mathcal{V}AS(0,0)A^*\mathcal{V}^*. \tag{7.121}$$

Now, the identity (7.67) takes the form

$$AS(0,0)A^* + S(0,0) = \Pi(0,0)\Pi(0,0)^*. \tag{7.122}$$

Taking into account (7.117) and (7.118), we introduce Π by the equations

$$\frac{\partial}{\partial x}\Pi = \frac{1}{4}\left(A\Pi\zeta - \mathrm{i}\left(\frac{\partial}{\partial t}\psi\right)\Pi j + A^{-1}\Pi\zeta^*\right), \tag{7.123}$$

$$\frac{\partial}{\partial t}\Pi = -\frac{\mathrm{i}}{4}\left(A\Pi\zeta - \left(\frac{\partial}{\partial x}\psi\right)\Pi j - A^{-1}\Pi\zeta^*\right). \tag{7.124}$$

The identity $\Pi_2 \equiv A^{-1}\Pi$ is an analog of formula (7.104) in the previous example, which is obtained in a similar way. Hence, formulas (7.55) and (7.70) take the form

$$\begin{aligned} \frac{\partial}{\partial x}S &= \frac{1}{4}\left(\Pi\zeta\Pi^*(A^*)^{-1} + A^{-1}\Pi\zeta^*\Pi^*\right), \\ \frac{\partial}{\partial t}S &= \frac{\mathrm{i}}{4}\left(A^{-1}\Pi\zeta^*\Pi^* - \Pi\zeta\Pi^*(A^*)^{-1}\right). \end{aligned} \tag{7.125}$$

It is easy to show that $\overline{\Pi} \equiv \mathcal{V}\Pi J$ and $\overline{S} \equiv \mathcal{V}ASA^*\mathcal{V}^*$. Hence, we have

$$I_2 - X_{-1} = I_2 - \Pi^*(A^*)^{-1}S^{-1}A^{-1}\Pi = I_2 - J\overline{\Pi^* S^{-1}\Pi}J. \tag{7.126}$$

Once the entry $(I_2 - X_{-1})_{22}$ is nonzero, the matrix $I_2 - X_{-1}$ is again a diagonal matrix. Moreover, it admits representation

$$I_2 - X_{-1} = I_2 - J\overline{\Pi^* S^{-1}\Pi}J = \operatorname{diag}\{a, a^{-1}\}, \quad a = \overline{a}. \tag{7.127}$$

The proof of formula (7.127) is similar to the proof of (7.112) and the following corollary is proved quite like Corollary 7.15 before.

Corollary 7.18. *Let an integer $n > 0$ and matrices A, $\Pi(0,0)$ and $S(0,0)$ be fixed and satisfy conditions (7.120)–(7.122). Let ψ satisfy the elliptic sinh-Gordon equation and be two times continuously differentiable. Then, in the points where $\det S \neq 0$ and $(I_2 - \Pi^* S^{-1}\Pi)_{11} \neq 0$, the function*

$$\hat{\psi} = \psi + 2\ln|a| \tag{7.128}$$

satisfies the elliptic sinh-Gordon equation too. Here, $a \in \mathbb{R}$ is given by (7.127) and (7.123)–(7.125).

8 Some further results on inverse problems and generalized Bäcklund-Darboux transformation (GBDT)

For the convenience of readers, we now offer several results closely related to some of those described in Chapters 1–7 and obtained by the same methods. Therefore, in this chapter, we don't present complete proofs, but restrict ourselves to precise references (and, sometimes, a scheme of the proof).

8.1 Inverse problems and the evolution of the Weyl functions

1. In several chapters, we used results from [273] (see Proposition 1.29 on the linear similarity of Volterra operators to operator of integration) in order to solve inverse problems. It is also possible to apply methods for solving inverse problems in order to obtain linear similarity results. In particular, let us consider operator

$$K = \mathrm{i}\tilde{\gamma}(x)\int_0^x \tilde{\gamma}(t)^* D\,\cdot\, dt \in B\left(L^2_m(0,l)\right), \tag{8.1}$$

where

$$D = \operatorname{diag}\{d_1, d_2, \dots, d_m\} > 0, \quad d_i \neq d_k \quad (i \neq k). \tag{8.2}$$

Section 15 in [243] is dedicated to the following statement.

Theorem 8.1. *Let an operator K have the form* (8.1)*, where the $m \times m$ matrix function $\tilde{\gamma}$ satisfies conditions*

$$\tilde{\gamma} \in B^1([0,l]), \quad \tilde{\gamma}\tilde{\gamma}^* \equiv I_m, \quad (\tilde{\gamma}'\tilde{\gamma}^*)_{ii} \equiv 0 \quad (1 \leq i \leq m). \tag{8.3}$$

Then, there is a boundedly invertible operator $E \in B(L^2_m(0,l))$ such that

$$K = EAE^{-1}, \quad A = \mathrm{i}D\int_0^x \cdot\, dt, \tag{8.4}$$

that is, K is similar to A. The operators $E^{\pm 1}$ here are lower triangular.

The scheme of the proof. It suffices to prove the theorem above for D satisfying (4.2). For that purpose, following (4.81), we introduce the potential

$$\zeta(x) = -\tilde{\gamma}'(x)\tilde{\gamma}(x)^* \quad (0 \leq x \leq l), \quad \zeta(x) \equiv 0 \quad (x > l). \tag{8.5}$$

For system (4.1) with such a potential, we consider its GW-function $\varphi(z)$ and show that, for some $M > 0$, this GW-function satisfies conditions

$$\varphi(\xi - \mathrm{i}\eta)^{-1} - I_m \in L^2_{m\times m}(-\infty,\infty) \quad (\eta > M), \quad \sup \|\varphi(z)^{-1}\| < \infty \quad (z \in \mathbb{C}^-_M).$$

Using the procedure from Chapter 4, we construct operator S, factorize it (like in (4.68)) and, finally, prove formula (4.72) for $y = D^{1/2}\tilde{y}$. Now, (8.4) is immediate.

2. The same methods as in Chapters 1–4 provide Weyl theory and a solution of the inverse problem for a system depending rationally on the spectral parameter [210, 265], namely, the system

$$y'(x,\lambda) = \mathrm{i}\sum_{k=1}^{m} b_k(\lambda - d_k)^{-1} h_k(x)^* h_k(x) y(x,\lambda), \quad b_k = \pm 1 \quad x \in [0,\infty), \tag{8.6}$$

where $h_k(x) = \begin{bmatrix} h_{k,1}(x) & h_{k,2}(x) \end{bmatrix}$ are $\mathbb{C}^2$-valued differentiable vector functions such that

$$\sup_{0<x<\infty} \|h_k'(x)\| < \infty, \quad h_k(x)h_k(x)^* \equiv 1, \quad 1 \le k \le m. \tag{8.7}$$

We always assume that

$$d_i = \overline{d_i} \neq d_k \quad \text{for } i \neq k, \quad 1 \le i,k \le m. \tag{8.8}$$

The normalized fundamental solution W of (8.6) is introduced in a standard way, that is,

$$W'(x,\lambda) = \mathrm{i}\sum_{k=1}^{m} b_k(\lambda - d_k)^{-1} h_k(x)^* h_k(x) W(x,\lambda), \quad W(0,\lambda) = I_2. \tag{8.9}$$

In order to consider the behavior of W in the neighborhood of $\lambda = d_k$, we introduce an additional variable z:

$$z = \frac{b_k}{2(\lambda - d_k)} \quad \text{i.e.} \quad \lambda = d_k + \frac{b_k}{2z}. \tag{8.10}$$

Definition 8.2. A function $\varphi_k(z)$ is called a GW_k-function of the system (8.6) with the properties (8.7) and (8.8) if and only if there exists an $M_k > 0$ such that φ_k is holomorphic on the half-plane $\Im z < -M/4$ and for all $x \in [0,\infty)$ we have

$$\sup_{\Im z < -M/4} \left\| W(x,\lambda(z)) \begin{bmatrix} \varphi_k(z) \\ 1 \end{bmatrix} \right\| < \infty. \tag{8.11}$$

According to Theorem 2.2 [210] and Theorem 2.1 [265], if only

$$h_{k,1}(0) \neq 0 \quad \text{for } 1 \le k \le m, \tag{8.12}$$

there exist unique GW_k-functions φ_k. Moreover, one can choose a sufficiently large value $M > 0$ so that φ_k are well-defined and bounded in the half-plane $\Im z < -M/4$. Then, the vector function Φ_1 given by

$$\Phi_1(x) = \frac{\mathrm{i}}{2\pi} \mathrm{e}^{-\eta x} \mathrm{l.i.m.}_{a\to\infty} \int_{-a}^{a} (\xi - \mathrm{i}\eta)^{-1} \mathrm{e}^{\mathrm{i}\xi x} \varphi\left(\frac{\xi - \mathrm{i}\eta}{2}\right) d\xi, \quad \eta > M/2, \tag{8.13}$$

$$\varphi(z) := \mathrm{col}\begin{bmatrix} \varphi_1(z) & \varphi_2(z) & \dots & \varphi_m(z) \end{bmatrix}, \quad \Phi_1 = \mathrm{col}\begin{bmatrix} \Phi_{1,1} & \Phi_{2,1} & \dots & \Phi_{m,1} \end{bmatrix} \tag{8.14}$$

is also well-defined. It is shown in the proof of the main Theorem 4.2 [265] that

$$\Phi_{k,1} \in B^1([0,\, l]) \quad \text{for any} \quad l > 0. \tag{8.15}$$

We note that the vector function φ of the form (8.14), where φ_k are the GW_k-functions, is called the GW-function.

Next, we introduce operators $\Pi = \Pi_l$ of multiplication by the $m \times 2$ matrix function $[\Phi_1(x) \quad \Phi_2(x)]$:

$$\Pi_l \begin{bmatrix} g_1 \\ g_2 \end{bmatrix} = g_1\Phi_1(x) + g_2\Phi_2(x), \quad \Pi_l \in B\left(\mathbb{C}^2,\, L^2_m(0,l)\right), \tag{8.16}$$

$$\Phi_2 = \operatorname{col}\left[\Phi_{1,2} \quad \Phi_{2,2} \quad \dots \quad \Phi_{m,2}\right], \quad \Phi_{k,2}(x) \equiv 1 \quad (1 \le k \le m). \tag{8.17}$$

Now, operators $S = S_l \in B(L^2_m(0,l))$ are uniquely defined by the operator identities

$$AS - SA^* = \mathrm{i}\Pi\Pi^*; \quad Af := Df + \mathrm{i}B\int_0^x f(t)dt, \tag{8.18}$$

$$D := \operatorname{diag}\{d_1, d_2, \dots, d_m\}, \quad B := \operatorname{diag}\{b_1, b_2, \dots, b_m\}, \quad A \in B\left(L^2_m(0,l)\right). \tag{8.19}$$

See Propositions 3.1 and 3.4 [265], where S is described in greater detail. It is essential that S is self-adjoint and boundedly invertible, its kernel is continuous, and S^{-1} admits factorization

$$S^{-1} = E^*BE, \quad (Ef)(x) = \widetilde{D}^{-\frac{1}{2}}f(x) + \int_0^x E(x,t)f(t)dt, \tag{8.20}$$

$$\widetilde{D} := I_m + \operatorname{diag}\{|\Phi_{1,1}(0)|^2, |\Phi_{2,1}(0)|^2, \dots, |\Phi_{m,1}(0)|^2\}, \tag{8.21}$$

$$E(\xi,t) = B\widetilde{D}^{\frac{1}{2}}T_\xi(\xi,t) \quad (\xi \ge t), \tag{8.22}$$

where $T_\xi(x,t)$ is the matrix kernel of the integral operator

$$T_\xi := S_\xi^{-1} = B\widetilde{D}^{-1} + \int_0^\xi T_\xi(x,t)\cdot dt. \tag{8.23}$$

Clearly, S_ξ $(\xi \le l)$ is connected with S_l by the simple relation $S_\xi = P_\xi S_l P_\xi^*$, where P_ξ is given by (1.99). Thus, there is no need to solve the operator identities (8.18) for each ξ in order to recover V.

Finally, we formulate the procedure for the solution of the inverse problem from Theorem 4.2 [265].

Theorem 8.3. *System* (8.6) *satisfying conditions* (8.7), (8.8) *and* (8.12) *is uniquely recovered from its* GW_k*-functions. In order to recover* (8.6) *on an arbitrary fixed interval*

$[0, l]$, *we first recover the vector function* $\Phi_1(x)$ *using formula* (8.13). *Thus,* Π *is also recovered since* Φ_2 *is given by* (8.17). *Now, the operator* S *is uniquely recovered from* (8.18) *and, taking into account* (8.20)–(8.23), *we recover* E. *The solution of the inverse problem is given by the equality*

$$h(x) = \begin{bmatrix} h_1(x) \\ h_2(x) \\ \dots \\ h_m(x) \end{bmatrix} = \big(E[\Phi_1 \quad \Phi_2]\big)J, \quad J = \begin{bmatrix} 0 & 1 \\ 1 & 0 \end{bmatrix}. \tag{8.24}$$

We note that a vector function φ, which is holomorphic in the half-plane $\Im z < -\frac{M}{4}$ and admits there an asymptotic representation

$$\varphi(z) = \phi_0 + \frac{\phi_1}{z} + O\left(\frac{1}{z^2}\right), \quad \Im z < -\frac{M}{4}, \quad z \to \infty,$$

is always a GW-function of some unique system (8.6) and the procedure to recover h works again, although the first inequality in (8.7) is substituted by the inequalities

$$\sup_{0<x<l} \|h_k'(x)\| < \infty \text{ for all } 0 < l < \infty$$

(see Theorem 4.3 [265]).

3. In view of the results from Chapters 2 and 3, the Inverse Spectral Transform approach from Chapter 6 is easily applied to integrable equations with rectangular matrix functions. In particular, the case of the focusing nonlinear Schrödinger equation (fNLS)

$$2\frac{\partial}{\partial t}v + \mathrm{i}\left(\frac{\partial^2}{\partial x^2}v + 2vv^*v\right) = 0 \tag{8.25}$$

is dealt with in Theorem 4.3 [108], which we present below. The fNLS equation with $m_1 \times m_2$ matrix function $v(x,t)$ is considered in the semistrip Ω_a given by (6.3). The auxiliary linear systems for fNLS have the form ([101, 337])

$$\frac{\partial}{\partial x}y = G(x,t,z)y, \quad \frac{\partial}{\partial t}y = F(x,t,z)y, \tag{8.26}$$

$$G = \mathrm{i}zj + jV, \quad F = \mathrm{i}\left(z^2 j - \mathrm{i}zjV - \frac{1}{2}\left(\frac{\partial}{\partial x}V + jV^2\right)\right), \tag{8.27}$$

$$j = \begin{bmatrix} I_{m_1} & 0 \\ 0 & -I_{m_2} \end{bmatrix}, \quad V(x,t) = \begin{bmatrix} 0 & v(x,t) \\ v(x,t)^* & 0 \end{bmatrix}. \tag{8.28}$$

Recall that $R(t,z)$ is determined by the boundary values $v(0,t)$ and $(\frac{\partial}{\partial x}v)(0,t)$ via equation $\frac{\partial}{\partial t}R = F(0,t,z)R$ and initial condition $R(0,z) = I_m$, where $m = m_1 + m_2$. We partition R similar to the partition of j in (8.28) and denote its blocks by R_{ik}.

Theorem 8.4. *Let an $m_1 \times m_2$ matrix function $v(x,t)$ be continuously differentiable on Ω_a and let $\frac{\partial^2}{\partial x^2}v$ exist. Assume that v satisfies the fNLS equation (8.25) as well as the following inequalities:*

$$\sup_{(x,t)\in\Omega_a} \|v(x,t)\| \leq M, \quad \sup_{x\in\mathbb{R}_+} \left\| \left(\frac{\partial}{\partial x}\right) v(x,t) \right\| < \infty \quad \text{for each} \quad 0 \leq t < a.$$

Then, the evolution $\varphi(t,z)$ of the Weyl functions of the skew-self-adjoint Dirac systems $\frac{\partial}{\partial x}y = Gy$, where G has the form (8.27), is given (for $\Im(z) > M$) by the equality

$$\varphi(t,z) = (R_{21}(t,z) + R_{22}(t,z)\varphi(0,z))\,(R_{11}(t,z) + R_{12}(t,z)\varphi(0,z))^{-1}.$$

8.2 GBDT for one and several variables

In this section, we formulate several GBDT results, which are closely related to the topics of this book, although they are not included in Chapter 7. We also consider a recent result on the GBDT for the time-dependent linear Schrödinger equation of dimension $k+1$ from [266].

1. The canonical system is one of the main systems discussed in this book. Since matrices J and $-\mathrm{i}\hat{J}$, where

$$J = \begin{bmatrix} 0 & I_p \\ I_p & 0 \end{bmatrix}, \qquad \hat{J} := \begin{bmatrix} 0 & -I_p \\ I_p & 0 \end{bmatrix}, \tag{8.29}$$

are unitarily equivalent, canonical system (0.2) is often written down in an equivalent form

$$y'(x,z) = z\hat{J}H(x)y(x,z), \quad H(x) = H(x)^* \in L^1_{m\times m}(0,l), \quad m = 2p, \tag{8.30}$$

$H(x) \geq 0$. The GBDT for system (8.30) is considered in [248, Theorem 1] even without the requirement $H(x) \geq 0$. As usual, we fix $n \in \mathbb{N}$ and a triple of parameter matrices $\{A,\, S(0) = S(0)^*,\, \Pi(0)\}$, where A and $S(0)$ are $n \times n$ matrices, $\Pi(0)$ is an $n \times m$ matrix and the identity

$$AS(0) - S(0)A^* = \Pi(0)\hat{J}\Pi(0)^* \tag{8.31}$$

holds. We define matrix functions $\Pi(x)$ and $S(x)$ by their initial values (at $x = 0$) and equations

$$\Pi' = -A\Pi\hat{J}H, \quad S' = \Pi\hat{J}H\hat{J}^*\Pi^*. \tag{8.32}$$

The transfer matrix function w_A is introduced in a standard way:

$$w_A(x,z) = I_m - \hat{J}\Pi(x)^*S(x)^{-1}(A - zI_n)^{-1}\Pi(x). \tag{8.33}$$

Matrix function $w_0(x)$ is determined by the relations

$$w_0' = (\hat{J}\Pi^* S^{-1}\Pi\hat{J}H - \hat{J}H\hat{J}\Pi^* S^{-1}\Pi)w_0, \quad w_0(0) = I_m. \tag{8.34}$$

It is easy to see that w_0 is invertible and $w_0^{-1} = \hat{J}w_0^*\hat{J}^*$. We fix some matrix function $W(x,z)$ satisfying the initial system (8.30) (in particular, W may be a vector function).

Theorem 8.5. *Let a triple $\{A,\ S(0) = S(0)^*,\ \Pi(0)\}$ be fixed, identity (8.31) hold and a matrix function $W(x,z)$ satisfy (8.30). Then, in the points of invertibility of $S(x)$, the matrix function*

$$\widetilde{W}(x,z) = w_0(x)^{-1}w_A(x,z)W(x,z), \tag{8.35}$$

where w_0 is determined by (8.34) and w_A is determined by (8.32) and (8.33), satisfies the transformed system

$$\widetilde{W}'(x,z) = z\hat{J}\widetilde{H}(x)\widetilde{W}(x,z), \quad \widetilde{H}(x) := w_0(x)^*H(x)w_0(x). \tag{8.36}$$

Remark 8.6. If $\det A \neq 0$ and $\det S(x) \neq 0$ for $0 \leq x \leq l$, formula (8) from [248] implies that $w_0(x) \equiv w_A(x,0)w_A(0,0)^{-1}$.

Remark 8.7. If $H \geq 0$ and $S(0) > 0$, then formula (8.32) yields $S(x) > 0$, and so $S(x)$ is invertible for $x \geq 0$. Furthermore, according to (8.36), in this case, we have $\widetilde{H} \geq 0$. Thus, the canonical system is transformed into a system which is also canonical.

2. Schrödinger and Dirac equations with singularities are of great interest and include, in particular, the well-known radial Schrödinger equation and radial Dirac system. The GBDT for these equations with singularities was studied in [111, 174, 253, 264] and below we present two such results.

First, following [111], we consider Dirac systems

$$\frac{d}{dx}y(x,z) = \mathrm{i}(zj + j\widetilde{V}(x))y(x,z) \quad (x \geq 0), \quad \widetilde{V} = \begin{bmatrix} 0 & \widetilde{v} \\ \widetilde{v}^* & 0 \end{bmatrix}, \tag{8.37}$$

with potentials $\widetilde{v}$ given by (2.58):

$$\widetilde{v}(x) = -2\mathrm{i}\varkappa_1^* \mathrm{e}^{\mathrm{i}xA^*}S(x)^{-1}\mathrm{e}^{\mathrm{i}xA}\varkappa_2. \tag{8.38}$$

The only difference with the generalized pseudoexponential potentials (Definition 2.28), which are considered in Subsection 2.2.2, is that we no longer assume the validity of (2.56). That is, instead of the condition $S(0) > 0$, we substitute a weaker condition

$$\det S(0) \neq 0. \tag{8.39}$$

This weaker condition allows one to consider the case of soliton-positon interactions (see [253] and see also an earlier paper [208] on positons).

It was shown in [133] that for the case of the pseudoexponential potentials (i.e. the case $S(0) > 0$ and $m_1 = m_2$), without loss of generality, we can assume

$$\sigma(A) \subset \overline{\mathbb{C}}_-, \quad \operatorname{span} \bigcup_{k=0}^{n-1} \operatorname{Im} A^k \varkappa_1 = \operatorname{span} \bigcup_{k=0}^{n-1} \operatorname{Im} A^k \varkappa_2 = \mathbb{C}^n. \tag{8.40}$$

The same relations (8.40) are assumed to hold now. Formula (2.8) in [253] and the proof of Theorem 2.1 in the same paper show that, under conditions (8.40), the matrix function $\mathrm{e}^{\mathrm{i}xA}S(x)\mathrm{e}^{-\mathrm{i}xA^*}$ is monotonic and tends to $+\infty$ for x tending to $+\infty$. Hence, for some $x_0 > 0$, we have

$$S(x) > 0, \quad x_0 \le x < \infty. \tag{8.41}$$

Moreover, it is easy to see that S given by (2.53) and (2.55) admits analytic continuation, and so inequality (8.41) implies that the set of points $x \in \mathbb{R}_+$, where $\det S(x) = 0$, is always finite. Like in Subsection 2.2.2, the matrix function

$$\widetilde{u}(x,z) = w_A(x,z)\mathrm{e}^{\mathrm{i}zxj}w_A(0,z)^{-1} \tag{8.42}$$

is nondegenerate, satisfies Dirac system (8.37) (excluding points of singularity, where $\det S(x) = 0$), and also admits the meromorphic (in x) continuation. Thus, in agreement with the standard for the singular case requirement of *one expression in the whole domain* [208], we call $\widetilde{u}$ the fundamental solution. Besides $\widetilde{u}$, we introduce a special (Jost) solution of (8.37):

$$Y(x,z) = \begin{bmatrix} Y_1(x,z) \\ Y_2(x,z) \end{bmatrix} = \mathrm{e}^{\mathrm{i}zx} \begin{bmatrix} I_{m_1} \\ 0 \end{bmatrix} + o(1), \quad x \to \infty. \tag{8.43}$$

An expression for Y is obtained in [111] using the representation (8.42) of the fundamental solution. The left reflection coefficient is introduced (via $Y(0,z)$) by the formula

$$R_L(z) := Y_2(0,z)Y_1(0,z)^{-1}. \tag{8.44}$$

According to [111, Theorem 3.3], the expression for R_L coincides with the right-hand side of (2.59), that is,

$$R_L(z) = \mathrm{i}\varkappa_2^* S(0)^{-1}(\alpha - zI_n)^{-1}\varkappa_1, \quad \alpha := A - \mathrm{i}\varkappa_1\varkappa_1^* S(0)^{-1}. \tag{8.45}$$

Unless $S(0) > 0$, we cannot state that R_L is nonexpansive in $\mathbb{C}_+$. However, the inverse problem to recover v from R_L can be solved in this *singular* case ([111, Theorem 4.1]).

Theorem 8.8. *Let R be a strictly proper rational $m_2 \times m_1$ matrix function. Then, R is the left reflection coefficient of a system* (8.37) *with $\widetilde{v}$ of the form* (8.38) *if and only if R is nonexpansive on $\mathbb{R}$. If R satisfies this condition, then $\widetilde{v}$ can be uniquely recovered* *from R in two steps. First, we recover a set of parameter matrices A, $S(0)$ and $\Pi(0)$, and next we construct $\widetilde{v}$ via formulas (2.53), (2.55) and (8.38).*

In order to recover the triple $\{A, S(0), \Pi(0)\}$, we consider a minimal realization

$$R(z) = -C(\mathcal{A} - zI_n)^{-1}B, \tag{8.46}$$

where n is the McMillan degree of R, that is, the order of $\mathcal{A}$. Then, there is an $n \times n$ Hermitian solution $X = X^*$ of the Riccati equation

$$\mathrm{i}(X\mathcal{A} - \mathcal{A}^*X) = C^*C + XBB^*X$$

such that

$$\sigma(\mathcal{A} + \mathrm{i}BB^*X) \subset \overline{\mathbb{C}}_-.$$

Finally, we put

$$A = \mathcal{A} + \mathrm{i}BB^*X; \quad S(0) = X^{-1}; \quad \varkappa_1 = B, \ \varkappa_2 = -\mathrm{i}S(0)C^*, \ \Pi(0) = \begin{bmatrix} \varkappa_1 & \varkappa_2 \end{bmatrix}.$$

3. It is of interest to consider also GBDT for the radial Dirac system, namely, the system

$$\left(-\mathrm{i}\sigma_2\frac{d}{dr} + \frac{\ell}{r}\sigma_1 + V(r)\right) y = zy \quad (r > 0), \tag{8.47}$$

or equivalently

$$\left(\frac{d}{dr} + zq_1 + q_0(r)\right) y(r,z) = 0 \quad (r > 0), \tag{8.48}$$

where ℓ is an integer, σ_i are Pauli matrices,

$$\sigma_1 = \begin{bmatrix} 0 & 1 \\ 1 & 0 \end{bmatrix}, \quad \sigma_2 = \begin{bmatrix} 0 & -\mathrm{i} \\ \mathrm{i} & 0 \end{bmatrix}, \quad \sigma_3 = \begin{bmatrix} 1 & 0 \\ 0 & -1 \end{bmatrix}, \tag{8.49}$$

$$V(r) = v_e(r)I_2 + v_a(r)\sigma_1 + v_s(r)\sigma_3, \tag{8.50}$$

v_e, v_a and v_s are real-valued functions which are locally integrable on intervals $[0, l)$ $(l \in \mathbb{R}_+)$,

$$q_1 = \hat{J} := (\mathrm{i}\sigma_2)^{-1}, \tag{8.51}$$

$$q_0(r) = v_*(r)\sigma_3 - \hat{J}\left(v_e(r)I_2 + v_s(r)\sigma_3\right), \quad v_*(r) = \frac{\ell}{r} + v_a(r). \tag{8.52}$$

The functions v_e, v_a and v_s represent the electrostatic potential, the anomalous magnetic moment, and the sum of the mass m and the scalar potential, respectively. In greater detail, the radial Dirac system is considered in Chapter 9 where a new half-inverse sliding problem for this system is formulated and solved. We note that, in Chapter 9, we have $v_a \equiv 0$, $v_s \equiv m$ (and V is defined in a somewhat different way).

The double commutation method was applied to (8.47) in an interesting paper by G. Teschl [319]. Using this method, S. Albeverio, R. Hryniv and Ya. Mykytyuk [14] proved that 1 is added to $\kappa = |\ell|$ when an eigenvalue is removed, and that 1 is subtracted from κ when an eigenvalue is inserted. The proof of the following theorem (i.e. [264, Theorem 4.1]) is based on GBDT results, factorization of the transfer matrix function w_A and some auxiliary results from [14].

Theorem 8.9. *In order to construct a class of explicit solutions of system* (8.47) *with some fixed integer ℓ, fix an integer $n > 0$ and $n \times n$ matrices $\mathcal{A}_1$ and $\mathcal{S}_1 > 0$. Also, fix a $\kappa \times \kappa$ lower triangular matrix $\mathcal{A}_2$ and an $(n + \kappa) \times 2$ matrix*

$$\Pi(0) = \begin{bmatrix} \Psi_1(0) \\ \Psi_2(0) \end{bmatrix},$$

where $\kappa = |\ell|$ and Ψ_1 and Ψ_2 are $n\times 2$ and $\kappa\times 2$, respectively, blocks of Π. It is required that

$$\mathcal{A}_1\mathcal{S}_1 - \mathcal{S}_1\mathcal{A}_1^* = -\Psi_1(0)\hat{J}\Psi_1(0)^*, \quad \Psi_2(0)\hat{J}\Psi_2(0)^* = 0.$$

Moreover, for $h := [1 \quad 0 \quad 0 \dots 0]\Psi_2$, we assume that $h(0) = c[0 \quad 1]$ when ℓ is positive and odd or negative and even, and we assume that $h(0) = c[1 \quad 0]$ when ℓ is positive and even or negative and odd ($c \neq 0$).

We introduce matrix A and vectors $\hat{\varkappa}_1, \hat{\varkappa}_2 \in \mathbb{C}^{n+\kappa}$

$$A := \begin{bmatrix} \mathcal{A}_1 & 0 \\ \mathcal{B}_1 & \mathcal{A}_2 \end{bmatrix}, \quad \mathcal{B}_1 = -\Psi_2(0)\hat{J}\Psi_1(0)^*\mathcal{S}_1^{-1};$$

$$[\hat{\varkappa}_1 \quad \hat{\varkappa}_2] := \Pi(0)\hat{\Theta}, \quad \hat{\Theta} := \frac{1}{\sqrt{2}}\begin{bmatrix} 1 & 1 \\ -\mathrm{i} & \mathrm{i} \end{bmatrix}.$$

Now, put

$$\Pi(r) := [\mathrm{e}^{\mathrm{i}rA}\hat{\varkappa}_1 \quad \mathrm{e}^{-\mathrm{i}rA}\hat{\varkappa}_2]\hat{\Theta}^*,$$

$$S(r) := \begin{bmatrix} \mathcal{S}_1 & 0 \\ 0 & 0 \end{bmatrix} + \int_0^r \left(\mathrm{e}^{\mathrm{i}tA}\hat{\varkappa}_1\hat{\varkappa}_1^*\mathrm{e}^{-\mathrm{i}tA^*} + \mathrm{e}^{-\mathrm{i}tA}\hat{\varkappa}_2\hat{\varkappa}_2^*\mathrm{e}^{\mathrm{i}tA^*}\right) dt. \tag{8.53}$$

Then, if $S(r) > 0$ for $r > 0$, the matrix function $\tilde{q}_0$ given by

$$\tilde{q}_0(r) := \hat{J}X(r)\hat{J}^* - X(r), \quad X = \{X_{ij}\}_{i,j=1}^2 := \Pi^*S^{-1}\Pi \tag{8.54}$$

admits representation

$$\tilde{q}_0(r) = \frac{\ell}{r}\sigma_3 + \tilde{v}_a(r)\sigma_3 - \tilde{v}_s(r)\sigma_1, \tag{8.55}$$

where $\tilde{v}_a$ and $\tilde{v}_s$ are bounded in the neighborhood of zero and

$$\tilde{v}_a(r) = X_{22}(r) - X_{11}(r) - \frac{\ell}{r}, \quad \tilde{v}_s(r) = X_{12}(r) + X_{21}(r). \tag{8.56}$$

The fundamental solution $\widetilde{u}(r,z)$ *of system*

$$\left(\frac{d}{dr} + zq_1 + \widetilde{q}_0(r)\right)\widetilde{u}(r,z) = 0 \quad (r > 0) \tag{8.57}$$

is given by the formula

$$\widetilde{u}(r,z) = w_A(r,z)\widehat{\Theta}\exp\{-izr\sigma_3\}, \tag{8.58}$$

where

$$w_A(z) = I_2 + \widehat{J}\Pi^* S^{-1}(A - zI_{n+\kappa})^{-1}\Pi.$$

Thus, we constructed explicit solutions $\widetilde{u}$ of the radial Dirac systems (8.57) where potentials are given by (8.54)–(8.56). By the neighborhood of zero, we mean the neighborhood of the form $(0,\,\varepsilon)$ or $[0,\,\varepsilon)$. Clearly, if $S(x) > 0$ for all x in some neighbourhood $(0,\,\varepsilon)$, then $S(x) > 0$ for all $x \in (0,\,\infty)$, the formulas in Theorem 8.9 are well-defined on $(0,\,\infty)$ and $\widetilde{q}_0$ is infinitely differentiable. The singularity for the system is connected with the fact that $\det S(0) = 0$ (see (8.53)).

4. Explicit solutions of multidimensional Schrödinger operators are actively studied (see, e.g. [69] and references therein). Like in many other topics, for BDT theory, the case of several variables is much more complicated than the case of one variable. Namely, it is much more difficult to introduce and study BDT for linear equations depending on several variables. The corresponding results are rare and usually incomplete. An interesting attempt to construct BDT for the important time-dependent Schrödinger (TDS), also called the nonstationary Schrödinger,

$$\left(\alpha\frac{\partial}{\partial t} + \Delta - q(x,t)\right)y = 0, \quad x \in \mathbb{C}^k, \quad \alpha \in \mathbb{C}, \quad \alpha \neq 0, \tag{8.59}$$

was undertaken in [233]. Singular as well as some stationary binary Darboux transformations of a scalar TDS equation into a matrix TDS equation (i.e. into a TDS equation, where the potential q is a matrix function) were treated in [233]. See [209] for a discussion on singular and binary Darboux transformations. Explicit GBDT solutions of TDS were constructed later in [266] (following, to some extent, the paper [255] on the matrix Kadomtsev–Petviashvili equation). In particular, solutions for the cases of nonsingular, rational and lump potentials were constructed in [266]. Matrices $\mathcal{R}$ simultaneously satisfying several operator (matrix) identities were used for that purpose. We present below the main definition and GBDT result from [266] for the case of matrix TDS (more precisely, for $y \in \mathbb{C}^p$, $p \in \mathbb{N}$, which also includes the scalar case).

Definition 8.10. A matrix S-multinode $\{k, A, B, \mathcal{R}, \nu, C_\Phi, C_\Lambda\}$ is a set of matrices consisting of $N \times N$ commuting matrices A_i $(1 \le i \le k)$, of $N \times N$ commuting matrices B_i $(1 \le i \le k)$, of $p \times p$ matrices ν_i $(1 \le i \le k)$ and of an $N \times N$ matrix $\mathcal{R}$, an $N \times p$ matrix C_Φ, and a $p \times N$ matrix C_Λ, such that the matrix identities

$$A_i\mathcal{R} - \mathcal{R}B_i = C_\Phi \nu_i C_\Lambda, \quad 1 \le i \le k \tag{8.60}$$

hold. An operator S-multinode is defined in the same way.

The condition that the matrices from $\{A_i\}$ mutually commute and the matrices from $\{B_i\}$ mutually commute:

$$A_iA_j = A_jA_i, \quad B_iB_j = B_jB_i \tag{8.61}$$

is important in Definition 8.10. For the case that $k = 1$, this definition coincides with the definition of the S-node (which is actively discussed and numerously applied in our book) and, for the case that $\mathcal{R} = I_N$ and $B_i = A_i^*$, Definition 8.10 coincides with the definition of a commutative colligation from [202].

Theorem 8.11. *Fix some $n \in \mathbb{N}$ and let an $n \times N$ matrix $\widehat{C}_\Phi$, an $N \times n$ matrix $\widehat{C}_\Lambda$, an $n \times n$ matrix $\mathcal{S}_0$ and a matrix S-multinode $\{k, A, B, \mathcal{R}, \nu, C_\Phi, C_\Lambda\}$ be given. Set*

$$\Phi(x,t) := \widehat{C}_\Phi e_A(x,t)C_\Phi, \quad e_A(x,t) := \exp\left\{\alpha^{-1}t\left(\sum_{i=1}^k A_i^2\right) + \sum_{i=1}^k x_iA_i\right\},$$

$$\Lambda(x,t)^* := C_\Lambda e_B(-x,-t)\widehat{C}_\Lambda, \quad S(x,t) := \mathcal{S}_0 + \widehat{C}_\Phi e_A(x,t)\mathcal{R}e_B(-x,-t)\widehat{C}_\Lambda.$$

Then, in the points of invertibility of $S(x,t)$, the matrix function

$$\widetilde{\Lambda}^* := \Lambda^* S^{-1} \tag{8.62}$$

satisfies the transformed TDS equation

$$\left(\alpha\frac{\partial}{\partial t} + \Delta - \widetilde{q}(x,t)\right)\widetilde{\Lambda}^* = 0, \tag{8.63}$$

$$\widetilde{q}(x,t) := -2\sum_{i=1}^k \frac{\partial}{\partial x_i}\left(\Lambda(x,t)^* S(x,t)^{-1}\Phi(x,t)\right)\nu_i. \tag{8.64}$$

Compare the theorem above with formula (7.61), which shows that $\Pi_2^* S^{-1}$ is a generalized eigenfunction of the transformed system (in the case of one variable).

Example 8.12. Finally, consider a simple example for the case that $1 < k \leq p$. Set

$$\begin{aligned}
&A_1 = \mathrm{diag}\,\{a_1, \ldots, a_N\}; \quad A_i = (c_iI_N - A_1)^{-1}, \quad c_i \notin \sigma(A_1) \quad (1 < i \leq k);\\
&B_1 = \mathrm{diag}\,\{b_1, \ldots, b_N\}; \quad B_i = (c_iI_N - B_1)^{-1}, \quad c_i \notin \sigma(B_1) \quad (1 < i \leq k);\\
&\mathcal{R} = \{(a_i - b_j)^{-1}\}_{i,j=1}^N \quad (a_i \neq b_j); \quad \nu_i = \{\delta_{i-j}\delta_{i-r}\}_{j,r=1}^p \quad (1 \leq i \leq k);\\
&C_\Phi = \begin{bmatrix} h & A_2h & \ldots & A_kh & \check{C}_\Phi \end{bmatrix}, \quad h = \mathrm{col}\begin{bmatrix} 1 & 1 & \ldots & 1 \end{bmatrix};\\
&C_\Lambda = \mathrm{col}\begin{bmatrix} h^* & h^*B_2 & \ldots & h^*B_k & \check{C}_\Lambda \end{bmatrix},
\end{aligned}$$

where δ_i is the Kronecker's δ, $\check{C}_\Phi$ is some $N \times (p-k)$ matrix, and $\check{C}_\Lambda$ is some $(p-k) \times N$ matrix. We see that $A_1\mathcal{R} - \mathcal{R}B_1 = hh^*$. Hence, the equalities

$$A_i\mathcal{R} - \mathcal{R}B_i = A_ihh^*B_i \quad (1 < i \leq k)$$

hold. Therefore, the matrices introduced in our example form an S-multinode

$$\{k, A, B, \mathcal{R}, \nu, C_\Phi, C_\Lambda\},$$

and the corresponding explicit expressions for solution and potential of the transformed TDS equation follow.

It is easy to see that for the case that $\tilde{q} = \{\tilde{q}_{ij}\}_{i,j=1}^{p}$ and $f = \{f_i\}_{i=1}^{p}$ are the potential and solution, respectively, of some matrix TDS equation, the functions

$$q_{sc} := \tilde{q}_{ii} + \Big(\sum_{j \neq i} \tilde{q}_{ij} f_j\Big) / f_i, \quad f_{sc} := f_i \tag{8.65}$$

are the potential and solution of a scalar TDS equation.

9 Sliding inverse problems for radial Dirac and Schrödinger equations

In this chapter, we introduce new inverse and half-inverse problems: *sliding problems*. In Section 9.1, we show that this approach can be used for the radial Schrödinger equation, radial Dirac system and the multidimensional Schrödinger equation. We note that inverse problems for the Dirac-type systems, which are considered in many of the previous chapters of the book (and are usually called Dirac systems), have been thoroughly investigated, see numerous references in Chapter 2 and the Introduction. The radial Dirac system was introduced before the Dirac-type system and differs from the Dirac-type system in two ways: the corresponding matrix function V has singularity and its nonzero diagonal entries are known ((9.6)). Therefore, we talk about a more difficult half-inverse problem. Here, we formulate the sliding half-inverse problem and recover the radial Dirac system from the quantum defect. Using the quantum defect, we also solve a sliding inverse problem for the case of the radial Schrödinger equation. In Section 9.2, we consider equations with Coulomb-type potentials. The chapter is based on the recent preprint [300].

9.1 Inverse and half-inverse sliding problems

9.1.1 Main definitions and results

1. The radial Schrödinger equation has the form

$$\frac{d^2y}{dr^2} + \left(z - \frac{\ell(\ell+1)}{r^2} - q(r)\right) y = 0, \quad 0 < r < \infty \quad (q = \overline{q}), \tag{9.1}$$

where $\ell = 0, 1, 2, \ldots$ Some recent results and references on this well-known equation can be seen, for instance, in [15, 173–175]. The radial Dirac system has the form

$$\left(\frac{d}{dr} + \frac{\ell}{r}\right) f_1 - (z + m - q(r)) f_2 = 0, \tag{9.2}$$

$$\left(\frac{d}{dr} - \frac{\ell}{r}\right) f_2 + (z - m - q(r)) f_1 = 0, \tag{9.3}$$

$$0 < r < \infty, \quad m > 0, \quad q = \overline{q}, \quad \ell = \pm 1, \pm 2, \ldots, \tag{9.4}$$

where m takes, in this chapter, positive (not necessarily integer) values and stands for mass. We easily rewrite (9.2) and (9.3) in the matrix form

$$\frac{d}{dr} f = (\mathrm{i} z \sigma_2 + V) f, \tag{9.5}$$

$f = \mathrm{col}\,[f_1, f_2]$, whereas the Pauli matrix σ_2 and matrix function V have the form

$$\sigma_2 = \begin{bmatrix} 0 & -\mathrm{i} \\ \mathrm{i} & 0 \end{bmatrix}, \quad V(r) = \begin{bmatrix} -\ell/r & m - q(r) \\ m + q(r) & \ell/r \end{bmatrix}. \tag{9.6}$$

The peculiarity of the radial Dirac system consists of the fact that the entry ℓ/r of the matrix $V(r)$ is known and $(\ell/r)\notin L^1(0,\infty)$. A well-posed inverse problem for the radial Dirac system was absent, which was discussed in [279]. In order to formulate a well-posed problem, we introduce the notion of the quantum defect $\delta(r)$ at infinity ($z=\infty$). The corresponding sliding half-inverse problem for the Dirac system is well-posed.

Without loss of generality, we only consider the case $\ell>0$ since the equality (9.5) yields the transformed equality, where Jf is substituted for f, $-z$ is substituted for z, and $-q$ and $-\ell$ are substituted for q and ℓ, respectively (in the expression for V). Here, J is given by (6.65).

First, we consider Schrödinger equation (9.1) with the boundary conditions

$$y(0)=y(a)=0, \tag{9.7}$$

and assume that

$$\int_0^a |q(t)|dt<\infty, \quad 0<a<\infty. \tag{9.8}$$

The corresponding eigenvalues (i.e. the values of z such that the solutions y satisfying (9.7) exist) are denoted by $z_n(\ell)$ $(n\in\mathbb{N})$. Then, we have (as a corollary of Theorem 9.17) the asymptotic equality

$$\sqrt{z_n(a,\ell)}=\frac{\pi}{a}(n+\ell/2)+\frac{a\delta(a)-\ell(\ell+1)}{2(n+\ell/2)\pi a}+o\left(n^{-1}\right) \tag{9.9}$$

where $\delta(a)$, which is defined by the formula

$$\delta(a)=\int_0^a q(t)dt, \tag{9.10}$$

is called the *quantum defect* of the equation (9.1). Now, we formulate a *sliding inverse problem*:

Problem 9.1. *Recover the potential q from the given quantum defect δ.*

It is immediate from (9.10) that the solution of Problem 9.1 is given by the equality

$$q(a)=\left(\frac{d}{da}\delta\right)(a). \tag{9.11}$$

We deal with the radial Dirac system in a similar way. In order to formulate our result, we introduce a parameter $\varepsilon(z)$, which for the case of the Dirac system is given by $\varepsilon=\sqrt{m^2-z^2}$.

Remark 9.2. More precisely, for all $z\in\mathbb{C}$, excluding the cuts along the semiaxes $(-\infty,-m)$ and (m,∞), we choose a branch of $\varepsilon=\sqrt{m^2-z^2}$ such that $\Re(\varepsilon)>0$

for $z \neq 0$. We introduce ε on the cuts as the product $\varepsilon = \sqrt{m+z}\,\sqrt{m-z}$, where

$$\arg(\sqrt{m+z}) = \begin{cases} 0, & \text{if } z > m, \\ -\pi/2, & \text{if } z < -m; \end{cases} \qquad \arg(\sqrt{m-z}) = \begin{cases} \pi/2, & \text{if } z > m, \\ 0, & \text{if } z < -m. \end{cases}$$

Hence, we have

$$\arg(\varepsilon) = \pi/2, \quad \text{if} \quad z > m; \quad \arg(\varepsilon) = -\pi/2, \quad \text{if} \quad z < -m. \tag{9.12}$$

The following theorem is proved in Subsection 9.1.3.

Theorem 9.3. *Let the condition*

$$\int_r^\infty |q(t)|dt < \infty, \quad 0 < r < \infty \tag{9.13}$$

be fulfilled. Then, there is a solution $\widetilde{F}(r, \varepsilon, \ell)$ of system (9.5) (where V is given by (9.6)) such that the relation

$$\widetilde{F}(r, \varepsilon, \ell) \sim -\frac{1}{2\sqrt{m+z}} e^{-i(r\varepsilon+\delta(r))} \begin{bmatrix} i \\ 1 \end{bmatrix}, \quad \varepsilon \to \infty, \quad z > m \tag{9.14}$$

holds and $\delta(r)$ (i.e. the quantum defect of this system) has the form

$$\delta(r) = \int_r^\infty q(t)dt. \tag{9.15}$$

So, for the Dirac system, we formulate a *sliding half-inverse problem*:

Problem 9.4. *Recover the matrix function V in (9.5) from the given value ℓ and quantum defect δ.*

Since V has the form (9.6), the solution of Problem 9.4 is immediate from (9.15) or, more precisely, from the equality

$$q(r) = -\left(\frac{d}{dr}\delta\right)(r). \tag{9.16}$$

Remark 9.5. The notion of the half-inverse problem and the first model of such a problem were introduced in the paper [272] (see also [292]).

Remark 9.6. The values of $\delta(r)$ are spectral characteristics of the radial Dirac system.

Remark 9.7. Sliding inverse Problem 9.1 and half-inverse Problem 9.4 are well-posed, that is, there is a one-to-one correspondence between the spectral data δ and potentials q (relations (9.10) and (9.16), respectively). Like all the classical inverse problems, these problems are not stable.

Remark 9.8. In the process of solving sliding Problems 9.1 and 9.4, we only use the spectral data for positive energy.

2. Next, we consider one-dimensional and multidimensional Schrödinger equations. Putting in (9.1) $l = 0$, we obtain the one-dimensional Schrödinger equation

$$-y'' + q(x)y = zy, \quad 0 \le x \le a, \quad y(0) = y(a) = 0. \tag{9.17}$$

Theorem 9.9 ([195]). *Let condition* (9.8) *be fulfilled. Then, the eigenvalues* $z_n(a) = z_n$ *of system* (9.17) *satisfy the relation*

$$\sqrt{z_n} = \frac{\pi}{a}n + \frac{1}{2n\pi}\delta(a) + o(n^{-1}), \tag{9.18}$$

where $\delta(a)$ *is given in* (9.10).

We note that formulas (9.9) and (9.18) coincide when $\ell = 0$. In Subsection 9.1.4, we prove the assertion:

Theorem 9.10. *Let condition* (9.8) *be fulfilled. Then, the corresponding statistical sum*

$$Z(T) = \sum_{n=1}^{\infty} \mathrm{e}^{-z_n/T} \tag{9.19}$$

satisfies the relation

$$Z(T) = (a/2)\sqrt{T/\pi} - 1/2 + \delta(a)/(2\sqrt{T\pi}) + o(1/\sqrt{T}), \quad T \to \infty. \tag{9.20}$$

Recall that T stands in physical applications for temperature.

The k-dimensional Schrödinger equation has the form

$$-\Delta y + q(x)y = zy, \quad x = (x_1, x_2, \dots, x_k) \in G \subset \mathbb{R}^k. \tag{9.21}$$

We deal here with a model case where

$$G = G(a_1, a_2, \dots) = \{x : 0 \le x_i \le a_i \quad (1 \le i \le k)\}, \tag{9.22}$$

q admits representation

$$q(x) = \sum_{i=1}^{k} q_i(x_i), \tag{9.23}$$

the functions q_i satisfy inequalities

$$\int_0^{a_i} |q_i(x_i)|dx_i < \infty, \quad 0 < a_i < \infty, \quad 1 \le i \le k, \tag{9.24}$$

and the boundary condition (on the boundary Γ of G) is given by

$$y|_\Gamma = 0. \tag{9.25}$$

Now, $z_{n,i}$ are the eigenvalues of the problem (9.17), where $x = x_i$ and $q(x) = q_i(x_i)$, the statistical sum Z_i is given by

$$Z_i(T) = \sum_{n=1}^{\infty} e^{-z_{n,i}/T}, \tag{9.26}$$

$\{z_n\}$ are the points of the spectrum of problem (9.21), (9.25), whereas the corresponding statistical sum $Z(T)$ is defined again by the relation (9.19). Representation (9.23) implies that

$$Z(T) = \prod_{i=1}^{k} Z_i(T). \tag{9.27}$$

From Theorem 9.10 and equality (9.27), we derive the following assertion.

Theorem 9.11. *Let the relations* (9.23) *and* (9.24) *hold. Then,*

$$Z(T) = \left(\frac{T}{4\pi}\right)^{k/2} \left(\mathcal{V}_k - \sqrt{\frac{\pi}{T}}\,\mathcal{V}_{k-1} + \frac{\pi}{T}\mathcal{V}_{k-2} + 2\frac{\delta(G)}{T} + o\left(\frac{1}{T}\right)\right), \tag{9.28}$$

where $T \to \infty$ *and* $\mathcal{V}_{-1} = 0, \quad \mathcal{V}_0 = 1,$

$$\mathcal{V}_k = \prod_{i=1}^{k} a_i, \quad \mathcal{V}_{k-1} = \mathcal{V}_k \sum_{i=1}^{k} 1/a_i, \quad \mathcal{V}_{k-2} = \mathcal{V}_k \sum_{1\le i<j\le k} 1/(a_i a_j), \tag{9.29}$$

$$\delta(G) = \int_G q(x)dx. \tag{9.30}$$

The coefficients $\mathcal{V}_i$ have clear geometric interpretation, namely,

$$\mathcal{V}_i = S_i/2^{k-i}, \quad i = k,\ k-1,\ k-2, \tag{9.31}$$

where S_k is the Lebesgue measure of the domain G, S_{k-1} is the Lebesgue measure of the boundary Γ, and S_{k-2} is the Lebesgue measure of the domain Γ_1 formed by the set of pairwise intersections of the facets of the boundary Γ.

Conjecture I. *Formulas* (9.28)–(9.30) *hold for a much wider class of potentials than the potentials of the form* (9.23).

Problem 9.12. *Let formulas* (9.28)–(9.30) *hold. Recover the potential q from the given* $\delta(G)$.

According to (9.22) and (9.30), the solution of Problem 9.12 is given by

$$q(a_1, a_2, \ldots, a_k) = \frac{\partial^k}{\partial a_1 \partial a_2 \cdots \partial a_k} \int_{G(a_1,a_2,\ldots,a_k)} q(x)dx. \tag{9.32}$$

3. Finally, we consider one-dimensional and multidimensional equations for anharmonic oscillators. The one-dimensional equation for the anharmonic oscillator (or

one-dimensional anharmonic oscillator, which is a shorter way to say it) has the form

$$-y'' + \left(x^2/4 + q(x)\right) y = zy, \quad 0 \le x < \infty \quad (q = \overline{q}), \tag{9.33}$$

and we introduce the boundary condition

$$y(0) = 0. \tag{9.34}$$

We assume that

$$\int_0^\infty |q(x)|(1 + x^{2+\varepsilon})dx < \infty, \quad \varepsilon > 0. \tag{9.35}$$

Theorem 9.13 ([278]). *Let condition (9.35) be fulfilled. Then, the eigenvalues z_n of problem (9.33), (9.34) satisfy the relation*

$$z_n = 2n - 1/2 + \frac{1}{\pi\sqrt{2n}}(\delta + o(1)), \quad n \to \infty, \tag{9.36}$$

where

$$\delta = \int_0^\infty q(x)dx. \tag{9.37}$$

Our next theorem easily follows from Theorem 9.13, see Subsection 9.1.4.

Theorem 9.14. *Let an anharmonic oscillator satisfy conditions of Theorem 9.13. Then, the statistical sum $Z(T)$, given by (9.19), satisfies the asymptotic relation*

$$Z(T) = T/2 - -1/4 - \delta \Big/ \left(2\sqrt{T\pi}\right) + o\left(1/\sqrt{T}\right). \tag{9.38}$$

The k-dimensional equation for the anharmonic oscillator has the form

$$-\Delta y + \left(\frac{|x|^2}{4} + q(x)\right) y = zy, \quad x \in G, \quad |x|^2 = \sum_{i=1}^{k} x_i{}^2. \tag{9.39}$$

We set

$$G = \{x : 0 \le x_i < \infty \quad (1 \le i \le k)\}, \tag{9.40}$$

and introduce the condition (9.25) on the boundary Γ of the domain G. We assume again that the potential $q(x)$ has the form (9.23) and that

$$\int_0^\infty |q_i(x_i)|(1 + x_i{}^{2+\varepsilon})dx_i < \infty, \quad \varepsilon > 0 \quad (1 \le i \le k). \tag{9.41}$$

Like in the case of the multidimensional Schrödinger equation, $z_{n,i}$ are the eigenvalues of the one-dimensional problem, this time regarding the problem (9.33), (9.34), where $x = x_i$ and $q(x) = q_i(x_i)$. By z_n, we denote the spectrum of the problem (9.39) with initial condition (9.25), where G is given by (9.40). The statistical sum $Z_i(T)$ has the form (9.26) and the statistical sum $Z(T)$ has the form (9.19). Like before, relation (9.23) yields (9.27). From (9.27) and (9.38) follows the next assertion.

Theorem 9.15. *Let a k-dimensional anharmonic oscillator, where the potential q admits representation (9.23) and the functions q_i satisfy the condition (9.41), be given. Then, we have*

$$Z(T) = \left(\frac{T}{2}\right)^m \left(1 - m/(2T) - \delta(G)\Big/\left(T\sqrt{T\pi}\right) + o\left(1\Big/T\sqrt{T}\right)\right), \tag{9.42}$$

where $T \to \infty$ and $\delta(G) = \int_G q(x)dx$.

Conjecture II. *Formula (9.42) holds for a much wider class of potentials than the potentials of the form (9.23).*

Remark 9.16. Theorem 9.15 is an analog of Theorem 9.11. Jointly with the results from [278], an easy modification of Theorem 9.15 can be used for the formulation and solution of a sliding inverse problem for the anharmonic oscillator.

9.1.2 Radial Schrödinger equation and quantum defect

We add variables $\varepsilon = \mathrm{i}\sqrt{z}$ and ℓ into the notation of the solutions of (9.1) and write $y(r, \varepsilon, \ell)$. Here, the parameter ε, continuously depending on z, is defined for all z in the complex plane with the cut along the negative part of the imaginary axis. Let us fix $\arg \varepsilon = \pi/2$ for $z > 0$. We start our study of the equation (9.1) from the case $q \equiv 0$. Two solutions $u_1(r, \varepsilon, \ell)$ and $u_2(r, \varepsilon, \ell)$ of (9.1), where $q \equiv 0$, are given by the relations

$$u_1(r, \varepsilon, \ell) = (2\varepsilon)^{-(\ell+1)} M_{0,\ell+1/2}(2r\varepsilon), \quad u_2(r, \varepsilon, \ell) = (2\varepsilon)^{\ell} W_{0,\ell+1/2}(2r\varepsilon), \tag{9.43}$$

and $M_{\kappa,\mu}(z)$ and $W_{\kappa,\mu}(z)$ in the formula above are Whittaker functions. The connections between Whittaker functions and confluent hypergeometric functions are well known (see, e.g. [98]), namely,

$$M_{0,\ell+1/2}(x) = \mathrm{e}^{-x/2} x^{c/2} \Phi(\alpha, c, x), \quad W_{0,\ell+1/2}(x) = \mathrm{e}^{-x/2} x^{c/2} \Psi(\alpha, c, x), \tag{9.44}$$

where $\alpha = \ell + 1$, $c = 2\ell + 2$. It follows from (9.44) and relations

$$\Phi(\alpha, c, x) \sim 1 \quad \text{and} \quad \Psi(\alpha, c, x) \sim \frac{\Gamma(c-1)}{\Gamma(\alpha)} x^{1-c} \quad \text{for } x \to 0 \tag{9.45}$$

that the solutions $u_1(r, \varepsilon, \ell)$ and $u_2(r, \varepsilon, \ell)$ satisfy the conditions

$$u_1(r, \varepsilon, \ell) \sim r^{\ell+1}, \quad u_2(r, \varepsilon, \ell) \sim \frac{\Gamma(2\ell+1)}{\Gamma(\ell+1)} r^{-\ell}, \quad r \to 0. \tag{9.46}$$

Hence, in view of (9.46), we have

$$|u_1(r, \varepsilon, \ell)| \le \mathcal{M} r^{\ell+1} \quad \text{for } r \le 1/|\varepsilon|. \tag{9.47}$$

We note that in this chapter, the letter $\mathcal{M}$ stands for different constants.

Let us write the following integral representation ([98, Section 6.11]), that is,

$$M_{0,\ell+1/2}(2r\varepsilon) = C(\ell)(2r\varepsilon)^{\ell+1} I(r,\varepsilon,\ell), \tag{9.48}$$

where

$$C(\ell) = 2^{-(2\ell+1)} \frac{\Gamma(2\ell+2)}{\Gamma^2(\ell+1)}, \quad I(r,\varepsilon,\ell) = \int_{-1}^{1} \mathrm{e}^{rt\varepsilon}(1-t^2)^{\ell} dt. \tag{9.49}$$

Integrating $I(r,\varepsilon,\ell)$ by parts $\ell+2$ times, we obtain the asymptotic formula

$$I(r,\varepsilon,\ell) \sim (r\varepsilon)^{-(\ell+1)} 2^{\ell}\Gamma(\ell+1)\left((-1)^{\ell+1}\mathrm{e}^{-r\varepsilon}\phi(r,\varepsilon,\ell) + \mathrm{e}^{r\varepsilon}\phi(r,-\varepsilon,\ell)\right), \tag{9.50}$$

where

$$\phi(r,\varepsilon,\ell) = 1 + \frac{\ell(\ell+1)}{2r\varepsilon} + o(1/\varepsilon), \quad \varepsilon \to \infty. \tag{9.51}$$

Here, we used the Leibnitz differentiation formula in order to obtain an expression for the $(\ell+1)$'th derivative of the product $(1-t)^{\ell}(1+t)^{\ell}$. Formulas (9.43), (9.44) and (9.48)–(9.50) imply that

$$u_1(r,\varepsilon,\ell) = C_1(\ell,\varepsilon)\left(-\mathrm{e}^{\mathrm{i}\pi\ell/2}\mathrm{e}^{-r\varepsilon}\phi(r,\varepsilon,\ell) + \mathrm{e}^{-\mathrm{i}\pi\ell/2}\mathrm{e}^{r\varepsilon}\phi(r,-\varepsilon,\ell)\right) \tag{9.52}$$

where $C_1(\ell,\varepsilon) = \mathrm{e}^{\mathrm{i}\pi\ell/2}\frac{\Gamma(2\ell+2)}{\Gamma(\ell+1)}(2\varepsilon)^{-(\ell+1)}$.

In order to study the asymptotic behavior of $u_2(r,\varepsilon,\ell)$, we need the following integral representation ([98, Section 6.11])

$$\Gamma(\ell+1)W_{0,\ell+1/2}(x) = \mathrm{e}^{-x/2}x^{\ell+1}\int_0^{\infty} \mathrm{e}^{-tx}t^{\ell}(1+t)^{\ell}dt, \quad x > 0. \tag{9.53}$$

Formula (9.53) leads us to the equality

$$\Gamma(\ell+1)W_{0,\ell+1/2}(x) = \mathrm{e}^{-x/2}x^{-\ell}\int_0^{\infty} \mathrm{e}^{-s}s^{\ell}(x+s)^{\ell}ds, \quad x > 0. \tag{9.54}$$

In view of the well-known integral representation

$$\Gamma(z) = \int_0^{\infty} \mathrm{e}^{-s}s^{z-1}ds, \quad \Re z > 0, \tag{9.55}$$

(9.54) takes the form

$$\Gamma(\ell+1)W_{0,\ell+1/2}(x) = \mathrm{e}^{-x/2}x^{-\ell}\sum_{p=0}^{\ell} C_{\ell}^{p}\Gamma(2\ell-p+1)x^{p}, \tag{9.56}$$

where C_ℓ^p are the binomial coefficients. Using analyticity, we rewrite (9.56) for the complex plane

$$\Gamma(\ell+1)W_{0,\ell+1/2}(z) = \mathrm{e}^{-z/2}z^{-\ell}\sum_{p=0}^{\ell} C_\ell^p\Gamma(2\ell-p+1)z^p, \quad z \neq 0. \tag{9.57}$$

Now, the asymptotic relation

$$u_2(r,\varepsilon,\ell) = (2\varepsilon)^\ell \mathrm{e}^{-r\varepsilon}\phi(r,\varepsilon,\ell), \quad \varepsilon\to\infty \tag{9.58}$$

directly follows from (9.43) and (9.57).

Next, we deal with the case $q \not\equiv 0$. We assume that q satisfies the condition

$$\int_0^r |q(t)|dt < \infty, \quad 0 < r < \infty. \tag{9.59}$$

Then, there is a solution u of (9.1) such that the equality

$$u(r,\varepsilon,\ell) = u_1(r,\varepsilon,\ell) + \int_0^r G(r,t,\varepsilon,\ell)q(t)u(t,\varepsilon,\ell)dt \tag{9.60}$$

holds for G of the form

$$G(r,t,\varepsilon,\ell) = \frac{\Gamma(\ell+1)}{\Gamma(2\ell+2)}\left(u_1(r,\varepsilon,\ell)u_2(t,\varepsilon,\ell) - u_2(r,\varepsilon,\ell)u_1(t,\varepsilon,\ell)\right). \tag{9.61}$$

Relations (9.46) and (9.61) imply that

$$(t/r)^{\ell+1}|G(r,t,\varepsilon,\ell)| \leq \mathcal{M}t, \quad t \leq r \leq 1/|\varepsilon|. \tag{9.62}$$

From (9.46), (9.60) and (9.62), we derive that the asymptotics of u is similar to the asymptotics of u_1, namely,

$$u(r,\varepsilon,\ell) \sim r^{\ell+1}, \quad r \leq 1/|\varepsilon|. \tag{9.63}$$

Moreover, according to (9.51), (9.52), (9.58) and (9.61), we have

$$G(r,t,\varepsilon,\ell) \sim \frac{1}{2\varepsilon}\left(\mathrm{e}^{(r-t)\varepsilon} - \mathrm{e}^{-(r-t)\varepsilon}\right), \quad \varepsilon\to\infty. \tag{9.64}$$

Using (9.46) and (9.60)–(9.63), we obtain the relation

$$\int_0^{1/|\varepsilon|} |G(r,t,\varepsilon,\ell)q(t)u(t,\varepsilon,\ell)|dt = o\left(|\varepsilon|^{-(\ell+2)}\right), \quad \varepsilon\to\infty. \tag{9.65}$$

It follows from (9.64) that, for some $\mathcal{M}$, we have

$$|G(r,t,\varepsilon,\ell)| \leq \mathcal{M}\mathrm{e}^{r|\sigma|}/|\varepsilon|, \quad 1/|\varepsilon| \leq t \leq r, \tag{9.66}$$

where $\sigma := \Re(\varepsilon)$. In view of (9.46), (9.51), (9.52), (9.58) and (9.63), the relation

$$\int_0^{1/|\varepsilon|} |G(r,t,\varepsilon,\ell)q(t)u(t,\varepsilon,\ell)|dt = \mathrm{e}^{r|\sigma|}o(1/|\varepsilon|), \quad r > 1/|\varepsilon|, \quad \varepsilon\to\infty \tag{9.67}$$

holds. Formulas (9.60), (9.66) and (9.67) imply the following equality, that is,

$$u(r,\varepsilon,\ell) = u_1(r,\varepsilon,\ell) + O\left(\mathrm{e}^{r|\sigma|}/|\varepsilon|\right), \quad r > 1/|\varepsilon|. \tag{9.68}$$

We denote by $z_n(a,\ell)$ the eigenvalues of problem (9.1), (9.7). It is well known that there is only one (up to a constant factor) solution of (9.1) which turns to zero at $r = 0$. Since u is such a solution, $z_n(a,\ell)$ is an eigenvalue if and only if

$$u(a, \mathrm{i}\sqrt{z_n(a,\ell)}, \ell) = 0. \tag{9.69}$$

By well-known methods (see, e.g. [177, Ch.1, Section 2]), using relations (9.52), (9.58) and (9.68), (9.69), we obtain that

$$\sqrt{z_n(a,\ell)} = \frac{\pi}{a}\left(n + \frac{\ell}{2}\right) + o(1). \tag{9.70}$$

Recall that $\sigma = \Re(\varepsilon)$. Putting

$$\Re(\varepsilon) = 0, \quad (2\varepsilon)^{\ell+1}u(r,\varepsilon,\ell) = \tilde{u}(r,\varepsilon,\ell), \quad (2\varepsilon)^{\ell+1}u_1(r,\varepsilon,\ell) = \tilde{u}_1(r,\varepsilon,\ell),$$

and taking into account estimates (9.64) and (9.68), we deduce from (9.60) that

$$\tilde{u}(r,\varepsilon,\ell) = \tilde{u}_1(r,\varepsilon,\ell) + \frac{1}{2\varepsilon}\int_0^r \left(\mathrm{e}^{(r-t)\varepsilon} - \mathrm{e}^{-(r-t)\varepsilon}\right)\tilde{u}_1(t,\varepsilon,\ell)q(t)dt + o\left(\frac{1}{\varepsilon}\right). \tag{9.71}$$

Thus, we obtain the following assertion.

Theorem 9.17. *Let condition* (9.59) *be fulfilled and equality* $\Re(\varepsilon) = 0$ *hold. Then, the solution* $u(r,\varepsilon,\ell)$ *of the equation* (9.1) *satisfies* (9.63) *and admits representation*

$$u(r,\varepsilon,\ell) = C_1(\ell,\varepsilon)\left(-\mathrm{e}^{\mathrm{i}\pi\ell/2}\mathrm{e}^{-r\varepsilon+\Delta(r)/(2\varepsilon)} + \mathrm{e}^{-\mathrm{i}\pi\ell/2}\mathrm{e}^{r\varepsilon-\Delta(r)/(2\varepsilon)} + o(1/|\varepsilon|)\right), \tag{9.72}$$

where $\Delta(r) = \frac{1}{r}\ell(\ell+1) - \delta(r)$ *and the quantum defect* $\delta(r)$ *has the form*

$$\delta(r) = \int_0^r q(t)dt. \tag{9.73}$$

Recall that (9.9) follows from Theorem 9.17. We denote by $z_n(\ell)$ the positive roots of the Whittaker function $M_{0,\ell+1/2}(\mathrm{i}z)$, which are not equal to zero. Formula (9.9) yields the next well-known result [154, Ch. XI]:

$$z_n(\ell) = \pi\left(n + \frac{\ell}{2}\right) - \frac{\ell(\ell+1)}{\pi(2n+\ell)} + o(n^{-1}). \tag{9.74}$$

9.1.3 Dirac equation and quantum defect

1. We assume that q satisfies (9.4) and (9.13), that is,

$$q(r) = \overline{q(r)}, \quad \int_r^\infty |q(t)|dt < \infty, \quad r > 0. \tag{9.75}$$

Given functions f_1 and f_2, it is easy to find functions Q_1 and Q_2 such that

$$f_1(r,\varepsilon,\ell) = \sqrt{m+z}\,\mathrm{e}^{-r\varepsilon}(2r\varepsilon)^{\ell-1}\left(Q_1(r,\varepsilon,\ell) + Q_2(r,\varepsilon,\ell)\right)r, \tag{9.76}$$

$$f_2(r,\varepsilon,\ell) = -\sqrt{m-z}\,\mathrm{e}^{-r\varepsilon}(2r\varepsilon)^{\ell-1}\left(Q_1(r,\varepsilon,\ell) - Q_2(r,\varepsilon,\ell)\right)r, \tag{9.77}$$

where $\varepsilon = \sqrt{m^2 - z^2}$. Recall that ε is defined more precisely in Remark 9.2. In order to prove Theorem 9.3, following [42], we consider solutions of (9.2)-(9.4) in the form (9.76) and (9.77). Substituting (9.76) and (9.77) into (9.2) and (9.3), we obtain

$$r(Q_1+Q_2)' + 2\ell(Q_1+Q_2) - 2r\varepsilon Q_2 + \sqrt{\frac{m-z}{m+z}}(Q_1-Q_2)rq(r) = 0, \tag{9.78}$$

$$r(Q_1-Q_2)' + 2r\varepsilon Q_2 - \sqrt{\frac{m+z}{m-z}}(Q_1+Q_2)rq(r) = 0, \tag{9.79}$$

where $Q' = \frac{d}{dr}Q$ in this chapter. It is easy to see that

$$\sqrt{\frac{m-z}{m+z}} - \sqrt{\frac{m+z}{m-z}} = -\frac{2}{\varepsilon}z, \quad \sqrt{\frac{m-z}{m+z}} + \sqrt{\frac{m+z}{m-z}} = \frac{1}{\varepsilon}m. \tag{9.80}$$

Hence, if $q(r) \equiv 0$, the equations (9.78) and (9.79) can be rewritten in the form

$$rQ_1' + \ell Q_1 + \ell Q_2 = 0, \quad rQ_2' + \ell Q_2 - 2r\varepsilon Q_2 + \ell Q_1 = 0. \tag{9.81}$$

We substitute into (9.81) functions

$$Q_i(r,\varepsilon,\ell) = Q_{i,0}\left(\rho/(2\varepsilon),\varepsilon,\ell\right) \quad (i = 1,2), \quad \rho = 2r\varepsilon, \tag{9.82}$$

and so switch to the new variable ρ. It follows from (9.81) that

$$\rho\frac{d^2}{d\rho^2}Q_{1,0} + (2\ell+1-\rho)\frac{d}{d\rho}Q_{1,0} - \ell Q_{1,0} = 0, \tag{9.83}$$

$$\rho\frac{d^2}{d\rho^2}Q_{2,0} + (2\ell+1-\rho)\frac{d}{d\rho}Q_{2,0} - (\ell+1)Q_{2,0} = 0. \tag{9.84}$$

The solutions of (9.83) and (9.84), which are regular at the point $\rho = 0$, are the confluent hypergeometric functions $a_1\Phi(\ell, 2\ell+1, \rho)$ and $a_2\Phi(\ell+1, 2\ell+1, \rho)$, respectively ([42, Section 36] and [47, Section 14]). Substitute $r = 0$ into (9.81), (9.82) to see that $a_1 = -a_2$. That is, up to the factor a_1, we have

$$Q_{1,0}(\rho) = \Phi(\ell, 2\ell+1, \rho) = \mathrm{e}^{\rho/2}\rho^{-\ell-1/2}M_{1/2,\ell}(\rho), \tag{9.85}$$

$$Q_{2,0}(\rho) = -\Phi(\ell+1, 2\ell+1, \rho) = -\mathrm{e}^{\rho/2}\rho^{-\ell-1/2}M_{-1/2,\ell}(\rho). \tag{9.86}$$

Now, we need an integral representation from [98, Section 6.11], that is,

$$M_{\pm 1/2,\ell}(\rho) = C(\ell)(\rho)^{\ell+1/2} I_\pm(\rho,\ell), \quad C(\ell) = 2^{-\ell}\frac{\Gamma(2\ell+1)}{\Gamma(\ell+1)\Gamma(\ell)}, \tag{9.87}$$

where

$$I_+(\rho,\ell) = \int_{-1}^{1} e^{t\rho/2}(1-t^2)^{\ell-1}(1-t)dt, \quad I_-(\rho,\ell) = \int_{-1}^{1} e^{t\rho/2}(1-t^2)^{\ell-1}(1+t)dt.$$

Integrating the expressions above by parts (ℓ times), we obtain

$$I_+(\rho,\ell) \sim (-1)^\ell(\rho/2)^{-\ell}\Gamma(\ell)e^{-\rho/2}, \quad \rho\to\infty \quad (\arg(\rho) = \pi/2), \tag{9.88}$$

$$I_-(\rho,\ell) \sim -(\rho/2)^{-\ell}\Gamma(\ell)e^{\rho/2}, \quad \rho\to\infty \quad (\arg(\rho) = \pi/2). \tag{9.89}$$

Using asymptotic formulas (9.88) and (9.89), we have

$$Q_{1,0}(\rho) \sim (-1)^\ell \rho^{-\ell}\Gamma(2\ell+1)/\Gamma(\ell+1), \quad \rho\to\infty \quad (\arg(\rho) = \pi/2), \tag{9.90}$$

$$Q_{2,0}(\rho) \sim -\rho^{-\ell}e^{\rho}\Gamma(2\ell+1)/\Gamma(\ell+1), \quad \rho\to\infty \quad (\arg(\rho) = \pi/2). \tag{9.91}$$

Formulas (9.90) and (9.91) show that the solutions of (9.83) and (9.84) of the form (9.85) and (9.86) are, indeed, regular at $\rho = 0$ and, moreover, we have

$$Q_{1,0} \sim 1, \quad Q_{2,0} \sim 1 \quad \text{for } \rho\to 0. \tag{9.92}$$

The functions

$$\tilde{Q}_{1,0}(\rho) = c_1\Psi(\ell, 2\ell+1, \rho) = c_1 e^{\rho/2}\rho^{-\ell-1/2}W_{1/2,\ell}(\rho), \tag{9.93}$$

$$\tilde{Q}_{2,0}(\rho) = c_2\Psi(\ell+1, 2\ell+1, \rho) = c_2 e^{\rho/2}\rho^{-\ell-1/2}W_{-1/2,\ell}(\rho), \tag{9.94}$$

where $\Psi(a,c,x)$ is the confluent hypergeometric function of the second kind, also satisfy (9.83) and (9.84), but are nonregular at $\rho = 0$ ([98, Ch.6]). According to [98, Section 6.11], we have the following integral representation for $W_{\pm 1/2,\ell}$:

$$W_{\pm 1/2,\ell}(\rho) = C_\pm(\ell)e^{-\rho/2}(\rho)^{\ell+1/2}J_\pm(\rho,\ell), \quad C_\pm(\ell) = 1/\Gamma(\ell+1/2\mp 1/2), \tag{9.95}$$

where

$$J_+(\rho,\ell) = \int_0^\infty e^{-t\rho}t^{\ell-1}(1+t)^\ell dt, \quad J_-(\rho,\ell) = \int_0^\infty e^{-t\rho}t^\ell(1+t)^{\ell-1}dt. \tag{9.96}$$

Using integral representation (9.55), we rewrite (9.96) as

$$J_+(\rho,\ell) = \rho^{-2\ell}\sum_{i=0}^{\ell} C_\ell^i\Gamma(2\ell-i)\rho^i, \quad J_-(\rho,\ell) = \rho^{-2\ell}\sum_{i=0}^{\ell-1} C_{\ell-1}^i\Gamma(2\ell-i)\rho^i, \tag{9.97}$$

where C_ℓ^i are again the binomial coefficients. The right-hand sides in (9.97) are analytic functions (for $\rho \neq 0$). Thus, (9.97) is valid for all $\rho \neq 0$ in the complex plane.

In view of (9.93)–(9.95) and (9.97), if $\rho \to 0$, we have

$$\widetilde{Q}_{1,0}(\rho) \sim c_1 \rho^{-2\ell}\Gamma(2\ell)/\Gamma(\ell), \quad \widetilde{Q}_{2,0}(\rho) \sim c_2 \rho^{-2\ell}\Gamma(2\ell)/\Gamma(\ell+1). \tag{9.98}$$

After the substitution $Q_{1,0} = \widetilde{Q}_{1,0}$ and $Q_{2,0} = \widetilde{Q}_{2,0}$ into (9.82), relations (9.81) and (9.98) imply that we can assume

$$c_1 = 1, \quad c_2 = \ell. \tag{9.99}$$

Using again formulas (9.93)–(9.95) and (9.97), we derive

$$\widetilde{Q}_{1,0}(\rho) \sim \rho^{-\ell}, \quad \widetilde{Q}_{2,0}(\rho) \sim \ell \rho^{-(\ell+1)}, \quad \rho \to \infty \quad (\arg(\rho) = \pi/2). \tag{9.100}$$

The constructed regular and nonregular (at $r = 0$) solutions of (9.2)–(9.3) (or, equivalently, (9.5)), where $q \equiv 0$, are denoted by

$$F_0(r) = F_0(r,\varepsilon,\ell) = \operatorname{col}\left[F_{1,0}(r,\varepsilon,\ell) \quad F_{2,0}(r,\varepsilon,\ell)\right] \tag{9.101}$$

and

$$\widetilde{F}_0(r) = \widetilde{F}_0(r,\varepsilon,\ell) = \operatorname{col}\left[\widetilde{F}_{1,0}(r,\varepsilon,\ell) \quad \widetilde{F}_{2,0}(r,\varepsilon,\ell)\right], \tag{9.102}$$

respectively. According to (9.76) and (9.82), we have

$$F_{1,0}(r) = \sqrt{m+z}\,\mathrm{e}^{-r\varepsilon}(2r\varepsilon)^{\ell-1}\left(Q_{1,0}(r) + Q_{2,0}(r)\right) r, \tag{9.103}$$

$$F_{2,0}(r) = -\sqrt{m-z}\,\mathrm{e}^{-r\varepsilon}(2r\varepsilon)^{\ell-1}\left(Q_{1,0}(r) - Q_{2,0}(r)\right) r; \tag{9.104}$$

$$\widetilde{F}_{1,0}(r) = \sqrt{m+z}\,\mathrm{e}^{-r\varepsilon}(2r\varepsilon)^{\ell-1}\left(\widetilde{Q}_{1,0}(r) + \widetilde{Q}_{2,0}(r)\right) r, \tag{9.105}$$

$$\widetilde{F}_{2,0}(r) = -\sqrt{m-z}\,\mathrm{e}^{-r\varepsilon}(2r\varepsilon)^{\ell-1}\left(\widetilde{Q}_{1,0}(r) - \widetilde{Q}_{2,0}(r)\right) r. \tag{9.106}$$

Hence, from (9.90) and (9.91), we obtain

$$\begin{aligned} F_0(r,\varepsilon,\ell) \sim{}& \frac{\Gamma(2\ell+1)}{2\varepsilon\Gamma(\ell+1)} \\ &\times \operatorname{col}\left[\left((-1)^\ell \mathrm{e}^{-r\varepsilon} - \mathrm{e}^{r\varepsilon}\right)\sqrt{m+z} \quad -\left((-1)^\ell \mathrm{e}^{-r\varepsilon} + \mathrm{e}^{r\varepsilon}\right)\sqrt{m-z}\right], \end{aligned} \tag{9.107}$$

where $z > m$ (i.e. $\arg(\varepsilon(z)) = \pi/2$) and either $\varepsilon \to \infty$ or $r \to \infty$. In a similar way, from (9.100), we derive that

$$\widetilde{F}_0(r,\varepsilon,\ell) \sim \frac{1}{2\varepsilon}\mathrm{e}^{-r\varepsilon}\operatorname{col}\left[\sqrt{m+z} \quad -\sqrt{m-z}\right], \quad z > m, \tag{9.108}$$

where either $\varepsilon \to \infty$ or $r \to \infty$.

2. Now, we consider the case when $q(r) \not\equiv 0$ and introduce 2×2 matrix functions

$$\hat{V}(r) = \begin{bmatrix} 0 & q(r) \\ -q(r) & 0 \end{bmatrix}, \quad U_0(r,\varepsilon,\ell) = \begin{bmatrix} F_{1,0}(r,\varepsilon,\ell) & \widetilde{F}_{1,0}(r,\varepsilon,\ell) \\ F_{2,0}(r,\varepsilon,\ell) & \widetilde{F}_{2,0}(r,\varepsilon,\ell) \end{bmatrix}. \tag{9.109}$$

It is easy to see that the solution $\widetilde{F}(r,\varepsilon,\ell)$ of the equation

$$\widetilde{F}(r) = \widetilde{F}_0(r) + \int_r^\infty U_0(r)U_0(t)^{-1}\hat{V}(t)\widetilde{F}(t)dt \tag{9.110}$$

satisfies the differential system (9.2), (9.3).

The following equality is valid, that is,

$$\det U_0(r,\varepsilon,\ell) = \Gamma(2\ell)/\left(\varepsilon\Gamma(\ell)\right). \tag{9.111}$$

Indeed, it follows from (9.2) and (9.3) that $(\det U_0(r,\varepsilon,\ell))' = 0$. Hence, we obtain the relation $\det U_0(r,\varepsilon,\ell) \equiv \text{const}$. Since the first and second columns of U_0 are F_0 and $\widetilde{F}_0$, respectively, formulas (9.107)–(9.109) imply that

$$\det U_0(r,\varepsilon,\ell) \sim \Gamma(2\ell)/\left(\varepsilon\Gamma(\ell)\right), \quad r\to\infty \quad (z > m), \tag{9.112}$$

which yields (9.111).

According to (9.107)–(9.109) and (9.111), the equality

$$U_0(r)U_0(t)^{-1} = \mathrm{e}^{(r-t)\varepsilon}\Theta_1 + \mathrm{e}^{(t-r)\varepsilon}\Theta_2 + o(1), \quad \varepsilon\to\infty \quad (z > m), \tag{9.113}$$

$$\Theta_1 := \frac{1}{2}\begin{bmatrix} 1 & \mathrm{i} \\ -\mathrm{i} & 1 \end{bmatrix}, \quad \Theta_2 := \frac{1}{2}\begin{bmatrix} 1 & -\mathrm{i} \\ \mathrm{i} & 1 \end{bmatrix} \tag{9.114}$$

holds. We multiply both sides of (9.110) by $-2\sqrt{m+z}\,\mathrm{e}^{r\varepsilon}$. Using (9.113) and passing to the limit $z \to +\infty$, we obtain

$$\widetilde{F}_\infty(r,\ell) = \begin{bmatrix} \mathrm{i} \\ 1 \end{bmatrix} + \int_r^\infty T_2\hat{V}(t)\widetilde{F}_\infty(t,\ell)dt, \tag{9.115}$$

where

$$\widetilde{F}_\infty(r,\ell) = -2\lim_{z\to+\infty}\left(\sqrt{m+z}\,\mathrm{e}^{r\varepsilon}\widetilde{F}(r,\varepsilon,\ell)\right). \tag{9.116}$$

The equality

$$\widetilde{F}_\infty(r,\ell) = \mathrm{e}^{-\mathrm{i}\int_r^\infty q(t)dt}\begin{bmatrix} \mathrm{i} \\ 1 \end{bmatrix} \tag{9.117}$$

follows directly from (9.108), (9.109), (9.114) and (9.115). From (9.116) and (9.117), we see that the constructed solution $\widetilde{F}(r,\varepsilon,\ell)$ has the properties stated in Theorem 9.3, that is, Theorem 9.3 is proved.

9.1.4 Proofs of Theorems 9.10 and 9.14

In order to show that the results on the multidimensional Schrödinger equation from Subsection 9.1.1 hold, we should prove Theorems 9.10 and 9.14. First, we prove the assertion below (see [294, 299]).

Lemma 9.18. *The following asymptotic relation*

$$\sum_{n=1}^{\infty} \mathrm{e}^{-n^2/z} = \frac{1}{2}\sqrt{z\pi} - \frac{1}{2} + o(1), \quad z \to \infty \tag{9.118}$$

is valid.

Proof. In the proof of (9.118), we use the Poisson formula (see, e.g. [100]):

$$\sum_{n=1}^{\infty} \mathrm{e}^{-n^2/z} = -\frac{1}{2} + \int_0^{\infty} \mathrm{e}^{-x^2/z} dx + 2\sum_{n=1}^{\infty} \int_0^{\infty} \mathrm{e}^{-x^2/z} \cos(2\pi n x) dx. \tag{9.119}$$

Since

$$\int_0^{\infty} \mathrm{e}^{-x^2/z} \cos(2\pi n x) dx = \frac{1}{2}\sqrt{z\pi}\, \mathrm{e}^{-zn^2\pi^2}, \tag{9.120}$$

formula (9.119) implies that

$$\sum_{n=1}^{\infty} \mathrm{e}^{-n^2/z} = -\frac{1}{2} + \frac{1}{2}\sqrt{z\pi} + \sqrt{z\pi} \sum_{n=1}^{\infty} \mathrm{e}^{-zn^2\pi^2}. \tag{9.121}$$

Relation (9.118) follows directly from (9.121). □

Proof of Theorem 9.10. In view of (9.18), we have

$$z_n = (n\pi/a)^2 + (1/a)\delta(a) + o(1), \quad n \to \infty. \tag{9.122}$$

Substituting $z = T(a/\pi)^2$ into (9.118) and taking into account (9.122), we derive (9.20), where Z is given by (9.19). □

In the proof of Theorem 9.14, the next well-known relations

$$\sum_{n=1}^{\infty} \mathrm{e}^{-2n/T} = \frac{1}{2\sinh(1/T)} = \frac{1}{2}\left(T + O(1/T)\right), \quad T \to \infty; \tag{9.123}$$

$$\sum_{n=1}^{\infty} \mathrm{e}^{-2n/T}/\sqrt{2n} = \int_0^{\infty} (\mathrm{e}^{-2x/T}/\sqrt{2x}) dx + O(1), \quad T \to \infty \tag{9.124}$$

are used instead of the Poisson formula, which was applied in the previous proof. Since

$$\int_0^{\infty} (\mathrm{e}^{-u}/\sqrt{u})\, du = \sqrt{\pi}, \tag{9.125}$$

the assertion of the Theorem 9.14 follows from (9.123) and (9.124).

9.1.5 Dirac system on a finite interval

The considerations from the previous subsections can also be used to deal with the Dirac system on a finite interval. In this final subsection of this section, we consider the Dirac system (9.2), (9.3) on the segment $[0, a]$ and assume that the inequality

$$\int_0^a |q(t)|dt < \infty \tag{9.126}$$

holds. The solution $F(r, \varepsilon, \ell)$ of the integral equation

$$F(r, \varepsilon, \ell) = F_0(r, \varepsilon, \ell) - \int_0^r U_0(r)U_0(t)^{-1}\widehat{V}(t)F(t, \varepsilon, \ell)dt, \tag{9.127}$$

where U_0 and $\widehat{V}$ are given by (9.109), satisfies the system (9.2), (9.3). Here, F_0 is a regular solution of the system (9.2), (9.3) with $q \equiv 0$ and is given by (9.101), (9.103) and (9.104). The parameter $\varepsilon = \varepsilon(z)$ above is the same as in Remark 9.2 and in Subsection 9.1.3. Recall that $B(0, l]$, or $B((0, l])$, stands for the class of bounded functions on $(0, l]$, that is, such functions the values of which are uniformly bounded in the norm. Below, we consider functions bounded with respect to the variable r. From formulas (9.85), (9.86), (9.103) and (9.104), we derive

$$F_0(r, \varepsilon, \ell)/\left(\sqrt{m+z}(2r\varepsilon)^{\ell-1}r\right) \in B(0, 1/|\varepsilon|] \quad (z > m). \tag{9.128}$$

In fact, the left-hand side of (9.128) is bounded when $z > m$ and $\rho = 2r\varepsilon$ is bounded. In a similar way, using (9.93), (9.94), (9.105) and (9.106), we have

$$\widetilde{F}_0(r, \varepsilon, \ell)/\left(\sqrt{m+z}(2r\varepsilon)^{-\ell-1}r\right) \in B(0, 1/|\varepsilon|] \quad (z > m). \tag{9.129}$$

According to (9.90), (9.91) and (9.98), the relation

$$U_0(r) \in B(0, 1/|\varepsilon|] \quad (z > m) \tag{9.130}$$

is valid. Relations (9.109), (9.128), (9.129) and (9.130) imply that

$$\sqrt{m+z}(2r\varepsilon)^{\ell-1}rU_0(r)^{-1} \in B(0, 1/|\varepsilon|] \quad (z > m) \tag{9.131}$$

and

$$\sup\left\|(t/r)^{2\ell}U_0(r)U_0(t)^{-1}\right\| < \infty, \quad 0 \le t \le r \le 1/|\varepsilon| \quad (z > m). \tag{9.132}$$

It follows from (9.127), (9.128) and (9.132) that

$$F(r, \varepsilon, \ell)/\left(\sqrt{m+z}(2r\varepsilon)^{\ell-1}r\right) \in [B(0, 1/|\varepsilon] \quad (z > m). \tag{9.133}$$

In view of (9.130), (9.131) and (9.133), the relation

$$\left\| \int_0^{1/|\varepsilon|} U_0(r)U_0(t)^{-1}\widehat{V}(t)F(t,\varepsilon,\ell)dt \right\| = o(1), \quad \varepsilon \to \infty \quad (z > m) \tag{9.134}$$

is valid. Substituting (9.90) and (9.91) into (9.103) and (9.104), we represent $F_0(r,\varepsilon,\ell)$ (for the case that $\varepsilon = \varepsilon(z)$, $z > m$) in the form

$$\mathrm{e}^{-\mathrm{i}\ell\pi/2}F_0(r,\varepsilon,\ell) = \mathrm{e}^{-r\varepsilon}\mathrm{e}^{\mathrm{i}\ell\pi/2}g_1 + \mathrm{e}^{r\varepsilon}\mathrm{e}^{-\mathrm{i}\ell\pi/2}g_2 + o(1), \quad \varepsilon \to \infty, \tag{9.135}$$

$$g_1 := -\frac{\Gamma(2\ell+1)}{2\Gamma(\ell+1)}\mathrm{e}^{\mathrm{i}\ell\pi/2}\begin{bmatrix} \mathrm{i} \\ 1 \end{bmatrix}, \quad g_2 := \frac{\Gamma(2\ell+1)}{2\Gamma(\ell+1)}\mathrm{e}^{\mathrm{i}\ell\pi/2}\begin{bmatrix} \mathrm{i} \\ -1 \end{bmatrix}. \tag{9.136}$$

Now, we can prove our next theorem.

Theorem 9.19. *Let condition* (9.126) *hold. Then (for the case that* $\varepsilon = \varepsilon(z)$, $z > m$*), we have*

$$\mathrm{e}^{-\mathrm{i}\ell\pi/2}F(r,\varepsilon,\ell) = \mathrm{e}^{-r\varepsilon}\mathrm{e}^{\mathrm{i}\ell\pi/2}h_1(r) + \mathrm{e}^{r\varepsilon}\mathrm{e}^{-\mathrm{i}\ell\pi/2}h_2(r) + o(1), \quad \varepsilon \to \infty, \tag{9.137}$$

$$h_1(r) := -\frac{\Gamma(2\ell+1)}{2\Gamma(\ell+1)}\mathrm{e}^{\mathrm{i}\delta(r)}\begin{bmatrix} \mathrm{i} \\ 1 \end{bmatrix}, \quad h_2(r) := \frac{\Gamma(2\ell+1)}{2\Gamma(\ell+1)}\mathrm{e}^{-\mathrm{i}\delta(r)}\begin{bmatrix} \mathrm{i} \\ -1 \end{bmatrix}, \tag{9.138}$$

where the quantum defect $\delta(r)$ *has the form*

$$\delta(r) = \int_0^r q(t)dt. \tag{9.139}$$

Proof. In order to prove the theorem, we represent $F(r,\varepsilon,\ell)$ in the form

$$\mathrm{e}^{-\mathrm{i}\ell\pi/2}F(r,\varepsilon,\ell) = \mathrm{e}^{-r\varepsilon}\mathrm{e}^{\mathrm{i}\ell\pi/2}h_1(r) + \mathrm{e}^{r\varepsilon}\mathrm{e}^{-\mathrm{i}\ell\pi/2}h_2(r) + \widehat{f}(r,\varepsilon,\ell) \tag{9.140}$$

and estimate $\widehat{f}$. Definitions (9.136) and (9.138) imply that

$$h_1(r) = g_1 - \int_0^r T_1\widehat{V}(t)h_1(t)dt, \quad h_2(r) = g_2 - \int_0^r T_2\widehat{V}(t)h_2(t)dt. \tag{9.141}$$

Let us multiply both sides of (9.140) by $\mathrm{e}^{\mathrm{i}\ell\pi/2}$ and substitute the result into (9.127). Then, formula (9.141) implies that the function

$$G(r,\varepsilon,\ell) = \widehat{f}(r,\varepsilon,\ell) + \int_0^r U_0(r)U_0(t)^{-1}\widehat{V}(t)\widehat{f}(t,\varepsilon,\ell)dt \tag{9.142}$$

satisfies the relation

$$\|G(r,\varepsilon,\ell)\| \to 0, \quad z \to +\infty. \tag{9.143}$$

Using (9.126), (9.142) and (9.143), we obtain

$$\|\widehat{f}(r, \varepsilon, \ell)\| \to 0, \quad z \to +\infty. \tag{9.144}$$

□

Recall that $F = \operatorname{col}[F_1 \quad F_2]$ is the regular solution of the radial Dirac system. Consider the case of the second boundary condition

$$F(a, \varepsilon_n, \ell) \sin \psi + F_2(a, \varepsilon_n, \ell) \cos \psi = 0, \quad -\pi/2 \le \psi \le \pi/2. \tag{9.145}$$

Here, $\varepsilon_n = \varepsilon(z_n) = \sqrt{m^2 - z_n^2}$ and (differently from other considerations of this section, where n always belongs $\mathbb{N}$) $n \in \mathbb{Z} - \{0\}$. Without loss of generality, we assume that $z_n > m$ for $n > 0$ and $z_n < -m$ for $n < 0$.

Corollary 9.20. *Let conditions* (9.126) *and* (9.145) *be fulfilled. Then, we have*

$$z_n(a) = \frac{\pi}{a}\left(n + \frac{\ell}{2}\right) + \frac{\psi - \pi/2 + \delta(a)}{a} + o(1). \tag{9.146}$$

Proof. It follows from Theorem 9.19 that

$$F_1(r, \varepsilon_n, \ell) = -\mathrm{e}^{\mathrm{i}\ell\pi/2} \frac{\Gamma(2\ell+1)}{\Gamma(\ell+1)} \sin(rs_n - \ell\pi/2 - \delta(r)) + o(1), \tag{9.147}$$

$$F_2(r, \varepsilon_n, \ell) = -\mathrm{e}^{\mathrm{i}\ell\pi/2} \frac{\Gamma(2\ell+1)}{\Gamma(\ell+1)} \cos(rs_n - \ell\pi/2 - \delta(r)) + o(1), \tag{9.148}$$

where $\varepsilon_n = \mathrm{i}\sqrt{z_n^2 - m^2} = \mathrm{i}s_n \ (n > 0,\ z_n > m,\ s_n > 0)$. In view of (9.145), (9.147) and (9.148), the relation

$$\cos(as_n - \ell\pi/2 - \delta(a) - \psi) = o(1) \tag{9.149}$$

is valid. Hence, the equality (9.146) is proved for $n > 0$. The case $n < 0$ can be dealt with in the same way. □

Remark 9.21. Formula (9.146) is essential for solving the corresponding inverse sliding problem.

Remark 9.22. When $\ell = 0$, formula (9.146) is well known ([195, Ch.VII]).

9.2 Schrödinger and Dirac equations with Coulomb-type potentials

The radial Schrödinger equation with the Coulomb-type potential has the form

$$\frac{d^2y}{dr^2} + \left(z + \frac{2a}{r} - \frac{\ell(\ell+1)}{r^2} - q(r)\right) y = 0, \quad 0 \le r < \infty, \quad a = \overline{a} \ne 0, \tag{9.150}$$

where $\ell = 0, 1, 2, \dots$ Its potential differs from the potential in (9.1) by the additional term $\frac{2a}{r}$. The radial Dirac system (relativistic case) with the Coulomb-type potential has the form

$$\left(\frac{d}{dr} + \frac{\ell}{r}\right) f_1 - \left(z + m + \frac{a}{r} - q(r)\right) f_2 = 0, \tag{9.151}$$

$$\left(\frac{d}{dr} - \frac{\ell}{r}\right) f_2 + \left(z - m + \frac{a}{r} - q(r)\right) f_1 = 0, \tag{9.152}$$

$$0 \le r < \infty, \quad m > 0, \quad q = \overline{q}, \quad a = \overline{a} \neq 0, \quad \ell = \pm 1, \pm 2, \dots, \quad \ell^2 > a^2. \tag{9.153}$$

Equations (9.151) and (9.152) differ from the equations (9.2) and (9.3) by the additional term $\frac{a}{r}$. Like in Section 9.1, without loss of generality, we only consider the case $\ell > 0$.

The scheme presented in Section 9.1 admits modification for the case of Coulomb-type potentials. Equation (9.150) and system (9.150), (9.151) are essential in the study of the spectrum of atoms and molecules [42, 47, 313].

We assume that

$$\int_r^\infty |q(t)|dt < \infty, \quad r > 0. \tag{9.154}$$

In the present section, we describe the asymptotic behavior of the solutions of equation (9.150) and of system (9.151), (9.152) for the energy tending to infinity (i.e. $z \to \infty$). Using asymptotics, we introduce the notion of the quantum defect and solve the corresponding sliding inverse problem for the relativistic case.

9.2.1 Asymptotics of the solutions: Schrödinger equation

If $q(r) \equiv 0$, then (9.150) takes the form

$$\frac{d^2 y}{dr^2} + \left(z + \frac{2a}{r} - \frac{\ell(\ell+1)}{r^2}\right) y = 0. \tag{9.155}$$

We denote by u_1 and u_2 the solutions of (9.155) such that

$$u_1(r, \varepsilon, \ell) = (2\varepsilon)^{-(\ell+1)} M_{a/\varepsilon, \ell+1/2}(2r\varepsilon), \quad u_2(r, \varepsilon, \ell) = (2\varepsilon)^{\ell} W_{a/\varepsilon, \ell+1/2}(2r\varepsilon), \tag{9.156}$$

where $M_{\kappa,\mu}(z)$ and $W_{\kappa,\mu}(z)$ are Whittaker functions. As in Subsection 9.1.2, the parameter $\varepsilon = \mathrm{i}\sqrt{z}$ is defined for all z in the complex plane with the cut along the negative part of the imaginary axis. We put $\arg \varepsilon = \pi/2$ for $z > 0$. Recall the connections between the Whittaker and confluent hypergeometric functions ([98]), namely,

$$M_{a/\varepsilon, \ell+1/2}(x) = \mathrm{e}^{-x/2} x^{c/2} \Phi(\alpha, c, x), \quad W_{a/\varepsilon, \ell+1/2}(x) = \mathrm{e}^{-x/2} x^{c/2} \Psi(\alpha, c, x), \tag{9.157}$$

where $\alpha = \ell + 1 - a/\varepsilon, \quad c = 2\ell + 2$. Instead of asymptotics in (9.45), we now need the asymptotics of Φ and Ψ for energy tending to infinity. More precisely, we need the following well-known formulas ([98, Section 6.13]), that is,

$$\Phi(\alpha, c, 2r\varepsilon) \sim \frac{\Gamma(2\ell+2)}{\Gamma(\ell+1)}(-2r\varepsilon)^{-\ell-1} + \frac{\Gamma(2\ell+2)}{\Gamma(\ell+1)}(2r\varepsilon)^{-\ell-1}\mathrm{e}^{2r\varepsilon}, \tag{9.158}$$

$$\Psi(\alpha, c, 2r\varepsilon) \sim (2r\varepsilon)^{-\ell-1}, \text{ where } \varepsilon \to \infty \text{ and } \arg(\varepsilon(z)) = \pi/2 \text{ (i.e. } z > 0). \tag{9.159}$$

Using (9.158) and (9.159), we obtain the relations

$$u_1(r, \varepsilon) \sim \frac{\Gamma(2\ell+2)}{\Gamma(\ell+1)}\left((-1)^{\ell+1}\mathrm{e}^{-r\varepsilon} + \mathrm{e}^{r\varepsilon}\right)(2\varepsilon)^{-(\ell+1)}, \quad \varepsilon \to \infty, \quad z > 0, \tag{9.160}$$

$$u_2(r, \varepsilon) \sim (2\varepsilon)^{\ell}\mathrm{e}^{-r\varepsilon}, \quad \varepsilon \to \infty, \quad z > 0. \tag{9.161}$$

Next, we consider the general-type Schrödinger equation (9.150) (where q is not necessarily trivial). The solution $u(r, \varepsilon, \ell)$ of the integral equation

$$u(r, \varepsilon, \ell) = u_2(r, \varepsilon, \ell) - \int_r^{\infty} k(r, t, \varepsilon, \ell)q(t)u(t, \varepsilon, \ell)dt, \tag{9.162}$$

where the kernel $k(r, t, \varepsilon, \ell)$ is defined by

$$k(r, t, \varepsilon, \ell) = \frac{\Gamma(\ell+1-a/\varepsilon)}{\Gamma(2\ell+2)}\left(u_1(r, \varepsilon, \ell)u_2(t, \varepsilon, \ell) - u_2(r, \varepsilon, \ell)u_1(t, \varepsilon, \ell)\right), \tag{9.163}$$

satisfies (9.150). It follows from (9.160), (9.161) and (9.163) that

$$k(r, t, \varepsilon, \ell) \sim \mathrm{e}^{(r-t)\varepsilon} - \mathrm{e}^{-(r-t)\varepsilon}, \quad \varepsilon \to \infty, \quad z > 0. \tag{9.164}$$

Using standard methods, we deduce from (9.161)–(9.164) the following statement.

Theorem 9.23. *Let* (9.154) *hold. Then, for* $u(r, \varepsilon, \ell)$ *satisfying* (9.162), *we have*

$$u(r, \varepsilon, \ell) \sim (2\varepsilon)^{-\ell}\mathrm{e}^{-r\varepsilon}, \quad \varepsilon \to \infty, \quad z > 0. \tag{9.165}$$

9.2.2 Asymptotics of the solutions: Dirac system

If $q(r) \equiv 0$, system (9.151), (9.152) takes the form

$$\left(\frac{d}{dr} + \frac{\ell}{r}\right)f_1 - (z+m+\frac{a}{r})f_2 = 0, \quad \left(\frac{d}{dr} - \frac{\ell}{r}\right)f_2 + (z-m+\frac{a}{r})f_1 = 0 \quad (\ell^2 > a^2). \tag{9.166}$$

We consider solutions of (9.166) in the form similar to (9.76) and (9.77), that is,

$$f_1(r, \varepsilon, \ell) = \sqrt{m+z}\,\mathrm{e}^{-r\varepsilon}(2r\varepsilon)^{\omega-1}\left(Q_1(r, \varepsilon, \ell) + Q_2(r, \varepsilon, \ell)\right)r, \tag{9.167}$$

$$f_2(r, \varepsilon, \ell) = -\sqrt{m-z}\,\mathrm{e}^{-r\varepsilon}(2r\varepsilon)^{\omega-1}\left(Q_1(r, \varepsilon, \ell) - Q_2(r, \varepsilon, \ell)\right)r, \tag{9.168}$$

where $\omega = \sqrt{\ell^2 - a^2} > 0$, $\varepsilon = \sqrt{m^2 - z^2}$ and the choice of arguments in $\sqrt{m \pm z}$ and $\sqrt{m^2 - z^2}$ is prescribed in Remark 9.2. For regular and nonregular solutions of (9.166) we use here the same notations as for regular and nonregular solutions of (9.2), (9.3), where $q \equiv 0$ in Section 9.1. For the case of regular (at $r = 0$) solutions of (9.166), the functions Q_1 and Q_2 can be expressed via the confluent hypergeometric functions $\Phi(\alpha, c, x)$ ([42, 47]). Thus, a regular solution F_0 of (9.166) is given by

$$F_0 = \operatorname{col}\begin{bmatrix} F_{1,0} & F_{2,0} \end{bmatrix}, \tag{9.169}$$

$$F_{1,0}(r,\varepsilon,\ell) = \sqrt{m+z}\,\mathrm{e}^{-r\varepsilon}(2r\varepsilon)^{\omega-1}\left(Q_{1,0}(r,\varepsilon,\ell) + Q_{2,0}(r,\varepsilon,\ell)\right) r, \tag{9.170}$$

$$F_{2,0}(r,\varepsilon,\ell) = -\sqrt{m-z}\,\mathrm{e}^{-r\varepsilon}(2r\varepsilon)^{\omega-1}\left(Q_{1,0}(r,\varepsilon,\ell) - Q_{2,0}(r,\varepsilon,\ell)\right) r, \tag{9.171}$$

where

$$Q_{1,0} = \alpha_1\Phi(\omega - az/\varepsilon, 2\omega+1, 2r\varepsilon), \quad Q_{2,0} = \alpha_2\Phi(\omega + 1 - az/\varepsilon, 2\omega+1, 2r\varepsilon), \tag{9.172}$$

and

$$\frac{\alpha_1 + \alpha_2}{\alpha_1 - \alpha_2} = -\frac{a}{\omega+\ell}\sqrt{\frac{m-z}{m+z}}. \tag{9.173}$$

Using asymptotic formulas for the confluent hypergeometric functions $\Phi(\alpha, c, x)$ ([98]), we obtain

$$Q_{1,0}(r,\varepsilon,\ell) \sim \hat{\alpha}_1\frac{\Gamma(2\omega+1)}{\Gamma(\omega+1-\mathrm{i}a)}(2r\varepsilon)^{-\omega-\mathrm{i}a}, \quad \varepsilon\to\infty \quad (z > m), \tag{9.174}$$

$$Q_{2,0}(r,\varepsilon,\ell) \sim \hat{\alpha}_2\frac{\Gamma(2\omega+1)}{\Gamma(\omega+1+\mathrm{i}a)}\mathrm{e}^{2r\varepsilon}(2r\varepsilon)^{-\omega+\mathrm{i}a}, \quad \varepsilon\to\infty \quad (z > m), \tag{9.175}$$

where

$$\frac{\hat{\alpha}_1 + \hat{\alpha}_2}{\hat{\alpha}_1 - \hat{\alpha}_2} = -\frac{\mathrm{i}a}{\omega+\ell}. \tag{9.176}$$

A nonregular at $r = 0$ solution $\widetilde{F}_0 = \operatorname{col}[\widetilde{F}_{1,0} \;\; \widetilde{F}_{2,0}]$ of (9.166) has the form [42, 47]

$$\widetilde{F}_{1,0}(r,\varepsilon,\ell) = \sqrt{m+z}\,\mathrm{e}^{-r\varepsilon}(2r\varepsilon)^{\omega-1}(\widetilde{Q}_{1,0}(r,\varepsilon,\ell) + \widetilde{Q}_{2,0}(r,\varepsilon,\ell))r, \tag{9.177}$$

$$\widetilde{F}_{2,0}(r,\varepsilon,\ell) = -\sqrt{m-z}\,\mathrm{e}^{-r\varepsilon}(2r\varepsilon)^{\omega-1}(\widetilde{Q}_{1,0}(r,\varepsilon,\ell) - \widetilde{Q}_{2,0}(r,\varepsilon,\ell))r, \tag{9.178}$$

where

$$\widetilde{Q}_{1,0}(r,\varepsilon,\ell) = \tilde{\alpha}_1\Psi(\omega - az/\varepsilon, 2\omega+1, 2r\varepsilon), \tag{9.179}$$

$$\widetilde{Q}_{2,0}(r,\varepsilon,\ell) = \tilde{\alpha}_2\Psi(\omega + 1 - az/\varepsilon, 2\omega+1, 2r\varepsilon). \tag{9.180}$$

Here, $\Psi(a, c, x)$ is the confluent hypergeometric function of the second kind. In view of (9.179) and (9.180), we have ([98, Ch.6])

$$\widetilde{Q}_{1,0} \sim \tilde{\alpha}_1(2r\varepsilon)^{-2\omega}, \quad \widetilde{Q}_{2,0} \sim \tilde{\alpha}_2(2r\varepsilon)^{-2\omega}, \quad r\to\infty; \quad \tilde{\alpha}_1 = \frac{\ell\varepsilon - am}{\omega\varepsilon + az}\tilde{\alpha}_2. \tag{9.181}$$

Using again asymptotic formulas for $\Phi(\alpha, c, x)$ ([98]), we obtain

$$\widetilde{Q}_{1,0} \sim \breve{\alpha}_1 (2r\varepsilon)^{-\omega-ia}, \quad \varepsilon \to \infty \quad (z > m), \tag{9.182}$$

$$\widetilde{Q}_{2,0} \sim \breve{\alpha}_2 (2r\varepsilon)^{-\omega-ia-1}, \quad \varepsilon \to \infty \quad (z > m), \tag{9.183}$$

where

$$\breve{\alpha}_1 = \frac{\ell}{\omega - ia} \breve{\alpha}_2. \tag{9.184}$$

From (9.170), (9.171), (9.174) and (9.175), we obtain

$$\begin{aligned} F_{1,0}(r,\varepsilon,\ell) \sim & \sqrt{m+z}\,\Gamma(2\omega+1) \\ & \times \left(e^{-r\varepsilon} \widehat{\alpha}_1 \frac{(2r\varepsilon)^{-ia}}{2\varepsilon\Gamma(\omega+1-ia)} + e^{r\varepsilon} \widehat{\alpha}_2 \frac{(2r\varepsilon)^{ia}}{2\varepsilon\Gamma(\omega+1+ia)} \right), \end{aligned} \tag{9.185}$$

$$\begin{aligned} F_{2,0}(r,\varepsilon,\ell) \sim & -\sqrt{m-z}\,\Gamma(2\omega+1) \\ & \times \left(e^{-r\varepsilon} \widehat{\alpha}_1 \frac{(2r\varepsilon)^{-ia}}{2\varepsilon\Gamma(\omega+1-ia)} - e^{r\varepsilon} \widehat{\alpha}_2 \frac{(2r\varepsilon)^{ia}}{2\varepsilon\Gamma(\omega+1+ia)} \right), \end{aligned} \tag{9.186}$$

where $\varepsilon \to \infty$, $z > m$. Formulas (9.177), (9.178) and (9.182)–(9.184) imply that

$$\widetilde{F}_0(r,\varepsilon,\ell) \sim \frac{1}{2\varepsilon} \breve{\alpha}_1 e^{-r\varepsilon} (2r\varepsilon)^{-ia} \operatorname{col}[\sqrt{m+z} \quad -\sqrt{m-z}], \quad \varepsilon \to \infty \quad (z > m). \tag{9.187}$$

Next, we consider the case $q(r) \not\equiv 0$. The solution $\widetilde{F}(r,\varepsilon,\ell)$ of the integral equation

$$\widetilde{F}(r,\varepsilon,\ell) = \widetilde{F}_0(r,\varepsilon,\ell) + \int_r^\infty U_0(r) U_0(t)^{-1} \widehat{V}(t) \widetilde{F}(t,\varepsilon,\ell)\,dt, \tag{9.188}$$

where $\widehat{V}$ and U_0 have the form (9.109), satisfies system (9.151), (9.152). We note that $F_{i,0}$ and $\widetilde{F}_{i,0}$ $(i = 1,2)$ in Section 9.1 are different from the entries $F_{i,0}$ and $\widetilde{F}_{i,0}$ of U_0 which are introduced in this section. For the present case, formulas (9.185)–(9.187) imply that

$$\det U_0(r,\varepsilon,\ell) \sim -\widehat{\alpha}_2 \breve{\alpha}_1 \frac{\Gamma(2\omega+1)}{2\varepsilon\Gamma(\omega+1+ia)}, \quad \varepsilon \to \infty \quad (z > m). \tag{9.189}$$

According to (9.185)–(9.187) and (9.189), the equality (9.113) from Section 9.1 is valid again. Hence, multiplying both sides of (9.188) by $-2\sqrt{m+z}\, e^{r\varepsilon}(2r\varepsilon)^{ia}$ and passing to the limit $z \to +\infty$, we obtain

$$\widetilde{F}_\infty(r,\ell) = \begin{bmatrix} i \\ 1 \end{bmatrix} + \int_r^\infty \Theta_1 \widehat{V}(t) \widetilde{F}_\infty(t,\ell)\,dt \tag{9.190}$$

where

$$\widetilde{F}_\infty(r,\ell) = -2 \lim_{z\to+\infty} \left(\sqrt{m+z}\, e^{r\varepsilon} (2r\varepsilon)^{ia} \widetilde{F}(r,\varepsilon,\ell) \right). \tag{9.191}$$

The equality

$$\widetilde{F}_{\infty}(r,\ell) = \mathrm{e}^{-\mathrm{i}\int_r^{\infty} q(t)dt}\begin{bmatrix}\mathrm{i}\\1\end{bmatrix} \tag{9.192}$$

follows directly from (9.190)–(9.192). Thus, we proved that $\widetilde{F}(r,\varepsilon,\ell)$ constructed above satisfies the requirements of the following statement.

Theorem 9.24. *Let condition* (9.154) *be fulfilled. Then, there is a solution* $\widetilde{F}(r,\varepsilon,\ell)$ *of system* (9.151), (9.152), *which satisfies the relation*

$$\widetilde{F}(r,\varepsilon,\ell) \sim -\frac{1}{2\sqrt{m+z}}\,\mathrm{e}^{-\mathrm{i}(r\varepsilon+\delta(r))}(2r\varepsilon)^{-\mathrm{i}a}\begin{bmatrix}\mathrm{i}\\1\end{bmatrix}, \quad \varepsilon\to\infty \quad (z>m), \tag{9.193}$$

where the quantum defect $\delta(r)$ *is given by the formula*

$$\delta(r) = \int_r^{\infty} q(t)dt. \tag{9.194}$$

Finally, we formulate a sliding half-inverse problem.

Problem 9.25. *Recover the pertubation potential* $q(r)$ *of the Dirac system* (9.151), (9.152) *from the given quantum defect* δ *and constant* a.

According to (9.194), the solution of Problem 9.25 has the form

$$q(r) = -\frac{d}{dr}\delta(r). \tag{9.195}$$

Appendices

A General-type canonical system: pseudospectral and Weyl functions

As shown in Chapter 1, important classical equations can be reduced to subclasses of the canonical system (0.2). However, general-type canonical system (0.2) is of independent interest. There are also many difficulties, especially, for the case where $\det H(x)$ turns to zero on a set of a nonzero measure. We already noted that $\det H(x) \equiv 0$ in the case of the canonical system (1.36) but, for that case, we have a simple transformation of (1.36) into the canonical system (1.8), (1.9), where $\det H(x) \neq 0$.

Recall that the fundamental solution W of the canonical system is normalized by the condition $W(0,z) = I_m$, that is, W is defined by the relations

$$dW(x,z)/dx = \mathrm{i}zJH(x)W(x,z), \quad W(0,z) = I_m, \tag{A.1}$$

$$m = 2p, \quad J = \begin{bmatrix} 0 & I_p \\ I_p & 0 \end{bmatrix}, \quad H(x) = H(x)^* \geq 0. \tag{A.2}$$

In this Appendix, we derive a description of the spectral and pseudospectral functions on the interval $[0, l]$. The description in Section A.1 is given in the general situation. Here, we do not assume (like in Chapters 1 and 2) the existence of the S-node.

In particular, we obtain simple sufficient requirements such that the important (refer to Subsection 1.2.3 and books [289, 290]) inequality

$$\int_{-\infty}^{\infty} \frac{d\tau(t)}{1+t^2} < \infty, \tag{A.3}$$

where τ is a spectral or pseudospectral function, holds.

In Section A.2, we consider system (A.1) under the "positivity" condition

$$\int_0^l H(x)dx > 0$$

which also includes cases where H is degenerate.

Orthogonal polynomials, their properties, asymptotics and spectral theory are of great current interest (see [310] and references therein). In Subsection A.2.2, we express asymptotics of the continuous analogs of the orthogonal polynomials in terms of the spectral functions.

Appendix A is an extension of [240]. The results of the main Section A.1 are closely related to the seminal book [85], where the subcase $p = 1$ was studied. See also an interesting paper [138], where J-theory and Potapov's fundamental (principal) matrix inequality (FMI) were applied in the new proof of the results from [85]. We use

FMI and Potapov's transformed matrix inequality (TFMI) in Subsection A.1.3. Usually, Potapov's inequalities are expressed in terms of the S-nodes (refer to the papers [145, 159, 225], a book [289, pp. 4–9] and formula (1.125) in Chapter 1), but one cannot require the existence of the S-nodes in the general situation considered in this Appendix. However, a certain modification of FMI and TFMI proves very useful here. Finally, we refer to [289, Ch4], where interesting connections between de Branges spaces and operator identities are studied, and to the paper [228] on Schrödinger operators and de Branges spaces.

Unlike some previous chapters, the notations α, ω and ζ stand in Appendix A for complex scalars (constants or variables).

A.1 Spectral and pseudospectral functions

A.1.1 Basic notions and results

We shall assume in what follows that the entries of the Hamiltonian $H(x)$ of the system (A.1) are summable on $[0, l]$. We denote by $L^2(H, l)$ or simply by $L^2(H)$ the space of vector functions on $(0, l)$ with the inner product

$$(f, \widetilde{f})_H = \int_0^l \widetilde{f}(x)^* H(x) f(x) dx. \tag{A.4}$$

We also introduce the operator

$$\widehat{U} f = F, \quad F(z) = \begin{bmatrix} F_1(z) \\ F_2(z) \end{bmatrix} = \int_0^l \mathfrak{A}(x, z) H(x) f(x) dx \tag{A.5}$$

where F_1 and F_2 are vector blocks ($F_1(z)$, $F_2(z) \in \mathbb{C}^p$) of F and

$$\mathfrak{A}(x, z) = W(x, \overline{z})^*. \tag{A.6}$$

We set

$$Uf = F_2(z) = \begin{bmatrix} 0 & I_p \end{bmatrix} \widehat{U} f. \tag{A.7}$$

If $\det H \neq 0$, then U diagonalizes operators given by the differential expression $-\mathrm{i}H^{-1}J\frac{d}{dx}$ on domains of functions satisfying boundary conditions

$$\begin{bmatrix} I_p & 0 \end{bmatrix} Y(0) = 0, \quad Y(l) = 0,$$

that is, we have

$$U\left(-\mathrm{i}H^{-1}J\frac{d}{dx}\right) Y = zUY.$$

Here, z is an operator of multiplication by z. The case of the more general boundary condition

$$\begin{bmatrix} D_2 & D_1 \end{bmatrix} Y(0) = 0 \quad \left(D_1 D_2^* + D_2 D_1^* = 0, \quad D_1 D_1^* + D_2 D_2^* = I_p\right)$$

can be reduced to the case (A.7) without loss of generality ([290], Section 4.1).

We consider $U \in B(L^2(H), L_p^2(d\tau))$, where $L_p^2(\tau)$ is the space of $p \times 1$ vector functions with the inner product

$$\langle f, \widetilde{f} \rangle_\tau = \int_{-\infty}^{\infty} \widetilde{f}(t)^* d\tau(t) f(t) < \infty, \tag{A.8}$$

and $\tau(t)$ is a nondecreasing $p \times p$ matrix function. The definition of the spectral functions τ below was already used in Chapter 2 (Definition 2.4).

Definition A.1. A nondecreasing matrix function $\tau(t)$ is called a spectral matrix function of the system (A.1) if U maps $L^2(H)$ isometrically into $L_p^2(d\tau)$.

Here, we define the space $\operatorname{Ker} U$ (and also $\operatorname{Ker} \widehat{U}$ several lines below) slightly differently from the standard definition of the operator kernel. Namely, we set

$$\operatorname{Ker} U = \left\{ f \in L^2(H) : \; Uf \equiv 0 \right\}, \quad L_1 = L^2(H) \ominus \operatorname{Ker} U. \tag{A.9}$$

Definition A.2. A nondecreasing matrix function $\tau(t)$ is called pseudospectral if U maps L_1 isometrically into $L_p^2(d\tau)$.

For the case $1 \le p < \infty$ that we study in this section, we consider the space of vector functions $B = \widehat{U} L^2(H)$ introduced by de Branges [86] and we need some of the results from [86] concerning this space. We set

$$\operatorname{Ker} \widehat{U} = \left\{ f \in L^2(H) : \; \widehat{U} f \equiv 0 \right\}, \quad L_0 = L^2(H) \ominus \operatorname{Ker} \widehat{U}. \tag{A.10}$$

Remark A.3. It is easy to see that $\operatorname{Ker} U$ and $\operatorname{Ker} \widehat{U}$ (given by (A.9) and (A.10), respectively) are closed in $L^2(H)$.

The inner product in B is given by

$$\left(\widehat{U} f, \widehat{U} \widetilde{f}\right)_B = \left(f, \widetilde{f}\right)_H; \quad f, \widetilde{f} \in L_0. \tag{A.11}$$

In Appendix A (differently from Chapter 4), ζ is a complex variable. We show that

$$\operatorname{Span}_1 := \operatorname{span} \left\{ \mathfrak{A}(x, \overline{\zeta})^* \begin{bmatrix} g_1 \\ g_2 \end{bmatrix} : \; \zeta \in \mathbb{C}; \; g_1, g_2 \in \mathbb{C}^p \right\} \tag{A.12}$$

belongs to L_0 and is dense everywhere, that is,

$$\operatorname{Span}_1 \subseteq L_0, \quad \overline{\operatorname{Span}_1} = L_0. \tag{A.13}$$

Indeed, if $f \in \operatorname{Ker} \widehat{U}$, then, by virtue of (A.5) and (A.10), we obtain $f \perp \mathfrak{A}(x, \overline{\zeta})^* \left[\begin{smallmatrix} g_1 \\ g_2 \end{smallmatrix}\right]$ for all $\zeta \in \mathbb{C}$ and $g_1, g_2 \in \mathbb{C}^p$, that is, $\mathrm{Span}_1 \subseteq L_0$. If the relation $f \perp \mathfrak{A}(x, \overline{\zeta})^* \left[\begin{smallmatrix} g_1 \\ g_2 \end{smallmatrix}\right]$ holds for all $\zeta \in \mathbb{C}$ and $g_1, g_2 \in \mathbb{C}^p$, then, according to (A.5), we have $\widehat{U} f \equiv 0$, and so $f \in \operatorname{Ker} \widehat{U}$, that is, either $f = 0$ or $f \notin L_0$. Thus, Span_1 is dense in L_0.

From (A.1) and (A.6), we see that

$$\int_0^l \mathfrak{A}(x, z) H(x) \mathfrak{A}(x, \overline{\zeta})^* dx = \mathrm{i} \frac{\mathfrak{A}(z) J \mathfrak{A}(\overline{\zeta})^* - J}{z - \zeta} \tag{A.14}$$

where $\mathfrak{A}(z) = \mathfrak{A}(l, z)$. Substitute $\zeta = \overline{z}$ into (A.14) to derive the important inequality

$$\mathrm{i}(z - \overline{z})^{-1} \left(\mathfrak{A}(z) J \mathfrak{A}(z)^* - J\right) \geq 0. \tag{A.15}$$

Since according to (A.1) and (A.6) we have

$$\mathfrak{A}(x, z) J \mathfrak{A}(x, \overline{z})^* = \mathfrak{A}(x, \overline{z})^* J \mathfrak{A}(x, z) = J, \tag{A.16}$$

the matrix function $\mathfrak{A}(z)$ is J-unitary on the real axis. Furthermore, relations (A.15) and (A.16) imply that

$$\mathfrak{A}(z) J \mathfrak{A}(z)^* \geq J, \quad \mathfrak{A}(z)^* J \mathfrak{A}(z) \geq J, \quad z \in \mathbb{C}_+. \tag{A.17}$$

From (A.13) and (A.14), it follows that

$$\mathrm{Span}_2 := \operatorname{span} \left\{ (z - \zeta)^{-1} \left(\mathfrak{A}(z) J \mathfrak{A}(\overline{\zeta})^* - J\right) \begin{bmatrix} g_1 \\ g_2 \end{bmatrix} : \zeta \in \mathbb{C};\ g_1, g_2 \in \mathbb{C}^p \right\} \tag{A.18}$$

is dense everywhere in B. By virtue of (A.11) and (A.14), we also have the equality

$$\begin{aligned} \left(F(z), \frac{\mathfrak{A}(z) J \mathfrak{A}(\omega)^* - J}{z - \overline{\omega}} \begin{bmatrix} \widetilde{g}_1 \\ \widetilde{g}_2 \end{bmatrix} \right)_B &= \begin{bmatrix} \widetilde{g}_1^* & \widetilde{g}_2^* \end{bmatrix} \int_0^l \mathfrak{A}(x, \omega) H(x) \mathfrak{A}(x, \overline{\zeta})^* dx \begin{bmatrix} g_1 \\ g_2 \end{bmatrix} \\ &= \mathrm{i} \begin{bmatrix} \widetilde{g}_1^* & \widetilde{g}_2^* \end{bmatrix} F(\omega) \end{aligned}$$

for $F(z)$ of the form

$$F(z) = (z - \zeta)^{-1} \left(\mathfrak{A}(z) J \mathfrak{A}(\overline{\zeta})^* - J\right) \begin{bmatrix} g_1 \\ g_2 \end{bmatrix}. \tag{A.19}$$

Thus, taking into account that Span_2 is dense in B, for all $F(z) \in B$, we derive the equality

$$\left(F(z), \frac{\mathfrak{A}(z) J \mathfrak{A}(\omega)^* - J}{z - \overline{\omega}} \begin{bmatrix} g_1 \\ g_2 \end{bmatrix} \right)_B = \mathrm{i} \begin{bmatrix} g_1^* & g_2^* \end{bmatrix} F(\omega), \tag{A.20}$$

that is, $\mathrm{i}(z - \overline{\omega})^{-1} (\mathfrak{A}(z) J \mathfrak{A}(\omega)^* - J)$ is a so called reproducing kernel.

From (A.16), we see that functions (A.19) can also be presented as

$$F(z) = (z - \zeta)^{-1} (\mathfrak{A}(z) - \mathfrak{A}(\zeta))\, h_\zeta, \quad h_\zeta \in \mathbb{C}^m. \tag{A.21}$$

Now, introduce operator R_α:

$$(R_\alpha F)(z) = (z - \alpha)^{-1} (F(z) - F(\alpha)). \tag{A.22}$$

The following equality is immediate for F of the form (A.21) and $\alpha \neq \zeta$:

$$(R_\alpha F)(z) = (\zeta - \alpha)^{-1} \left(\frac{\mathfrak{A}(z) - \mathfrak{A}(\zeta)}{z - \zeta} - \frac{\mathfrak{A}(z) - \mathfrak{A}(\alpha)}{z - \alpha} \right) h_\zeta. \tag{A.23}$$

Using (A.20) and (A.23), for $F,\ G \in \mathrm{Span}_2$, we derive an important de Branges' "basic identity"

$$\begin{aligned} &(F(z), R_{\alpha_1} G(z))_B - (R_{\alpha_2} F(z), G(z))_B \\ &+ (\alpha_2 - \overline{\alpha_1})\left(R_{\alpha_2} F(z), R_{\alpha_1} G(z)\right)_B = \mathrm{i} G(\alpha_1)^* J F(\alpha_2). \end{aligned} \tag{A.24}$$

Identity (A.24) also holds for arbitrary $F,\ G \in B$, which is immediate from the next statement given in [86].

Proposition A.4. *The operator R_α is well-defined everywhere on B and bounded.*

Scheme of the proof. The proof consists of two steps. First, we prove that $\|(\alpha - \overline{\alpha}) R_\alpha\|$ is bounded on Span_2 uniformly with respect to α taken from some bounded domain. For this purpose, notice that by putting $\alpha_1 = \alpha_2 = \alpha$ and $G = F \in \mathrm{Span}_2$ in the basic identity (A.24), we obtain the equality

$$(F + (\alpha - \overline{\alpha}) R_\alpha F, F + (\alpha - \overline{\alpha}) R_\alpha F)_B = (F, F)_B - \mathrm{i}(\alpha - \overline{\alpha}) F(\alpha)^* J F(\alpha). \tag{A.25}$$

According to (A.20), we have

$$\left(F(z), \frac{\mathfrak{A}(z) J \mathfrak{A}(\alpha)^* - J}{z - \overline{\alpha}} F(\alpha) \right)_B = \mathrm{i} F(\alpha)^* F(\alpha). \tag{A.26}$$

Taking into account (A.11), (A.14) and (A.26), we see that

$$\begin{aligned} \|F(\alpha)\|^4 &\le \|F(z)\|_B^2 \|\mathfrak{A}(x, \alpha)^* F(\alpha)\|_H^2 \\ &\le \|F(z)\|_B^2 \left\| \int_0^l \mathfrak{A}(x, \alpha) H(x) \mathfrak{A}(x, \alpha)^* dx \right\| \|F(\alpha)\|^2, \end{aligned} \tag{A.27}$$

where $\|\cdot\|$ is the usual matrix or vector norm in $\mathbb{C}^m$, $\|\cdot\|_B$ is the norm in B and $\|\cdot\|_H$ is the norm in $L^2(H)$. In view of (A.14), it follows from (A.27) that

$$|\alpha - \overline{\alpha}|\, \|F(\alpha)\|^2 \le \|F(z)\|_B^2 \|\mathfrak{A}(\alpha) J \mathfrak{A}(\alpha)^* - J\|. \tag{A.28}$$

Relations (A.25) and (A.28) imply the boundedness of $\|(\alpha - \overline{\alpha})R_\alpha\|$ on Span_2.

Our next step is an integral representation of $R_\alpha F$ as a residue:

$$\frac{F(z) - F(\alpha)}{z - \alpha} = \frac{1}{2\pi \mathrm{i}} \int\limits_{\Gamma} \frac{(1 - \zeta^2)\,(F(z) - F(\zeta + \alpha))}{\zeta(z - \zeta - \alpha)} d\zeta$$
$$= \frac{1}{\pi \mathrm{i}} \int\limits_{0}^{2\pi} \frac{\sin\theta\,\left(F(z) - F(\alpha + \mathrm{e}^{\mathrm{i}\theta})\right)}{\mathrm{e}^{-\mathrm{i}\theta}(z - \alpha - \mathrm{e}^{\mathrm{i}\theta})} d\theta,$$

where $\Gamma = \{z : |z| = 1\}$ is the unit circle. From this representation and the uniform boundedness of $\|(\alpha - \overline{\alpha})R_\alpha\|$, the boundedness of R_α follows for all α (including $\alpha \in \mathbb{R}$).

A.1.2 Description of the pseudospectral functions

Clearly, spectral functions can exist only when $\mathrm{Ker}\, U = 0$. In fact, if $\mathrm{Ker}\, U = 0$, then the set of spectral functions coincides with the set of pseudospectral functions of system (A.1) (compare Definitions A.1 and A.2). Below, we study the general case where $\mathrm{Ker}\, U = 0$ does not necessarily hold.

First, let us construct one pseudospectral matrix function of the system. For that, we partition the matrix $\mathfrak{A}(z)$ into blocks of order p:

$$\mathfrak{A}(z) = \begin{bmatrix} a(z) & b(z) \\ c(z) & d(z) \end{bmatrix}. \tag{A.29}$$

An essential role in our study is played by the Möbius transformation (0.8)

$$\varphi(z) = \mathrm{i}\,(a(z)\mathcal{P}_1(z) + b(z)\mathcal{P}_2(z))\,(c(z)\mathcal{P}_1(z) + d(z)\mathcal{P}_2(z))^{-1}, \quad z \in \mathbb{C}_+, \tag{A.30}$$

where the pair $\mathcal{P}_1$, $\mathcal{P}_2$ is nonsingular with property-J (Definition 0.1). Only nonstrict inequalities (A.17) hold for $\mathfrak{A}$ in the general case and a nonstrict inequality

$$c(z)d(z)^* + d(z)c(z)^* \geq 0, \quad z \in \mathbb{C}_+ \tag{A.31}$$

follows from (A.17). Hence, in order to be able to use Proposition 1.43 and show that

$$\det\,(c(z)\mathcal{P}_1(z) + d(z)\mathcal{P}_2(z)) \neq 0, \quad z \in \mathbb{C}_+, \tag{A.32}$$

we need the strict inequality

$$\begin{bmatrix} \mathcal{P}_1(z)^* & \mathcal{P}_2(z)^* \end{bmatrix} J \begin{bmatrix} \mathcal{P}_1(z) \\ \mathcal{P}_2(z) \end{bmatrix} > 0. \tag{A.33}$$

Taking into account (A.33) and applying Proposition 1.43 to $\vartheta = \mathrm{col}\,[\mathcal{P}_1 \;\; \mathcal{P}_2]$ and $\widetilde{\vartheta} = J\mathrm{col}\,[c^* \;\; d^*]$, we derive (A.32).

The standard (see [289]) reasoning implies that if (A.32) holds at some $z \in \mathbb{C}_+$ and the pair $\mathcal{P}_1(z)$, $\mathcal{P}_2(z)$ is well-defined and satisfies (0.6), then we have

$$\mathrm{i}(\varphi(z)^* - \varphi(z)) \geq 0. \tag{A.34}$$

Indeed, rewriting (A.30) in the form

$$\begin{bmatrix} -\mathrm{i}\varphi \\ I_p \end{bmatrix} = \mathfrak{A} \begin{bmatrix} \mathcal{P}_1 \\ \mathcal{P}_2 \end{bmatrix} (c\mathcal{P}_1 + d\mathcal{P}_2)^{-1}, \tag{A.35}$$

and using (0.6) and (A.17), we obtain $[\mathrm{i}\varphi(z)^* \;\; I_p]J\mathrm{col}\,[-\mathrm{i}\varphi(z) \;\; I_p] \geq 0$, which is equivalent to (A.34). Clearly, the strict inequality (A.33) implies the strict inequality $[\mathrm{i}\varphi(z)^* \;\; I_p]J\mathrm{col}\,[-\mathrm{i}\varphi(z) \;\; I_p] > 0$. The next proposition is immediate.

Proposition A.5. *Let φ have the form* (A.30)*, where the pair $\mathcal{P}_1$, $\mathcal{P}_2$ is nonsingular with property-J, and assume that* $\det(c\mathcal{P}_1 + d\mathcal{P}_2) \not\equiv 0$. *Then, φ belongs to Nevanlinna class (i.e.* (A.34) *holds everywhere in $\mathbb{C}_+$). Relation* $\det(c\mathcal{P}_1 + d\mathcal{P}_2) \not\equiv 0$ *holds automatically if, at some $z \in \mathbb{C}_+$, the strict inequality* (A.33) *is valid. Moreover, at this z, we have* $\mathrm{i}(\varphi(z)^* - \varphi(z)) > 0$.

We start with the case $\mathcal{P}_1(z) \equiv \mathcal{P}_2(z) \equiv I_p$ and the Herglotz representation of the corresponding $\varphi = \varphi_0$:

$$\begin{aligned} \varphi_0(z) &= \mathrm{i}\,(a(z) + b(z))\,(c(z) + d(z))^{-1} \\ &= \mu_0 z + \nu_0 + \int_{-\infty}^{\infty} \left(\frac{1}{t-z} - \frac{t}{1+t^2}\right) d\tau_0(t), \end{aligned} \tag{A.36}$$

where $\mu_0 \geq 0$, $\nu_0 = \nu_0^*$. Clearly, our (constant) pair is well defined on $\mathbb{R}$ and (A.33) holds for this pair on $\mathbb{R}$ also. Therefore, quite like in the proof of Proposition A.5, we derive that $\det(c(t) + d(t)) \neq 0$ for $t \in \mathbb{R}$. Thus, we can continuously extend φ (given by (A.30)) on $\mathbb{R}$ and use the Stieltjes–Perron inversion formula (1.189) to see that $\tau_0(t)$ is absolutely continuous and

$$\tau_0'(t) = \frac{\mathrm{i}}{2\pi}(\varphi_0(t)^* - \varphi_0(t)). \tag{A.37}$$

In view of (A.16) and (A.35), we have

$$\begin{aligned} \mathrm{i}(\varphi_0(t)^* - \varphi_0(t)) &= \begin{bmatrix} \mathrm{i}\varphi_0(t)^* & I_p \end{bmatrix} J \begin{bmatrix} -\mathrm{i}\varphi_0(t) \\ I_p \end{bmatrix} \\ &= 2\,(c(t)^* + d(t)^*)^{-1}\,(c(t) + d(t))^{-1}. \end{aligned} \tag{A.38}$$

From (A.37) and (A.38), we obtain

$$\tau_0'(t) = \frac{1}{\pi}\,(c(t)^* + d(t)^*)^{-1}\,(c(t) + d(t))^{-1}. \tag{A.39}$$

According to [86], the equality

$$(f,\widetilde{f})_H = \left\langle Uf, U\widetilde{f}\right\rangle_{\tau_0}; \quad f,\widetilde{f} \in L_1 \tag{A.40}$$

is valid for our absolutely continuous $\tau_0(t)$, satisfying (A.39). That is, Theorem 9 from [86, p.90] yields our next lemma.

Lemma A.6. *The absolutely continuous matrix function $\tau_0(t)$ given by* (A.39) *is a pseudospectral function of the canonical system.*

Without essential restrictions on the Hamiltonian, we sometimes require one of the two following hypotheses.

Hypothesis I. *If $c(z)h \equiv 0$ $(h \in \mathbb{C}^p)$, then $h = 0$.*

Hypothesis II. *If $c(z)^*h \equiv 0$ and $d(z)^*h \equiv h$, then $h = 0$.*

From (A.16), we see that $\begin{bmatrix} 0 & I_p \end{bmatrix} \mathfrak{A}(z) J \mathfrak{A}(\overline{z})^* \begin{bmatrix} 0 \\ I_p \end{bmatrix} = 0$, that is,

$$c(z)d(\overline{z})^* + d(z)c(\overline{z})^* = 0. \tag{A.41}$$

Thus, $c(z)h \equiv 0$ is immediate from $c(z)^*h \equiv 0$ and $d(z)^*h \equiv h$, and so Hypothesis II is even weaker than Hypothesis I.

According to (A.14), we have the relation

$$z^{-1}\left(\mathfrak{A}(z) - I_m\right)\begin{bmatrix} g_1 \\ g_2 \end{bmatrix} \in B. \tag{A.42}$$

When Hypothesis II holds, it follows from (A.42) that

$$\operatorname{span}\{F_2(\omega) : F(z) \in B,\ \omega \in \mathbb{C}\} = \mathbb{C}^p. \tag{A.43}$$

Similar to the proof of its analog (for $p = 1$) from [85], we prove the next theorem.

Theorem A.7. *Let $\tau(t)$ be a pseudospectral matrix function of the system* (A.1) *and let Hypothesis* II *hold. Then, inequality* (A.3) *is valid.*

Proof. Let $F = \widehat{U}f \in B$ (i.e. let $f \in L^2(H)$). Then, f admits decomposition $f = f_L + f_U$, where $f_L \in L_1$, $f_U \in \operatorname{Ker} U$ (see (A.9)). In view of (A.7) and (A.9), the lower block F_2 of the vector function F satisfies equality $F_2 = Uf = Uf_L$, that is, F_2 is also the lower block of the vector function $\widehat{U}f_L \in B_1$. Thus, the pseudospectral property of τ yields $F_2(z) \in L^2_p(d\tau)$. From Proposition A.4, we derive that $(z-\omega)^{-1}(F(z)-F(\omega)) \in B$, and using the arguments from above, we obtain $(z-\omega)^{-1}(F_2(z)-F_2(\omega)) \in L^2_p(d\tau)$. Clearly, relations

$$F_2(z),\ (z-\omega)^{-1}(F_2(z)-F_2(\omega)) \in L^2_p(d\tau)$$

also imply

$$(z-\omega)^{-1}F_2(\omega) \in L_p^2(d\tau) \quad (\omega \neq \overline{\omega}). \tag{A.44}$$

By virtue of (A.43) and (A.44), the theorem holds. □

Recall notation $\mathcal{N}(\mathfrak{A})$ from the Introduction for the class of Möbius transformations (A.30), where $\det(c\mathcal{P}_1 + d\mathcal{P}_2) \not\equiv 0$ and the pairs $\mathcal{P}_1(z), \mathcal{P}_2(z)$ are nonsingular with property-J. Put $B_1 = \widehat{U}L_1$. Using reasoning as in [85], we obtain

Proposition A.8.

(a) *If $\mu_0 = 0$ and $F \in B_1$, then also $R_\alpha F \in B_1$ $(\alpha \in \mathbb{C})$.*

(b) *If $\mu_0 = 0$ and Hypothesis* I *holds, then* $\operatorname{Ker}\widehat{U} = \operatorname{Ker} U$ *and $B = B_1$.*

Proposition A.9. *If* $\operatorname{Ker}\widehat{U} = \operatorname{Ker} U$ *and Hypothesis* II *holds, then $\mu_0 = 0$.*

Theorem A.10. *Let system* (A.1) *satisfy Hypothesis* I, *let $\mu_0 = 0$ and let $\tau(t)$ be a pseudospectral matrix function. Then, there exist a matrix $\nu = \nu^*$ such that*

$$\varphi(z) = \nu + \int_{-\infty}^{\infty} \left(\frac{1}{t-z} - \frac{t}{1+t^2}\right) d\tau(t) \in \mathcal{N}(\mathfrak{A}). \tag{A.45}$$

Proof of Proposition A.8. Step 1. It is useful to rewrite (A.40) as

$$(F, \widetilde{F})_B = \left\langle F_2, \widetilde{F}_2 \right\rangle_{\tau_0}; \quad F \in B_1, \ \widetilde{F} \in B. \tag{A.46}$$

Below, we consider lower blocks of $(\mathfrak{A}(z)J\mathfrak{A}(\omega)^* - J)/(z-\overline{\omega})$, namely,

$$\aleph_1(z,\omega) := (c(z)b^*(\omega) + d(z)a^*(\omega) - I_p)/(z-\overline{\omega}), \tag{A.47}$$

$$\aleph_2(z,\omega) := (c(z)d^*(\omega) + d(z)c^*(\omega))/(z-\overline{\omega}). \tag{A.48}$$

Putting $g_1 = 0$ in (A.20) and using (A.46), we derive

$$\langle F_2(z), \aleph_2(z,\omega)g_2 \rangle_{\tau_0} = \mathrm{i}g_2^* F_2(\omega)$$

for $F \in B_1$, which immediately yields a more general result, that is,

$$\langle F_2(z), \aleph_2(z,\omega)g_2 \rangle_{\tau_0} = \mathrm{i}g_2^* F_2(\omega), \quad F \in B. \tag{A.49}$$

Putting $g_2 = 0$ in (A.20) and using (A.46), we obtain

$$\langle F_2(z), \aleph_1(z,\omega)g_1 \rangle_{\tau_0} = \mathrm{i}g_1^* F_1(\omega) \tag{A.50}$$

for $F \in B_1$. It is also apparent that if $F \in B$ and (A.50) holds for all $g_1 \in \mathbb{C}^p$ and $\omega \in \mathbb{C}$, then $F \in B_1$. In the next two steps, we prove that

$$\langle F_2(z), \aleph_1(z,\alpha)\widetilde{g}_1 \rangle_{\tau_0} = \widetilde{g}_1^* \left(\varphi_0(\alpha)F_2(\alpha) - \int_{-\infty}^{\infty} \tau_0'(t) \frac{F_2(t)dt}{(t-\alpha)} \right), \tag{A.51}$$

$$F \in B, \quad \alpha \in \mathbb{C}_+.$$

Step 2. In view of (A.16), we have $[I_p \quad -I_p]\mathfrak{A}(\overline{z})^* J\mathfrak{A}(z)\left[\begin{smallmatrix} I_p \\ I_p \end{smallmatrix}\right] = 0$. The last equality can be rewritten in the form

$$(a(\overline{z})^* - b(\overline{z})^*)(c(z) + d(z)) = (d(\overline{z})^* - c(\overline{z})^*)(a(z) + b(z)). \tag{A.52}$$

The inequalities

$$\det(c(z) + d(z)) \neq 0 \quad (z \in \mathbb{C}_+ \cup \mathbb{R}), \quad \det(c(z) - d(z)) \neq 0 \quad (z \in \mathbb{C}_- \cup \mathbb{R}) \tag{A.53}$$

are valid. Indeed, using (A.15) and (A.33) (where we put $\mathcal{P}_1 = \mathcal{P}_2 = I_p$), we already obtained the first inequality in (A.53). In the same way (dealing in $\mathbb{C}_- \cup \mathbb{R}$ with the pair $\mathcal{P}_1 = -\mathcal{P}_2 = I_p$), we derive the second inequality in (A.53) from

$$\begin{bmatrix} I_p & -I_p \end{bmatrix} J \operatorname{col} \begin{bmatrix} I_p & -I_p \end{bmatrix} < 0.$$

Hence, formulas (A.30) and (A.52) yield

$$\varphi_0(z) = \mathrm{i}(d(\overline{z})^* - c(\overline{z})^*)^{-1}(a(\overline{z})^* - b(\overline{z})^*). \tag{A.54}$$

From (A.16) and (A.54), we see that

$$\frac{\mathfrak{A}(z)J\mathfrak{A}(\omega)^* - J}{z - \overline{\omega}} \begin{bmatrix} I_p \\ -\mathrm{i}\varphi_0(\omega)^* \end{bmatrix} = \frac{\mathfrak{A}(z) - \mathfrak{A}(\overline{\omega})}{z - \overline{\omega}} \begin{bmatrix} I_p \\ -I_p \end{bmatrix} (c(\overline{\omega}) - d(\overline{\omega}))^{-1}. \tag{A.55}$$

Recall that the lower blocks of $(\mathfrak{A}(z)J\mathfrak{A}(\omega)^* - J)/(z - \overline{\omega})$ are denoted by $\aleph_1$ and $\aleph_2$. Thus, (A.55) implies that

$$\aleph_1(z, \omega) - \mathrm{i}\aleph_2(z, \omega)\varphi_0(\omega)^* = (z - \overline{\omega})^{-1}(e_1(z)e_1(\overline{\omega})^{-1} - I_p), \tag{A.56}$$

$$e_k(z) := c(z) + (-1)^k d(z), \quad k = 1, 2. \tag{A.57}$$

An easy calculation also shows that

$$2\begin{bmatrix} 0 & I_p \end{bmatrix} (\mathfrak{A}(z)J\mathfrak{A}(\omega)^* - J) \begin{bmatrix} 0 \\ I_p \end{bmatrix} = e_2(z)e_2(\omega)^* - e_1(z)e_1(\omega)^*. \tag{A.58}$$

According to (A.16) and (A.58), the matrix functions e_k introduced in (A.57) have the following property:

$$e_1(z)e_1(\overline{z})^* = e_2(z)e_2(\overline{z})^*. \tag{A.59}$$

Substituting in (A.58) z for ω and taking into account (A.15) and (A.53), we obtain inequalities

$$e_2(z)^{-1}e_1(z)e_1(z)^*(e_2(z)^*)^{-1} \leq I_p \quad (z \in \mathbb{C}_+ \cup \mathbb{R}), \tag{A.60}$$

$$e_1(z)^{-1}e_2(z)e_2(z)^*(e_1(z)^*)^{-1} \leq I_p \quad (z \in \mathbb{C}_- \cup \mathbb{R}). \tag{A.61}$$

Using (A.57) and (A.59), we rewrite (A.39)

$$\tau_0'(t) = (e_1(t)^*)^{-1} e_1(t)^{-1}/\pi = (e_2(t)^*)^{-1} e_2(t)^{-1}/\pi. \tag{A.62}$$

We apply formulas (A.49), (A.56) and (A.62) in the following calculations. First, because of (A.56), we have

$$\begin{aligned}\langle F_2(z), \aleph_1(z,\alpha)\widetilde{g}_1\rangle_{\tau_0} = &- \mathrm{i}\langle F_2(z), \aleph_2(z,\alpha)\varphi_0(\alpha)^* \widetilde{g}_1\rangle_{\tau_0} \\ &+ \langle F_2(z), (z-\overline{\alpha})^{-1}(e_1(z)e_1(\overline{\alpha})^{-1} - I_p)\widetilde{g}_1\rangle_{\tau_0}.\end{aligned} \tag{A.63}$$

Hence, from (A.49), (A.62) and (A.63), we derive

$$\begin{aligned}\langle F_2(z), \aleph_1(z,\alpha)\widetilde{g}_1\rangle_{\tau_0} = &\widetilde{g}_1^* \varphi_0(\alpha) F_2(\alpha) \\ &+ \widetilde{g}_1^* \int_{-\infty}^{\infty} \left(\frac{1}{\pi}(e_1(\overline{\alpha})^*)^{-1} e_1(t)^{-1} - \tau_0'(t)\right) \frac{F_2(t)dt}{(t-\alpha)}.\end{aligned} \tag{A.64}$$

Step 3. In order to proceed with the calculations above, we consider F of the form

$$F(z) = F(z,\omega) = (z-\overline{\omega})^{-1}\left(\mathfrak{A}(z) J \mathfrak{A}(\omega)^* - J\right) \begin{bmatrix} 0 \\ g_2 \end{bmatrix}. \tag{A.65}$$

That is, according to (A.47) and (A.48), we consider

$$F_1(z) = -\aleph_1(\omega, z)^* g_2, \quad F_2(z) = \aleph_2(z,\omega) g_2. \tag{A.66}$$

Formula (A.49) implies

$$\langle F_2(z), \aleph_1(z,\alpha) g_1\rangle_{\tau_0} = \overline{\langle \aleph_1(z,\alpha) g_1, \aleph_2(z,\omega) g_2\rangle_{\tau_0}} = -\mathrm{i} g_1^* \aleph_1(\omega,\alpha)^* g_2. \tag{A.67}$$

Using the first relation in (A.66), we rewrite (A.67) as

$$\langle F_2(z), \aleph_1(z,\alpha) g_1\rangle_{\tau_0} = \mathrm{i} g_1^* F_1(\alpha), \quad g_1 \in \mathbb{C}^p, \quad \alpha \in \mathbb{C}. \tag{A.68}$$

Identities (A.50) and (A.68) coincide (up to α instead of ω). Therefore, as explained in Step 1, (A.68) leads us to $F(z,\omega) \in B_1$. Moreover, (A.20) shows that any $\widetilde{F}$ such that $\widetilde{F}(z) \perp F(z,\omega)$ for all $\omega \in \Omega$, where Ω is a set with a limit point in $\mathbb{C}$, satisfies the identity $\widetilde{F}_2(z) \equiv 0$. That is,

$$\overline{\mathrm{span}}\,\{F(z,\omega) : \omega \in \Omega, \ g_2 \in \mathbb{C}^p\} = B_1. \tag{A.69}$$

For F of the form (A.65), formula (A.58) also implies

$$F_2(z) = \left(e_2(z)e_2(\omega)^* - e_1(z)e_1(\omega)^*\right) g_2/2(z-\overline{\omega}). \tag{A.70}$$

It follows from (A.61) and (A.70) that the function $(z-\overline{\omega})e_1(z)^{-1}F_2(z)$ is bounded in $\mathbb{C}_- \cup \mathbb{R}$. This function is also analytic in $\mathbb{C}_- \cup \mathbb{R}$ and so we obtain

$$\int_{-\infty}^{\infty} (e_1(\overline{\alpha})^*)^{-1} e_1(t)^{-1} \frac{F_2(t)dt}{(t-\alpha)} = 0 \quad \text{for } \alpha \in \mathbb{C}_+, \quad \omega \in \mathbb{C}_-. \tag{A.71}$$

Substituting (A.71) into (A.64), we see that

$$\langle F_2(z), \aleph_1(z,\alpha)\widetilde{g}_1\rangle_{\tau_0} = \widetilde{g}_1^* \left(\varphi_0(\alpha)F_2(\alpha) - \int_{-\infty}^{\infty} \tau_0'(t)\frac{F_2(t)dt}{(t-\alpha)} \right),$$

where F_2 has the form (A.70). In view of (A.69), this equality also holds for the lower blocks of all $F \in B_1$ and thus of $F \in B$, that is, (A.51) is proved.

Step 4. From the definition (A.22) of R_α, we see that

$$(R_{\alpha_1}\widetilde{F})(\alpha_2) = (R_{\alpha_2}\widetilde{F})(\alpha_1), \quad \frac{(R_{\alpha_1}\widetilde{F})(t)}{t-\alpha_2} - \frac{(R_{\alpha_2}\widetilde{F})(t)}{t-\alpha_1} = \frac{\widetilde{F}(\alpha_2)-\widetilde{F}(\alpha_1)}{(t-\alpha_1)(t-\alpha_2)}.$$

Therefore, substituting first $R_{\alpha_1}\widetilde{F}$ for F and α_2 for α in (A.51), and then substituting $R_{\alpha_2}\widetilde{F}$ for F and α_1 for α in (A.51) and taking the difference of the results, we derive

$$\begin{aligned} &\langle (R_{\alpha_1}\widetilde{F})_2(z), \aleph_1(z,\alpha_2)\widetilde{g}_1\rangle_{\tau_0} - \langle (R_{\alpha_2}\widetilde{F})_2(z), \aleph_1(z,\alpha_1)\widetilde{g}_1\rangle_{\tau_0} \\ &= \widetilde{g}_1^* \left((\varphi_0(\alpha_2)-\varphi_0(\alpha_1))(R_{\alpha_1}\widetilde{F})(\alpha_2) - \int_{-\infty}^{\infty} \tau_0'(t)\frac{\widetilde{F}(\alpha_2)-\widetilde{F}(\alpha_1)}{(t-\alpha_1)(t-\alpha_2)}dt \right). \end{aligned} \tag{A.72}$$

Taking into account the Herglotz representation (A.36) of φ_0, we simplify (A.72). Namely, for all $\widetilde{F}$ such that $R_{\alpha_1}\widetilde{F}$, $R_{\alpha_2}\widetilde{F} \in B$, we have

$$\begin{aligned} &\langle (R_{\alpha_1}\widetilde{F})_2(z), \aleph_1(z,\alpha_2)\widetilde{g}_1\rangle_{\tau_0} - \langle (R_{\alpha_2}\widetilde{F})_2(z), \aleph_1(z,\alpha_1)\widetilde{g}_1\rangle_{\tau_0} \\ &= \widetilde{g}_1^*\mu_0(\widetilde{F}_2(\alpha_2) - \widetilde{F}_2(\alpha_1)), \quad \alpha_1, \alpha_2 \in \mathbb{C}_+. \end{aligned} \tag{A.73}$$

For the case where $\mu_0 = 0$, formula (A.73) yields the *permutability* relation

$$\langle (R_{\alpha_1}\widetilde{F})_2(z), \aleph_1(z,\alpha_2)\widetilde{g}_1\rangle_{\tau_0} = \langle (R_{\alpha_2}\widetilde{F})_2(z), \aleph_1(z,\alpha_1)\widetilde{g}_1\rangle_{\tau_0}, \quad \alpha_1, \alpha_2 \in \mathbb{C}_+. \tag{A.74}$$

We set

$$\widetilde{F}(z) = zF(z), \quad F \in B \tag{A.75}$$

and easily obtain that

$$R_\alpha\widetilde{F} = \alpha R_\alpha F + F. \tag{A.76}$$

Next, we substitute equalities $\mu_0 = 0$ and (A.76) into (A.73) and take into account the permutability relation (A.74) (for $\widetilde{F} = F$) in order to obtain

$$(\alpha_1 - \alpha_2)\langle (R_{\alpha_1}F)_2(z), \aleph_1(z,\alpha_2)\widetilde{g}_1\rangle_{\tau_0} = \langle F_2(z), (\aleph_1(z,\alpha_1) - \aleph_1(z,\alpha_2))\widetilde{g}_1\rangle_{\tau_0},$$
$$\text{for } \alpha_1, \alpha_2 \in \mathbb{C}_+. \tag{A.77}$$

Considering $\langle F_2(z), \aleph_1(z,\alpha_2)\widetilde{g}_1\rangle_{\tau_0}$ as a function depending on α_2 and applying operator R_{α_1} to this function, we rewrite (A.77) in the form

$$\langle (R_{\alpha_1}F)_2(z), \aleph_1(z,\alpha_2)\widetilde{g}_1\rangle_{\tau_0} = R_{\alpha_1}(\langle F_2(z), \aleph_1(z,\alpha_2)\widetilde{g}_1\rangle_{\tau_0}) \tag{A.78}$$

for $\alpha_1,\ \alpha_2 \in \mathbb{C}_+,\ \alpha_1 \neq \alpha_2$. From the equality

$$R_{\alpha_1} - R_{\alpha_2} = (\alpha_1 - \alpha_2)R_{\alpha_1}R_{\alpha_2}, \tag{A.79}$$

we see that R_{α_1} depends on α_1 analytically, and so the left-hand side of (A.78) is analytic with respect to α_1. Since the right-hand side of (A.78) is also analytic with respect to α_1 and both sides of (A.78) are analytic with respect to α_2, equality (A.78) holds for all $\alpha_1,\ \alpha_2 \in \mathbb{C}$.

When $F \in B_1$, equality (A.50) is valid, and so formula (A.78) yields

$$\langle (R_{\alpha_1}F)_2(z), \aleph_1(z,\alpha_2)\widetilde{g}_1\rangle_{\tau_0} = \mathrm{i}\widetilde{g}_1^*(R_{\alpha_1}F)(\alpha_2). \tag{A.80}$$

On the other hand (as discussed in Step 1), identity (A.80) implies $R_{\alpha_1}F \in B_1$, that is, statement (a) of proposition is proved.

Step 5. In order to prove statement (b), we rewrite (A.65) in the form

$$F(z,\omega) = (z-\overline{\omega})^{-1}\left(\mathfrak{A}(z) - \mathfrak{A}(\overline{\omega})\right) J\mathfrak{A}(\omega)^* \begin{bmatrix} 0 \\ g_2 \end{bmatrix} \tag{A.81}$$

and use (A.23) and (A.81) in order to obtain

$$\begin{aligned} R_\alpha F(z,\omega) &= (\overline{\omega}-\alpha)^{-1}\left(\frac{\mathfrak{A}(z)-\mathfrak{A}(\overline{\omega})}{z-\overline{\omega}} - \frac{\mathfrak{A}(z)-\mathfrak{A}(\alpha)}{z-\alpha}\right) J\mathfrak{A}(\omega)^* \begin{bmatrix} 0 \\ g_2 \end{bmatrix} \\ &= (\overline{\omega}-\alpha)^{-1}\left(F(z) - \frac{\mathfrak{A}(z)-\mathfrak{A}(\alpha)}{z-\alpha} J\mathfrak{A}(\omega)^* \begin{bmatrix} 0 \\ g_2 \end{bmatrix}\right). \end{aligned} \tag{A.82}$$

Since formula (A.69) and statement (a) of the proposition yield

$$F(z,\omega),\ R_\alpha F(z,\omega) \in B_1,$$

it is immediate from (A.82) that

$$\frac{\mathfrak{A}(z)-\mathfrak{A}(\alpha)}{z-\alpha} J\mathfrak{A}(\omega)^* \begin{bmatrix} 0 \\ g_2 \end{bmatrix} \in B_1 \quad (\omega \neq \alpha). \tag{A.83}$$

Next, we show that

$$\operatorname{span}\left\{\mathfrak{A}(\omega)^* \begin{bmatrix} 0 \\ g_2 \end{bmatrix} : \quad \omega \neq \alpha,\ g_2 \in \mathbb{C}^p\right\} = \mathbb{C}^m. \tag{A.84}$$

Indeed, if (A.84) does not hold, then there is a row vector $h = \begin{bmatrix} h_1 & h_2 \end{bmatrix} \in \mathbb{C}^m$, $h_k \in \mathbb{C}^p$ such that

$$h\mathfrak{A}(\omega)^* \begin{bmatrix} 0 \\ g_2 \end{bmatrix} \equiv 0 \quad (\omega \in \mathbb{C},\ g_2 \in \mathbb{C}^p), \quad h \neq 0. \tag{A.85}$$

However, from (A.85) at $\omega = 0$, we derive $h_2 = 0$. Thus, using block representation (A.29) of $\mathfrak{A}$, we can rewrite (A.85) in the form

$$h_1 c(\omega)^* \equiv 0 \quad (\omega \in \mathbb{C}), \quad h_1 \neq 0, \tag{A.86}$$

which contradicts Hypothesis I. Since Hypothesis I is assumed in statement (b), formula (A.84) is proved by contradiction.

From (A.83) and (A.84), we see that

$$\overline{\operatorname{span}} \left\{ \frac{\mathfrak{A}(z) - \mathfrak{A}(\alpha)}{z - \alpha} g : \ \alpha \in \mathbb{C}, \, g \in \mathbb{C}^m \right\} \subseteq B_1. \tag{A.87}$$

On the other hand, recalling that Span_2 (given in (A.18)) is dense in B and recalling also representation (A.21) of the functions generating Span_2, we obtain

$$\overline{\operatorname{span}} \left\{ \frac{\mathfrak{A}(z) - \mathfrak{A}(\alpha)}{z - \alpha} g : \ \alpha \in \mathbb{C}, \, g \in \mathbb{C}^m \right\} = B. \tag{A.88}$$

Compare (A.87) and (A.88) in order to derive $B = B_1$. Recall that

$$B = \widehat{U}(L^2(H) \ominus \operatorname{Ker} \widehat{U}) = \widehat{U} L^2(H), \quad B_1 = \widehat{U}(L^2(H) \ominus \operatorname{Ker} U). \tag{A.89}$$

Hence, $B = B_1$ implies $\operatorname{Ker} \widehat{U} = \operatorname{Ker} U$. □

Proof of Proposition A.9. From (A.89), we see that $\operatorname{Ker} \widehat{U} = \operatorname{Ker} U$ yields $B = B_1$, and so (A.50) is valid for $F \in B$. Hence, for the case where $R_{\alpha_1} \widetilde{F}, R_{\alpha_2} \widetilde{F} \in B$, the left-hand side of (A.73) vanishes. Therefore, the right-hand side of (A.73) vanishes too, that is,

$$\widetilde{g}_1^* \mu_0 (\widetilde{F}_2(\alpha_2) - \widetilde{F}_2(\alpha_1)) = 0 \quad (\alpha_1, \alpha_2 \in \mathbb{C}_+; \quad \widetilde{g}_1 \in \mathbb{C}^p). \tag{A.90}$$

It is easy to see that

$$\operatorname{span} \left\{ \widetilde{F}_2(\alpha_2) - \widetilde{F}_2(\alpha_1) : \ \alpha_1, \alpha_2 \in \mathbb{C}_+; \ \widetilde{F}(z) = zF(z), \ F \in B \right\} = \mathbb{C}^p. \tag{A.91}$$

Indeed, if some vector $h \neq 0$ is orthogonal to the span above, in view of analyticity of $\widetilde{F}$, we also have

$$h^* \operatorname{span} \left\{ \widetilde{F}_2(\alpha_2) - \widetilde{F}_2(0) : \ \alpha_2 \in \mathbb{C}; \ \widetilde{F}(z) = zF(z), \ F \in B \right\} = 0, \quad h \neq 0.$$

According to (A.43), this is impossible.

Finally, the equality $\mu_0 = 0$ is immediate from (A.90) and (A.91). □

Proof of Theorem A.10. Under conditions of the theorem, the statement (b) of Proposition A.8, that is, $B = B_1$, is valid. Since τ is pseudospectral and $B = B_1$, we rewrite

(A.24) in the form

$$
\begin{aligned}
\mathrm{i}G(\alpha_1)^*JF(\alpha_2) &= \int_{-\infty}^{\infty}\left(\frac{1}{t-\alpha_2}-\frac{1}{t-\overline{\alpha_1}}\right)G_2(\alpha_1)^*d\tau(t)F_2(\alpha_2) \qquad \text{(A.92)}\\
&\quad -\int_{-\infty}^{\infty}G_2(\alpha_1)^*\frac{d\tau(t)}{t-\alpha_2}F_2(t)+\int_{-\infty}^{\infty}G_2(t)^*\frac{d\tau(t)}{t-\overline{\alpha_1}}F_2(\alpha_2)\\
&= G_2(\alpha_1)^*(\widetilde{\varphi}(\alpha_2)-\widetilde{\varphi}(\alpha_1)^*)F_2(\alpha_2)\\
&\quad -\int_{-\infty}^{\infty}G_2(\alpha_1)^*\frac{d\tau(t)}{t-\alpha_2}F_2(t)+\int_{-\infty}^{\infty}G_2(t)^*\frac{d\tau(t)}{t-\overline{\alpha_1}}F_2(\alpha_2),
\end{aligned}
$$

where

$$
F,G\in B,\quad \widetilde{\varphi}(z)=\int_{-\infty}^{\infty}\left(\frac{1}{t-z}-\frac{t}{1+t^2}\right)d\tau(t). \tag{A.93}
$$

Notice that, in view of Theorem A.7, the integral in (A.93) (and in (A.45) also) converges. From (A.92), we see that functions $\widetilde{F}$ and $\widetilde{G}$ given by the relations

$$
\widetilde{F}=\begin{bmatrix}\widetilde{F}_1\\ F_2\end{bmatrix},\quad \widetilde{G}=\begin{bmatrix}\widetilde{G}_1\\ G_2\end{bmatrix}, \tag{A.94}
$$

$$
\widetilde{F}_1(z)=F_1(z)+\mathrm{i}\widetilde{\varphi}(z)F_2(z)-\mathrm{i}\int_{-\infty}^{\infty}\frac{d\tau(t)}{t-z}F_2(t), \tag{A.95}
$$

$$
\widetilde{G}_1(z)=G_1(z)+\mathrm{i}\widetilde{\varphi}(z)G_2(z)-\mathrm{i}\int_{-\infty}^{\infty}\frac{d\tau(t)}{t-z}G_2(t) \tag{A.96}
$$

satisfy the identity

$$
\widetilde{G}(\alpha_1)^*J\widetilde{F}(\alpha_2)=0,\quad \alpha_1,\alpha_2\in\mathbb{C}. \tag{A.97}
$$

According to (A.97), span $\{\widetilde{F}(\omega)\,:\,F\in B,\ \omega\in\mathbb{C}\}$ is a J-neutral subspace of $\mathbb{C}^m$, and so its dimension is less or equal to p (in fact, it is equal to p). Hence, formula (A.43) implies that there is a matrix ν such that

$$
\widetilde{F}_1(\omega)=-\mathrm{i}\nu F_2(\omega). \tag{A.98}
$$

By substituting (A.98) into (A.97), we obtain $G_2(\alpha_1)^*(\nu-\nu^*)F_2(\alpha_2)=0$, which (in view of (A.43)) yields

$$
\nu=\nu^*. \tag{A.99}
$$

We set

$$
\varphi(z)=\nu+\int_{-\infty}^{\infty}\left(\frac{1}{t-z}-\frac{t}{1+t^2}\right)d\tau(t), \tag{A.100}
$$

where ν satisfies (A.98). Using (A.93), (A.98) and (A.100), we rewrite (A.95) as

$$\begin{bmatrix} I_p & i\varphi(z) \end{bmatrix} F(z) - i \int_{-\infty}^{\infty} \frac{d\tau(t)}{t-z} F_2(t) = \widetilde{F}_1(z) + i\nu F_2(z) = 0,$$

that is,

$$\langle F_2, \breve{g}/(z-\overline{\omega})\rangle_\tau = -i\breve{g}^* \begin{bmatrix} I_p & i\varphi(\omega) \end{bmatrix} F(\omega), \quad \omega \neq \overline{\omega}. \tag{A.101}$$

Consider again functions F of the form (A.19), more precisely, of the form

$$F(z) = (z-\overline{\omega})^{-1}(\mathfrak{A}(z)J\mathfrak{A}(\omega)^* - J)\begin{bmatrix} I_p & i\varphi(\omega) \end{bmatrix}^* \breve{g}, \quad \breve{g} \in \mathbb{C}^p. \tag{A.102}$$

From $B = B_1$, (A.20) and (A.102), we have

$$\langle F_2, F_2\rangle_\tau = (F, F)_B = i\breve{g}^* \begin{bmatrix} I_p & i\varphi(\omega) \end{bmatrix} F(\omega). \tag{A.103}$$

Taking into account (A.101) and (A.103), for $\omega \neq \overline{\omega}$, we obtain

$$\begin{aligned} 0 \le \langle F_2 + \breve{g}/(z-\overline{\omega}), F_2 + \breve{g}/(z-\overline{\omega})\rangle_\tau &= i\breve{g}^* \begin{bmatrix} I_p & i\varphi(\omega) \end{bmatrix} F(\omega) \\ - i\breve{g}^* \begin{bmatrix} I_p & i\varphi(\omega) \end{bmatrix} F(\omega) + iF(\omega)^* \begin{bmatrix} I_p & i\varphi(\omega) \end{bmatrix}^* \breve{g} &+ \int_{-\infty}^{\infty} \frac{\breve{g}^* d\tau(t)\breve{g}}{(t-\omega)(t-\overline{\omega})}. \end{aligned} \tag{A.104}$$

In view of (A.100) and (A.102), we rewrite the equality in (A.104) in the form

$$\begin{aligned} \langle F_2 + \breve{g}/(z-\overline{\omega}), F_2 + \breve{g}/(z-\overline{\omega})\rangle_\tau &= \breve{g}^*(\overline{\omega}-\omega)^{-1} \\ \times \Big(i\begin{bmatrix} I_p & i\varphi(\omega) \end{bmatrix} (\mathfrak{A}(\omega)J\mathfrak{A}(\omega)^* - J)&\begin{bmatrix} I_p & i\varphi(\omega) \end{bmatrix}^* + \varphi(\omega)^* - \varphi(\omega)\Big)\breve{g} \\ = i\breve{g}^*(\overline{\omega}-\omega)^{-1}\begin{bmatrix} I_p & i\varphi(\omega) \end{bmatrix} \mathfrak{A}(\omega)J\mathfrak{A}(\omega)^* &\begin{bmatrix} I_p & i\varphi(\omega) \end{bmatrix}^* \breve{g}. \end{aligned} \tag{A.105}$$

Clearly, relations (A.104) and (A.105) imply

$$\begin{bmatrix} I_p & i\varphi(\omega) \end{bmatrix} \mathfrak{A}(\omega)J\mathfrak{A}(\omega)^* \begin{bmatrix} I_p & i\varphi(\omega) \end{bmatrix}^* \le 0 \quad \text{for } \omega \in \mathbb{C}_+. \tag{A.106}$$

It is apparent from (A.16) that

$$\begin{bmatrix} I_p & i\varphi(\omega) \end{bmatrix} \mathfrak{A}(\omega)J\mathfrak{A}(\overline{\omega})^* \begin{bmatrix} I_p & i\varphi(\omega)^* \end{bmatrix}^* \equiv 0. \tag{A.107}$$

From the relations (A.106) and (A.107), we see that for

$$\vartheta^* = \begin{bmatrix} I_p & i\varphi(\omega) \end{bmatrix} \mathfrak{A}(\omega)j, \quad \widetilde{\vartheta} = j\mathfrak{A}(\overline{\omega})^* \begin{bmatrix} I_p & i\varphi(\omega)^* \end{bmatrix}^*, \quad j = \operatorname{diag}\{I_p, -I_p\},$$

the conditions of Remark 1.45 hold, that is,

$$\begin{bmatrix} I_p & i\varphi(\omega)^* \end{bmatrix} \mathfrak{A}(\overline{\omega})J\mathfrak{A}(\overline{\omega})^* \begin{bmatrix} I_p & i\varphi(\omega)^* \end{bmatrix}^* \ge 0, \quad \omega \in \mathbb{C}_+. \tag{A.108}$$

Here, we used the equality $jJj = -J$. According to (A.108), the pair

$$\begin{bmatrix} \mathcal{P}_1(z) \\ \mathcal{P}_2(z) \end{bmatrix} = J\mathfrak{A}(\overline{z})^* \begin{bmatrix} I_p & \mathrm{i}\varphi(z)^* \end{bmatrix}^*, \quad z \in \mathbb{C}_+ \tag{A.109}$$

is nonsingular with property-J. Using (A.109), we rewrite (A.107) in the form

$$\begin{bmatrix} I_p & \mathrm{i}\varphi(z) \end{bmatrix} \mathfrak{A}(z) \begin{bmatrix} \mathcal{P}_1(z) \\ \mathcal{P}_2(z) \end{bmatrix} \equiv 0 \tag{A.110}$$

and it is clear from (A.109) also that $c\mathcal{P}_1 + d\mathcal{P}_2 = I_p$. Thus, $c\mathcal{P}_1 + d\mathcal{P}_2$ is invertible and (A.110) yields representation (0.8) of φ. In other words, $\varphi \in \mathcal{N}(\mathfrak{A})$. □

A.1.3 Potapov's inequalities and description of the pseudospectral functions

For $\omega \in \mathbb{C}_-$ and $R_\omega F \in B$, we set

$$\mathfrak{p}(z,\omega) := \frac{\mathfrak{A}(z)J\mathfrak{A}(\overline{\omega})^* - J}{z-\omega} \begin{bmatrix} I_p \\ -\mathrm{i}\varphi(\overline{\omega})^* \end{bmatrix} F_2(\omega), \tag{A.111}$$

that is, $\mathfrak{p}$ has the form (A.102), where ω is substituted for $\overline{\omega}$ and $F_2(\omega)$ for $\breve{g}$. Then, (A.103) yields

$$\begin{aligned} &(\mathfrak{p}(z,\omega), \mathfrak{p}(z,\omega))_B \\ &= \mathrm{i}F_2(\omega)^* \begin{bmatrix} I_p & \mathrm{i}\varphi(\overline{\omega}) \end{bmatrix} \frac{\mathfrak{A}(\overline{\omega})J\mathfrak{A}(\overline{\omega})^* - J}{\overline{\omega}-\omega} \begin{bmatrix} I_p \\ -\mathrm{i}\varphi(\overline{\omega})^* \end{bmatrix} F_2(\omega). \end{aligned} \tag{A.112}$$

It follows from (A.106) and (A.112) that

$$(\mathfrak{p}(z,\omega), \mathfrak{p}(z,\omega))_B \le F_2(\omega)^* \frac{\varphi(\overline{\omega}) - \varphi(\overline{\omega})^*}{\overline{\omega}-\omega} F_2(\omega), \quad \omega \in \mathbb{C}_-. \tag{A.113}$$

Note that an inequality like (A.106) yields Potapov's FMI (see, e.g. [289, p. 10]), and so (A.113) can be considered as an analog of FMI for the case of the general-type canonical system. Using FMI, the transformed fundamental matrix inequality (TFMI) is constructed. In our case, function N given by

$$N(\zeta) = \left(F, R_{\overline{\zeta}}F - \mathfrak{p}(z,\overline{\zeta})\right)_B, \quad \zeta \in \mathbb{C}_+ \tag{A.114}$$

is an analog of the transformed block B_T in TFMI considered in [289, p. 6]. Our analog of TFMI is introduced in the next lemma.

Lemma A.11. *For $F, \widetilde{F} \in B$ and N given by* (A.114), *we have*

$$\begin{aligned} &(\widetilde{F}, \widetilde{F})_B + \left(R_\omega F - \mathfrak{p}(z,\omega), \widetilde{F}\right)_B + \left(\widetilde{F}, R_\omega F - \mathfrak{p}(z,\omega)\right)_B \\ &\quad + (\overline{\omega}-\omega)^{-1}\left(N(\overline{\omega}) - \overline{N(\overline{\omega})}\right) \ge 0, \quad \omega \in \mathbb{C}_-. \end{aligned} \tag{A.115}$$

Proof. For $\widetilde{F} \in B$ and F such that $R_\omega F \in B$, we clearly have

$$\Big(\widetilde{F}(z) + R_\omega F(z) - \mathfrak{p}(z,\omega),\ \widetilde{F}(z) + R_\omega F(z) - \mathfrak{p}(z,\omega)\Big)_B \geq 0. \tag{A.116}$$

Replacing in (A.116) the left term of the inequality (A.113) by the greater right term, we obtain

$$\begin{aligned}
&(\widetilde{F},\widetilde{F})_B + \Big(R_\omega F - \mathfrak{p}(z,\omega),\widetilde{F}\Big)_B + \Big(\widetilde{F}, R_\omega F - \mathfrak{p}(z,\omega)\Big)_B \\
&- (\mathfrak{p}(z,\omega), R_\omega F)_B - (R_\omega F, \mathfrak{p}(z,\omega))_B \\
&+ (R_\omega F, R_\omega F)_B + F_2(\omega)^* \frac{\varphi(\overline{\omega}) - \varphi(\overline{\omega})^*}{\overline{\omega} - \omega} F_2(\omega) \geq 0, \quad \omega \in \mathbb{C}_-.
\end{aligned} \tag{A.117}$$

It remains to show that

$$\begin{aligned}
\frac{N(\overline{\omega}) - \overline{N(\overline{\omega})}}{\overline{\omega} - \omega} &= (R_\omega F, R_\omega F)_B - (R_\omega F, \mathfrak{p}(z,\omega))_B \\
&\quad - (\mathfrak{p}(z,\omega), R_\omega F)_B + F_2(\omega)^* \frac{\varphi(\overline{\omega}) - \varphi(\overline{\omega})^*}{\overline{\omega} - \omega} F_2(\omega).
\end{aligned} \tag{A.118}$$

Indeed, according to (A.24), we have

$$\begin{aligned}
&(R_\omega F, R_\omega F)_B + (\omega - \overline{\omega})^{-1}((F, R_\omega F)_B - (R_\omega F, F)_B) \\
&= \mathrm{i}(\omega - \overline{\omega})^{-1} F(\omega)^* J F(\omega).
\end{aligned} \tag{A.119}$$

Also, by virtue of (A.20) and (A.111), we have

$$\begin{aligned}
&\big(R_\omega F, \mathfrak{p}(z,\omega)\big)_B + \big(\mathfrak{p}(z,\omega), R_\omega F\big)_B + (\omega - \overline{\omega})^{-1}\Big(\big(F, \mathfrak{p}(z,\omega)\big)_B - \big(\mathfrak{p}(z,\omega), F\big)_B\Big) \\
&= (\omega - \overline{\omega})^{-1}\left(\mathrm{i}F(\omega)^* J F(\omega) - F_2(\omega)^*(\varphi(\overline{\omega}) - \varphi(\overline{\omega})^*)F_2(\omega)\right).
\end{aligned} \tag{A.120}$$

Finally, (A.118) follows from (A.114), (A.119) and (A.120). □

Recall that, according to Proposition A.5, the relation $\varphi \in \mathcal{N}(\mathfrak{A})$ implies that φ belongs to the Nevanlinna class and so it admits the Herglotz representation

$$\varphi(z) = \mu z + \nu + \int_{-\infty}^{\infty} \left(\frac{1}{t - z} - \frac{t}{1 + t^2}\right) d\tau(t), \tag{A.121}$$

where $\mu \geq 0$, $\nu = \nu^*$, and τ is a nondecreasing $p \times p$ matrix function. Some simple conditions under which the pseudospectrality of the distribution function τ in (A.121) yields $\varphi \in \mathcal{N}(\mathfrak{A})$ are given in Theorem A.10. Now, we start with the assumption that $\varphi \in \mathcal{N}(\mathfrak{A})$ and study (using TFMI (A.115)) the distribution function τ from the Herglotz representation of φ.

Theorem A.12. *If $\varphi \in \mathcal{N}(\mathfrak{A})$, then the distribution function τ satisfies inequalities*

$$\langle F_2, F_2 \rangle_\tau \leq \langle F_2, F_2 \rangle_{\tau_0} \tag{A.122}$$

for all $F \in B$.

Proof. According to (A.113), the right-hand side of (A.118) is nonnegative. This means that $N(z)$ is a Nevanlinna function. We denote the distribution function in the Herglotz representation of $N(z)$ by $\tilde{\tau}$ and the set of discontinuities of $\tilde{\tau}$ by $M(\tilde{\tau})$. For $t_1, t_2 \notin M(\tilde{\tau})$, the Stieltjes–Perron formula (1.189) yields

$$\tilde{\tau}(t_2) - \tilde{\tau}(t_1) = \frac{\mathrm{i}}{2\pi} \lim_{\eta \to 0} \int_{t_1}^{t_2} \left(\overline{N(z)} - N(z)\right) d\xi; \quad z = \xi + \mathrm{i}\eta, \quad \eta > 0. \tag{A.123}$$

Using (A.118) and (A.123), we shall express $\tilde{\tau}$ via τ. First, recall that, in view of (A.79), the operator R_ω is analytic with respect to ω ($\omega \in \mathbb{C}$), and so the function $(R_\omega F, R_\omega F)_B$ is uniformly bounded on compact sets. It follows from (A.20) and (A.111) that

$$(R_\omega F, \mathfrak{p}(z,\omega))_B = \mathrm{i}(\overline{\omega} - \omega)^{-1} F_2(\omega)^* \begin{bmatrix} I_p & \mathrm{i}\varphi(\overline{\omega}) \end{bmatrix} (F(\overline{\omega}) - F(\omega)).$$

Therefore, it is easy to see that $|\Im(\omega)(R_\omega F, \mathfrak{p}(z,\omega))_B|$ is uniformly bounded when the values of $|\omega|$ ($\omega \in \mathbb{C}_-$) are uniformly bounded and tends to 0 when $\Im(\omega)$ tends to 0, whereas $\Re(\omega) \notin M(\tau)$. Hence, taking into account (A.118), we rewrite (A.123) in the form

$$\tilde{\tau}(t_2) - \tilde{\tau}(t_1) = \frac{\mathrm{i}}{2\pi} \lim_{\eta \to 0} \int_{t_1}^{t_2} F_2(\overline{z})^* \left(\varphi(z)^* - \varphi(z)\right) F_2(\overline{z})\, d\xi. \tag{A.124}$$

Using the Stieltjes–Perron formula (for τ and φ) and the analyticity of $F_2(z)$, we derive

$$\frac{\mathrm{i}}{2\pi} \lim_{\eta \to 0} \int_{t_1}^{t_2} F_2(\overline{z})^* \left(\varphi(z)^* - \varphi(z)\right) F_2(\overline{z}) d\xi = \int_{t_1}^{t_2} F_2(t)^* d\tau(t) F_2(t) \tag{A.125}$$

for $t_1, t_2 \notin M(\tilde{\tau})$. From (A.124) and (A.125), we obtain

$$\tilde{\tau}(t_2) - \tilde{\tau}(t_1) = \int_{t_1}^{t_2} F_2(t)^* d\tau(t) F_2(t). \tag{A.126}$$

Now, setting $\tilde{F} = \mathrm{i}\zeta F$ ($\zeta = \overline{\zeta}$), we rewrite (A.115) as

$$\zeta^2 (F,F)_B + \mathrm{i}\zeta \left(N(z) - \overline{N(z)}\right) + \frac{N(z) - \overline{N(z)}}{z - \overline{z}} \geq 0, \quad z \in \mathbb{C}_+. \tag{A.127}$$

Since the quadratic polynomial in ζ above is nonnegative for all real ζ, we have

$$(F,F)_B \geq \Im(z)\, \Im\left(N(z)\right). \tag{A.128}$$

In particular, formula (A.128) implies that μ (denoted by μ_N) in the Herglotz representation of N equals zero. Taking also into account (A.126), we derive representation

$$N(z) = \nu_N + \int_{-\infty}^{\infty} \left(\frac{1}{t - z} - \frac{t}{1 + t^2}\right) F_2(t)^*\, d\tau(t)\, F_2(t) \quad (\nu_N \in \mathbb{R}). \tag{A.129}$$

Putting $z = \mathrm{i}\eta$ $(\eta = \overline{\eta})$, we deduce from (A.128) and (A.129) that

$$\int_{-\infty}^{\infty} F_2(t)^* \, d\tau(t) \, F_2(t) = \lim_{\eta \to \infty} \eta \, \Im \left(N(\mathrm{i}\eta) \right) \leq (F, F)_B. \tag{A.130}$$

When $F \in B_1$, formula (A.130) is equivalent to (A.122). Since (A.122) is proved for $f \in B_1$, it also holds for all $F \in B$. □

TFMI yields also another property of N. Setting $\widetilde{F} = \zeta F (\zeta = \overline{\zeta})$, we rewrite (A.115) as

$$\zeta^2 (F, F)_B + \zeta \left(N(z) + \overline{N(z)} \right) + \frac{N(z) - \overline{N(z)}}{z - \overline{z}} \geq 0, \quad z \in \mathbb{C}_+.$$

Hence, using (A.128), we obtain

$$\left| \eta \left(N(z) + \overline{N(z)} \right) \right| \leq 2(F, F)_B, \quad z = \xi + \mathrm{i}\eta. \tag{A.131}$$

In view of (A.130) and (A.131), representation (A.129) takes the form

$$N(z) = \int_{-\infty}^{\infty} (t - z)^{-1} F_2(t)^* \, d\tau(t) \, F_2(t), \tag{A.132}$$

which, again taking into account (A.130), implies

$$\mathrm{i} \langle F_2, F_2 \rangle_\tau = \lim_{\eta \to \infty} \eta N(\mathrm{i}\eta). \tag{A.133}$$

Next, we impose an additional condition in order to achieve equality (for all $F \in B$) in formula (A.122) from Theorem A.12.

Theorem A.13.

(a) *Let $\varphi \in \mathcal{N}(\mathfrak{A})$ and let the condition*

$$\lim_{\eta \to \infty} \eta^{-1} \left(c(-\mathrm{i}\eta)^* - d(-\mathrm{i}\eta)^* \right) \left(\varphi(\mathrm{i}\eta) - \varphi_0(\mathrm{i}\eta) \right) \left(c(\mathrm{i}\eta) - d(\mathrm{i}\eta) \right) = 0 \tag{A.134}$$

hold. Then, the distribution function τ from the Herglotz representation of φ is pseudospectral.

(b) *Let τ be pseudospectral, let $\mu_0 = 0$ and let Hypothesis* I *be valid. Then, there exists $\varphi \in \mathcal{N}(\mathfrak{A})$ with distribution function τ. For this function φ, equality* (A.134) *holds.*

Proof. In order to prove statement (a), we first show that

$$\lim_{\eta \to \infty} \left(c(-\mathrm{i}\eta)^* - d(-\mathrm{i}\eta)^* \right) \left(\varphi(\mathrm{i}\eta) - \varphi_0(\mathrm{i}\eta) \right) \widetilde{F}_2(\mathrm{i}\eta) = 0 \quad (\widetilde{F} \in B). \tag{A.135}$$

For that purpose, we set

$$\begin{aligned} \psi(\omega) &= (R_\omega F, R_\omega F)_B - (\mathfrak{p}(z, \omega), R_\omega F)_B - (R_\omega F, \mathfrak{p}(z, \omega))_B \\ &\quad + (\overline{\omega} - \omega)^{-1} F_2(\omega)^* \left(\varphi(\overline{\omega}) - \varphi(\overline{\omega})^* \right) F_2(\omega), \end{aligned} \tag{A.136}$$

$$F(z) = \mathfrak{A}(z) \begin{bmatrix} I_p \\ -I_p \end{bmatrix} \breve{g} \quad (\breve{g} \in \mathbb{C}^p). \tag{A.137}$$

Here, ψ coincides with the right-hand side of (A.118) for the case that $F \in B$. However, the function F of the form (A.137) does not necessarily belong B. Nevertheless, like in the case where $F \in B$, it can be shown that

$$\lim_{\eta\to\infty} \psi(-\mathrm{i}\eta) = 0. \tag{A.138}$$

Taking into account (A.20), (A.136), (A.137) and the representation

$$R_\omega F = \frac{\mathfrak{A}(z)J\mathfrak{A}(\overline{\omega})^* - J}{z-\omega} J\mathfrak{A}(\omega)\begin{bmatrix} I_p \\ -I_p \end{bmatrix}\breve{g} \tag{A.139}$$

of $R_\omega F$ (for F given by (A.137)), we have

$$\psi(\omega) = \psi_1(\omega) + \psi_1(\omega)^* + 2\mathrm{i}\,(\omega-\overline{\omega})^{-1}\breve{g}^*\breve{g}, \tag{A.140}$$

$$\psi_1(\omega) := \mathrm{i}\,(\omega-\overline{\omega})^{-1}F_2(\omega)^*\begin{bmatrix} I_p & \mathrm{i}\varphi(\overline{\omega}) \end{bmatrix}F(\overline{\omega}). \tag{A.141}$$

In view of (A.137), we rewrite (A.141) in the form

$$\begin{aligned}\psi_1(\omega) &= (\overline{\omega}-\omega)^{-1}\breve{g}^*\,(c(\omega)^*-d(\omega)^*)\\ &\quad\times\left(\varphi(\overline{\omega}) - \mathrm{i}\,(a(\overline{\omega})-b(\overline{\omega}))\,(c(\overline{\omega})-d(\overline{\omega}))^{-1}\right)\\ &\quad\times(c(\overline{\omega})-d(\overline{\omega}))\,\breve{g}.\end{aligned} \tag{A.142}$$

Since (A.16) implies

$$\begin{bmatrix} I_p & -I_p \end{bmatrix}\mathfrak{A}(\omega)^*J\mathfrak{A}(\overline{\omega})\begin{bmatrix} I_p & -I_p \end{bmatrix}^* = -2I_p,$$

we have

$$(c(\omega)-d(\omega))^*\,(a(\overline{\omega})-b(\overline{\omega})) = -2I_p - (a(\omega)-b(\omega))^*\,(c(\overline{\omega})-d(\overline{\omega}))\,. \tag{A.143}$$

Using (A.54), we rewrite (A.143) in the form

$$\begin{aligned}\mathrm{i}\,(a(\overline{\omega})-b(\overline{\omega}))\,(c(\overline{\omega})-d(\overline{\omega}))^{-1} &= \varphi_0(\overline{\omega}) - 2\mathrm{i}\,(c(\omega)^*-d(\omega)^*)^{-1}\\ &\quad\times(c(\overline{\omega})-d(\overline{\omega}))^{-1}.\end{aligned} \tag{A.144}$$

Substituting (A.144) into (A.142), we obtain

$$\begin{aligned}\psi_1(\omega) &= (\overline{\omega}-\omega)^{-1}\breve{g}^*\\ &\quad\times\left((c(\omega)^*-d(\omega)^*)\,(\varphi(\overline{\omega})-\varphi_0(\overline{\omega}))\,(c(\overline{\omega})-d(\overline{\omega})) + 2\mathrm{i}I_p\right)\breve{g}.\end{aligned} \tag{A.145}$$

From (A.134) and (A.145), we derive

$$\lim_{\eta\to\infty} \psi_1(-\mathrm{i}\eta) = 0. \tag{A.146}$$

Because of (A.140) and (A.146), we see that (A.138) holds.

We note that (A.117) is valid. Therefore, relations (A.136) and (A.138) imply

$$\lim_{\eta\to-\infty}(\widetilde{F}, R_{\mathrm{i}\eta}F - \mathfrak{p}(z,\mathrm{i}\eta))_B = 0, \quad \widetilde{F}\in B, \tag{A.147}$$

since otherwise, substituting $\zeta\widetilde{F}$ for $\widetilde{F}$ in (A.117), we can choose a small enough $\zeta\in\mathbb{C}$ such that (A.117) does not hold for sufficiently large values of $|\eta|$.

Taking into account (A.111), (A.137) and (A.139), we immediately verify that

$$R_\omega F - \mathfrak{p}(z,\omega) = (z-\omega)^{-1}\left(\mathfrak{A}(z)J\mathfrak{A}(\overline{\omega})^* - J\right)\begin{bmatrix}0\\ \mathfrak{g}(\omega)\end{bmatrix}, \tag{A.148}$$

$$\mathfrak{g}(\omega) := \left(a(\omega) - b(\omega) + \mathrm{i}\varphi(\overline{\omega})^*\,(c(\omega)-d(\omega))\right)\breve{g}. \tag{A.149}$$

Therefore, formula (A.20) yields

$$\left(\widetilde{F}, R_\omega F - \mathfrak{p}(z,\omega)\right)_B = \mathrm{i}\mathfrak{g}(\omega)^*\,\widetilde{F}_2(\overline{\omega}). \tag{A.150}$$

From (A.54), (A.72) and (A.150), we derive

$$\left(\widetilde{F}, R_\omega F - \mathfrak{p}(z,\omega)\right)_B = (c(\omega)^* - d(\omega)^*)\,(\varphi(\overline{\omega}) - \varphi_0(\overline{\omega}))\,\widetilde{F}_2(\overline{\omega}). \tag{A.151}$$

Clearly, relations (A.147) and (A.151) imply (A.135).

Finally, we switch to functions F from the set

$$\mathrm{Span}_3 = \mathrm{span}\,\{F(z,\omega):\ \omega\in\mathbb{C},\ g_2\in\mathbb{C}^p\}, \tag{A.152}$$

where $F(z,\omega)$ are given in (A.65). Then, relations (A.61) and (A.70) yield

$$\eta e_1(-\mathrm{i}\eta)^{-1}F_2(-\mathrm{i}\eta) = O(1) \quad \text{for } \eta\to\infty. \tag{A.153}$$

Recall that, according to (A.57), $e_1 = c - d$. Therefore, from (A.135) and (A.153), we obtain

$$\lim_{\eta\to\infty}\eta F_2(-\mathrm{i}\eta)^*\,(\varphi(\mathrm{i}\eta) - \varphi_0(\mathrm{i}\eta))\,F_2(\mathrm{i}\eta) = 0. \tag{A.154}$$

For $\varphi = \varphi_0$, we set $N(\zeta) = N_0(\zeta) = (F, R_{\overline{\zeta}}F - \mathfrak{p}_0(z,\overline{\zeta}))_B$, where

$$\mathfrak{p}_0(z,\omega) := \frac{\mathfrak{A}(z)J\mathfrak{A}(\overline{\omega})^* - J}{z-\omega}\begin{bmatrix}I_p\\ -\mathrm{i}\varphi_0(\overline{\omega})^*\end{bmatrix}F_2(\omega). \tag{A.155}$$

Then, by virtue of (A.20), (A.111) and (A.155), we have

$$N(\zeta) - N_0(\zeta) = \left(F, \mathfrak{p}_0(z,\overline{\zeta}) - \mathfrak{p}(z,\overline{\zeta})\right)_B = F_2(\overline{\zeta})^*\,(\varphi(\zeta) - \varphi_0(\zeta))\,F_2(\zeta). \tag{A.156}$$

Since formulas (A.154) and (A.156) yield $\lim\eta(N(\mathrm{i}\eta) - N_0(\mathrm{i}\eta)) = 0$ for $\eta\to\infty$, we deduce from (A.133) that

$$\langle F_2, F_2\rangle_\tau - \langle F_2, F_2\rangle_{\tau_0} = 0. \tag{A.157}$$

According to (A.69), Span_3 is dense in B_1, that is, relation (A.157) holds on a dense set in B_1. Recall that, by virtue of Theorem A.12, inequality (A.122) is valid everywhere in B_1 and therefore the passage to the limit is possible. Thus, (A.157) holds for all $F \in B_1$. Taking into account that τ_0 is pseudospectral, we see that τ is also pseudospectral.

In order to prove statement (b), we recall that the existence of φ satisfying (A.45), where τ is the given pseudospectral function, follows from Theorem A.10. Furthermore, $\varphi \in \mathcal{N}(\mathfrak{A})$ is uniquely determined by the distribution function τ. Indeed, if there are two such functions, φ_1 and φ_2, then, clearly,

$$\varphi_1(z) - \varphi_2(z) = \breve{\mu}z + \breve{\nu} \quad (\breve{\mu} = \breve{\mu}^*, \quad \breve{\nu} = \breve{\nu}^*). \tag{A.158}$$

In the same way as (A.156), we obtain the equality

$$N_1(z) - N_2(z) = F_2(\overline{z})^* \left(\varphi_1(z) - \varphi_2(z)\right) F_2(z), \tag{A.159}$$

where N_1 and N_2 are functions N corresponding to φ_1 and φ_2, respectively. According to (A.132), N depends on τ only, that is, $N_1 - N_2 = 0$, and so (A.159) yields $F_2(\overline{z})^*(\breve{\mu}z + \breve{\nu})F_2(z) \equiv 0$ for all $F \in B$. We immediately verify that $\breve{\mu} = \breve{\nu} = 0$.

From the considerations above, we see that φ exists and is unique. Notice that for $F \in B$, formulas (A.133), (A.156) and (A.157) hold. Hence, we have

$$\lim_{\eta\to\infty} \eta F_2(-\mathrm{i}\eta)^* \left(\varphi(\mathrm{i}\eta) - \varphi_0(\mathrm{i}\eta)\right) F_2(\mathrm{i}\eta) = 0. \tag{A.160}$$

Following [86], we introduce two families of vector-valued functions

$$\mathfrak{L}_k = \left\{ R_0 \left(F(z) - \mathfrak{A}(z) \begin{bmatrix} (-1)^k I_p \\ I_p \end{bmatrix} F_2(0) \right) : F \in \mathrm{Span}_3 \right\}.$$

For $\widetilde{F} \in \mathfrak{L}_k$ $(k = 1, 2)$, we have a representation

$$\widetilde{F}_2(z,k) = z^{-1} \left(F_2(z) - (-1)^k e_k(z) F_2(0) \right). \tag{A.161}$$

It follows from (A.153) and (A.161) that

$$\lim_{\eta\to\infty} \eta e_1(-\mathrm{i}\eta)^{-1} \widetilde{F}_2(-\mathrm{i}\eta, 1) = \mathrm{i}F_2(0). \tag{A.162}$$

Using (A.60) and (A.70), we obtain an analog of (A.153) for $\mathbb{C}_+$:

$$\eta e_2(\mathrm{i}\eta)^{-1} F_2(\mathrm{i}\eta) = O(1) \quad (\eta \to \infty), \quad F \in \mathrm{Span}_3, \tag{A.163}$$

which yields

$$\lim_{\eta\to\infty} \eta e_2(\mathrm{i}\eta)^{-1} \widetilde{F}_2(\mathrm{i}\eta, 2) = \mathrm{i}F_2(0). \tag{A.164}$$

Taking into account (A.162), (A.164) and the result of substitution of $\widetilde{F}$ for F into (A.160), we derive

$$\begin{aligned}
&\lim_{\eta\to\infty} \eta^{-1} F_2(0)^* e_1(-\mathrm{i}\eta)^* \left(\varphi(\mathrm{i}\eta) - \varphi_0(\mathrm{i}\eta)\right) e_2(\mathrm{i}\eta) F_2(0) \\
&= \lim_{\eta\to\infty} \eta \widetilde{F}_2(-\mathrm{i}\eta, 1)^* \left(\varphi(\mathrm{i}\eta) - \varphi_0(\mathrm{i}\eta)\right) \widetilde{F}_2(\mathrm{i}\eta, 2) = 0, \quad F \in \mathrm{Span}_3.
\end{aligned} \tag{A.165}$$

According to (A.65) and (A.155), we have

$$\{F_2(0):\ F \in \mathrm{Span}_3\} \supseteq \mathrm{span}\ \{-c(\omega)^* g_2/\overline{\omega}:\ \omega \neq 0,\ g_2 \in \mathbb{C}^p\}. \tag{A.166}$$

Hypothesis I yields

$$\mathrm{span}\ \{-c(\omega)^* g_2/\overline{\omega}:\ \omega \neq 0,\ g_2 \in \mathbb{C}^p\} = \mathbb{C}^p. \tag{A.167}$$

Clearly, formulas (A.165)–(A.167) imply that φ satisfies condition (A.134). □

A.1.4 Description of the spectral functions

If $\mu_0 = 0$ and Hypothesis I holds, Theorems A.10 and A.13 completely describe the set of pseudospectral functions of the canonical system in terms of the linear-fractional (Möbius) transformations. The simplest description occurs in the case where (A.134) follows from $\varphi(z) \in \mathcal{N}(\mathfrak{A})$ automatically. We set

$$\mathfrak{B} = \{F_2(z):\ F(z) \in B\}, \quad \mathfrak{B}_1 = \{F_2(z):\quad zF_2(z),\, F_2(z)\ \in \mathfrak{B}\}, \tag{A.168}$$

whereas $\overline{\mathfrak{B}}_1$ is the closure of $\mathfrak{B}_1$. We define the inner product in $\mathfrak{B}$ via the equality $(F_2, \widetilde{F}_2)_{\mathfrak{B}} = \langle F_2, \widetilde{F}_2\rangle_{\tau_0}$. Like in [85], we deduce the next two propositions.

Proposition A.14. *If* $\overline{\mathfrak{B}}_1 = \mathfrak{B}$ *and* $\varphi(z) \in \mathcal{N}(\mathfrak{A})$, *then the distribution function* $\tau(t)$ *is pseudospectral.*

Proof. Let $F_2 \in \mathfrak{B}_1$. Then, putting $\widetilde{F}_2 = zF_2$ and substituting $\widetilde{F}_2$ for F_2 in (A.133) and (A.156), we obtain

$$\lim_{\eta\to\infty} \mathrm{i}\eta \widetilde{F}_2(-\mathrm{i}\eta)^* \left(\varphi(\mathrm{i}\eta) - \varphi_0(\mathrm{i}\eta)\right) \widetilde{F}_2(i\eta) = \langle \widetilde{F}_2, \widetilde{F}_2\rangle_{\tau_0} - \langle \widetilde{F}_2, \widetilde{F}_2\rangle_{\tau} < \infty. \tag{A.169}$$

For $F_2 \in \mathfrak{B}_1$, we derive from (A.169) that

$$\lim_{\eta\to\infty} \eta F_2(-\mathrm{i}\eta)^* \left(\varphi(\mathrm{i}\eta) - \varphi_0(\mathrm{i}\eta)\right) F_2(i\eta) = 0, \quad \text{i.e.}$$

$$\langle F_2, F_2\rangle_{\tau_0} = \langle F_2, F_2\rangle_{\tau}. \tag{A.170}$$

Since (A.170) holds on a set, which is dense in $\mathfrak{B}$, taking into account (A.122), we obtain the validity of (A.170) for all $F \in B$. □

Remark A.15. When Hypothesis I is valid, $\mu_0 = 0, \overline{\mathfrak{B}}_1 = \mathfrak{B}$, and $\varphi(z) \in \mathcal{N}(\mathfrak{A})$, the asymptotic relation (A.134) is immediate from Theorem A.12 and Proposition A.14.

In the proposition below, we study $\mathfrak{B}_1$ in somewhat greater detail.

Proposition A.16. *If* $\mathrm{Ker}\,\widehat{U} = \mathrm{Ker}\,U$ *and* $\widetilde{F}_2 = c(z)g_1 + d(z)g_2 \in \mathfrak{B}$ *for some* $g_1, g_2 \in \mathbb{C}^p$, *then*

$$\widetilde{F}_2 \perp \mathfrak{B}_1. \tag{A.171}$$

Proof. In order to prove the statement above, we note that the equality $\operatorname{Ker}\widehat{U} = \operatorname{Ker} U$ implies that F_2 uniquely determines $F \in B$ (and $B = B_1$). Therefore, since

$$(R_\alpha \widetilde{F})_2 = (R_\alpha \mathfrak{A}(z)g)_2 \quad \left(g := \operatorname{col}\begin{bmatrix} g_1 & g_2 \end{bmatrix}\right), \quad R_\alpha \mathfrak{A}(z)g \in B,$$

we derive $R_\alpha \widetilde{F} = R_\alpha(\mathfrak{A}(z)g)$. In particular, using (A.20), we obtain

$$(R_0\widetilde{F}, \breve{F})_B = -\mathrm{i}\breve{F}(0)^* Jg \quad \text{for } \breve{F} \in B. \tag{A.172}$$

The equality $R_\alpha \widetilde{F} = R_\alpha(\mathfrak{A}(z)g)$ also yields that $\widetilde{F}$ has the form

$$\widetilde{F}(z) = \mathfrak{A}(z)g + \operatorname{col}\begin{bmatrix} g_0 & 0 \end{bmatrix}, \quad g_0 \in \mathbb{C}^p. \tag{A.173}$$

Now, in a similar way, we obtain a representation of any function $\breve{F} \in B$ such that $\breve{F}_2 = zF_2$, $F \in B$. Indeed, taking into account (A.76), we see that

$$R_\alpha\left(zF(z)\right) \in B, \quad \left(R_\alpha\left(zF(z)\right)\right)_2 = (R_\alpha\breve{F})_2,$$

that is, $R_\alpha(zF) = R_\alpha\breve{F}$, and so

$$\breve{F}(z) = zF(z) + \operatorname{col}\begin{bmatrix} \breve{g}_0 & 0 \end{bmatrix}, \quad \breve{g}_0 \in \mathbb{C}^p. \tag{A.174}$$

In particular, (A.174) yields $R_0\breve{F} = F$. Hence, substituting $\widetilde{F}$ for F and $\breve{F}$ for G into (A.24) and setting $\alpha_1 = \alpha_2 = 0$, we see that

$$(\widetilde{F}, F)_B = (R_0\widetilde{F}, \breve{F})_B + \mathrm{i}\breve{F}(0)^* J\widetilde{F}(0). \tag{A.175}$$

Using (A.172)–(A.174), we rewrite (A.175) as $(\widetilde{F}, F)_B = 0$, that is, (A.171) holds. □

Finally, we study under what restrictions on the Hamiltonian $H(t)$ important hypotheses and assumptions of the preceding statements hold. We set

$$Kf = J\int_x^l H(t)f(t)\,dt, \quad K \in B(L^2(H)). \tag{A.176}$$

Lemma A.17. *A vector-valued function $f \in L^2(H)$ belongs* $\operatorname{Ker}\widehat{U}$ *if and only if $Kf = 0$ (i.e. $f \in \operatorname{Ker} K$) and*

$$\int_0^l H(t)f(t)\,dt = 0. \tag{A.177}$$

Proof. From (A.176), we see that

$$K^*f = J\int_0^x H(t)f(t)\,dt, \quad (K+K^*)f = J\int_0^l H(t)f(t)\,dt. \tag{A.178}$$

It can be checked directly (see also [201]) that

$$\mathfrak{A}(x,\overline{z})^* = I + \mathrm{i}zJ \int_0^x H(t)\left((I - \mathrm{i}zK^*)^{-1} I_p\right)(t)dt. \tag{A.179}$$

The relation $f \in \operatorname{Ker} \hat{U}$ is equivalent to

$$\int_0^l f(x)^* H(x)\mathfrak{A}(x,\overline{z})^* dx \equiv 0. \tag{A.180}$$

The necessity of the conditions of Lemma A.17 for $f \in \operatorname{Ker} \hat{U}$ are shown in the following way. First, substitute $z = 0$ into (A.180) in order to derive (A.177). Next, note that formulas (A.177), (A.179) and (A.180) yield

$$\begin{aligned} &\int_0^l f(x)^* H(x) J \int_0^x H(t)\left((I - \mathrm{i}zK^*)^{-1} I_p\right)(t)dt\,dx \\ &= \int_0^l f(x)^* H(x)\left(K^*(I - \mathrm{i}zK^*)^{-1} I_p\right)(x)dx = 0. \end{aligned} \tag{A.181}$$

According to (A.177), (A.178) and (A.181), we obtain

$$f(x) \perp \bigcup_{j=0}^{\infty} (K^*)^j (K + K^*) L^2(H), \quad \text{i.e.}$$

$$L_f \perp (K + K^*) L^2(H), \quad L_f := \overline{\operatorname{span}} \bigcup_{j=0}^{\infty} K^j f. \tag{A.182}$$

By virtue of (A.182), L_f is an invariant subspace of K on which $K = -K^*$. Note that K is an integral Volterra operator and its spectrum, in particular, is concentrated at zero. Therefore, we have $KL_f = 0$ and so $Kf = 0$. Thus, the equalities $Kf = 0$ and (A.177) follow from (A.180).

The converse is also valid – formula (A.181) follows from $Kf = 0$ and formula (A.180) follows from (A.177) and (A.181). □

Taking into account representation (A.179), we prove another useful lemma.

Lemma A.18. *A vector-valued function $f \in L^2(H)$ belongs to* Ker U *if and only if*

$$Kf = 0, \quad \int_0^l \begin{bmatrix} 0 & I_p \end{bmatrix} H(t) f(t)\, dt = 0. \tag{A.183}$$

Proof. The relation $f \in \operatorname{Ker} U$ is equivalent to

$$\int_0^l f(x)^* H(x) \mathfrak{A}(x, \overline{z})^* dx \begin{bmatrix} 0 \\ I_p \end{bmatrix} \equiv 0, \tag{A.184}$$

which, in view of (A.179), is equivalent to

$$K^j f \perp \operatorname{Im} \begin{bmatrix} 0 \\ I_p \end{bmatrix} \quad (j = 0, 1, 2, \ldots) \tag{A.185}$$

and thus (A.183) is a sufficient condition for $f \in \operatorname{Ker} U$.

Next, we show that (A.183) is a necessary condition for $f \in \operatorname{Ker} U$. Let $f \in \operatorname{Ker} U$. Putting $j = 0$ in (A.185), we derive the second condition in (A.183). From the second equalities in (A.178) and (A.183), we obtain

$$(K + K^*) \operatorname{Ker} U \in \operatorname{Im} \begin{bmatrix} 0 \\ I_p \end{bmatrix}. \tag{A.186}$$

Recall that $f \in \operatorname{Ker} U$ is equivalent to (A.184). Since relations (A.184) for f yield similar relations for Kf, the subspace $\operatorname{Ker} U$ is an invariant subspace of K. Denote the reduction of K on $\operatorname{Ker} U$ by $\widehat{K}$ and consider the block representation of K corresponding to the representation $L^2(H) = \operatorname{Ker} U \oplus L_1$, namely,

$$K = \begin{bmatrix} \widehat{K} & \cdots \\ 0 & \cdots \end{bmatrix}. \tag{A.187}$$

In view of (A.185) (for $j = 0$), we have $\operatorname{Im} \left[\begin{smallmatrix} 0 \\ I_p \end{smallmatrix} \right] \subseteq L_1$ and so formulas (A.186) and (A.187) imply $\widehat{K} + \widehat{K}^* = 0$. As in the previous proof, we use the last equality and the equality $\sigma(\widehat{K}) = \sigma(K) = 0$ in order to derive $\widehat{K} = 0$. Thus, we derive the necessity of the first equality in (A.183) for $f \in \operatorname{Ker} U$. □

Lemmas A.17 and A.18 can also be obtained without difficulty from the results of [86]. In the case $p = 1$, a description of $\operatorname{Ker} U$ in terms of the so called distorting (H-indivisible) intervals, where $\det H = 0$, is given in [157], [158].

Consider the simplest example.

Example A.19. Let $p = 1$ and

$$H(x) = \begin{bmatrix} 1 \\ 0 \end{bmatrix} \begin{bmatrix} 1 & 0 \end{bmatrix}, \quad f(x) = \begin{bmatrix} 1 \\ 0 \end{bmatrix} \quad (0 < x < l_0 < l). \tag{A.188}$$

Assume that on the interval $l_0 < x < l$, we have $f(x) = 0$ and H is chosen so that "positivity type" condition $\int_0^l H(x) dx > 0$ holds. Then, the relations

$$(Kf)(x) = (l_0 - x) \begin{bmatrix} 0 \\ 1 \end{bmatrix} \quad \text{for } x < l_0, \quad (Kf)(x) = 0 \quad \text{for } x > l_0 \tag{A.189}$$

easily follow. In other words, $Kf = 0$ in $L^2(H)$. Moreover, one can see that

$$\int_0^l H(x)f(x)dx = l_0 \begin{bmatrix} 1 \\ 0 \end{bmatrix} \neq 0, \quad \int_0^l \begin{bmatrix} 0 & 1 \end{bmatrix} H(x)f(x)dx = 0. \tag{A.190}$$

Using Lemmas A.17 and A.18, we finally obtain

$$f \notin \operatorname{Ker} \hat{U}, \quad f \in \operatorname{Ker} U, \quad \operatorname{Ker} \hat{U} \neq \operatorname{Ker} U. \tag{A.191}$$

It is proved in Section A.2 that inequality $\int_0^l H(x)dx > 0$ implies Hypothesis I. Hence, according to Lemma A.8, inequality $\mu_0 \neq 0$ follows.

Consider again the subspace $\mathfrak{B}_1$ introduced in (A.168). Denote by $\hat{L}$ the subspace of vectors in $L^2(H)$, which are orthogonal to $\operatorname{Im} [0 \quad I_p]^*$. The relation $f \in \hat{L}$ is equivalent to

$$F_2(0) = (\hat{U}f)_2(0) = 0. \tag{A.192}$$

From the definitions (A.5) and (A.176) of $\hat{U}$ and K, respectively, we deduce that

$$\hat{U}Kf = \mathrm{i}z^{-1}(F(z) - F(0)). \tag{A.193}$$

Thus, formulas (A.192) and (A.193) yield

$$z(UKf) = \mathrm{i}F_2(z) \in \mathfrak{B} \quad \text{for } f \in \hat{L},$$

and so

$$\mathfrak{B}_1 \supseteq UK\hat{L}. \tag{A.194}$$

If $\widetilde{F}_2 \in \mathfrak{B}_1$, then there exists an f such that

$$\hat{U}f = F = \begin{bmatrix} F_2 \\ z\widetilde{F}_2 \end{bmatrix} \quad \text{and} \quad f \in \hat{L}.$$

Then, according to (A.193), we have $UKf = \mathrm{i}\widetilde{F}_2$, that is, $\mathfrak{B}_1 \subseteq UK\hat{L}$. Therefore, taking (A.194) into account, we obtain

$$\mathfrak{B}_1 = UK\hat{L}. \tag{A.195}$$

Lemma A.20. *The relation* $\overline{\mathfrak{B}_1} = \mathfrak{B}$ *holds if and only if the relations*

$$K^*f = \begin{bmatrix} 0 \\ g_2 \end{bmatrix}, \quad f \in L^2(H), \quad g_2 \in \mathbb{C}^p \tag{A.196}$$

imply that

$$(K + K^*)f = \begin{bmatrix} 0 \\ g_2 \end{bmatrix}, \quad \int_0^l \begin{bmatrix} 0 & I_p \end{bmatrix} H(t)f(t)\, dt = 0. \tag{A.197}$$

Proof. From (A.195) and definitions of $\mathfrak{B}$ and $\mathfrak{B}_1$ in (A.168), we see that $\overline{\mathfrak{B}_1} = \mathfrak{B}$ if and only if

$$\overline{\mathrm{span}}\left(K\hat{L} \cup \mathrm{Ker}\, U\right) = L^2(H). \tag{A.198}$$

Presenting K in the form $K = (K + K^*) - K^*$ and using (A.178) and Lemma A.18, we obtain $K\hat{L} \perp \mathrm{Ker}\, U$, that is,

$$K\hat{L} \subseteq L_1. \tag{A.199}$$

By virtue of (A.199), relation (A.198) is equivalent to $\overline{K\hat{L}} = L_1$. However, this, in turn, is equivalent to the statement that $f \perp K\hat{L}$ implies $f \in \mathrm{Ker}\, U$. Hence, we should only show that the condition that (A.196) implies (A.197) is equivalent to the condition that $f \perp K\hat{L}$ implies $f \in \mathrm{Ker}\, U$.

Finally, we note that $f \perp K\hat{L}$ is equivalent to $K^* f \perp \hat{L}$, which can be written in the form (A.196). In case that (A.196) holds, taking into account Lemma A.18, we see that (A.197) is equivalent to $f \in \mathrm{Ker}\, U$. □

The next lemma can be used to derive Hypothesis I.

Lemma A.21. *If $c(z)h \equiv 0$ for some $h \in \mathbb{C}^p$, then for each $\omega \in \mathbb{C}$ almost everywhere in x, we have*

$$H(x)\begin{bmatrix} 0 & I_p \end{bmatrix}^* a(\omega)h = 0 \quad (0 < x < l). \tag{A.200}$$

Proof. By virtue of (A.14), we derive

$$\begin{aligned} &\int_0^l \begin{bmatrix} c(t,z) & d(t,z) \end{bmatrix} H(t) \begin{bmatrix} c(t,\omega)^* \\ d(t,\omega)^* \end{bmatrix} dt\, a(\overline{\omega})h \\ &= \frac{\mathrm{i}}{z - \overline{\omega}}(c(z)d(\omega)^* + d(z)c(\omega)^*)a(\overline{\omega})h. \end{aligned} \tag{A.201}$$

From (A.16), we see that

$$\begin{aligned} a(\omega)^* c(\overline{\omega}) + c(\omega)^* a(\overline{\omega}) &= 0 \text{ and} \\ b(\omega)^* c(\overline{\omega}) + d(\omega)^* a(\overline{\omega}) &= I_p, \quad \text{i.e.} \\ c(\omega)^* a(\overline{\omega}) &= -a(\omega)^* c(\overline{\omega}), \\ d(\omega)^* a(\overline{\omega}) &= I_p - b(\omega)^* c(\overline{\omega}). \end{aligned} \tag{A.202}$$

Let $c(z)h \equiv 0$. Then, in view of (A.201) and (A.202), we obtain

$$\int_0^l \begin{bmatrix} c(t,z) & d(t,z) \end{bmatrix} H(t) \begin{bmatrix} c(t,\omega)^* \\ d(t,\omega)^* \end{bmatrix} dt\, a(\overline{\omega})h = 0. \tag{A.203}$$

For any $\omega \in \mathbb{C}$, formula (A.203) yields (almost everywhere in t) the equality

$$H(t)\begin{bmatrix} c(t,\omega)^* \\ d(t,\omega)^* \end{bmatrix} a(\overline{\omega})\, h = 0. \tag{A.204}$$

From (A.1) and (A.6), we see that equation (A.204) is equivalent to

$$\frac{d}{dt}\begin{bmatrix} c(t,\omega)^* \\ d(t,\omega)^* \end{bmatrix} a(\overline{\omega})\,h = 0, \quad \text{i.e.} \quad \begin{bmatrix} c(t,\omega)^* \\ d(t,\omega)^* \end{bmatrix} a(\overline{\omega})\,h = \begin{bmatrix} 0 \\ a(\overline{\omega})h \end{bmatrix}.$$

Substituting the last equality into (A.204) (and substituting also ω for $\overline{\omega}$), we deduce (A.200). □

It could be useful to describe also spectral functions under stronger but simpler conditions. For this purpose, recall that spectral functions exist (and coincide with pseudospectral functions) if and only if $\operatorname{Ker} U = 0$. Recall also that Hypothesis I implies Hypothesis II and that $a(0) = I_p$. Hence, the theorem below is immediate from Proposition A.9, Theorem A.13 and Lemma A.21.

Theorem A.22. *Let* $\operatorname{Ker} U = 0$ *and assume that the equality* $H(x)\begin{bmatrix} 0 & I_p \end{bmatrix}^* h = 0$ $(h \in \mathbb{C}^p)$ *holds almost everywhere in* x *only for* $h = 0$. *Then, the set of spectral functions of the canonical system coincides with the set of distribution functions from Herglotz representations of functions* $\varphi \in \mathcal{N}(\mathfrak{A})$ *satisfying* (A.134).

The last theorem in this section also easily follows from our previous results.

Theorem A.23. *Let* $\det H(x) \neq 0$ *almost everywhere. Then,* $\operatorname{Ker} U = 0$ *and the set of spectral functions of system* (A.1) *coincides with the set of distribution functions from Herglotz representations of functions* $\varphi \in \mathcal{N}(\mathfrak{A})$.

Proof. First, note that in view (A.178) and $\det H(x) \neq 0$, relations (A.196) imply $\int_0^x H(t)f(t)dt = 0$ pointwise. Hence, we obtain $f(x) = 0$ almost everywhere and the condition of Lemma A.20 is fulfilled. Therefore, we have $\overline{\mathfrak{B}}_1 = \mathfrak{B}$. In a similar way, $\det H(x) \neq 0$ yields that $f = 0$ if only $Kf = 0$ and using Lemma A.18 we derive $\operatorname{Ker} U = 0$. Clearly, for the case that $\det H(x) \neq 0$, the equality $H(x)\begin{bmatrix} 0 & I_p \end{bmatrix}^* h = 0$ means that $h = 0$, and applying Lemma A.21, we see that Hypothesis I holds. Since $\operatorname{Ker} U = 0$ and Hypothesis I holds, from Proposition A.9, we obtain $\mu_0 = 0$. From Remark A.15, taking into account that $\overline{\mathfrak{B}}_1 = \mathfrak{B}$, Hypothesis I holds and $\mu_0 = 0$, we derive that condition (A.134) is fulfilled for $\varphi \in \mathcal{N}(\mathfrak{A})$ automatically. Thus, our proof shows that all the conditions of Theorem A.22 are satisfied and the statement of the theorem easily follows from Theorem A.22. □

The case $\det H(x) \neq 0$ permits reformulation in terms of expansions of symmetric operators. We refer to [167], where the spectral functions of a canonical system, which are generated by self-adjoint expansions, are described for the case $l = \infty$. Also, see the additional references and results in [230].

A.2 Special cases

A.2.1 Positivity-type condition

The “positivity-type” (or simply *positivity*) condition

$$\int_0^l H(t)dt > 0 \tag{A.205}$$

is often assumed to be valid (see, e.g. [137]). Then, according to [137, p. 249], we obtain

$$\int_0^l \mathfrak{A}(x,z)H(x)\mathfrak{A}(x,z)^*dx = \int_0^l W(x,\overline{z})^*H(x)W(x,\overline{z})dx > 0. \tag{A.206}$$

Indeed, if (A.206) is not valid, then, for some $g \neq 0$ (and some $z \in \mathbb{C}$), we have $H(x)^{\frac{1}{2}}W(x,\overline{z})g \equiv 0$ $(x \in [0, l])$. It follows that $H(x)W(x,\overline{z})g \equiv 0$, and hence, from (A.1), we obtain $W(x,\overline{z})g \equiv g$. Therefore, the equality $H(x)W(x,\overline{z})g \equiv 0$ yields $H(x)g \equiv 0$, contradicting (A.205).

In view of (A.14) and Potapov’s theorem (see Corollary E.4), formula (A.206) yields

$$\mathfrak{A}(z)J\mathfrak{A}(z)^* - J > 0, \quad \mathfrak{A}(z)^*J\mathfrak{A}(z) - J > 0 \quad (z \in \mathbb{C}_+). \tag{A.207}$$

In particular, we have

$$a(z)^*c(z) + c(z)^*a(z) > 0, \quad \det(c(z)\mathcal{P}_1(z) + d(z)\mathcal{P}_2(z)) \not\equiv 0 \tag{A.208}$$

for $\mathcal{P}_1$ and $\mathcal{P}_2$ satisfying (0.6). From the first relation in (A.208), we see that Hypothesis I in Section A.1.2 is satisfied.

Recall Definition A.1 of the spectral functions of a canonical system, which is given on $[0, l]$ (see also Definition 2.4).

Remark A.24. Note that if Hypothesis I for system (A.1) is fulfilled, then according to Theorem A.7, the spectral functions of system (A.1) satisfy inequality (A.3). Thus, assuming that Hamiltonian H satisfies either the positivity-type condition or sufficient conditions from Lemma A.21, we obtain (A.3).

Definition A.25. A nondecreasing $p \times p$ matrix function τ on $\mathbb{R}$ is called a spectral function of the system (A.1) on $[0, \infty)$ if this τ is a spectral function of (A.1) considered on $[0, l]$ for all $0 < l < \infty$.

In other words, τ is called a spectral function if operator

$$U_\infty := [0 \quad I_p] \int_0^\infty \mathfrak{A}(x,z)H(x) \cdot dx \tag{A.209}$$

defined on functions from $L^2(H, \infty)$ with compact support extends to isometry, also denoted by U_∞, on the entire space $L^2(H, \infty)$.

As mentioned in the Introduction, we sometimes use notation $\mathcal{N}(l)$ instead of $\mathcal{N}(\mathfrak{A})$. To show the existence of a spectral function for system (A.1) on $[0, \infty)$ under positivity condition, we first prove the following theorem.

Theorem A.26. *Let Hamiltonian H of system* (A.1) *be locally summable on* $[0, \infty)$ *and satisfy* (A.205) *for some* $\breve{l} > 0$. *Then,* $\bigcap_{l<\infty} \mathcal{N}(l) \neq \emptyset$.

Proof. Consider bounded closed domains $\mathbb{D}_+$ in $\mathbb{C}_+$. Since these domains are compact, formula (A.207) implies the existence of constants $\varepsilon = \varepsilon(l, \mathbb{D}_+) > 0 \quad (l \geq \breve{l})$ such that

$$J - (\mathfrak{A}(z)^*)^{-1} J \mathfrak{A}(z)^{-1} \geq \varepsilon I_m \quad (z \in \mathbb{D}_+). \tag{A.210}$$

From (A.210), we see that functions $\varphi \in \mathcal{N}(\breve{l})$ are uniformly bounded in $\mathbb{D}_+$. Indeed, if this is not valid, there exist sequences $\{\varphi_k\}$, $\{z_k\}$ and $\{g_k\}$ $(g_k \in \mathbb{C}^p)$ such that

$$\varphi_k \in \mathcal{N}(\breve{l}), \quad \|\varphi_k(z_k)\| \uparrow \infty, \quad \|\varphi_k(z_k) g_k\| = 1, \quad \|g_k\| = \|\varphi_k(z_k)\|^{-1}. \tag{A.211}$$

Notice that, according to (A.30), the pairs of $p \times p$ blocks (upper and lower) of matrix functions

$$\mathfrak{A}(\breve{l}, z)^{-1} \begin{bmatrix} -\mathrm{i}\varphi(z) \\ I_p \end{bmatrix} = \begin{bmatrix} \mathcal{P}_1(z) \\ \mathcal{P}_2(z) \end{bmatrix} (c(\breve{l}, z)\mathcal{P}_1(z) + d(\breve{l}, z)\mathcal{P}_2(z))^{-1}, \tag{A.212}$$

where $\varphi \in \mathcal{N}(\breve{l})$, are nonsingular with property-J. Thus, inequality (A.210) and the first relation in (A.211) yield

$$\begin{aligned} \mathrm{i} g_k^* (\varphi_k(z_k)^* - \varphi_k(z_k)) g_k &= g_k^* \begin{bmatrix} \mathrm{i}\varphi_k(z_k)^* & I_p \end{bmatrix} J \begin{bmatrix} -\mathrm{i}\varphi_k(z_k) \\ I_p \end{bmatrix} g_k \\ &\geq \varepsilon g_k^* \begin{bmatrix} \mathrm{i}\varphi_k(z_k)^* & I_p \end{bmatrix} \begin{bmatrix} -\mathrm{i}\varphi_k(z_k) \\ I_p \end{bmatrix} g_k \geq \varepsilon, \end{aligned}$$

which contradicts the last three relations in (A.211). Therefore, the functions $\varphi \in \mathcal{N}(\breve{l})$ are uniformly bounded in $\mathbb{D}_+$.

Recall now that $\varphi \in \mathcal{N}(\breve{l})$ are Nevanlinna functions and $\mathcal{N}(l_k) \subseteq \mathcal{N}(l_j)$ for $l_k > l_j$. Therefore, by Montel's theorem (i.e. Theorem E.9), there is a sequence

$$\{(\varphi_k - \mathrm{i}I_p)(\varphi_k + \mathrm{i}I_p)^{-1}\}, \quad \varphi_k \in \mathcal{N}(l_k), \quad l_k \uparrow \infty$$

which tends to an analytic function $\widetilde{\psi}$. Taking into account the boundedness of φ_k, we see that $\|\widetilde{\psi}(z) - I_p\| \geq \widetilde{\varepsilon}(\mathbb{D}_+)$. Thus, the functions φ_k tend to an analytic limit, namely,

$$\varphi_k \to \widetilde{\varphi} := \mathrm{i}(I_p + \widetilde{\psi})(I_p - \widetilde{\psi})^{-1}. \tag{A.213}$$

For arbitrary $l > 0$ and $k \in \mathbb{N}$ such that $l_k > l$, the blocks of $\mathfrak{A}^{-1}\left[\begin{smallmatrix} -i\varphi_k \\ I_p \end{smallmatrix}\right]$ form (according to (A.212)) a nonsingular pair with property-J. Hence, (A.213) implies that $\widetilde{\mathcal{P}}_1$ and $\widetilde{\mathcal{P}}_2$ given by

$$\begin{bmatrix} \widetilde{\mathcal{P}}_1(z) \\ \widetilde{\mathcal{P}}_2(z) \end{bmatrix} = \mathfrak{A}(l,z)^{-1} \begin{bmatrix} -i\widetilde{\varphi}(z) \\ I_p \end{bmatrix} \tag{A.214}$$

satisfy (0.6) too. Substitute now $\widetilde{\mathcal{P}}_1$ and $\widetilde{\mathcal{P}}_2$ instead of $\mathcal{P}_1$ and $\mathcal{P}_2$ into the right-hand side of (A.30) to show that $\widetilde{\varphi} \in \mathcal{N}(l)$. Therefore,

$$\widetilde{\varphi} \in \bigcap_{l<\infty} \mathcal{N}(l), \tag{A.215}$$

and the theorem is proved. □

Taking into account Theorem A.23, we obtain a corollary of Theorem A.26.

Corollary A.27. *Let* $\det H(x) \neq 0$ *almost everywhere. Then, there exists a spectral function of system* (A.1) *on* $[0, \infty)$.

Notice that formulas (A.14), (A.16) and (A.212) yield

$$\begin{aligned} &\int_0^l \begin{bmatrix} i\varphi(z)^* & I_p \end{bmatrix} J\mathfrak{A}(x,\overline{z})H(x)\mathfrak{A}(x,\overline{z})^* J \begin{bmatrix} -i\varphi(z) \\ I_p \end{bmatrix} dx \\ &= i\begin{bmatrix} i\varphi(z)^* & I_p \end{bmatrix} J \frac{\mathfrak{A}(\overline{z})J\mathfrak{A}(\overline{z})^* - J}{\overline{z} - z} J \begin{bmatrix} -i\varphi(z) \\ I_p \end{bmatrix} \\ &= \frac{i}{\overline{z} - z}\Big(\big((c(z)\mathcal{P}_1(z) + d(z)\mathcal{P}_2(z))^{-1}\big)^* (\mathcal{P}_1(z)^*\mathcal{P}_2(z) + \mathcal{P}_2(z)^*\mathcal{P}_1(z)) \\ &\quad \times (c(z)\mathcal{P}_1(z) + d(z)\mathcal{P}_2(z))^{-1} + i(\varphi(z) - \varphi(z)^*)\Big) \quad \text{for} \quad \varphi \in \mathcal{N}(l). \end{aligned} \tag{A.216}$$

According to (0.6) and (A.216), we obtain

$$\begin{aligned} &\int_0^\infty \begin{bmatrix} i\varphi(z)^* & I_p \end{bmatrix} J\mathfrak{A}(x,\overline{z})H(x)\mathfrak{A}(x,\overline{z})^* J \begin{bmatrix} -i\varphi(z) \\ I_p \end{bmatrix} dx \\ &\leq (\varphi(z) - \varphi(z)^*)/(z - \overline{z}) < \infty \quad \Big(\varphi \in \bigcap_{l<\infty} \mathcal{N}(l) \Big). \end{aligned} \tag{A.217}$$

Below, we recall Definition 0.4.

Definition A.28. Holomorphic functions φ such that

$$\int_0^\infty \begin{bmatrix} I_p & i\varphi(z)^* \end{bmatrix} W(x,z)^* H(x) W(x,z) \begin{bmatrix} I_p \\ -i\varphi(z) \end{bmatrix} dx < \infty \quad (z \in \mathbb{C}_+) \tag{A.218}$$

are called Weyl functions of system (A.1) on $[0, \infty)$.

Remark A.29. From (A.217), we see that the functions $\varphi \in \bigcap_{l<\infty} \mathcal{N}(l)$ satisfy (A.218) and so they are Weyl functions.

Corollary A.30. *Let Hamiltonian H of system* (A.1) *be locally summable on* $[0, \infty)$ *and satisfy* (A.205) *for some* $\breve{l} > 0$. *Then, there exists a Weyl function of system* (A.1) *on* $[0, \infty)$.

Corollary A.30 and Theorem A.31 below are wide generalizations of the Corollary 2.9 on Weyl functions of those systems (A.1) which are equivalent to Dirac systems.

Theorem A.31. *Let Hamiltonian H of system* (A.1) *be locally summable on* $[0, \infty)$ *and satisfy* (A.205) *for some* $\breve{l} > 0$. *Suppose additionally that* $H(x) \geq -\delta J$ *for some* $\delta > 0$. *Then, there exists a unique Weyl function of system* (A.1) *on* $[0, \infty)$.

Proof. From Corollary A.30, we see that there exists a Weyl function $\widetilde{\varphi} \in \bigcap_{l<\infty} \mathcal{N}(l)$. Suppose there exists also another Weyl function $\hat{\varphi}$, where $\hat{\varphi} \neq \widetilde{\varphi}$. Denote by $\mathbf{G}_1$ a subspace of $\mathbb{C}^m$ such that

$$\int_0^\infty g^* W(x,z)^* H(x) W(x,z) g dx < \infty \quad (g \in \mathbf{G}_1). \tag{A.219}$$

Since $\widetilde{\varphi}$ and $\hat{\varphi}$ are Weyl functions and $\hat{\varphi} \neq \widetilde{\varphi}$, Definition A.28 yields $\dim \mathbf{G}_1 > p$. On the other hand, conditions of the theorem imply that

$$W(x,z)^* H(x) W(x,z) \geq -\delta W(x,z)^* J W(x,z).$$

From (A.14), we derive $-W(x,z)^* J W(x,z) \geq -J$ for $z \in \mathbb{C}_+$. Hence, we see that

$$W(x,z)^* H(x) W(x,z) \geq -\delta J. \tag{A.220}$$

Taking into account that $-[-I_p \quad I_p] J \left[\begin{smallmatrix} -I_p \\ I_p \end{smallmatrix} \right] = 2I_p$, from (A.220), we derive

$$\int_0^\infty g^* W(x,z)^* H(x) W(x,z) g dx = \infty \quad \text{for } g \in \mathbf{G}_2 := \operatorname{Im} \begin{bmatrix} -I_p \\ I_p \end{bmatrix}. \tag{A.221}$$

Relations (A.219) and (A.221) show that $\mathbf{G}_1 \bigcap \mathbf{G}_2 = 0$. Since

$$\dim \mathbf{G}_1 + \dim \mathbf{G}_2 > 2p$$

and $\mathbf{G}_1, \mathbf{G}_2 \in \mathbb{C}^{2p}$, the equality $\mathbf{G}_1 \bigcap \mathbf{G}_2 = 0$ is impossible. We arrive at a contradiction. Thus, the Weyl function is unique and the theorem is proved. □

One can find similar uniqueness (and uniqueness up to simple transformations) results for nonclassical systems, for instance, in Chapter 4 and Section 8.1, see also papers [96, 142, 175, 210] and the references therein.

Remark A.32. Assuming $\det H(x) \neq 0$ almost everywhere, one can introduce the symmetric in $L^2(H)$ operator $\mathcal{L}$:

$$\mathcal{L}f = -\mathrm{i}H(x)^{-1}Jf', \quad f(0) = 0,$$

acting on the compactly supported absolutely continuous vector functions. The deficiency index p_+ of $\mathcal{L}$ coincides with the number of the linear independent solutions of system (A.1) on $[0, \infty)$ for $\Im z > 0$ (refer to Proposition 2.1 in [290, p. 123]). It is immediate now from Corollary A.30 that $p_+ \geq p$. Moreover, if $H(x) \geq -\delta J$, according to the proof of Theorem A.31, we obtain $p_+ = p$. Our conditions for $p_+ = p$ are essentially weaker than those in Theorem 2.2 from [290, p. 124], whereas Theorem 2.2 generalizes, in turn, a celebrated theorem by B. M. Levitan [194].

A.2.2 Asymptotics of the continuous analogs of orthogonal polynomials

There are close connections between orthogonal polynomials and solutions of linear differential equations (see, e.g. [21, 102, 139, 178, 308, 310] and references therein). In particular, important equality (A.14) and equality (5.50) for the discrete Dirac system are analogs of the well-known Christoffel–Darboux formula. One can find various results connected with the Christoffel–Darboux formula (including interesting asymptotical relations) and further references in [116, 309, 310, 316]. The matrix functions

$$\widetilde{\aleph}(l, z, \zeta) = c(l, z)\, d(l, \zeta)^* + d(l, z)\, c(l, \zeta)^*$$

are analogs of polynomial kernels. (Actually, in the case of discrete S-nodes, polynomial kernels are described in this form.) In this subsection, using some methods from the theory of orthogonal polynomials (see [21]) and Theorem A.12, we shall derive asymptotics of $\widetilde{\aleph}(l, z, \zeta)$, when l tends to infinity.

Consider now a locally summable Hamiltonian $H(x) \geq 0$ given on the semiaxis $0 \leq x < \infty$. The sets $\mathcal{N}(\mathfrak{A}(l, z))$ are included in one another, that is,

$$\mathcal{N}(\mathfrak{A}(l_1, z)) \subseteq \mathcal{N}(\mathfrak{A}(l_2, z)) \quad \text{if} \quad l_1 > l_2.$$

We require that there exists a unique matrix function $\varphi(z)$ which belongs to every $\mathcal{N}(\mathfrak{A}(l, z))$:

$$\varphi(z) = \bigcap_{l<\infty} \mathcal{N}(\mathfrak{A}(l, z)) \tag{A.222}$$

(the Weyl point case), and we also require that the distribution function $\tau(t)$ from the Herglotz representation of φ satisfies the Szegö condition

$$-\infty < \int_{-\infty}^{\infty} \left(1 + t^2\right)^{-1} \ln\left(\det \tau'(t)\right) dt\,. \tag{A.223}$$

According to the Krein–Zasukhin theorem [232] (see also a closely related Theorem 7.13 from [331]), τ' satisfying (A.223) admits the factorization

$$\tau'(t) = \upsilon(t)^*\upsilon(t), \tag{A.224}$$

where the analytic (in $\mathbb{C}_+$) $p \times p$ matrix function $\upsilon(z)$ having the limit $\upsilon(t)$ on $\mathbb{R}$ belongs $\widetilde{D}_S$. Here, $\widetilde{D}_S$ denotes the class of matrix functions f such that the entries of $f\left((\overline{\omega}\lambda - \omega)(\lambda - 1)^{-1}\right)^{\pm 1}$ belong to Smirnov class D_S, where $\omega \in \mathbb{C}_+$ and D_S is the class of the functions on the unit disc, which can be represented in the form of the quotient of a function from the Hardy class H_∞ and an exterior function from H_∞. Therefore, for $\Im\omega > 0$, the definition of the class $\widetilde{D}_S$ does not depend on the choice of ω. (See, e.g. [161] for the main facts and references on the Smirnov class.)

It follows from (A.14) that matrix functions $\widetilde{\aleph}(l, z, z)$ are monotonically nondecreasing in l for $\Im z > 0$. According to (A.14) and (A.40), we have

$$\frac{i\widetilde{\aleph}(l, z, z)}{z - \overline{z}} = \int_{-\infty}^{\infty} \frac{\widetilde{\aleph}(l, t, z)^*\, d\tau_0(t)\widetilde{\aleph}(l, t, z)}{(t - z)(t - \overline{z})}. \tag{A.225}$$

Hence, when $\Im z > 0$ and $\det c(l, z) \neq 0$, we obtain $\widetilde{\aleph}(l, z, z) > 0$. If the relation

$$c(l, z) \in \widetilde{D}_S \tag{A.226}$$

holds, then [27, Theorems 1 and 2] yield

$$|\det(\upsilon(z))|^2 \geq \lim_{l \to \infty} \det\left(\widetilde{\aleph}(l, z, z)^{-1}/(2\pi)\right). \tag{A.227}$$

According to (A.226) and [27, Theorem 3], we also have

$$\upsilon(z)^*\upsilon(z) \leq \widetilde{\aleph}(l, z, z)^{-1}/(2\pi). \tag{A.228}$$

From (A.227) and (A.228), we derive the equality

$$\lim_{l \to \infty} \widetilde{\aleph}(l, z, z)^{-1} = 2\pi\upsilon(z)^*\upsilon(z), \tag{A.229}$$

which was announced in [235]. In view of the monotonicity of $\widetilde{\aleph}$, the above limit is uniform on compact subsets of the open upper half-plane. Taking (A.229) into account, we prove the next theorem.

Theorem A.33. *Let z and ζ lie in a compact subset of $\mathbb{C}_+$ and let* (A.222), (A.223) *and* (A.226) *hold. Then,*

$$\lim_{l \to \infty} \widetilde{\aleph}(l, z, \zeta) = \frac{1}{2\pi}\upsilon(z)^{-1}\left(\upsilon(\zeta)^*\right)^{-1}, \tag{A.230}$$

where υ is the factoring multiplier from (A.224)*. The limit is uniform in z and ζ.*

Proof. We denote by Γ_r the curve $|(z-\zeta)(z-\overline{\zeta})^{-1}| = r \le 1$, choose anticlockwise orientation for Γ_r and put

$$\psi(l,r,\zeta) := \frac{1}{2\pi\mathrm{i}} \int_{\Gamma_r} \left(2\pi\upsilon(\zeta)\widetilde{\aleph}(l,z,\zeta)^*\upsilon(z)^* - I_p\right)$$

$$\times \left(2\pi\upsilon(z)\widetilde{\aleph}(l,z,\zeta)\upsilon(\zeta)^* - I_p\right) \frac{\left(\zeta-\overline{\zeta}\right)dz}{(z-\zeta)\left(z-\overline{\zeta}\right)};$$

$$\widetilde{\psi}(l,r,\zeta) := 2\pi\mathrm{i} \int_{\Gamma_r} \upsilon(\zeta)\widetilde{\aleph}(l,z,\zeta)^*\upsilon(z)^*\upsilon(z)\widetilde{\aleph}(l,z,\zeta) \frac{\upsilon(\zeta)^*\left(\overline{\zeta}-\zeta\right)dz}{(z-\zeta)\left(z-\overline{\zeta}\right)}.$$

Clearly, for $r < 1$, we have

$$\psi(l,r,\zeta) = I_p + \widetilde{\psi}(l,r,\zeta) - 4\pi\upsilon(\zeta)\widetilde{\aleph}(l,\zeta,\zeta)\,\upsilon(\zeta)^*. \tag{A.231}$$

From formula (A.225) (one could also use Theorem A.12), we have

$$\widetilde{\psi}(l,1,\zeta) \le 2\pi\,\upsilon(\zeta)\,\widetilde{\aleph}(l,\zeta,\zeta)\,\upsilon(\zeta)^*. \tag{A.232}$$

This implies that the entries of $\upsilon(z)\widetilde{\aleph}(z,\zeta)$ belong, after substitution

$$z = (\overline{\omega}\lambda - \omega)\,(\lambda-1)^{-1},$$

to the space H_2 of analytic functions in the unit disc. Then, using the theorem of F. Riesz [229] (see formula (4.1.1) in [226]) and properties of subharmonic functions, we have

$$\widetilde{\psi}(l,r,\zeta) \le \widetilde{\psi}(l,1,\zeta). \tag{A.233}$$

Since $\psi(l,r,\zeta) \ge 0$, it follows from (A.229)–(A.233) that

$$\lim_{l\to\infty} \psi(l,r,\zeta) = 0, \tag{A.234}$$

uniformly in ζ and r. Taking into account (A.234), we obtain the assertion of the theorem by expanding

$$\upsilon\left(\frac{\overline{\omega}\lambda-\omega}{\lambda-1}\right)\widetilde{\aleph}\left(\frac{\overline{\omega}\lambda-\omega}{\lambda-1},\zeta\right)$$

in series in λ. □

Theorem A.33 yields the equality

$$\lim_{l\to\infty} \mathfrak{A}(l,z)\,J\,\mathfrak{A}(l,\zeta)^* = \frac{1}{2\pi}\begin{bmatrix} -\mathrm{i}\varphi(z) \\ I_p \end{bmatrix} \upsilon(z)^{-1}\left(\upsilon(\zeta)^*\right)^{-1}\begin{bmatrix}\mathrm{i}\varphi(\zeta)^* & I_p\end{bmatrix}.$$

B Mathematical system theory

In the considerations of this book connected with explicit solutions of the direct and inverse problems and with Bäcklund–Darboux transformations, we often use some results and notations from mathematical system theory. This material has its roots in the Kalman theory [153], and can be found in various books (see, e.g. [34, 81, 185]). See also interesting historical remarks in [160].

For the convenience of the readers, we shall shortly present some basic facts below. The rational matrix functions appearing in this book are mostly *proper*, that is, analytic at infinity. Such an $m_2 \times m_1$ matrix function M can be represented in the form

$$M(\lambda) = D - C(A - \lambda I_n)^{-1}B, \tag{B.1}$$

where A is a square matrix of some order n, the matrices B and C are of sizes $n \times m_1$ and $m_2 \times n$, respectively, and $D = M(\infty)$. The representation (B.1) is called a *realization* or a *transfer matrix representation* of M, and the number $\mathrm{ord}(A)$ (order of the matrix A) is called the *state space dimension* of the realization.

If $M(\infty) = 0$, then the proper matrix function M is called *strictly proper*.

Definition B.1. The realization (B.1) is said to be *minimal* if its state space dimension n is minimal among all possible realizations of M. This minimal n is called the *McMillan degree* of M.

The realization (B.1) of M is minimal if and only if

$$\mathrm{span} \bigcup_{k=0}^{n-1} \mathrm{Im}\, A^k B = \mathbb{C}^n, \quad \mathrm{span} \bigcup_{k=0}^{n-1} \mathrm{Im}\, (A^*)^k C^* = \mathbb{C}^n, \quad n = \mathrm{ord}(A), \tag{B.2}$$

where Im stands for image. If for a pair of matrices $\{A, B\}$ the first equality in (B.2) holds, then the pair $\{A, B\}$ is called *controllable* or *full range*. If the second equality in (B.2) is fulfilled, then the pair $\{C, A\}$ is said to be *observable*. If the pair $\{A, B\}$ is full range, and K is an $m_1 \times n$ matrix, then the pair $\{A - BK, B\}$ is also full range.

If the pair $\{A, B\}$ is controllable, it is c-stabilizable, that is, there exists a matrix K such that $A-BK$ has all its eigenvalues in the open left half-plane. Some useful results on Riccati equations are formulated in terms of controllability and observability. The following theorem is a reduction of Theorem 16.3.3 [185] (see also [152]).

Theorem B.2. *Let the pair $\{A, B\}$ be c-stabilizable and the pair $\{Q, A\}$ (where Q is an $n \times n$ matrix) be observable. Then, there is a unique solution $X \geq 0$ of the Riccati equation*

$$XBB^*X - XA - A^*X - Q = 0. \tag{B.3}$$

Moreover, this X is positive (i.e. $X > 0$).

Minimal realizations are unique up to a basis transformation, that is, if (B.1) is a minimal realization of M and if $M(\lambda) = D - \widetilde{C}(\widetilde{A} - \lambda I_n)^{-1}\widetilde{B}$ is a second minimal realization of M, then there exists an invertible matrix $\mathcal{S}$ such that

$$\widetilde{A} = \mathcal{S}A\mathcal{S}^{-1}, \quad \widetilde{B} = \mathcal{S}B, \quad \widetilde{C} = C\mathcal{S}^{-1}. \tag{B.4}$$

In this case, (B.4) is called a *similarity transformation.*

Finally, if M is a square matrix ($m_1 = m_2 = p$) and $D = I_p$, then M^{-1} admits representation

$$M(\lambda)^{-1} = I_p + C(A^{\times} - \lambda I_n)^{-1}B, \quad A^{\times} = A - BC. \tag{B.5}$$

C Krein's system

An operator S with a scalar difference kernel

$$S_\xi f = f(x) + \int_0^\xi k(x-t)f(t)dt, \quad 0 < \xi < \infty \tag{C.1}$$

was used by M. G. Krein to solve an inverse spectral problem, that is, the problem to recover the self-adjoint Dirac system from its spectral function, in his classical work [176]. See also [20, 64, 169, 214, 227, 252, 295], where this approach is discussed and developed, and the references therein. It was the first time that structured operators, to which also operators with difference kernels belong, were applied in order to solve an inverse spectral problem. (Different classes of structured operators are dealt with in Appendix D.) Here, we write down Krein's system (formula (C.4)), which was the main object of study in [176], discuss some basic results on Krein's system from this paper and give further references. We follow the lines of [295] (and preserve some of the notations from [176]). Note that Krein's system is closely related to the Dirac system (see, e.g. [176] and [290, Section 10.4]).

We suppose that the operator S_ξ is positive and that the function $k(t)$ is continuous and satisfies the equality

$$k(t) = \overline{k(-t)}, \quad -\xi \le t \le \xi.$$

Then, there exists the Hermitian resolvent $\Gamma(t,s,\xi) = \overline{\Gamma(s,t,\xi)}$, satisfying the relation

$$\Gamma(t,s,\xi) + \int_0^\xi k(t-r)\Gamma(r,s,\xi)dr = k(t-s), \quad 0 \le s,\, t \le \xi.$$

We set

$$P(\xi,z) = \mathrm{e}^{\mathrm{i}z\xi}\left(1 - \int_0^\xi \mathrm{e}^{-\mathrm{i}zr}\Gamma(r,0,\xi)dr\right), \tag{C.2}$$

$$P_\star(\xi,z) = 1 - \int_0^\xi \mathrm{e}^{\mathrm{i}zr}\Gamma(0,r,\xi)dr. \tag{C.3}$$

M. G. Krein [176] deduced the following differential system

$$\frac{dP(\xi,z)}{d\xi} = \mathrm{i}zP(\xi,z) - \overline{A(\xi)}P_\star(\xi,z), \quad \frac{dP_\star(\xi,z)}{d\xi} = -A(\xi)P(\xi,z), \tag{C.4}$$

where $A(\xi) = \Gamma(0,\xi,\xi)$.

It was proved that there exists a nondecreasing function $\tau(x)$ (spectral function) such that the operator

$$Uf = \int_0^\infty P(\xi,x)f(\xi)d\xi, \quad -\infty < x < \infty$$

is an isometry from $L^2(0,\infty)$ to $L^2(d\tau)$.

The important theorem below was formulated in [176] without proof.

Theorem C.1. *The following statements are equivalent:*

(a) *The integral*

$$K(z_0) = \int_0^{\infty} |P(\xi, z_0)|^2 d\xi \tag{C.5}$$

converges for at least one point $z_0 \in \mathbb{C}_+$*;*

(b) *The function* $P_\star(\xi, z_0)$, $\quad 0 \le \xi < \infty$ *is bounded for at least one* $z_0 \in \mathbb{C}_+$*;*

(c) *The integral* $K(z)$*, given by* (C.5)*, converges uniformly at any bounded closed set in* $\mathbb{C}_+$*;*

(d) *There exists a limit*

$$P(z) = \lim P_\star(\xi, z), \quad \xi \to \infty, \tag{C.6}$$

which converges uniformly at any bounded closed set in $\mathbb{C}_+$*;*

(e) *The integral* $\int_{-\infty}^{\infty} \frac{\log(\tau'(x))}{(1+x^2)} dx$ *is finite.*

If the conditions (a)–(e) hold, then $P(z)$ given by (C.6) admits a representation

$$P(z) = \frac{1}{\sqrt{2\pi}} \exp\left\{ \frac{1}{2\pi \mathrm{i}} \int_{-\infty}^{\infty} \frac{(1+tz)\log(\tau'(t))}{(z-t)(1+t^2)} dt + \mathrm{i}\alpha \right\}, \quad \alpha = \overline{\alpha}. \tag{C.7}$$

Several interesting results stated in [176] and their generalizations for a wide class of canonical differential systems were proved in [288, 290, 291, 295]. Some open problems remain unsolved.

Open problem C.1. *Is Krein's theorem (Theorem C.1) valid in its original formulation?*

Open problem C.2. *Is there a case when the limit function* $P(z) \equiv \infty$*?*

D Operator identities corresponding to inverse problems

The operator S with difference kernel, which was discussed in Appendix C, is one of the most well-known so called structured operators. Structured operators usually satisfy some simple operator identities and the operator S with a scalar difference kernel satisfies identity (D.2), where j is a 2×2 diagonal matrix $\mathrm{diag}\{1, \ -1\}$ [275, 277, 280]. Based on the method of operator identities and its applications to inverse spectral problems for canonical systems, as presented in the review [280] and books [287, 290], many other systems were treated in the same way using other operators satisfying somewhat different operator identities (see, e.g. [112–114, 238, 262, 265, 289, 290]). Various facts about operator identities that are required in our book are collected in this Appendix, which is based on the paper [107]. We always assume in Sections D.1–D.4 that

$$j = \begin{bmatrix} I_{m_1} & 0 \\ 0 & -I_{m_2} \end{bmatrix}, \qquad A = -\mathrm{i}\int_0^x \cdot\, dt \in B\left(L^2_{m_2}(0, l)\right), \qquad m_1 m_2 \neq 0. \tag{D.1}$$

In particular, we study operators which are used in Chapter 2 on a self-adjoint Dirac system and in Chapter 3 on a skew-self-adjoint Dirac system. Operator identities corresponding to a self-adjoint Dirac system have the form

$$AS - SA^* = \mathrm{i}\Pi j \Pi^*, \quad S \in B\left(L^2_{m_2}(0, l)\right), \tag{D.2}$$

$$\Pi := \begin{bmatrix} \Phi_1 & \Phi_2 \end{bmatrix} \in B\left(\mathbb{C}^m, L^2_{m_2}(0, l)\right), \quad m := m_1 + m_2, \tag{D.3}$$

$$\Phi_i \in B\left(\mathbb{C}^{m_i}, L^2_{m_2}(0, l)\right), \quad (\Phi_1 f)(x) = \Phi_1(x) f, \quad \Phi_2 f = I_{m_2} f \equiv f. \tag{D.4}$$

The operator identities, which appear in the case of a skew-self-adjoint Dirac system, have the form

$$\mathrm{i}(AS - SA^*) = \Pi\Pi^*, \quad S \in B\left(L^2_{m_2}(0, l)\right), \tag{D.5}$$

where Π is given by formulas (D.3) and (D.4).

Structured operators (that satisfy operator identities) are also of interest on their own. For applications of structured operators to probability theory and other fields, see [92, 136, 149, 287, 296, 330] and various references therein. In particular, we show that for the case that $\Phi_1(x)$ is continuously differentiable, the operators S satisfying (D.2) have *close to displacement kernels*. Operators with close to displacement kernels were considered in [151] (see also [135, Section 2.4] and references therein) in connection with slightly nonhomogeneous processes and an algebra generated by Toeplitz operators. We also derive explicit inversion formulas for operators satisfying (D.2)–(D.4). Explicit inversion formulas for convolution integral operators on a finite interval are presented in [134].

The next Section D.1 is dedicated to the operator identities for the case of a self-adjoint Dirac system. Operator identities for a skew-self-adjoint Dirac system are studied in Section D.2. The positivity of operators S is discussed in Section D.3. Section D.4 is dedicated to explicit inversion of structured operators with semiseparable kernels. Operator identities for operators with D-difference kernels are the subject of the last Section D.5.

D.1 Operator identity: the case of self-adjoint Dirac system

The following theorem will be proved in this section.

Theorem D.1. *Assume that the entries of the $m_2 \times m_1$ matrix function $\Phi_1(x)$ are boundedly differentiable on the interval $0 \le x \le l$. Then, there exists a unique operator $S \in B(L^2_{m_2}(0,l))$ satisfying the identity* (D.2) *where Π is expressed via formulas* (D.3) *and* (D.4)*. This operator S is given by*

$$(Sf)(x) = \left(I_{m_2} - \Phi_1(0)\Phi_1(0)^*\right) f(x) - \int_0^l s(x,t)f(t)dt, \tag{D.6}$$

$$s(x,t) := \int_0^{\min(x,t)} \Phi_1'(x-\zeta)\Phi_1'(t-\zeta)^* d\zeta + \begin{cases} \Phi_1'(x-t)\Phi_1(0)^*, & x > t; \\ \Phi_1(0)\Phi_1'(t-x)^*, & t > x. \end{cases} \tag{D.7}$$

Furthermore, S is self-adjoint.

As mentioned in the introduction, for the case that $\Phi_1(x)$ is continuously differentiable, the kernel s of the form (D.7) is called a "close to displacement" kernel [135, 151]. To proceed with the proof of Theorem D.1, we need the lemma below, which easily follows from [171, Theorem 1.1] (see also [113, Proposition 3.2]).

Lemma D.2. *Let $\Phi(x)$ and $\widehat{\Phi}(x)$ be $m_2 \times m_1$ and $m_1 \times m_2$ matrix functions, respectively, and assume that these functions are boundedly differentiable on the interval $[0,l]$ and satisfy equalities $\Phi(0) = 0$, $\widehat{\Phi}(0) = 0$. Then, the operator S given by*

$$(Sf)(x) = -\frac{1}{2}\int_0^l \int_{|x-t|}^{x+t} \Phi'\left(\frac{\xi + x - t}{2}\right)\widehat{\Phi}'\left(\frac{\xi + t - x}{2}\right) d\xi f(t)dt \tag{D.8}$$

satisfies the operator identity

$$AS - SA^* = \mathrm{i}\Phi(x)\int_0^l \widehat{\Phi}(t) \cdot dt. \tag{D.9}$$

The following useful proposition is a special case of Theorem 3.1 in [270] (and a simple generalization of a subcase of Theorem 1.3 in [277, Ch. 1], which concerns the scalar case; see also [275, 287]).

Proposition D.3. *Let an operator $T \in B\left(L^2_{m_2}(0,l)\right)$ satisfy the operator identity*

$$TA - A^*T = \mathrm{i}\int_0^l Q(x,t) \cdot dt, \quad Q(x,t) = Q_1(x)Q_2(t), \tag{D.10}$$

where Q, Q_1 and Q_2 are $m_2 \times m_2$, $m_2 \times p$ and $p \times m_2$ $(p > 0)$ matrix functions, respectively. Then, T has the form

$$Tf = \frac{d}{dx}\int_0^l \left(\frac{\partial}{\partial t}\Upsilon(x,t)\right) f(t)dt, \tag{D.11}$$

where Υ is absolutely continuous in t and

$$\Upsilon(x,t) := -\frac{1}{2}\int_{x+t}^{2l-|x-t|} Q_1\left(\frac{\xi + x - t}{2}\right) Q_2\left(\frac{\xi - x + t}{2}\right) d\xi. \tag{D.12}$$

In particular, the identity $TA - A^*T = 0$ implies $T = 0$.

Proof of Theorem D.1. We split the proof into two parts. The first part shows that S given by (D.6) and (D.7) satisfies (D.2). Then, we deal with the uniqueness of S.

Step 1. Rewrite (D.6) as

$$S = \sum_{i=1}^4 S_i, \quad (S_1 f)(x) = \left(I_{m_2} - \Phi_1(0)\Phi_1(0)^*\right) f(x), \tag{D.13}$$

$$S_2 = -\int_0^x \Phi_1'(x-t)\Phi_1(0)^* \cdot dt, \quad S_3 = -\int_x^l \Phi_1(0)\Phi_1'(t-x)^* \cdot dt,$$

$$S_4 = -\int_0^l \int_0^{\min(x,t)} \Phi_1'(x-\zeta)\Phi_1'(t-\zeta)^* d\zeta \cdot dt.$$

It is immediately clear that

$$AS_1 - S_1A^* = \mathrm{i}\left(\Phi_1(0)\Phi_1(0)^* - I_{m_2}\right)\int_0^l \cdot dt. \tag{D.14}$$

Changing the order of integration and integrating by parts, we easily derive

$$AS_2 - S_2A^* = \mathrm{i}\left(\Phi_1(x) - \Phi_1(0)\right)\Phi_1(0)^*\int_0^l \cdot dt, \tag{D.15}$$

$$AS_3 - S_3A^* = \mathrm{i}\Phi_1(0)\int_0^l \left(\Phi_1(t) - \Phi_1(0)\right)^* \cdot dt. \tag{D.16}$$

Because of (D.13)–(D.16), it remains to show that

$$AS_4 - S_4A^* = \mathrm{i}\,(\Phi_1(x) - \Phi_1(0)) \int_0^l (\Phi_1(t) - \Phi_1(0))^* \cdot dt \tag{D.17}$$

in order to prove (D.2). After substitution

$$\xi = x + t - 2\zeta, \quad \Phi(x) = \Phi_1(x) - \Phi_1(0), \quad \widehat{\Phi}(t) = (\Phi_1(t) - \Phi_1(0))^*,$$

it follows that operator S in Lemma D.2 equals S_4, and formula (D.9) yields (D.17). Thus, (D.2) is proved.

Step 2. First, let us show that $S = 0$ is the unique operator from the class $B(L^2_{m_2}(0, l))$, which satisfies the operator identity $AS - SA^* = 0$. We show it by contradiction. Let $S_0 \neq 0$ ($S_0 \in B(L^2_{m_2}(0, l))$) satisfy the identity $AS_0 - S_0A^* = 0$. From the definition of A in (D.1), we have

$$\mathfrak{U}A\mathfrak{U} = A^*, \quad \mathfrak{U}A^*\mathfrak{U} = A \quad \text{for} \quad (\mathfrak{U}f)(x) := \overline{f(l - x)}. \tag{D.18}$$

It directly follows from the identity $AS_0 - S_0A^* = 0$ and equality (D.18) that

$$T_0A - A^*T_0 = 0 \quad \text{for } T_0 := \mathfrak{U}S_0\mathfrak{U} \neq 0, \tag{D.19}$$

where $T_0 \in B(L^2_{m_2}(0, l))$. Proposition D.3 and formula (D.19) imply $T_0 = 0$ and we arrive at a contradiction. Now, the uniqueness of the operator S, which satisfies (D.2), is apparent.

Note that the operator S^* satisfies (D.2) (or (D.5)) together with S, and so $S = S^*$ follows directly from the uniqueness of the solution of the corresponding operator identity. □

The next statement is apparent from the fact that $S = 0$ is the unique operator from the class $B(L^2_{m_2}(0, l))$, which satisfies the operator identity $AS - SA^* = 0$ (see the proof above, Step 2) and from [287, Ch.1, Theorem 1.2].

Corollary D.4. *Let* $m_1 = m_2 = p$, $\Pi = \begin{bmatrix} \Phi_1 & \Phi_2 \end{bmatrix}$, $\Phi_1, \Phi_2 \in B(\mathbb{C}^p, L^2_p(0, l))$,

$$\Phi_1 g = -s(x)g, \quad \Phi_2 g = g; \quad J = \begin{bmatrix} 0 & I_p \\ I_p & 0 \end{bmatrix}.$$

Then, the solution S of the operator identity $AS - SA^* = \mathrm{i}\Pi J\Pi^*$ *is given by*

$$S = \frac{d}{dx} \int_0^l s(x - t) \cdot dt, \quad s(-x) := -s(x)^* \tag{D.20}$$

if only the operator in (D.20) *is bounded.*

D.2 Operator identity for skew-self-adjoint Dirac system

Let

$$S = 2I - \check{S}, \quad A\check{S} - \check{S}A^* = \mathrm{i}\Pi j \Pi^*. \tag{D.21}$$

In view of (D.1) and (D.4), we have

$$\mathrm{i}(A - A^*) = \Phi_2 \Phi_2^*. \tag{D.22}$$

Therefore, the first equality in (D.21) yields the equivalence between the second equality in (D.21) and identity (D.5). In other words, we can rewrite Theorem D.1 in the following way.

Theorem D.5. *Assume that the entries of the $m_2 \times m_1$ matrix function $\Phi_1(x)$ are boundedly differentiable on the interval $0 \le x \le l$. Then, there exists a unique operator $S \in B(L^2_{m_2}(0, l))$ satisfying the operator identity* (D.5), *where Π is expressed via formulas* (D.3) *and* (D.4). *This operator S has the form*

$$(Sf)(x) = \left(I_{m_2} + \Phi_1(0)\Phi_1(0)^*\right) f(x) + \int_0^l s(x,t) f(t) dt, \tag{D.23}$$

where $s(x,t)$ is given by (D.7). *Furthermore, S is strictly positive (in fact, $S \ge I$).*

Proof. The first statements of the theorem present a reformulation of Theorem D.1, and this reformulation easily follows from the considerations at the beginning of the section.

It remains to prove the positivity statements. In order to prove $S \ge I$, it suffices to show that the inequalities $\mathcal{S}_\varepsilon \ge 0$, where $\mathcal{S}_\varepsilon$ is given by

$$(\mathcal{S}_\varepsilon f)(x) = \left(\varepsilon I_{m_2} + \Phi_1(0)\Phi_1(0)^*\right) f(x) + \int_0^l s(x,t) f(t) dt, \tag{D.24}$$

hold for all $0 < \varepsilon < 1$. For this reason, we note that $\mathcal{S}_\varepsilon = S - (1-\varepsilon)I$. Identities (D.5) and (D.22) thus lead us to the formula

$$\mathrm{i}(A\mathcal{S}_\varepsilon - \mathcal{S}_\varepsilon A^*) = \Phi_1\Phi_1^* + \varepsilon \Phi_2 \Phi_2^* \ge 0. \tag{D.25}$$

Formula (D.25) implies the equality of the scalar products (in $L^2_{m_2}(0, l)$) below:

$$\mathrm{i}((A\mathcal{S}_\varepsilon f, f) - (\mathcal{S}_\varepsilon A^* f, f)) = (\Phi_1\Phi_1^* f, f) + (\varepsilon \Phi_2\Phi_2^* f, f). \tag{D.26}$$

For the case that $f \in \mathrm{Ker}\, \mathcal{S}_\varepsilon$, equality (D.26) yields $\Phi_1^* f = 0$ and $\Phi_2^* f = 0$, and so the relation

$$A^* \mathrm{Ker}\, \mathcal{S}_\varepsilon \subseteq \mathrm{Ker}\, \mathcal{S}_\varepsilon \tag{D.27}$$

is immediate from (D.25). We note that according to (D.25), the operator A^* is $\mathcal{S}_\varepsilon$-dissipative and inequality (D.27) for an $\mathcal{S}_\varepsilon$-dissipative operator is presented in [28, statement 9°]. (At the end of the proof, we use Theorem 1 from the paper [28], where earlier results on operators in the space $\Pi_\varkappa$, from [180, 186], are developed for the case that we are interested in.)

Since the integral part of $\mathcal{S}_\varepsilon$ is a compact operator, we see that $\operatorname{Ker} \mathcal{S}_\varepsilon$ is finite-dimensional. However, A^* does not have eigenvectors and finite-dimensional invariant subspaces. Therefore, in view of (D.27), we derive $\operatorname{Ker} \mathcal{S}_\varepsilon = 0$. Hence, $\mathcal{S}_\varepsilon$ admits the representation

$$\mathcal{S}_\varepsilon = -K\mathcal{J}K, \quad K > 0, \quad \mathcal{J} = P_1 - P_2 \quad \left(P_i, K, K^{-1} \in B\left(L^2_{m_2}(0, l)\right)\right), \tag{D.28}$$

where P_1 and P_2 are orthoprojectors, $P_1 + P_2 = I$. Furthermore, since $\varepsilon > 0$ and the integral part of $\mathcal{S}_\varepsilon$ is a compact operator, we see that P_1 is a finite-dimensional orthoprojector. In other words, $\mathcal{J}$ determines some space $\Pi_\varkappa$, where $\varkappa < \infty$ is the dimension of $\operatorname{Im} P_1$. According to (D.25) and (D.28), the operator $-KA^*K^{-1}$ is $\mathcal{J}$-dissipative. From [28, Theorem 1], we see that there is a $\varkappa$-dimensional invariant subspace of $-KA^*K^{-1}$ (i.e. there is a $\varkappa$-dimensional invariant subspace of A^*), which leads us to $\varkappa = 0$ and $\mathcal{J} = -I$. Now, the inequality $\mathcal{S}_\varepsilon \geq 0$ follows directly from the first relation in (D.28). □

D.3 Families of positive operators

In this section, the operator $S \in B(L^2_{m_2}(0, \xi))$, which is given by (D.6) (after substitution ξ for l there), is denoted by S_ξ, that is, index ξ is added since we consider different values of ξ simultaneously. (Correspondingly, A is denoted by A_ξ, and Π by Π_ξ.) The family of orthoprojectors P_ξ $(\xi \leq l)$ from $L^2_{m_2}(0, l)$ on $L^2_{m_2}(0, \xi)$

$$\left(P_\xi f\right)(x) = f(x) \quad (0 < x < \xi), \quad f \in L^2_{m_2}(0, l) \tag{D.29}$$

is the same as in Chapters 1–3 (and formula (D.29) coincides with(1.99)). Clearly, for $\xi < l$, we have

$$A_\xi = P_\xi A_l P_\xi^*, \quad S_\xi = P_\xi S_l P_\xi^*. \tag{D.30}$$

The case of strictly positive operators S_l, which satisfy (D.2) (as well as strictly positive operators S_l, which satisfy (D.5) and are dealt with in Section D.2), is of special interest. Such operators are invertible and admit the factorization

$$S_l^{-1} = E^*_{\Phi,l} E_{\Phi,l}, \quad E_{\Phi,l} = I + \int_0^x E_\Phi(x,t) \cdot dt \in B\left(L_2^{m_2}(0, l)\right). \tag{D.31}$$

Below, we assume that $\varepsilon_\xi > 0$ in our notations. The following result on positivity of S_l satisfying (D.2) holds.

Proposition D.6. *Let $\Phi_1(x)$ be an $m_2 \times m_1$ matrix function, which is boundedly differentiable on the interval $[0, l]$ and satisfies the inequality*

$$(I_{m_2} - \Phi_1(0)\Phi_1(0)^*) > 0. \tag{D.32}$$

Furthermore, let operators S_ξ of the form (D.6)*, where we put ξ instead of l and s is expressed (via $\Phi_1(x)$) in* (D.7)*, be boundedly invertible for all $0 < \xi \le l$. Then, the operators S_ξ are strictly positive (i.e. $S_\xi \ge \varepsilon_\xi I$).*

Proof. Taking into account that (D.32) is valid and operators S_l are given by (D.6), substituting ξ for l, we obtain $S_\xi \ge \varepsilon_\xi I$ for some values $\varepsilon_\xi > 0$ and small values of ξ. We proceed by negation and suppose that some operators S_ξ are not strictly positive. Then, there is a value $0 < \xi_0 \le l$ such that $S_\xi \ge \varepsilon_\xi I$ for all $\xi < \xi_0$ and the inequality does not hold for all ξ such that $\xi > \xi_0$ (or $\xi_0 = l$ and the inequality does not hold for $\xi = \xi_0$). This is impossible since by passing to the limit, we obtain $S_{\xi_0} \ge 0$ and, moreover, the invertibility of S_{ξ_0} and the inequality $S_{\xi_0} \ge 0$ yield $S_{\xi_0} \ge \varepsilon_{\xi_0} I$. We also note that in the case that $\xi_0 < l$, formula (D.32) and inequality $S_{\xi_0} \ge \varepsilon_{\xi_0} I$ imply $S_{\xi_0+\delta} \ge \varepsilon_{\xi_0+\delta} I$ for some values $\varepsilon_{\xi_0+\delta} > 0$ and small values of δ. □

Another useful result easily follows from Proposition 1.38.

Theorem D.7. *Let the $m_2 \times m_1$ matrix function $\Phi_1(x)$ be boundedly differentiable on each finite interval $[0, l]$ and satisfy equality $\Phi_1(0) = 0$. Assume that operators S_l, which are expressed via Φ_1 in* (D.6) *and* (D.7)*, are boundedly invertible for all $0 < l < \infty$. Then, the operators S_l^{-1} admit factorizations* (D.31)*, where $E_\Phi(x, t)$ is continuous with respect to x, t and does not depend on l. Furthermore, all the factorizations* (D.31) *with continuous $E_\Phi(x, t)$ are unique.*

Proof. Since $\Phi_1(0) = 0$, formulas (D.6) and (D.7) take the form

$$S_l = I - \int_0^l s(x,t) \cdot dt, \quad s(x,t) = \int_0^{\min(x,t)} \Phi_1'(x-\zeta)\Phi_1'(t-\zeta)^* d\zeta. \tag{D.33}$$

Because of (D.33), we see that the kernel $s(x,t)$ of S_l is continuous. Hence, we can apply the factorization "result 2" from [137, pp. 185–186], which is included in Proposition 1.38 from Chapter 1. It follows that operators S_l^{-1} admit unique upper-lower triangular factorizations (1.164), where the kernels of the corresponding triangular operators $E_\pm$ are continuous and don't depend on l. From the equality $S_l = S_l^*$ (i.e. $S_l^{-1} = (S_l^{-1})^*$) and the uniqueness of the factorization, we derive $E_{\Phi,l} := E_- = E_+^*$. Hence, (1.164) leads us to (D.31). □

D.4 Semiseparable operators S

The inversion of structured operators S is used in inverse problems and various other applications. In this section, we discuss the inversion of S of the form (D.6), (D.7), where $\Phi_1(0) = 0$ and S is semiseparable, and focus on the case in which explicit inversion is possible. More precisely, we consider the subcase in which an explicit solution of the inverse problem for the self-adjoint Dirac system can be found. Thus, the inversion of the semiseparable operators S is an alternative to GBDT approach to the explicit solution of inverse problems (compare with Subsection 2.2.2, where GBDT is applied to direct and inverse problems for Dirac system). Note that the subcase of semiseparable operators corresponding to the skew-self-adjoint Dirac system, where $m_1 = m_2$, was discussed in [113] and similar results are easily derived in the same way for $m_1 \neq m_2$.

Recall (see, e.g. [128, Ch. IX]) that the operator

$$S = I + \int_0^l k(x,t) \cdot dt \tag{D.34}$$

is called semiseparable when k admits representation

$$k(x,t) = F_1(x)G_1(t) \quad \text{for } x > t, \quad k(x,t) = F_2(x)G_2(t) \quad \text{for } x < t, \tag{D.35}$$

where F_1 and F_2 are $p \times \widetilde{p}$ matrix functions and G_1 and G_2 are $\widetilde{p} \times p$ matrix functions. When the kernel k of the operator S is given by (D.35), the kernel of the operator $T = S^{-1}$ is expressed in terms of the $2\widetilde{p} \times 2\widetilde{p}$ solution U of the differential equation

$$\left(\frac{d}{dx}U\right)(x) = H(x)U(x), \quad x \geq 0, \quad U(0) = I_{2\widetilde{p}}, \tag{D.36}$$

where

$$\begin{aligned} &H(x) := B(x)C(x), \\ &B(x) := \begin{bmatrix} -G_1(x) \\ G_2(x) \end{bmatrix}, \quad C(x) := \begin{bmatrix} F_1(x) & F_2(x) \end{bmatrix}. \end{aligned} \tag{D.37}$$

More precisely, we have

$$T = S^{-1} = I + \int_0^l T(x,t) \cdot dt, \tag{D.38}$$

$$T(x,t) = \begin{cases} C(x)U(x)\left(I_{2\widetilde{p}} - P^{\times}\right)U(t)^{-1}B(t), & x > t, \\ -C(x)U(x)P^{\times}U(t)^{-1}B(t), & x < t. \end{cases} \tag{D.39}$$

Here, $P^{\times}$ is given in terms of the $\widetilde{p} \times \widetilde{p}$ blocks $U_{21}(l)$ and $U_{22}(l)$ of $U(l)$, that is,

$$P^{\times} = \begin{bmatrix} 0 & 0 \\ U_{22}(l)^{-1}U_{21}(l) & I_{\widetilde{p}} \end{bmatrix}. \tag{D.40}$$

Furthermore, the invertibility of $U_{22}(l)$ is a necessary and sufficient condition for the invertibility of S.

Remark D.8. Note that the operator K given by (2.92) is a triangular semiseparable operator. This property of K is actively used in [20].

As in the subcases from [19, 20, 107, 113], where explicit inversion results were dealt with, we assume that Φ_1' has the form

$$\Phi_1'(x) = 2\vartheta_2^* e^{2ix\theta}\vartheta_1, \quad \Phi_1(0) = 0, \tag{D.41}$$

where θ is an $n \times n$ matrix. Here, we present results from [107], where the matrices ϑ_1 and ϑ_2 are of different order (i.e. ϑ_i is an $n \times m_i$ matrix), which is essential for applications to Dirac systems with rectangular potentials. Because of (D.6), (D.7) and (D.41), the kernel k of S admits the representation (D.35) where

$$F_1(x) = 2\vartheta_2^* e^{2ix\theta}, \quad G_1(t) = -2\int_0^t e^{-2i\xi\theta}\vartheta_1\vartheta_1^* e^{2i\xi\theta^*}d\xi e^{-2it\theta^*}\vartheta_2, \tag{D.42}$$

$$F_2(x) = 2\vartheta_2^* e^{2ix\theta}\int_0^x e^{-2i\xi\theta}\vartheta_1\vartheta_1^* e^{2i\xi\theta^*}d\xi, \quad G_2(t) = -2e^{-2it\theta^*}\vartheta_2. \tag{D.43}$$

Clearly, if there is a solution X of the matrix identity

$$\theta X - X\theta^* = i\vartheta_1\vartheta_1^*, \tag{D.44}$$

then there is also a self-adjoint solution of (D.44) and it is often convenient to choose $X = X^*$. For X satisfying (D.44), we have

$$\frac{d}{d\xi}\left(e^{-2i\xi\theta}Xe^{2i\xi\theta^*}\right) = 2e^{-2i\xi\theta}\vartheta_1\vartheta_1^* e^{2i\xi\theta^*}, \quad \text{i.e.}$$

$$2\int_0^r e^{-2i\xi\theta}\vartheta_1\vartheta_1^* e^{2i\xi\theta^*}d\xi = e^{-2ir\theta}Xe^{2ir\theta^*} - X. \tag{D.45}$$

In view of (D.45), the matrix function U satisfying (D.36) can be constructed explicitly, and thus, so can $T = S^{-1}$.

Theorem D.9. *Let a structured operator S be given by* (D.6) *and* (D.7), *where Φ_1 has the form* (D.41). *Assume that the matrix identity* (D.44) *admits some solution X. Then, S is a semiseparable operator (of the form* (D.34)*) with the kernel k given by* (D.35), (D.42) *and* (D.43). *The matrix functions B, C, and H defined in* (D.37) *admit representations*

$$B(x) = \begin{bmatrix} e^{-2ix\theta}X - Xe^{-2ix\theta^*} \\ -2e^{-2ix\theta^*} \end{bmatrix}\vartheta_2, \quad C(x) = \vartheta_2^*\begin{bmatrix} 2e^{2ix\theta} & Xe^{2ix\theta^*} - e^{2ix\theta}X \end{bmatrix},$$

$$H(x) = B(x)C(x) = -2\Omega e^{-x\mathcal{A}}\begin{bmatrix} X \\ I_n \end{bmatrix}\vartheta_2\vartheta_2^*\begin{bmatrix} -I_n & X \end{bmatrix}e^{x\mathcal{A}}\Omega^{-1}, \tag{D.46}$$

where

$$\Omega := \begin{bmatrix} I_n & -X \\ 0 & -2I_n \end{bmatrix}, \quad \mathcal{A} := 2\mathrm{i}\begin{bmatrix} \theta & 0 \\ 0 & \theta^* \end{bmatrix}. \tag{D.47}$$

Moreover, the matrix function U satisfying (D.36)*, where $\widetilde{p} = n$, is given explicitly by the formula*

$$U(x) = \Omega \mathrm{e}^{-x\mathcal{A}}\mathrm{e}^{x\mathcal{A}^\times}\Omega^{-1}, \quad \mathcal{A}^\times := \mathcal{A} - 2\begin{bmatrix} X \\ I_n \end{bmatrix}\vartheta_2\vartheta_2^*\begin{bmatrix} -I_n & X \end{bmatrix}. \tag{D.48}$$

The operator S is invertible if and only if

$$\det\begin{bmatrix} 0 & I_n \end{bmatrix}\mathrm{e}^{l\widehat{\mathcal{A}}}\begin{bmatrix} 0 \\ I_n \end{bmatrix} \neq 0, \quad \widehat{\mathcal{A}} := \begin{bmatrix} 2\mathrm{i}\theta & \vartheta_1\vartheta_1^* \\ -4\vartheta_2\vartheta_2^* & 2\mathrm{i}\theta^* \end{bmatrix}. \tag{D.49}$$

If the inequality in (D.49) *holds, then $T = S^{-1}$ is given by the formulas* (D.38)–(D.40)*, where U is explicitly determined in* (D.48).

Proof. It was already shown above that S is a semiseparable operator. The representations of B, C and H easily follow from (D.37), (D.42) and (D.43). The U of the form (D.48) satisfies the relation

$$\frac{d}{dx}U = \Omega \mathrm{e}^{-x\mathcal{A}}\left(\mathcal{A}^\times - \mathcal{A}\right)\mathrm{e}^{x\mathcal{A}^\times}\Omega^{-1} = \Omega \mathrm{e}^{-x\mathcal{A}}\left(\mathcal{A}^\times - \mathcal{A}\right)\mathrm{e}^{x\mathcal{A}}\Omega^{-1}U(x), \tag{D.50}$$

which can be verified directly. Formulas (D.46), (D.50) and the definitions of $\mathcal{A}$ and $\mathcal{A}^\times$ yield (D.36). Thus our U is the same as in (D.39) and (D.40).

Recall that the invertibility of S is equivalent to the invertibility of $U_{22}(l)$. Because of (D.48), it is also equivalent to the invertibility of the right lower block of $\mathrm{e}^{l\mathcal{A}^\times}\Omega^{-1} = \Omega^{-1}\mathrm{e}^{l\Omega\mathcal{A}^\times\Omega^{-1}}$, which is equivalent to the invertibility of the right lower block of $\mathrm{e}^{l\Omega\mathcal{A}^\times\Omega^{-1}}$. According to (D.44), (D.47), (D.48) and the definition of $\widehat{\mathcal{A}}$ in (D.49), we have

$$\Omega\mathcal{A}^\times\Omega^{-1} = \begin{bmatrix} 2\mathrm{i}\theta & -\mathrm{i}(\theta X - X\theta^*) \\ -4\vartheta_2\vartheta_2^* & 2\mathrm{i}\theta^* \end{bmatrix} = \widehat{\mathcal{A}}. \tag{D.51}$$

Therefore, the invertibility of S is equivalent to the invertibility of the right lower block of $\mathrm{e}^{l\widehat{\mathcal{A}}}$, that is, to the inequality in (D.49). When S is invertible, $T = S^{-1}$ is given (as already mentioned before) by the formulas (D.38)–(D.40) (see [128]). □

D.5 Operators with D-difference kernels

Operator $S \in B(L^2_p(0,l))$ with D-difference kernel is an operator of the form

$$Sf = \frac{d}{dx}\int_0^l \mathfrak{s}(x,t)f(t)dt, \quad \mathfrak{s}(x,t) = \left\{\mathfrak{s}_{ij}(x,t)\right\}_{i,j=1}^p, \tag{D.52}$$

$$\mathfrak{s}_{ij}(x,t) = \mathfrak{s}_{ij}(d_ix - d_jt), \quad \mathfrak{s}_{ij}(x) \in L^2(-d_jl, d_il), \tag{D.53}$$

where we assume that

$$D = \operatorname{diag}\{d_1, d_2, \dots, d_p\}, \quad d_1 \geq d_2 \geq \dots \geq d_p > 0. \tag{D.54}$$

According to [287, Ch. 6], a bounded in $L_p^2(0, l)$ operator S of the form (D.52), (D.53) satisfies the operator identity

$$AS - SA^* = \mathrm{i}\Pi J \Pi^*, \tag{D.55}$$

where J is introduced in (0.2), $A \in B(L_p^2(0, l))$, $\Pi = [\Phi_1 \quad \Phi_2] \in B(\mathbb{C}^{2p}, L_p^2(0, l))$, $\Phi_i \in B(\mathbb{C}^p, L_p^2(0, l))$ and

$$A = \mathrm{i}D \int_0^x \cdot\, dt, \quad \Phi_1 g = D\mathfrak{s}(x, 0)g, \quad (\Phi_2 g)(x) \equiv g. \tag{D.56}$$

The identity (D.55) is easily proved directly. Following the lines of [270, Section 3], it is easy to show that (D.55) implies, in turn, that S is an operator with a D-difference kernel, and so the inverse statement is valid (see also [290, p. 104, Ex.1.2]).

Next, we formulate a generalization of Proposition D.3, the proof of which coincides (up to evident modifications) with the proof of Theorem 1.3 [287, p. 11].

Proposition D.10. *Let an operator $T \in B(L_p^2(0, l))$ satisfy the operator identity*

$$TA - A^*T = \mathrm{i} \int_0^l Q(x, t) \cdot dt, \quad Q(x, t) = Q_1(x) Q_2(t), \tag{D.57}$$

where A is given in (D.56) and Q, Q_1 and Q_2 are $p \times p$, $p \times \hat{p}$ and $\hat{p} \times p$ ($\hat{p} > 0$), respectively, matrix functions. Then, T has the form

$$Tf = \frac{d}{dx} \int_0^l \frac{\partial}{\partial t} Y(x, t) f(t) dt, \tag{D.58}$$

where $Y(x, t) = \{Y_{ij}(x, t)\}_{i,j=1}^p$ is absolutely continuous in t and

$$Y_{ij}(x, t) := (2 d_i d_j)^{-1} \int_{d_i x + d_j t}^{\chi} Q_{ij}\left(\frac{\xi + d_i x - d_j t}{2 d_i}, \frac{\xi - d_i x + d_j t}{2 d_j} \right) d\xi, \tag{D.59}$$

$$\chi := \min\Big(d_i(2l - x) + d_j t,\ d_i x + d_j(2l - t) \Big). \tag{D.60}$$

In particular, the identity $TA - A^*T = 0$ implies $T = 0$.

In fact, Proposition D.10 holds for a much wider class of functions Q than the one given in (D.57). Like Theorem D.1 is derived using Proposition D.3, the next corollary follows from a more general Proposition D.10 and from formula (D.18), which is also valid for A of the form (D.56).

Corollary D.11. *Suppose $S \in B(L_p^2(0,l))$ satisfies the operator identity*

$$AS - SA^* = \mathrm{i}\int_0^l \left(\Phi_1(x) + \widehat{\Phi}_1(t)\right) \cdot dt, \quad \Phi_1(x), \widehat{\Phi}_1(t) \in L^2_{p\times p}(0,l).$$

Then, S is an operator of the form (D.52), (D.53) *and*

$$\mathfrak{s}(x,0) = D^{-1}\Phi_1(x), \quad \mathfrak{s}(0,t) = -D^{-1}\widehat{\Phi}_1(t).$$

Moreover, if (D.55) *holds, that is, $\widehat{\Phi}_1(t) = \Phi_1(t)^*$, we have*

$$\mathfrak{s}(x,t) = -D^{-1}\mathfrak{s}(t,x)^*D, \quad S = S^*. \tag{D.61}$$

Corollary D.11 was essential in the study of a subclass of canonical systems from [270]. In Chapter 4, we use another direct corollary of Proposition D.10.

Corollary D.12. *Suppose $S \in B(L_p^2(0,l))$ satisfies the operator identity*

$$AS - SA^* = \mathrm{i}\Pi\Pi^*, \quad (\Pi g)(x) = \Pi(x)g, \tag{D.62}$$

where A is given in (D.56) *and $\Pi_1(x)$ is a $p \times p$ matrix function. Then, S has the form*

$$Sf = \frac{d}{dx}\int_0^l \frac{\partial}{\partial t}\widehat{\Upsilon}(x,t)f(t)dt, \tag{D.63}$$

where $\widehat{\Upsilon}(x,t) = \{\widehat{\Upsilon}_{ij}(x,t)\}_{i,j=1}^p$ is absolutely continuous in t and

$$\widehat{\Upsilon}_{ij}(x,t) := (2d_id_j)^{-1}\int_{|d_ix-d_jt|}^{d_ix+d_jt} \widehat{Q}_{ij}\left(\frac{\xi + d_ix - d_jt}{2d_i}, \frac{\xi - d_ix + d_jt}{2d_j}\right)d\xi, \tag{D.64}$$
$$\widehat{Q}(x,t) := \Pi(x)\Pi(t)^*.$$

In particular, the identity $AS - SA^* = 0$ implies $S = 0$.

E Some basic theorems

Here, we formulate several basic theorems which are essential for our proofs and are usually applied several times in the text. Although these theorems are mostly well known, it is convenient to formulate them in the book.

1. We start with a useful linear algebraic result by V. P. Potapov ([223, Ch.2, Theorem 7]).

Theorem E.1. *For* $j = \begin{bmatrix} I_{m_1} & 0 \\ 0 & -I_{m_2} \end{bmatrix}$ *and* $m \times m$ *matrix* u $(m = m_1 + m_2)$, *the inequalities*

$$uju^* \leq j \quad \text{and} \quad u^*ju \leq j \tag{E.1}$$

are equivalent.

Using [223, Ch.2, Theorem 6], like in the proof in [223] of Theorem E.1, we easily obtain the equivalence of the corresponding strict inequalities.

Corollary E.2. *For* $j = \begin{bmatrix} I_{m_1} & 0 \\ 0 & -I_{m_2} \end{bmatrix}$ *and* $m \times m$ *matrix* u $(m = m_1 + m_2)$, *the inequalities* $uju^* < j$ *and* $u^*ju < j$ *are equivalent.*

Since the equalities $J = J^* = J^{-1}$ imply (for matrix J) the existence of representation $J = UjU^*$, where U is a unitary matrix, and also the existence of representation $J = -UjU^*$ for another choice of j and U, we can substitute J or $-J$ instead of j in Theorem E.1.

Corollary E.3. *For* $J = J^* = J^{-1}$ *and* $m \times m$ *matrix* u $(m = m_1 + m_2)$, *inequality* $uJu^* \leq J$ *is equivalent to inequality* $u^*Ju \leq J$ *and inequality* $uJu^* \geq J$ *is equivalent to inequality* $u^*Ju \geq J$.

In the same way, Corollary E.2 yields our next corollary.

Corollary E.4. *For* $J = J^* = J^{-1}$ *and* $m \times m$ *matrix* u $(m = m_1 + m_2)$, *inequality* $uJu^* < J$ *is equivalent to inequality* $u^*Ju < J$ *and inequality* $uJu^* > J$ *is equivalent to inequality* $u^*Ju > J$.

2. A simple first Liouville theorem (see, e.g. [170, Ch. 9]) is of use in our considerations.

Theorem E.5. *A nonconstant entire function assumes arbitrarily large values outside every circle.*

It is immediate from the Liouville theorem that an entire function whose absolute values are bounded on $\mathbb{C}$ is a constant.

3. We also need Phragmen–Lindelöf theorems for an angle and for a strip. These theorems follow from the general Phragmen–Lindelöf theorem (see, e.g. [80, Theorem VI.4.1]).

Theorem E.6. *Let Ω be a simply connected open set and let f be an analytic function on Ω. Suppose there is an analytic function ψ which never vanishes and is bounded (on Ω). Assume that the extended boundary $\partial\Omega \cup \{\infty\}$ can be split into components ∂_1 and ∂_2 $(\partial\Omega \cup \{\infty\} = \partial_1 \cup \partial_2)$ so that for some $M > 0$ we have*

$$\limsup_{z\to\lambda} |f(z)| \leq M \tag{E.2}$$

for every $\lambda \in \partial_1$, and

$$\limsup_{z\to\lambda} |\psi(z)|^{\eta} |f(z)| \leq M \tag{E.3}$$

for every $\lambda \in \partial_2$ and $\eta > 0$. Then, the inequality $|f(z)| \leq M$ holds for all $z \in \Omega$.

For the case where Ω is an angle ($\Omega = \{z : |\arg(z)| < \pi/a\}$) and $\psi(z)$ are functions of the form $\exp\{-z^{(a-\varepsilon)/2}\}$ $(0 < \varepsilon < a)$, we obtain the next corollary.

Corollary E.7. *Suppose that $\Omega = \{z : |\arg(z)| < \pi/a\}$ and $a > 1$. Assume that f is analytic on Ω and that*

$$\limsup_{z\to\lambda} |f(z)| \leq M \quad \text{for } \lambda \in \partial\Omega; \tag{E.4}$$

$$|f(z)| \leq M_1 \exp\{|z|^{(a/2)-\varepsilon}\} + M_2 \tag{E.5}$$

for some $\varepsilon, M_1, M_2 > 0$ and all $z \in \Omega$. Then, $|f(z)| \leq M$ for all $z \in \Omega$.

Clearly, we can rotate the angle and consider $e^{i\xi}\Omega$ in the same way as Ω. Theorem E.6 also yields a corollary for a semistrip.

Corollary E.8. *Suppose that $\Omega = \{z : \eta_2 < \Im(z) < \eta_1\}$. Assume that f is analytic on Ω and that*

$$\limsup_{z\to\lambda} |f(z)| \leq M \quad \text{for } \lambda \in \partial\Omega; \tag{E.6}$$

$$|f(z)| \leq \exp\{M_1 \exp\{\pi a|\Re(z)|/(\eta_1 - \eta_2)\}\} + M_2 \tag{E.7}$$

for some $M_1, M_2 > 0$, $a < 1$ and all $z \in \Omega$. Then, $|f(z)| \leq M$ for all $z \in \Omega$.

In our study of Weyl discs, we apply a well-known Montel's theorem on analytic functions (see [80, Section VII.2]).

Theorem E.9. *A family $\mathcal{F}$ of the functions analytic on an open set $\Omega \subset \mathbb{C}$ is normal (i.e. each sequence in $\mathcal{F}$ has a subsequence which converges to an analytic function) if $\mathcal{F}$ is locally bounded.*

4. We use Theorem V [219, §4] and Theorem VIII [219, §5] (see also [324]) on the Fourier transform in the complex domain. Since those results are used (in the book) in the domain $\mathbb{C}_M$, mostly, we reformulate them in this way.

Theorem E.10. *The class of all functions F, which are analytic in the half-plane $\mathbb{C}_M = \{z : \Im(z) > M \geq 0\}$ and satisfy condition*

$$\sup_{\eta > M} \int_{-\infty}^{\infty} |F(\xi + \mathrm{i}\eta)|^2 d\xi < \infty, \tag{E.8}$$

coincides with the class of functions F in $\mathbb{C}_M$ such that

$$F(\xi + \mathrm{i}\eta) = \mathrm{l.i.m.}_{a\to\infty} \int_0^a \mathrm{e}^{\mathrm{i}(\xi+\mathrm{i}\eta)x} f(x) dx \quad \left(\mathrm{e}^{-xM} f(x) \in L^2(0, \infty)\right). \tag{E.9}$$

Moreover, we have

$$\mathrm{l.i.m.}_{\eta\to M} F(\xi + \mathrm{i}\eta) = \mathrm{l.i.m.}_{a\to\infty} \int_0^a \mathrm{e}^{\mathrm{i}(\xi+\mathrm{i}M)x} f(x) dx. \tag{E.10}$$

It is easy to see that (E.9) implies the pointwise equality

$$F(\xi + \mathrm{i}\eta) = \int_0^{\infty} \mathrm{e}^{\mathrm{i}(\xi+\mathrm{i}\eta)x} f(x) dx, \quad \eta > M. \tag{E.11}$$

Theorem E.11. *If F is analytic and bounded in $\overline{\mathbb{C}_M} = \{z : \Im(z) \geq M \geq 0\}$ and*

$$\int_{-\infty}^{\infty} |F(\xi + \mathrm{i}M)|^2 d\xi < \infty, \tag{E.12}$$

then (E.8) *holds, and so F admits representation* (E.9).

Bibliography

[1] M. J. Ablowitz, G. Biondini and B. Prinari. Inverse scattering transform for the integrable discrete nonlinear Schrödinger equation with nonvanishing boundary conditions. *Inverse Problems*, 23(4):1711–1758, 2007.

[2] M. J. Ablowitz, S. Chakravarty, A. D. Trubatch and J. Villarroel. A novel class of solutions of the non-stationary Schrödinger and the Kadomtsev–Petviashvili I equations. *Phys. Lett. A*, 267(2–3):132–146, 2000.

[3] M. J. Ablowitz and R. Haberman. Resonantly coupled nonlinear evolution equations. *J. Math. Phys.*, 16:2301–2305, 1975.

[4] M. J. Ablowitz, D. J. Kaup, A. C. Newell and H. Segur. Method for solving the sine-Gordon equation. *Phys. Rev. Lett.*, 30(25):1262–1264, 1973.

[5] M. J. Ablowitz, D. J. Kaup, A. C. Newell and H. Segur. The inverse scattering transform – Fourier analysis for nonlinear problems. *Stud. Appl. Math.*, 53:249–315, 1974.

[6] M. J. Ablowitz, B. Prinari and A. D. Trubatch. *Discrete and continuous nonlinear Schrödinger systems*. London Mathematical Society Lecture Note Series. 302. Cambridge: Cambridge University Press, 257 p., 2004.

[7] M. J. Ablowitz and H. Segur. *Solitons and the inverse scattering transform*. SIAM Stud. Appl. Math. 4. Philadelphia: SIAM, 425 p, 1981.

[8] V. M. Adamyan and D. Z. Arov. A class of scattering operators and of characteristic operator-functions of contractions. *Sov. Math. Dokl.*, 6:1–5, 1965.

[9] H. Aden and B. Carl. On realizations of solutions of the KdV equation by determinants on operator ideals. *J. Math. Phys.*, 37:1833–1857, 1996.

[10] M. Adler and P. van Moerbeke. Birkhoff strata, Bäcklund transformations, and regularization of isospectral operators. *Adv. Math.*, 108(1):140–204, 1994.

[11] N. I. Akhiezer. *The classical moment problem and some related questions in analysis*. Moscow: Fizmatgiz, 1961. Translated in: London: Oliver & Boyd and New York: Hafner, 263 p., 1965.

[12] S. A. Akhmanov, V. A. Vysloukh and A. S. Chirkin. *Optics of femtosecond laser pulses*. New York: American Institute of Physics, 366 p., 1992.

[13] A. Aksoy and M. Martelli. Mixed partial derivatives and Fubini's theorem. *Coll. Math. J.*, 33:126–130, 2002.

[14] S. Albeverio, R. Hryniv and Ya. Mykytyuk. Reconstruction of radial Dirac operators. *J. Math. Phys.*, 48(4):043501, 14 pp., 2007.

[15] S. Albeverio, R. Hryniv and Ya. Mykytyuk. Reconstruction of radial Dirac and Schrödinger operators from two spectra. *J. Math. Anal. Appl.*, 339(1):45–57, 2008.

[16] D. Alpay and I. Gohberg. Inverse spectral problems for difference operators with rational scattering matrix function. *Integral Equations Operator Theory*, 20:125–170, 1994.

[17] D. Alpay and I. Gohberg. Inverse spectral problem for differential operators with rational scattering matrix functions. *J. Differential Equations*, 118:1–19, 1995.

[18] D. Alpay and I. Gohberg. Discrete analogs of canonical systems with pseudo-exponential potential. Inverse Problems. In: *Interpolation, Schur Functions and Moment Problems*, pp. 31–65. Oper. Theory Adv. Appl. 165. Basel: Birkhäuser, 2006.

[19] D. Alpay, I. Gohberg, M. A. Kaashoek, L. Lerer and A. Sakhnovich. Krein systems. In: V. Adamyan et al. (eds). *Modern Analysis and Applications*, pp. 19–36. Oper. Theory Adv. Appl. 191. Basel: Birkhäuser, 2009.

[20] D. Alpay, I. Gohberg, M. A. Kaashoek, L. Lerer and A. Sakhnovich. Krein systems and canonical systems on a finite interval: accelerants with a jump discontinuity at the origin and continuous potentials. *Integral Equations Operator Theory*, 68(1):115–150, 2010.

[21] A. I. Aptekarev and E. M. Nikishin. The scattering problem for a discrete Sturm–Liouville operator. *Mat. Sb.*, 121/163(3/7):327–358, 1983. Translated in: *Math. USSR Sb.*, 49(2):325–355, 1984.

[22] D. Z. Arov. On monotone families of J-contractive matrix-valued functions. *St. Petersburg Math. J.*, 9(6):1025–1051, 1997.

[23] D. Z. Arov and H. Dym. J-inner matrix functions, interpolation and inverse problems for canonical systems. IV: Direct and inverse bitangential input scattering problems. *Integral Equations Operator Theory*, 43:68–129, 2002.

[24] D. Z. Arov and H. Dym. The bitangential inverse spectral problem for canonical systems. *J. Funct. Anal.*, 214(2):312–385, 2004.

[25] D. Z. Arov and H. Dym. Direct and inverse problems for differential systems connected with Dirac systems and related factorization problems. *Indiana Univ. Math. J.*, 54(6):1769–1815, 2005.

[26] D. Z. Arov and H. Dym. *J-contractive matrix valued functions and related topics.* Encyclopedia of Mathematics and its Applications. 116. Cambridge: Cambridge University Press, 575 p., 2008.

[27] D. Z. Arov and M. G. Krein. Problem of search of the minimum of entropy in indeterminate extension problems. *Funktsional. Anal. i Prilozhen.*, 15(2):61–64, 1981. Translated in: *Funct. Anal. Appl.*, 15:123–126, 1981.

[28] T. Ja. Azizov. Dissipative operators in a Hilbert space with an indefinite metric. *Math. USSR Izv.*, 7:639–660, 1973.

[29] A. V. Bäcklund. Zur Theorie der partiellen Differentialgleichungen erster Ordnung. *Math. Ann.*, 17:285–328, 1880.

[30] I. Bakas. Conservation laws and geometry of perturbed coset models. *Internat. J. Modern Phys. A*, 9:3443–3472, 1994.

[31] J. A. Ball. Conservative dynamical systems and nonlinear Livsic–Brodskii nodes. In: *Nonselfadjoint operators and related topics (Beer Sheva, 1992)*, pp. 67–95. Oper. Theory Adv. Appl. 73. Basel: Birkhäuser, 1994.

[32] J. A. Ball, C. Sadosky and V. Vinnikov. Scattering systems with several evolutions and multidimensional input/state/output systems. *Integral Equations Operator Theory*, 52:323–393, 2005.

[33] I. V. Barashenkov and D. E. Pelinovsky. Exact vortex solutions of the complex sine-Gordon theory on the plane. *Phys. Lett. B*, 436:117–124, 1998.

[34] H. Bart, I. Gohberg and M. A. Kaashoek. *Minimal factorization of matrix and operator functions.* Oper. Theory Adv. Appl. 1. Basel/Boston: Birkhäuser, 227 p., 1979.

[35] F. G. Bass and V. G. Sinitsyn. Non-stationary theory of second-harmonic generation. *Ukr. Fiz. Zh.*, 17:124, 1972.

[36] R. Beals and R. R. Coifman. Scattering and inverse scattering for first order systems. *Comm. Pure Appl. Math.*, 37:39–90, 1984.

[37] R. Beals and R. R. Coifman. Scattering and inverse scattering for first-order systems. II. *Inverse Problems*, 3:577–593, 1987.

[38] R. Beals and R. R. Coifman. Scattering and inverse scattering for first order systems. In: *Surv. Differ. Geom. IV*, pp. 467–519. Boston, MA: IntPress, 1998.

[39] R. Beals, P. Deift and C. Tomei. *Direct and inverse scattering on the line.* Mathematical Surveys and Monographs. 28. Providence, RI: Amer. Math. Soc., 209 p., 1988.

[40] R. Beals, P. Deift and X. Zhou. The inverse scattering transform on the line. In: A. S. Fokas and V. E. Zakharov (eds). *Important Developments in Soliton Theory*, pp. 7–32. Berlin: Springer, 1993.

[41] C. M. Bender and S. Böttcher. Real spectra in non-Hermitian Hamiltonians having $\mathcal{PT}$-symmetry. *Phys. Rev. Lett.*, 80:5243–5246, 1998.

[42] V. P. Berestetskii, L. P. Pitaevskii and E. M. Lifshitz. *Relativistic quantum theory.* Oxford: Pergamon Press, 1974.

[43] H. Berestycki et al. (eds). *Perspectives in nonlinear partial differential equations*. Contemporary Mathematics. 446. Providence, RI: Amer. Math. Soc., 2007.

[44] Yu. M. Berezanskii. Integration of non-linear difference equations by means of inverse problem technique. *Dokl. Akad. Nauk SSSR*, 281(1):16–19, 1985.

[45] Yu. M. Berezanskij. On the direct and inverse spectral problems for Jacobi fields. *St. Petersburg Math. J.*, 9(6):1053–1071, 1998.

[46] Yu. M. Berezanskii and M. I. Gekhtman. Inverse problem of spectral analysis and nonabelian chains of nonlinear equations. *Ukr. Math. J.*, 42(6):645–658, 1990.

[47] H. A. Bethe and E. E. Salpeter. *Quantum mechanics of one and two electron atoms*. Berlin: Springer, 1957.

[48] P. A. Binding, P. Drábek and Y. X. Huang. Existence of multiple solutions of critical quasilinear elliptic Neumann problems. *Nonlinear Anal. TMA*, 42:613–629, 2000.

[49] D. A. Bini and B. Iannazzo. A note on computing matrix geometric means. *Adv. Comput. Math.*, 35(2–4):175–192, 2011.

[50] A. I. Bobenko and Yu. B. Suris (eds). *Discrete differential geometry. Integrable structure.* Graduate Studies in Mathematics. 98. Providence, RI: Amer. Math. Soc., 404 p., 2008.

[51] J. Bognar. *Indefinite inner product spaces*. Ergebnisse der Mathematik und ihrer Grenzgebiete. 78. New York–Heidelberg–Berlin: Springer, 224 p., 1974.

[52] J. L. Bona and A. S. Fokas. Initial-boundary-value problems for linear and integrable nonlinear dispersive equations. *Nonlinearity*, 21(10):T195–T203, 2008.

[53] J. L. Bona, S. M. Sun and B.-Yu. Zhang. Boundary smoothing properties of the Korteweg–de Vries equation in a quarter plane and applications. *Dyn. Partial Differ. Equ.*, 3:1–69, 2006.

[54] J. L. Bona and R. Winther. The Korteweg–de Vries equation, posed in a quarter-plane. *SIAM J. Math. Anal.*, 14:1056–1106, 1983.

[55] A. B. Borisov and V. V. Kiseliev. Inverse problem for an elliptic sine-Gordon equation with an asymptotic behaviour of the cnoidal-wave type. *Inverse Problems*, 5:959–982, 1989.

[56] A. Boutet de Monvel and V. Marchenko. Generalization of the Darboux transform. *Mat. Fiz. Anal. Geom.*, 1:479–504, 1994.

[57] P. Bowcock and G. Tzamtzis. Quantum complex sine-Gordon model on a half line. *J. High Energy Phys.*, 11, 018, 22 pp., 2007.

[58] A. Bressan, G. Crasta and B. Piccoli. *Well-posedness of the Cauchy problem for $n \times n$ systems of conservation laws.* Mem. Am. Math. Soc. 694. Providence, RI: Amer. Math. Soc., 134 p., 2000.

[59] M. S. Brodskii. *Triangular and Jordan representations of linear operators*. Transl. Math. Monographs. 32., Providence, RI: Amer. Math. Soc., 246 p., 1971.

[60] I. Brunner, M. Herbst, W. Lerche and J. Walcher. Matrix factorizations and mirror symmetry: the cubic curve. *J. High Energy Phys.*, 0611: 006, 2006.

[61] F. Calogero and A. Degasperis. *Spectral transforms and solitons*. Studies in Mathematics and its Applications. 13. Amsterdam–New York–Oxford: North-Holland Publishing Company, 516 p., 1982.

[62] B. Carl and C. Schiebold. Nonlinear equations in soliton physics and operator ideals. *Nonlinearity*, 12:333–364, 1999.

[63] S. Carl and S. Heikkilä. Existence of solutions for discontinuous functional equations and elliptic boundary-value problems. *Electron. J. Differential Equations*, Paper No. 61, 10 pp., 2002.

[64] R. Carroll. Some remarks on orthogonal polynomials and transmutation methods. *Boll. Unione Mat. Ital. VI. B.*, 5(2):465–486, 1986.

[65] R. Carroll and Q. Bu. Solution of the forced nonlinear Schrödinger (NLS) equation using PDE techniques. *Appl. Anal.*, 41:33–51, 1991.

[66] R. C. Cascaval, F. Gesztesy, H. Holden and Yu. Latushkin. Spectral analysis of Darboux transformations for the focusing NLS hierarchy. *J. Anal. Math.*, 93:139–197, 2004.

[67] J. Čepička, P. Drábek and P. Girg. Quasilinear boundary value problems: existence and multiplicity results. *Contemp. Math.*, 357:111–139, 2004.

[68] K. Chadan and P. C. Sabatier. *Inverse problems in quantum scattering theory.* Texts and Monographs in Physics. New York etc.: Springer, 499 p., 1989.

[69] O. Chalykh. Algebro-geometric Schrödinger operators in many dimensions. *Philos. Trans. R. Soc. Lond. Ser. A Math. Phys. Eng. Sci.*, 366(1867):947–971, 2008.

[70] D. V. Chudnovsky and G. V. Chudnovsky. Bäcklund transformation as a method of decomposition and reproduction of two-dimensional nonlinear systems. *Phys. Lett. A*, 87(7):325–329, 1982.

[71] J. L. Cieslinski. An effective method to compute N-fold Darboux matrix and N-soliton surfaces. *J. Math. Phys.*, 32:2395–2399, 1991.

[72] J. L. Cieslinski. An algebraic method to construct the Darboux matrix. *J. Math. Phys.*, 36(10):5670–5706, 1995.

[73] J. L. Cieslinski. Algebraic construction of the Darboux matrix revisited. *J. Phys. A*, 42(40):404003, 40 pp., 2009.

[74] S. Clark and F. Gesztesy. Weyl–Titchmarsh M-function asymptotics, local uniqueness results, trace formulas, and Borg-type theorems for Dirac operators. *Trans. Amer. Math. Soc.*, 354:3475–3534, 2002.

[75] S. Clark and F. Gesztesy. On Weyl–Titchmarsh theory for singular finite difference Hamiltonian systems. *J. Comput. Appl. Math.*, 171:151–184, 2004.

[76] S. Clark and F. Gesztesy, On Self-adjoint and J-self-adjoint Dirac-type Operators: A Case Study. *Contemp. Math.*, 412:103–140, 2006.

[77] S. Clark, F. Gesztesy and W. Renger. Trace formulas and Borg-type theorems for matrix-valued Jacobi and Dirac finite difference operators. *J. Differential Equations*, 219(1):144–182, 2005.

[78] S. Clark, F. Gesztesy and M. Zinchenko. Weyl–Titchmarsh theory and Borg–Marchenko-type uniqueness results for CMV operators with matrix-valued Verblunsky coefficients. *Oper. Matrices*, 1:535–592, 2007.

[79] S. Clark, F. Gesztesy and M. Zinchenko. Borg–Marchenko-type uniqueness results for CMV operators. *Skr. K. Nor. Vidensk. Selsk.*, 1:1–18, 2008.

[80] J. B. Conway. *Functions of one complex variable.* 2nd ed. Graduate Texts in Mathematics. 11. New York: Springer, 1978.

[81] M. J. Corless and A. E. Frazho. *Linear Systems and Control – An Operator Perspective*. Pure and Applied Mathematics. 254. New York: Marcel Dekker, 339 p., 2003.

[82] M. M. Crum. Associated Sturm–Liouville systems. *Quart. J. Math., Oxford II Ser.*, 6:121–127, 1955.

[83] G. Darboux. Sur les équations aux dérivées partielles du second ordre. *Ann. de l'Éc. Norm.*, VII:163–173, 1870.

[84] G. Darboux. *Lecons sur la Theorie Generale de Surface et les Applications Geometriques du Calcul Infinitesimal. II: Les congruences et les équations linéaires aux dérivées partielles. Des lignes tracées sur les surfaces.* Paris: Gauthier–Villars & Fils, 1889.

[85] L. de Branges. *Hilbert spaces of entire functions*. Englewood Cliffs, N.J.: Prentice-Hall, 326 p., 1968.

[86] L. de Branges. The expansion theorem for Hilbert spaces of entire functions. In: *Entire Functions and Related Parts of Analysis*, pp. 79–148. Proc. Sympos. Pure Math. 11. Providence, RI: Amer. Math. Soc., 1968.

[87] P. A. Deift. Applications of a commutation formula. *Duke Math. J.*, 45:267–310, 1978.

[88] P. Deift and E. Trubowitz. Inverse scattering on the line. *Commun. Pure Appl. Math.*, 32:121–251, 1979.

[89] Ph. Delsarte, Y. Genin and Y. Kamp. Orthogonal polynomial matrices on the unit circles. *IEEE Trans. Circuits and Systems*, 25:149–160, 1978.

[90] Ph. Delsarte, Y. Genin and Y. Kamp. Schur parametrization of positive definite block-Toeplitz systems. *SIAM J. Appl. Math.*, 36(1):34–46, 1979.

[91] N. Dorey and T. J. Hollowood. Quantum scattering of charged solitons in the complex sine-Gordon model. *Nuclear Phys. B*, 440:215–233, 1995

[92] V. K. Dubovoj, B. Fritzsche and B. Kirstein. *Matricial version of the classical Schur problem.* Teubner-Texte zur Mathematik [Teubner Texts in Mathematics]. 129. Stuttgart: B. G. Teubner Verlagsgesellschaft mbH, 355 p., 1992.

[93] J. J. Duistermaat and F. A. Grünbaum. Differential equations in the spectral parameter. *Commun. Math. Phys.*, 103:177–240, 1986.

[94] H. Dym. J contractive matrix functions, reproducing kernel Hilbert spaces and interpolation. *CBMS Reg. Conf. Ser. in Math.*, 71:1–147, 1989.

[95] H. Dym and A. Iacob. Positive definite extensions, canonical equations and inverse problems. In: *Topics in operator theory and systems and networks*, pp. 141–240. Oper. Theory Adv. Appl. 12. Basel–Boston: Birkhäuser, 1984.

[96] J. Eckhardt, F. Gesztesy, R. Nichols and G. Teschl. Weyl–Titchmarsh Theory for Sturm–Liouville Operators with Distributional Coefficients. arXiv:1208.4677.

[97] J. C. Eilbeck and M. Johansson. *The discrete nonlinear Schrödinger equation – 20 years on*. In: L. Vázquez et al. (eds). *Proceedings of the 3rd conference on localization and energy transfer in nonlinear systems*, pp. 44–67. River Edge, NJ: World Scientific, 2003.

[98] A. Erdelyi et al. *Higher transcendental functions.* Bateman Manuscript Project. New York/Toronto/London: McGraw-Hill Book Co., 396 p., 1953.

[99] J. Escher and Z. Yin. Well-posedness, blow-up phenomena, and global solutions for the b-equation. *J. Reine Angew. Math.*, 624:51–80, 2008.

[100] M. A. Evgrafov. *Asymptotic Estimates and Entire Functions.* New York: Gordon and Breach, 1961.

[101] L. D. Faddeev and L. A. Takhtajan. *Hamiltonian methods in the theory of solitons.* Springer Series in Soviet Mathematics. Berlin etc.: Springer, 592 p., 1987.

[102] L. Faddeev, P. Van Moerbeke and F. Lambert (eds). *Bilinear integrable systems. From classical to quantum, continuous to discrete.* Dordrecht: Springer, 390 p., 2006.

[103] A. S. Fokas. Integrable nonlinear evolution equations on the half-line. *Comm. Math. Phys.* 230:1–39, 2002.

[104] A. S. Fokas and C. R. Menyuk. Integrability and self-similarity in transient stimulated Raman scattering. *J. Nonlinear Sci.* 9:1–31, 1999.

[105] B. Fritzsche, B. Kirstein, I. Ya. Roitberg and A. L. Sakhnovich. Weyl matrix functions and inverse problems for discrete Dirac type self-adjoint system: explicit and general solutions. *Oper. Matrices*, 2:201–231, 2008.

[106] B. Fritzsche, B. Kirstein, I. Ya. Roitberg and A. L. Sakhnovich. Recovery of Dirac system from the rectangular Weyl matrix function. *Inverse Problems*, 28(1), 015010, 18 p., 2012.

[107] B. Fritzsche, B. Kirstein, I. Ya. Roitberg and A. L. Sakhnovich. Operator identities corresponding to inverse problems. *Indag. Math., New Ser.*, 23(4):690–700, 2012.

[108] B. Fritzsche, B. Kirstein, I. Ya. Roitberg and A. L. Sakhnovich. Skew-self-adjoint Dirac systems with a rectangular matrix potential: Weyl theory, direct and inverse problems. *Integral Equations Operator Theory*, 74(2):163–187, 2012.

[109] B. Fritzsche, B. Kirstein, I. Ya. Roitberg and A. L. Sakhnovich. Discrete Dirac system: rectangular Weyl functions, direct and inverse problems. arXiv:1206.2915.

[110] B. Fritzsche, B. Kirstein, I. Ya. Roitberg and A. L. Sakhnovich. Weyl theory and explicit solutions of direct and inverse problems for a Dirac system with rectangular matrix potential. *Oper. Matrices*, 7(1):183–196, 2013.

[111] B. Fritzsche, B. Kirstein and A. L. Sakhnovich. Completion problems and scattering problems for Dirac type differential equations with singularities. *J. Math. Anal. Appl.*, 317:510–525, 2006.

[112] B. Fritzsche, B. Kirstein and A. L. Sakhnovich. On a new class of structured matrices related to the discrete skew-self-adjoint Dirac systems. *Electron. J. Linear Algebra*, 17:473–486, 2008.

[113] B. Fritzsche, B. Kirstein and A. L. Sakhnovich. Semiseparable integral operators and explicit solution of an inverse problem for the skew-self-adjoint Dirac-type system. *Integral Equations Operator Theory*, 66:231–251, 2010.

[114] B. Fritzsche, B. Kirstein and A. L. Sakhnovich. Weyl functions of Dirac systems and of their generalizations: integral representation, inverse problem and discrete interpolation. *J. Anal. Math.*, 116(1):17–51, 2012.

[115] V. S. Gerdjikov and P. P. Kulish. The generating operator for the $n \times n$ linear system *Phys. D*, 3:549–564, 1981

[116] Ya. L. Geronimus. *Polynomials orthogonal on a circle and interval.* International Series of Monographs on Pure and Applied Mathematics. 18. Oxford–London–New York–Paris: Pergamon Press, 218 p., 1961.

[117] F. Gesztesy. A complete spectral characterization of the double commutation method. *J. Funct. Anal.*, 117(2):401–446, 1993.

[118] F. Gesztesy and H. Holden. *Soliton equations and their algebro-geometric solutions.* I : $(1+1)$*-dimensional continuous models*. Cambridge Studies in Advanced Mathematics. 79. Cambridge: Cambridge University Press, 505 p., 2003.

[119] F. Gesztesy, H. Holden, J. Michor and G. Teschl. *Soliton equations and their algebro-geometric solutions.* II : $(1+1)$*-dimensional discrete models*. Cambridge Studies in Advanced Mathematics. 114. Cambridge: Cambridge University Press, 438 p., 2008.

[120] F. Gesztesy, W. Schweiger and B. Simon. Commutation methods applied to the mKdV-equation. *Trans. Amer. Math. Soc.*, 324:465–525, 1991.

[121] F. Gesztesy and B. Simon. On local Borg–Marchenko uniqueness results. *Commun. Math. Phys.*, 211:273–287, 2000.

[122] F. Gesztesy and B. Simon. A new approach to inverse spectral theory. II: General real potentials and the connection to the spectral measure. *Ann. of Math. (2)*, 152(2):593–643, 2000.

[123] F. Gesztesy and G. Teschl. On the double commutation method. *Proc. Amer. Math. Soc.*, 124(6):1831–1840, 1996.

[124] F. Gesztesy and E. Tsekanovskii. On matrix-valued Herglotz functions. *Math. Nachr.*, 218:61–138, 2000.

[125] F. Gesztesy and M. Zinchenko. A Borg-type theorem associated with orthogonal polynomials on the unit circle. *J. Lond. Math. Soc., II. Ser.*, 74(3):757–777, 2006.

[126] F. Gesztesy and M. Zinchenko. Weyl–Titchmarsh theory for CMV operators associated with orthogonal polynomials on the unit circle. *J. Approx. Theory*, 139:172–213, 2006.

[127] W. H. Glenn. Second-harmonic generation by picosecond optical pulses. *IEEE J. Quantum Electron.*, QE-5:284–290, 1969.

[128] I. Gohberg, S. Goldberg and M. A. Kaashoek. *Classes of Linear Operators. I.* Oper. Theory Adv. Appl. 49. Basel etc.: Birkhäuser, 468 p., 1990.

[129] I. Gohberg, M. A. Kaashoek and A. L. Sakhnovich. Canonical systems with rational spectral densities: explicit formulas and applications. *Math. Nachr.*, 194:93–125, 1998.

[130] I. Gohberg, M. A. Kaashoek and A. L. Sakhnovich. Pseudocanonical systems with rational Weyl functions: explicit formulas and applications. *J. Differential Equations*, 146(2):375–398, 1998.

[131] I. Gohberg, M. A. Kaashoek and A. L. Sakhnovich. Sturm–Liouville systems with rational Weyl functions: explicit formulas and applications. *Integral Equations Operator Theory*, 30(3):338–377, 1998.

[132] I. Gohberg, M. A. Kaashoek and A. L. Sakhnovich. Bound states for canonical systems on the half and full line: explicit formulas. *Integral Equations Operator Theory*, 40(3):268–277, 2001.

[133] I. Gohberg, M. A. Kaashoek and A. L. Sakhnovich. Scattering problems for a canonical system with a pseudo-exponential potential. *Asymptotic Analysis*, 29(1):1–38, 2002.

[134] I. Gohberg, M. A. Kaashoek and F. van Schagen. On inversion of convolution integral operators on a finite interval. In: I. Gohberg et al. (eds). *Operator theoretical methods and applications to mathematical physics*, pp. 277–285. Oper. Theory Adv. Appl. 147. Basel: Birkhäuser, 2004.

[135] I. Gohberg and I. Koltracht. Numerical solution of integral equations, fast algorithms and Krein–Sobolev equations. *Numer. Math.*, 47:237–288, 1985.

[136] I. Gohberg and M. G. Krein. Systems of integral equations on a half line with kernels depending on the difference of arguments. *Amer. Math. Soc. Transl. (2)*, 14:217–287, 1960.

[137] I. Gohberg and M. G. Krein. *Theory and applications of Volterra operators in Hilbert space*. Moscow: Nauka, 1967. Translated in: Transl. of math. monographs. 24. Providence, RI: Amer. Math. Soc., 430 p., 1970.

[138] L. B. Golinskii and I. V. Mikhailova (edited by V. P. Potapov). *Hilbert Spaces of entire functions as an object of J-theory*. Preprint No. 28–80. Kharkov: Inst. Low-Temperature Phys. Engineering, Acad. Sci. Ukrain. SSR, 1980. Translated in: H. Dym et al. (eds). *Topics in interpolation theory*, pp. 205–251. Oper. Theory Adv. Appl. 95. Basel: Birkhäuser, 1997.

[139] F. A. Grünbaum, I. Pacharoni and J. Tirao. Matrix valued orthogonal polynomials of the Jacobi type. *Indag. Math., New Ser.*, 14(3–4):353–366, 2003.

[140] C. H. Gu, H. Hu and Z. Zhou. *Darboux transformations in integrable systems. Theory and their applications to geometry*. Mathematical Physics Studies. 26. Dordrecht: Springer, 308 p., 2005.

[141] D. B. Hinton and J. K. Shaw. On Titchmarsh–Weyl $M(\lambda)$-functions for linear Hamiltonian systems. *J. Differential Equations*, 40:316–342, 1981.

[142] D. B. Hinton and J. K. Shaw. Hamiltonian systems of limit point or limit circle type with both endpoints singular. *J. Differential Equations*, 50:444–464, 1983.

[143] J. Holmer. The initial-boundary value problem for the Korteweg–de Vries equation. *Comm. Partial Differential Equations*, 31:1151–1190, 2006.

[144] E. Infeld and G. Rowlands. *Nonlinear waves, solitons and chaos*. 2-nd ed. Cambridge: Cambridge University Press, 2000.

[145] T. S. Ivanchenko and L. A. Sakhnovich. *An operator approach to the study of interpolation problems* (Russian). Manuscript No. 701Uk-85. Deposited at Ukrainian NIINTI, 1985.

[146] C. G. T. Jacobi. Über eine neue Methode zur Integration der hyperelliptischen Differentialgleichungen und über die rationale Form ihrer vollständigen algebraischen Integralgleichungen. *J. Reine Angew. Math.*, 32:220–226, 1846.

[147] M. Jaworski and D. Kaup. Direct and inverse scattering problem associated with the elliptic sinh-Gordon equation. *Inverse Problems*, 6:543–556, 1990.

[148] M. A. Kaashoek and A. L. Sakhnovich. Discrete skew self-adjoint canonical system and the isotropic Heisenberg magnet model. *J. Funct. Anal.*, 228:207–233, 2005.

[149] M. Kac. On some connections between probability theory and differential and integral equations. In: Proc. Berkeley Sympos. Math. Statist. Probability, pp. 189–215. Berkley, Calif.: Univ. of Calif. Press, 1951.

[150] M. Kac and P. van Moerbeke. A complete solution of the periodic Toda problem. *Proc. Natl. Acad. Sci. USA*, 72:2879–2880, 1975.

[151] T. Kailath, B. Levy, L. Ljung and M. Morf. The factorization and representation of operators in the algebra generated by Toeplitz operators. *SIAM J. Appl. Math.*, 37:467–484, 1979.

[152] R. E. Kalman. Contributions to the theory of optimal control. *Bol. Soc. Mat. Mex.*, 5:102–199, 1960.

[153] R. E. Kalman, P. Falb and M. Arbib. *Topics in mathematical system theory*. International Series in Pure and Applied Mathematics. New York etc.: McGraw-Hill Book Company, 358 p., 1969.

[154] J. Kampe de Feriet. *Fonctions de la physique mathematique*. Paris: Paris Editions du CNRS, 1957.

[155] A. Kapustin and Y. Li. Topological correlators in Landau–Ginzburg models with boundaries. *Adv. Theor. Math. Phys.*, 7:727–749, 2004.

[156] A. Kasman and M. Gekhtman. Solitons and almost-intertwining matrices. *J. Math. Phys.*, 42:3540–3551, 2001.

[157] I. S. Kats. Some general theorems on the density of the spectrum of a string. *Dokl. Akad. Nauk SSSR*, 238(4):785–788, 1978.

[158] I. S. Kats. Linear relations generated by a canonical differential equation of dimension 2, and eigenfunction expansions. *St. Petersburg Math. J.*, 14(3):429–452, 2003.

[159] V. E. Katsnelson. Continuous analogues of the Hamburger–Nevanlinna theorem and fundamental matrix inequalities for classical problems. IV: Problems on integral representation as continuous interpolation problems; from the fundamental matrix inequality (FMI) to the asymptotic relation. *Teor. Funktsii, Funktsional. Anal. i Prilozhen.*, 40:79–90, 1983. Translated in: *Amer. Math. Soc. Transl. (2)*, 136:97–108, 1987.

[160] V. E. Katsnelson. Right and left joint system representation of a rational matrix function in general position. In: D. Alpay, et al. (eds). *Operator theory, system theory and related topics*, pp. 337–400. Oper. Theory Adv. Appl. 123. Basel: Birkhäuser, 2001.

[161] V. E. Katsnelson and B. Kirstein. On the theory of matrix-valued functions belonging to the Smirnov class. In: H. Dym et al. (eds). *Topics in interpolation theory*, pp. 299–350. Oper. Theory Adv. Appl. 95. Basel: Birkhäuser, 1997.

[162] D. J. Kaup. Simple harmonic generation: an exact method of solution. *Stud. Appl. Math.*, 59:25–35, 1978.

[163] D. J. Kaup. The forced Toda lattice: an example of an almost integrable system. *J. Math. Phys.*, 25(2):277–281, 1984.

[164] D. J. Kaup and A. C. Newell. An exact solution for a derivative nonlinear Schrödinger equation. *J. Math. Phys.*, 19:798–801, 1978.

[165] D. J. Kaup and A. C. Newell. The Goursat and Cauchy problems for the sine-Gordon equation. *SIAM J. Appl. Math.*, 34(1):37–54, 1978.

[166] D. J. Kaup and H. Steudel. Recent results on second harmonic generation. *Contemp. Math.*, 326:33–48, 2003.

[167] A. M. Khol'kin. Description of self-adjoint extensions of differential operators of arbitrary order on an infinite interval in the absolutely indeterminate case. *Teor. Funktsii, Funktsional. Anal. i Prilozhen.*, 44:112–122, 1985. Translated in: *J. Soviet Math.*, 48(3):337–345, 1990.

[168] K. R. Khusnutdinova and H. Steudel. Second harmonic generation: Hamiltonian structures and particular solutions. *J. Math. Phys.*, 39:3754–3764, 1998.

[169] R. Killip and B. Simon. Sum rules and spectral measures of Schrödinger operators with L^2 potentials. *Ann. of Math. (2)*, 170(2):739–782, 2009.

[170] K. Knopp. *Theory of Functions. I: Elements of the General Theory of Analytic Functions*. New York: Dover Publications, 1945.

[171] I. Koltracht, B. Kon and L. Lerer. Inversion of structured operators. *Integral Equations Operator Theory*, 20:410–448, 1994.

[172] B. G. Konopelchenko and C. Rogers. Bäcklund and reciprocal transformations: gauge connections. In: W. F. Ames and C. Rogers (eds.) *Nonlinear equations in applied sciences*, pp. 317–362. San Diego: Academic Press, 1992.

[173] A. Kostenko, A. Sakhnovich and G. Teschl. Inverse eigenvalue problems for perturbed spherical Schrödinger operators. *Inverse Problems*, 26:105013, 14 p., 2010.

[174] A. Kostenko, A. Sakhnovich and G. Teschl. Commutation Methods for Schrödinger Operators with Strongly Singular Potentials. *Math. Nachr.*, 285(4):392–410, 2012.

[175] A. Kostenko, A. Sakhnovich and G. Teschl. Weyl–Titchmarsh theory for Schrödinger operators with strongly singular potentials. *Int. Math. Res. Not.*, 8:1699–1747, 2012.

[176] M. G. Krein. Continuous analogues of propositions on polynomials orthogonal on the unit circle (Russian). *Dokl. Akad. Nauk SSSR*, 105:637–640, 1955.

[177] M. G. Krein. On the theory of accelerants and S-matrices of canonical differential systems. *Dokl. Akad. Nauk SSSR*, 111:1167–1170, 1956.

[178] M. G. Krein. On a continuous analogue of a Christoffel formula from the theory of orthogonal polynomials (Russian). *Dokl. Akad. Nauk SSSR*, 113:970–973, 1957.

[179] M. G. Krein. *Topics in differential and integral equations and operator theory*. (Edited by I. Gohberg). Oper. Theory Adv. Appl. 7. Basel–Boston–Stuttgart: Birkhäuser, 1983.

[180] M. G. Krein and H. Langer. Defect subspaces and generalized resolvents of an Hermitian operator in the space Π_κ. *Funct. Anal. Appl.*, 5:136–146, 1971.

[181] M. G. Krein and Ju. L. Šmul'jan. On linear-fractional transformations with operator coefficients. *Amer. Math. Soc. Transl. (2)*, 103:125–152, 1974.

[182] I. M. Krichever. The analog of the d'Alembert formula for the main field equation and for the sine-Gordon equation. *Dokl. Akad. Nauk. SSSR*, 253(2):288–292, 1980.

[183] I. Kukavica, R. Temam, V. C. Vicol and M. Ziane. Local existence and uniqueness for the hydrostatic Euler equations on a bounded domain. *J. Differential Equations*, 250:1719–1746, 2011.

[184] V. B. Kuznetsov, M. Salerno and E. K. Sklyanin. Quantum Bäcklund transformation for the integrable DST model. *J. Phys. A*, 33(1):171–189, 2000.

[185] P. Lancaster and L. Rodman. *Algebraic Riccati equations*. Oxford: Clarendon Press, 480 p., 1995.

[186] H. Langer. Zur Spektraltheorie J-selbstadjungierter Operatoren. *Math. Ann.*, 146:60–85, 1962.

[187] M. Langer and H. Woracek. A local inverse spectral theorem for Hamiltonian systems. *Inverse Problems*, 27:055002, 17 pp., 2011.

[188] P. D. Lax. Integrals of nonlinear equations of evolution and solitary waves. *Comm. Pure Appl. Math.*, 21:467–490, 1968.

[189] J. Leon and A. Spire. The Zakharov–Shabat spectral problem on the semi-line: Hilbert formulation and applications. *J. Phys. A* 34:7359–7380, 2001.

[190] M. Lesch and M. Malamud. The inverse spectral problem for first order systems on the half-line. In: V. M. Adamyan et al. (eds). *Differential operators and related topics. I*, pp. 199–238. Oper. Theory Adv. Appl. 117. Basel: Birkhäuser, 2000.

[191] M. Lesch and M. Malamud. *On the number of square integrable solutions and self-adjointness of symmetric first order systems of differential equations*. Los Alamos preprint, 2000.

[192] D. Levi and O. Ragnisco (eds). *SIDE III – Symmetries and integrability of difference equations*. CRM Proceedings and Lecture Notes. 25. Providence, RI: Amer. Math. Soc., 2000.

[193] D. Levi, O. Ragnisco and A. Sym. Dressing method vs. classical Darboux transformation. *Nuovo Cimento B*, 83:34–41, 1984.

[194] B. M. Levitan and I. S. Sargsjan. *Introduction to the spectral theory. Selfadjoint differential operators*. Transl. Math. Monographs. 34. Providence, RI: Amer. Math. Soc., 1975.

[195] B. M. Levitan and I. S. Sargsjan. *Sturm–Liouville and Dirac operators*. Mathematics and its Applications (Soviet Series). 59. Dordrecht: Kluwer, 1990.

[196] B. M. Levitan and I. S. Sargsjan. *Spectrum and trace of ordinary differential operators* (Russian). Moscow: MGU, 2003.

[197] A. N. Leznov and M. V. Saveliev. *Group-theoretical methods for integration of nonlinear dynamical systems*. Progress in Physics. 15. Basel: Birkhäuser, 1992.

[198] J. Liouville. Sur l'equation aux differences partielles. *J. Math. Pure Appl.*, 18:71, 1853.

[199] Q. P. Liu and M. Manas. Vectorial Darboux transformations for the Kadomtsev–Petviashvili hierarchy. *J. Nonlinear Sci.*, 9(2):213–232, 1999.

[200] M. S. Livšic. On a class of linear operators in Hilbert space. *Rec. Math. [Mat. Sbornik] N.S.* 19(61):239–262, 1946). Translated in: *Amer. Math. Soc. Transl. (2)*, 13:85–103, 1960.

[201] M. S. Livšic. *Operators, oscillations, waves (open systems)*. Moscow: Nauka, 1966. Translated in: Providence, RI: Amer. Math. Soc., 274 p., 1973.

[202] M. S. Livšic. Operator waves in Hilbert space and related partial differential equations. *Integral Equations Operator Theory*, 2:25–47, 1979.

[203] M. S. Livšic, N. Kravitsky, A. S. Markus and V. Vinnikov. *Theory of commuting nonselfadjoint operators*. Mathematics and its Applications. 332. Dordrecht: Kluwer Acad. Publ., 1995.

[204] F. Lund and T. Regge. Unified approach to strings and vortices with soliton solutions. *Phys. Rev. D (3)*, 14:1524–1535, 1976.

[205] V. A. Marchenko. *Sturm–Liouville operators and applications*. Oper. Theory Adv. Appl. 22. Basel/Boston/Stuttgart: Birkhäuser, 367 p., 1986.

[206] V. A. Marchenko, *Nonlinear equations and operator algebras*. Mathematics and Its Applications (Soviet Series). 17. Dordrecht etc.: D. Reidel Publishing Company, 157 p., 1988.

[207] J. E. Marsden. *Elementary classical analysis*. San Francisco: W. H. Freeman and Co., 1974.

[208] V. B. Matveev. Positons: slowly decaying soliton analogs. *Teoret. Mat. Fiz.*, 131(1):44–61, 2002.

[209] V. B. Matveev and M. A. Salle. *Darboux transformations and solitons*. Berlin: Springer, 120 p., 1991.

[210] R. Mennicken, A. L. Sakhnovich and C. Tretter. Direct and inverse spectral problem for a system of differential equations depending rationally on the spectral parameter. *Duke Math. J.*, 109(3):413–449, 2001.

[211] F. E. Melik-Adamjan. Canonical differential operators in a Hilbert space. *Izv. Akad. Nauk Arm. SSR Math.*, 12:10–31, 1977.

[212] I. V. Mikhailova and V. P. Potapov. On a criterion of positive definiteness In: *Topics in interpolation theory*, pp. 419–451. Oper. Theory Adv. Appl. 95. Basel: Birkhäuser, 1997.

[213] R. Miura (ed.). *Bäcklund Transformations*. Lecture Notes in Math. 515. Berlin–Heidelberg–New York: Springer, 295 p., 1976.

[214] Ya. V. Mykytyuk and D. V. Puyda. Inverse spectral problems for Dirac operators on a finite interval. *J. Math. Anal. Appl.*, 386(1):177–194, 2012.

[215] G. Neugebauer and D. Kramer. Einstein–Maxwell solitons. *J. Phys. A*, 16:1927–1936, 1983.

[216] S. P. Novikov. A periodic problem for the Korteweg–de Vries equation. *Funct. Anal. Appl.*, 8(3):236–246, 1974.

[217] S. Novikov, S. V. Manakov, L. P. Pitaevskij and V. E. Zakharov. *Theory of solitons*. Contemporary Soviet Mathematics. New York–London: Plenum Publishing Corporation. Consultants Bureau, 286 p, 1984.

[218] S. A. Orlov. Nested matrix balls, analytically depending on a parameter, and theorems on invariance of ranks of radii of limit matrix balls. *Math. USSR Izv.*, 10:565–613, 1976.

[219] R. E. A. C. Paley and N. Wiener. *Fourier transforms in the complex domain*. Amer. Math. Soc. Colloquium Publications. 19. Providence, RI: Amer. Math. Soc., 1987.

[220] Q-H. Park and H. J. Shin. Duality in complex sine-Gordon theory. *Phys. Lett. B*, 359:125–132, 1995.

[221] Q-H. Park and H. J. Shin. Complex sine-Gordon equation in coherent optical pulse propagation. *J. Korean Phys. Soc.*, 30:336–340, 1997.

[222] K. Pohlmeyer. Integrable Hamiltonian systems and interactions through quadratic constraints. *Commun. Math. Phys.*, 46:207–221, 1976.

[223] V. P. Potapov. The multiplicative structure of J-contractive matrix functions. *Amer. Math. Soc. Transl.*, 15:131–243, 1960.

[224] V. P. Potapov. *Collected papers of V. P. Potapov*. Hokkaido University, Research Institute of Applied Electricity, Division of Applied Mathematics, Sapporo, 1982.

[225] V. P. Potapov. Basic facts in the theory of J-nonexpansive analytic matrix-valued functions (Russian). In: All-Union Conf. Theory of Functions of a Complex Variable, Summaries of Reports, pp. 170–181. Kharkov: Fiz.-Tekhn. Inst. Nizkikh Temperatur Akad. Nauk Ukrain. SSR, 1971.

[226] I. I. Privalov. *Boundary properties of analytic functions*. 2nd ed. Moscow: GITTL, 1950. German transl. in: VEB Deutscher Verlag Wiss., Berlin, 1956.

[227] D. V. Puyda. Inverse spectral problems for Dirac operators with summable matrix-valued potentials. *Integral Equations Operator Theory*, 74(3):417–450, 2012.

[228] C. Remling. Schrödinger operators and de Branges spaces. *J. Funct. Anal.*, 196:323–394, 2002.

[229] F. Riesz. Über die Randwerte einer analitischen Funktion. *Math. Zs.*, 18:87–95, 1923.

[230] F. S. Rofe-Beketov and A. M. Hol'kin. *Spektral'nyj analiz differentzial'nyh operatorov* (Russian). Mariupol': Azov State University, 2001.

[231] C. Rogers and W. K. Schief. *Bäcklund and Darboux transformations. Geometry and modern applications in soliton theory*. Cambridge Texts in Applied Mathematics. Cambridge: Cambridge University Press, 413 p., 2002.

[232] Yu. A. Rosanov. *Stationary stochastic processes* (Russian). Moscow: Fizmatgiz, 284 p., 1963.

[233] P. C. Sabatier. On multidimensional Darboux transformations. *Inverse Problems*, 14:355–366, 1998.

[234] Yu. Safarov and D. Vassiliev. *The asymptotic distribution of eigenvalues of partial differential operators*. Translations of Mathematical Monographs. 155. Providence, RI: Amer. Math. Soc., 354 p., 1996.

[235] A. L. Sakhnovich. On a class of extremal problems. *Izv. Akad. Nauk SSSR, Ser. Mat.*, 51:436–443, 1987. Translated in: *Math. USSR Izv.*, 30, 1988.

[236] A. L. Sakhnovich. Asymptotics of spectral functions of an S-colligation. *Izv. Vyssh. Uchebn. Zaved., Mat.*, 9:62–72, 1988. Translated in: *Soviet Math. (Iz. VUZ)*, 32:92–105, 1988.

[237] A. L. Sakhnovich. *The mixed problem for nonlinear Shrödinger equation and the inverse spectral problem*. Archives of VINITI AN SSSR, no. 3255-B89, 1989.

[238] A. L. Sakhnovich. Nonlinear Schrödinger equation on a semi-axis and an inverse problem associated with it. *Ukr. Mat. Zh.* 42(3):356–363, 1990. Translated in: *Ukr. Math. J.*, 42(3):316–323, 1990.

[239] A. L. Sakhnovich. The N-wave problem on the semiaxis. *Russ. Math. Surveys* 46(4):198–200, 1991.

[240] A. L. Sakhnovich. Spectral functions of the canonical systems of the $2n$-th order. *Mat. Sb.*, 181(11):1510–1524, 1990. Translated in: *Math. USSR Sb.*, 71(2):355–369, 1992.

[241] A. L. Sakhnovich. The N-wave problem on the semiaxis (Russian). In: *16 All-Union school on the operator theory in functional spaces. Lecture materials*, pp. 95–114, Nyzhni Novgorod, 1992.
[242] A. L. Sakhnovich. The Goursat problem for the sine-Gordon equation and the inverse spectral problem. *Russ. Math. Iz. VUZ*, 36(11):42–52, 1992.
[243] A. L. Sakhnovich. *Spectral theory for the systems of differential equations and applications.* Thesis for the secondary doctorship. Kiev: Institute of Mathematics, 1992.
[244] A. L. Sakhnovich. Exact solutions of nonlinear equations and the method of operator identities. *Lin. Alg. Appl.*, 182:109–126, 1993.
[245] A. L. Sakhnovich. Dressing procedure for solutions of nonlinear equations and the method of operator identities. *Inverse Problems*, 10:699–710, 1994.
[246] A. L. Sakhnovich. Iterated Darboux transform (the case of rational dependence on the spectral parameter). *Dokl. Natz. Akad. Nauk Ukrain.*, 7:24–27, 1995.
[247] A. L. Sakhnovich. Iterated Bäcklund–Darboux transformation and transfer matrix-function (nonisospectral case). *Chaos, Solitons and Fractals*, 7:1251–1259, 1996.
[248] A. L. Sakhnovich. Iterated Bäcklund–Darboux transform for canonical systems. *J. Funct. Anal.*, 144:359–370, 1997.
[249] A. L. Sakhnovich. Inverse spectral problem related to the N-wave equation. In: V. M. Adamyan et al. (eds). *Differential operators and related topics. I*, pp. 323–338. Oper. Theory Adv. Appl. 117. Basel: Birkhäuser, 2000.
[250] A. L. Sakhnovich. Toeplitz matrices with an exponential growth of entries and the first Szegö limit theorem. *J. Functional Anal.*, 171:449–482, 2000.
[251] A. L. Sakhnovich. Generalized Bäcklund–Darboux transformation: spectral properties and nonlinear equations. *J. Math. Anal. Appl.*, 262:274–306, 2001.
[252] A. L. Sakhnovich. Dirac type and canonical systems: spectral and Weyl–Titchmarsh functions, direct and inverse problems. *Inverse Problems*, 18:331–348, 2002.
[253] A. L. Sakhnovich. Dirac type system on the axis: explicit formulas for matrix potentials with singularities and soliton-positon interactions. *Inverse Problems*, 19:845–854, 2003.
[254] A. L. Sakhnovich. Non-Hermitian matrix Schrödinger equation: Bäcklund–Darboux transformation, Weyl functions, and $\mathcal{PT}$ symmetry. *J. Phys. A*, 36:7789–7802, 2003.
[255] A. L. Sakhnovich. Matrix Kadomtsev–Petviashvili equation: matrix identities and explicit non-singular solutions. *J. Phys. A*, 36:5023–5033, 2003.
[256] A. L. Sakhnovich. Second harmonic generation: Goursat problem on the semi-strip, Weyl functions and explicit solutions. *Inverse Problems*, 21(2):703–716, 2005.
[257] A. L. Sakhnovich. Non-self-adjoint Dirac-type systems and related nonlinear equations: wave functions, solutions, and explicit formulas. *Integral Equations Operator Theory*, 55:127–143, 2006.
[258] A. L. Sakhnovich. Harmonic maps, Bäcklund–Darboux transformations and “line solution” analogues. *J. Phys. A, Math. Gen.*, 39:15379–15390, 2006.
[259] A. L. Sakhnovich. Skew-self-adjoint discrete and continuous Dirac-type systems: inverse problems and Borg–Marchenko theorems. *Inverse Problems*, 22:2083–2101, 2006.
[260] A. L. Sakhnovich. Bäcklund–Darboux transformation for non-isospectral canonical system and Riemann–Hilbert problem. *Symmetry Integrability Geom. Methods Appl.*, 3:054, 11 p., 2007.
[261] A. L. Sakhnovich. Discrete canonical system and non-Abelian Toda lattice: Bäcklund–Darboux transformation and Weyl functions. *Math. Nachr.*, 280(5–6):1–23, 2007.
[262] A. L. Sakhnovich. Weyl functions, inverse problem and special solutions for the system auxiliary to the nonlinear optics equation. *Inverse Problems*, 24:025026, 2008.

[263] A. L. Sakhnovich. Nonisospectral integrable nonlinear equations with external potentials and their GBDT solutions. *J. Phys. A, Math. Theor.*, 41(15):155204, 2008.

[264] A. L. Sakhnovich. On the GBDT version of the Bäcklund–Darboux transformation and its applications to the linear and nonlinear equations and spectral theory. *Mathematical Modelling of Natural Phenomena*, 5(4):340–389, 2010.

[265] A. L. Sakhnovich. Construction of the solution of the inverse spectral problem for a system depending rationally on the spectral parameter, Borg–Marchenko-type theorem, and sine-Gordon equation. *Integral Equations Operator Theory*, 69:567–600, 2011.

[266] A. L. Sakhnovich. Time-dependent Schrödinger equation in dimension $k+1$: explicit and rational solutions via GBDT and multinodes. *J. Phys. A, Math. Theor.*, 44:475201, 2011.

[267] A. L. Sakhnovich. On the factorization formula for fundamental solutions in the inverse spectral transform. *J. Differential Equations*, 252:3658–3667, 2012.

[268] A. L. Sakhnovich. Sine-Gordon theory in a semi-strip. *Nonlinear Analysis*, 75(2):964–974, 2012.

[269] A. L. Sakhnovich. KdV equation in the quarter–plane: evolution of the Weyl functions and unbounded solutions. *Math. Model. Nat. Phenom.* 7(2):131–145, 2012.

[270] A. L. Sakhnovich, A. A. Karelin, J. Seck-Tuoh-Mora, G. Perez-Lechuga and M. Gonzalez-Hernandez. On explicit inversion of a subclass of operators with D-difference kernels and Weyl theory of the corresponding canonical systems. *Positivity*, 14:547–564, 2010.

[271] A. L. Sakhnovich and J. P. Zubelli. Bundle bispectrality for matrix differential equations. *Integral Equations Operator Theory*, 41:472–496, 2001.

[272] L. A. Sakhnovich. A semi-inverse problem. *Uspehi Mat. Nauk*, 18(3):199–206, 1963.

[273] L. A. Sakhnovich. Spectral analysis of Volterra's operators defined in the space of vector-functions $L^2_m(0,l)$. *Ukr. Mat. Zh.*, 16(2): 259–268, 1964. Translated in: *Amer. Math. Soc. Transl. (2)*, 61:85–95, 1967.

[274] L. A. Sakhnovich. Dissipative operators with absolutely continuous spectrum. *Trans. Moskow Math. Soc.*, 19:233–297, 1968.

[275] L. A. Sakhnovich. An integral equation with a kernel dependent on the difference of the arguments. *Mat. Issled.*, 8:138–146, 1973.

[276] L. A. Sakhnovich. On the factorization of the transfer matrix function. *Dokl. Akad. Nauk SSSR* , 226:781–784, 1976. Translated in: *Sov. Math. Dokl.*, 17:203–207, 1976.

[277] L. A. Sakhnovich. Equations with a difference kernel on a finite interval. *Russian Math. Surveys* 35(4):81–152, 1980.

[278] L. A. Sakhnovich. The asymptotic behavior of the spectrum of an anharmonic oscillator. *Theoret. and Math. Phys.* 47(2):449–456, 1981.

[279] L. A. Sakhnovich. Scattering theory for Coulomb type problem. In: V. P. Havin et al. (eds). *Linear and Complex Analysis Problem Book*, pp. 116–120. Lecture Notes in Mathematics 1043. Berlin: Springer, 1984.

[280] L. A. Sakhnovich. Factorization problems and operator identities. *Uspekhi Mat. Nauk*, 41(1):3–55, 1986. Translated in: *Russian Math. Surveys*, 41(1):1–64, 1986.

[281] L. A. Sakhnovich. *The non-linear equations and the inverse problems on the half-axis*. Preprint 30, Inst. Mat. AN Ukr.SSR. Kiev: Izd-vo Inst. Matem. AN Ukr.SSR, 1987.

[282] L. A. Sakhnovich. Evolution of spectral data and nonlinear equations. *Ukr. Mat. Zh.* 40(4):533–535, 1988. Translated in: *Ukr. Math. J.*, 40(4):459–461, 1988.

[283] L. A. Sakhnovich. The explicit formulas for the spectral characteristics and for solution of the $\sinh$-Gordon equation. *Ukr. Math. J.* 42(11):1359–1365, 1990.

[284] L. A. Sakhnovich. Integrable nonlinear equations on the halfaxis. *Ukr. Mat. Zh.*, 43, 1991. Translated in: *Ukr. Math. J.*, 43:1578–1584, 1991.

[285] L. A. Sakhnovich. Interpolation problems, inverse spectral problems and nonlinear equations. In: T. Ando et al. (eds). *Operator theory and complex analysis*, pp. 292–304. Oper. Theory Adv. Appl. 59. Basel: Birkhäuser, 1992.

[286] L. A. Sakhnovich. The method of operator identities and problems in analysis. *St. Petersburg Math. J.*, 5(1):1–69, 1994.

[287] L. A. Sakhnovich. *Integral equations with difference kernels on finite intervals*. Oper. Theory Adv. Appl. 84. Basel: Birkhäuser, 175 p, 1996.

[288] L. A. Sakhnovich. On a class of canonical systems on half-axis. *Integral Equations Operator Theory*, 31:92–112, 1998.

[289] L. A. Sakhnovich. *Interpolation theory and its applications*. Mathematics and its Applications. 428. Dordrecht: Kluwer Academic Publishers, 197 p., 1997.

[290] L. A. Sakhnovich. *Spectral theory of canonical differential systems, method of operator identities*. Oper. Theory Adv. Appl. 107. Basel: Birkhäuser, 202 p., 1999.

[291] L. A. Sakhnovich. Spectral theory of a class of canonical differential systems. *Funct. Anal. Appl.* 34(2):119–128, 2000.

[292] L. A. Sakhnovich. Half-inverse problems on the finite interval. *Inverse problems*, 17(3):527–532, 2001.

[293] L. A. Sakhnovich. Matrix finite-zone Dirac-type equations. *J. Funct. Anal.* 193(2):385–408, 2002.

[294] L. A. Sakhnovich. Comparison of thermodynamic characteristics of a potential well under quantum and classical approaches. *Funct. Anal. Appl.*, 36(3):205–211, 2002.

[295] L. A. Sakhnovich. On Krein's differential system and its generalization. *Integral Equations Operator Theory*, 55: 561–572, 2006.

[296] L. A. Sakhnovich. Effective construction of a class of positive operators in Hilbert space, which do not admit triangular factorization. *J. Funct. Anal.*, 263(3):803–817, 2012.

[297] L. A. Sakhnovich. On the solutions of Knizhnik–Zamolodchikov system. *Cent. Eur. J. Math.*, 7(1):145–162, 2009.

[298] L. A. Sakhnovich. Rationality of the Knizhnik–Zamolodchikov equation solution. *Theoret. and Math. Phys.*, 163(1):472–478, 2010.

[299] L. A. Sakhnovich. *Levy Processes, Integral Equations, Statistical Physics: Connections and Interactions*. Oper. Theory Adv. Appl. 225. Basel: Birkhäuser, 2012.

[300] L. A. Sakhnovich. Sliding inverse problems for radial Dirac and Schrödinger equations. *ArXiv*: 1302.2078.

[301] D. S. Sattinger and V. D. Zurkowski. Gauge theory of Bäcklund transformations. II. *Physica D*, 26:225–250, 1987.

[302] C. Schiebold. An operator-theoretic approach to the Toda lattice equation. *Physica D*, 122(1–4):37–61, 1998.

[303] C. Schiebold. Solitons of the sine-Gordon equation coming in clasters. *Revista Mat. Compl.*, 15:262–325, 2002.

[304] C. Schiebold. Explicit solution formulas for the matrix-KP. *Glasg. Math. J.*, 51A:147–155, 2009.

[305] A. C. Scott, F. Y F. Chu and D. W. McLaughlin. The soliton: a new concept in applied science. *Proc. IEEE* 61:1443–1483, 1973.

[306] R. T. Seeley. Fubini Implies Leibniz Implies $F_{yx} = F_{xy}$. *Amer. Math. Monthly*, 68:56–57, 1961.

[307] B. Simon. A new approach to inverse spectral theory. I: Fundamental formalism. *Ann. of Math.*, 150:1029–1057, 1999.

[308] B. Simon. Analogs of the m-function in the theory of orthogonal polynomials on the unit circle. *J. Comput. Appl. Math.*, 171:411–424, 2004.

[309] B. Simon. Ratio asymptotics and weak asymptotic measures for orthogonal polynomials on the real line. *J. Approximation Theory*, 126(2):198–217, 2004.

[310] B. Simon. *Orthogonal polynomials on the unit circle. 1, 2.* Colloquium Publications. 51, 54. Providence, RI: Amer. Math. Soc., 2005.

[311] E. K. Sklyanin. Some algebraic structures connected with the Yang–Baxter equation. *Funct. Anal. Appl.*, 16:263–270, 1983.

[312] V. I. Smirnov. *A course of higher mathematics, IV.* Oxford/New York: Pergamon Press, 1964.

[313] A. Sommerfeld. *Atombau und Spektrallinien, I.* Braunschweig: Friedr. Vieweg & Sohn, 1939.

[314] H. Steudel, C. F. de Morisson Faria, A. M. Kamchatnov and M. Paris. The inverse problem for second harmonic generation with an amplitude-modulated initial pulse. *Phys. Lett. A*, 276:267–271, 2000.

[315] C. Sulem and P. Sulem. *The nonlinear Schrödinger equation. Self-focusing and wave collapse.* Applied Mathematical Sciences 139. New York, NY: Springer, 350 p., 1999.

[316] G. Szegö. *Orthogonal polynomials. Revised ed.* Colloquium Publ. 23. New York: Amer. Math. Soc., 1939.

[317] B. Sz.-Nagy and C. Foias. *Harmonic analysis of operators on Hilbert space.* Budapest: Akadémiai Kiadó; Amsterdam–London: North-Holland Publishing Company, 387 p., 1970.

[318] C. L. Terng and K. Uhlenbeck. Bäcklund transformations and loop group actions. *Commun. Pure Appl. Math.*, 53:1–75, 2000.

[319] G. Teschl. Deforming the point spectra of one-dimensional Dirac operators. *Proc. Amer. Math. Soc.*, 126(10):2873–2881, 1998.

[320] G. Teschl. *Jacobi operators and completely integrable nonlinear lattices.* Mathematical Surveys and Monographs. 72. Providence, RI: Amer. Math. Soc., 351 p., 2000.

[321] G. Teschl. Almost everything you always wanted to know about the Toda equation. *Jahresber. Dtsch. Math.-Ver.*, 103(4):149–162, 2001.

[322] G. Teschl. *Mathematical Methods in Quantum Mechanics. With Applications to Schrödinger Operators.* Graduate Studies in Mathematics. 99. Providence, RI: Amer. Math. Soc., 305 p., 2009.

[323] E. C. Titchmarsh. *Eigenfunction expansions associated with second-order differential equations.* Oxford: Clarendon Press, 1946.

[324] E. C. Titchmarsh. *The theory of functions.* 2nd ed. London: Oxford University Press, 454 p., 1975.

[325] B. A. Ton. Initial boundary value problems for the Korteweg–de Vries equation. *J. Differential Equations*, 25:288–309, 1977.

[326] P. L. Vu. The Dirichlet initial-boundary-value problems for sine and sinh-Gordon equations on a half-line. *Inverse Problems*, 21:1225–1248, 2005.

[327] M. Wadati. The exact solution of the modified Korteweg–de Vries equation. *J. Phys. Soc. Japan*, 32:1681, 1972.

[328] J. R. L. Webb and K. Q. Lan. Eigenvalue criteria for existence of multiple positive solutions of nonlinear boundary value problems of local and nonlocal type. *Topol. Methods Nonlinear Anal.*, 27:91–115, 2006.

[329] R. Weikard. A local Borg–Marchenko theorem for difference equations with complex coefficients. *Contemp. Math.*, 362:403–410, 2004.

[330] N. Wiener. *Extrapolation, interpolation, and smoothing of stationary time series. With engineering applications.* Cambridge, Mass.: The M. I. T. Press, Massachusetts Institute of Technology, 163 p., 1966.

[331] N. Wiener and P. Masani. The prediction theory of multivariate stochastic processes. I: The regularity condition. *Acta Math.*, 98:111–150, 1957.

[332] O. C. Wright and M. G. Forest. On the Bäcklund-gauge transformation and homoclinic orbits of a coupled nonlinear Schrödinger system. *Physica D*, 141:104–116, 2000.

[333] A. E. Yagle and B. C. Levy. The Schur algorithm and its applications. *Acta Appl. Math.*, 3:255–284, 1985.

[334] V. E. Zakharov and S. V. Manakov. Theory of resonance interaction of wave packages in nonlinear medium. *Soviet Phys. JETP*, 69(5):1654–1673, 1975.

[335] V. E. Zakharov and A. V. Mikhailov. On the integrability of classical spinor models in two-dimensional space-time. *Commun. Math. Phys.*, 74:21–40, 1980.

[336] V. E. Zakharov and A. V. Mikhailov. Relativistically invariant two-dimensional models of field theory which are integrable by means of the inverse scattering problem method (Russian). *Soviet Phys. JETP*, 74(6):1953–1973, 1978.

[337] V. E. Zakharov and A. B. Shabat. On soliton interaction in stable media. *Soviet Phys. JETP*, 64:1627–1639, 1973.

[338] V. E. Zakharov, L. A. Takhtadzhyan and L. D. Faddeev. Complete description of solutions of the 'sine-Gordon' equation. *Soviet Phys. Dokl.*, 19:824–826, 1974.

Index

Bei Fragen zur Produktsicherheit wenden Sie sich bitte an:
If you have any questions regarding product safety,
please contact:

Walter de Gruyter GmbH
Genthiner Straße 13
10785 Berlin
productsafety@degruyterbrill.com